Springer Series on Polymer and Composite Materials

Series editor

Susheel Kalia, Dehradun, India

More information about this series at http://www.springer.com/series/13173

Susheel Kalia
Editor

Lignocellulosic Composite Materials

Springer

Editor
Susheel Kalia
Army Cadet College Wing
Indian Military Academy
Dehradun
India

ISSN 2364-1878 ISSN 2364-1886 (electronic)
Springer Series on Polymer and Composite Materials
ISBN 978-3-319-68695-0 ISBN 978-3-319-68696-7 (eBook)
https://doi.org/10.1007/978-3-319-68696-7

Library of Congress Control Number: 2017954285

Printed on acid-free paper

This Springer imprint is published by Springer Nature
The registered company is Springer International Publishing AG
The registered company address is: Gewerbestrasse 11, 6330 Cham, Switzerland

Preface

Lignocellulosic materials refer to the organic matter produced by trees, shrubs, and agricultural crops and are major feedstock for the pulp & paper industry, composite industry and packaging material. Renewable sources of lignocellulosic materials are natural fibers, agricultural residues and forest products. These agricultural wastes and forest feedstocks are sufficiently abundant and generate very low net greenhouse emissions. Forest wood products contain more lignin and less ash content, which makes them attractive to cost-effective transportation in comparison to agricultural residues. Lignocellulosics are being used in food packaging, composites and textile industries due to their advantages over other traditional materials which include low cost, renewability, non-toxicity, biodegradability etc. Lignocellulosic biomass is a main raw material for many industrial processes, but it usually displays a very poor microbial and moisture resistance. Lignocellulosic materials are attractive materials for a variety of potential applications such as soil conservation, textile applications, as alternate materials especially wood substitutes in the construction market and as reinforcement in composite materials to produce automotive structural components. Lignocellulosic materials possess sufficient strength and stiffness but are difficult to use in load bearing applications by themselves because of their fibrous structure. In fiber-reinforced composites, fibers gave strength and stiffness to the structure while the plastic matrix serves as the binder to hold the fibers in place.

Physical, chemical and biological pretreatments of lignocellulosic fibers stop moisture absorption, increase surface roughness and improve other properties also. The main objective of this book is to explicate some important features of lignocellulosic materials, their characterization, properties, and applications, pretreatments of lignocellulosics and lignocellulosics reinforced composite materials. First chapter of this book discusses about lignocellulosic materials, their characterization and applications in polymer composites. Chapter "Retting Process as a Pretreatment of Natural Fibers for the Development of Polymer Composites" aims to provide a classification and an overview of the retting process that have been developed during years and are applied to extract mainly bast fibers. Various physical, chemical, mechanical, enzymatic and microbiological retting techniques are discussed in this chapter. Physical, chemical and biological pretreatments of natural fibers and their effect on properties of natural fibers are discussed in Chapter "Pretreatments of Natural Fibers for Polymer Composite Materials".

Chapter "Mechanical and Thermal Properties of Less Common Natural Fibres and Their Composites" addresses with morphological, chemical and physical properties of less common natural fibers. Thermoplastic and thermosetting composites reinforced with the less common natural fibres are reviewed and compared with the traditional and common natural fibres. The development of cellulose-based nanocomposites is also discussed here.

Chapter "Lignocellulosic Fibres Reinforced Thermoset Composites: Preparation, Characterization, Mechanical and Rheological Properties" highlights the origin, structure and chemical composition of lignocellulosic fibers. This chapter also includes the preparation, characterization, mechanical and rheological properties of lignocellulosic fiber reinforced thermoset composites. Applications of lignocellulosic fiber composites in automobile industry and building sector are also discussed here. Chapter "Pineapple Leaf Fiber: From Waste to High Performance Green Reinforcement for Plastics and Rubbers" deals with extraction methods of pineapple leaf fibers. The reinforcing potential of pineapple leaf fibers in polymer matrix composites is also discussed in this chapter.

Chapter "Lightweight Wood Composites: Challenges, Production and Performance" discusses the production and performance of lightweight wood composites. Fiberboards, particleboards, extruded particleboards, and sandwich panels are included in this chapter. The fabrication techniques of natural fiber polymer composites are discussed in Chapter "Design and Fabrication of Kenaf Fibre Reinforced Polymer Composites for Portable Laptop Table". This chapter also deals with design and fabrication of kenaf fiber reinforced polymer composites for portable laptop table. Chapter "Lignocellulosic Materials for Geotextile and Geocomposites for Engineering Applications" presents an extensive overview of potential methods for enhancement of durability of lignocellulosic materials. Designing and testing parameters for lignocellulosic materials based geotextiles and geocomposites are also discussed. Chapter "Lignocellulosic Fibres-based Biocomposites Materials for Food Packaging" highlights the applications of lignocellulosic fibers-based biocomposites materials for food packaging. Chapter "Lignocellulosic Fibres Reinforced Polymer Composites for Acoustical Applications" includes the acoustical applications of lignocellulosic fiber reinforced polymer composites. Factors affecting the sound absorption coefficients of lignocellulosic fiber reinforced composites are also discussed in this chapter.

All the chapters in this book are contributed by renowned researchers from academia and research laboratories across the world. This book will prove to be a very useful for scientists, academicians, research scholars, material engineers and industries. This book will be supportive for undergraduate and post graduate students in Institutes of Materials Science & other Technical Institutes, and Technologists & Researchers from R&D laboratories working in this area.

The Editor would like to express their gratitude to all contributors, who have provided excellent contributions in this book. The Editor would like to thank his research team, who helped him in the editorial work. Finally, I gratefully acknowledge permissions to reproduce copyright materials from a number of sources.

Dehradun, India

Susheel Kalia

Contents

About the Editor

Susheel Kalia is Associate Professor & Head in Department of Chemistry at Army Cadet College Wing of Indian Military Academy Dehradun, India. He was visiting researcher in the Department of Civil, Chemical, Environmental and Materials Engineering at University of Bologna, Italy in 2013 and held a position as Assistant Professor in the Department of Chemistry, Bahra University and Shoolini University, India until 2015. Kalia has around 80 research articles in international journals along with 11 books and 11 book chapters in his academic career. His research interests include polymeric composites, bio- and nanocomposites, conducting polymers, nanofibers, nanoparticles, hybrid materials and hydrogels. Kalia is an experienced book editor, and he has edited a number of successful books (with Springer & Wiley), such as "Cellulose Fibers: Bio- and Nano-Polymer Composites", "Polymers at Cryogenic Temperatures", "Polysaccharide Based Graft Copolymers", "Organic–Inorganic Hybrid Nanomaterials", "Polymeric Hydrogels as Smart Biomaterials", "Biopolymers: Environmental and Biomedical Applications", "Polymer Nanocomposites based on Organic and Inorganic Nanomaterials", "Biodegradable Green Composites", etc. Kalia is the main editor of the "Springer Series on Polymer and Composite Materials". In addition, he is a member of a number of professional organizations, including the Asian

Polymer Association, Indian Cryogenics Council, the Society for Polymer Science, Indian Society of Analytical Scientists, and the International Association of Advanced Materials.

Lignocellulosic Materials of Brazil—Their Characterization and Applications in Polymer Composites and Art Works

Kestur G. Satyanarayana, Thais H. S. Flores-Sahagun and Pamela Bowman

Abstract Growing environmental concerns in recent years have led to finding new resources to replace the synthetic materials such as glass and carbon fibers in the development of polymeric and cementitious matrices-based composites. One such material is lignocellulosic fibers, due to their inherent characteristics such as abundant availability and renewability, low weight, biodegradability, interesting specific properties, etc. Studies have been carried out throughout the world in the utilization of these fibers particularly in the development of light weight, but high-performance composites. Some of their products are already in use in many applications. Although these fibers have been used even in sculptures or in arts in general, there is no report on the characteristics required for such applications. This chapter focuses on the exploration of these aspects particularly about the appearance of some plant fibers as well as their texture and potential use in polymeric sculptures or in arts. While doing so, the chapter will also present availability, methodologies used for the characterization as well as the reported properties (chemical, physical, mechanical thermal, electrical, environmental, and moisture absorption) of various Brazilian-based lignocellulosic fibers. It is hoped this will be applicable to fibers from other parts of the world to open up new areas of their

K. G. Satyanarayana (✉)
Formerly PIPE & Department of Chemistry, Federal University of Parana, Curitiba, Brazil
e-mail: gundsat42@hotmail.com; kgs_satya@yahoo.co.in

K. G. Satyanarayana
Poornaprajna Scientific Research Institute (PPISR), Sy. No. 167, Poornaprajnapura, Bidalur Post, Devanahalli, Bangalore, Karnataka 562 110, India

T. H. S. Flores-Sahagun
Department of Mechanical Engineering, Federal University of Parana, P.O. Box. 19011, Curitiba, CEP 81531-990, Brazil
e-mail: tsydenstricker@gmail.com

P. Bowman
3033 Navajo Lane, 84604 Provo, UT, USA
e-mail: pam@pambowman.com

S. Kalia (ed.), *Lignocellulosic Composite Materials*, Springer Series on Polymer and Composite Materials, https://doi.org/10.1007/978-3-319-68696-7_1

utilization in a scientific manner enabling new value-added application for these agro-industrial byproducts, which may otherwise go waste.

Keywords Lignocellulosic materials · Chemical characterization
Thermal characterization · Textile properties · Tensile properties
Flexural and impact properties · Surface treatments · Polymer composites

Contents

1 Introduction

Increasing global pollution and climate changes have been the focus areas for environmental safety leading to increased environmental awareness during the past few decades. One of the solutions to overcome these is by increasing the use of biodegradable and locally available materials in the development of new materials. The use of such renewable materials also helps in increasing the economic conditions of the farmers all over the world when such materials are used due to the value addition they make. Thus, these natural resources contribute significantly to a country's gross domestic product (GDP) in its economic activities, besides contributing to the social and economic development particularly in the case of developing countries such as Brazil. In fact, this trend of transition of substitution of hitherto used and largely produced materials based on dwindling and expensive fossil (petrochemical) or mineral resources by materials from renewable resources (plant and animal based) would lead to a bio-based economy and sustainable developments. This is due to the Kyoto protocols on greenhouse gas reduction and CO_2 neutral production, which would lead to high perspectives for plant-based fiber markets. Besides, increasing quest for new materials since 90s supporting global sustainability and diversification of non-nutrient areas addressing global environment and stable economy through agricultural industry, and successful use of their composites in many sectors (automotive, construction, etc.) and increased use of such materials becomes important for these countries through the development of new processes and products.

One such resource, which has attracted the attention of researchers all over the world, is plant-based materials, particularly the fibers (also called "lignocellulosic fibers") obtained from different parts of plants. This is particularly so in them being used to replace the synthetic fibers such glass or carbon fibers in the development of polymer or cement/ceramic-based composites (Abdul Khalil et al. 2012; Dicker et al. 2014; La Mantia and Morreale 2011; Satyanarayana et al. 2009a, b; Thakur et al. 2014) due to two reasons. First, these fibers present attractive characteristics. These include abundant availability, low cost, eco-friendly and bio degradability, better stiffness, comparable specific strength, Young's modulus, etc. Besides, use of these renewable and abundantly available materials would contribute to sustainable development as well as generating employment opportunities to the rural community, where these are available/produced (Satyanarayana et al. 2009a). Such replacement calls for complete knowledge of the characteristics of these materials.

Second reason is that greater attention being paid in recent years for the use of natural materials in view of the increasing cost of nonrenewable petroleum-based matrix and reinforcements in the development of composites. In addition, longer time (about 106 years-geological time) is taken for producing fossil-based materials such as fibers from biomass when compared to the shorter time (1–10 years) to convert them (Narayan 2006). Besides, the petroleum-based materials are continuously becoming expensive in addition to their persisting environmental degradation due to the increasing release of CO_2.

Plants that are of interest to the readers of this book are those normally referred to as "fiber crops", which are the crops grown for their fibers and are generally harvestable after a single growing season. These are distinct from other lignocellulosic materials such as trees. The latter ones are typically grown for many years before being harvested for other purposes (named as wood) such as making furniture, building materials (doors, windows, etc), pulp for paper making fiber, etc. Various types of plant fibers are available throughout the world. Names of various plants, the area under cultivation for these and amount of fibers obtained from the plants annually all over the world are normally published by FAO. These can be seen from the annual reports of FAO. It is reported that of the annual production of more than 13 million tons of various fibers, major fiber yielding plants ones are cotton and flax, while hemp with ramie, jute, and sisal, being less important plants grown in small numbers (http://lifeofplant.blogspot.in/2011/02/plant-fibers.html-accessed). Another report has estimated average global production of different plant fibers of about 29.61 million tons of major plant fibers per average in recent years compared to that of 3.3 million tons manmade cellulosic fibers (Proceedings of the Symposium on Natural Fibers, Rome 20 October 2008. Available at ftp://ftp.fao.org/docrep/fao/011/i0709e/i0709e.pdf).

Since fibers can be obtained from different parts of the plants, they are broadly classified as (a) bast or bark or stem-skin fibers (fibers from different types of grasses such as esparto, banana, flax, hemp, jute, kenaf, malva, ramie, sugarcane bagasse, etc); (b) leaf fibers (curauá, pineapple, sisal, etc.); (c) fruit or seed fibers (coconut, cotton, Luffa cylindrica, etc) and finally others such as bamboo and different types of straws (rice, wheat, etc). Short fibers of various types of woods can also be considered under this. Besides, there is also economic classification of various fibers such as brush fibers (stiff tough fibers, small stems, and twigs are used in making brooms and brushes), reinforcing or filling fibers (for mattresses, cushions, upholstery, packing and in polymer composites as reinforcing or fillers), fibers for paper industry includes textile fibers and wood fibers, natural fabrics (From barks in layers or sheets made of tough interlacing fibers as possible rough substitutes for cloth; For e.g., Luffa cylindrica), rough weaving and plaiting fibers (used in matting, baskets, hats, footwear, seat chairs, etc.) and finally the textile fibers (possible for twisting, knitting spinning, and weaving) (http://www.faculty.ucr.edu/~legneref/botany/fibers.htm).

These fibers are obtained from different parts of the plants and by different extraction methods for different fibers practiced by different countries, which include manual, machine, and microbial techniques. More details on this aspect can be seen in the earlier publications, particularly from the authors group (Satyanarayana and Wypych 2007; Satyanarayana et al. 2009b). These methods include (a) mechanical extractor (coir), (b) decorticator (banana and pineapple fibers); (c) periquita (curauá and sisal fibers), and (d) splitting using knife followed by beating them by wooden mallet (malva and piassava). Some of the other fibers such as sugarcane bagasse (byproduct of sugar industry), bamboo and straws of rice and wheat, are mostly available without resorting to any extraction methods. Accordingly, the characteristics of these fibers vary considerably depending on the part from which they are extracted, the quality of their origin, age of their source, and their preconditioning (Satyanarayana and Wypych 2007). Thus, it can be said that the properties of plant fibers depend not only on the location where they are produced, but also on the method of extraction.

Utilization of these fibers is important in order to avoid these natural resources to go waste and hence studies have been carried out throughout the world toward this end particularly in the development of light weight, but high-performance composites, besides their conventional applications in making ropes, mats, mattresses, etc. Based on the characteristics, many of the products so developed are already in use in many applications, some of which have become very attractive due to the natural appearance. For example, less-resistant plant fibers may be efficiently used in applications such as packaging, interior components, etc, while fibers exhibiting high strength properties can be used in load-bearing applications. One of the other nonconventional uses of these fibers is in the field of art works considering the great diversity and often low cost, which become incentives for their use in nets, packaging threads, blankets, and carpets. It may be noted that this is possible including in making sculptures or in arts in general, even in the absence of any report on the characteristics required for such applications.

Considering the above and publication of various reports on the sources and amount of these fibers, characteristics, and applications in various fields, this chapter will present availability of plant-based fibers, their characterization on these in respect of their reported properties (chemical, physical, mechanical thermal, electrical, environmental, and moisture absorption) and morphology of various Brazilian-based lignocellulosic fibers and methodologies used for the same. In addition, the chapter also presents some attempts made in the country by various researchers in the area of lignocellulosic fibers to prepare polymeric composites using these fibers and use of these as well as those of fibers themselves in arts.

2 Plant Fibers of Brazil

2.1 Plants and Fibers—Availability and Production

Brazil, one of the fast developing countries, is the fifth in terms of area in the world and the largest country among the South American countries with about 50% of this region, having about 5–6% arable lands, 22% permanent pastures, and 58% forests and woodlands totaling to an area of about 8.5 million km^2 (Satyanarayana et al. 2007). It has a large number of natural resources, which play a dominant role in the economic activities contributing substantially to its GDP. Besides, this also contributes to the social and economic development in the case of developing countries such as Brazil. One of these natural resources is the lignocellulosic materials-based ones, which are available to the extent of about 2.5 billion tons and have been used since more than 6000 BC (Kozlowski et al. 2004). Table 1a lists harvested area for various types of plants (fiber related) in Brazil, quantity of fibers produced, average yield, value of the production, variation of the production, and the value of the production in relation to the previous year of the main products during 2013 [IBGE], while Table 1b shows names of some of the fibers and their botanical names. Figure 1a shows the resource plants of some of these fibers, particularly not so commonly reported, while Fig. 1b shows some of the fibers. The numbers shown in Fig. 1b (vii and viii) are the lengths of fibers (Naldony et al. 2016). It is reported that Brazil produces about 56% of the total global production of sisal thus becoming the world's leading producer of sisal (http://www.ipsnews.net/2006/10/ecobreves-brazil-plant-fibers-to-replace-asbestos/).

Table 1 **a** List of harvested area for various types of plants (fiber related) in Brazil, quantity of fibers produced, average yield, value of the production, variation of the production, and the value of the production in relation to the previous year of the main products during 2013 (IBGE 2013), **b** some of the Lignocellulosic fibers available in Brazil and their botanical names

Name of plant	Harvested area (ha)	Quantity produced (tons)	Medium yield (Kg/ha)	Value of production (in R$ 1000)	Variation in production in relation to the previous year (%)	Variation in the value of the production in relation to the previous year (%)
Banana	485,075	6,892,622	14,209	5,114,223	−(0.1)	16.3
Cotton	975,046	3378.502	3465		(−)6.9	
Curauá						
Jute	396	490	1237	822	(−)50.7	(−)50.1
Malva (Fiber)	6612	9502	1437	16,384	12.2	21.8
Piassava						
Pineapple	63,204	1,655,887	26,199	1,854,309	(−)2.5	7.3
Ramie	150	495	3300	990	(−)20.7	(−)11.8
Sisal (Fiber)	176,739	150,584	852	188,120	69.0	89.8
Sponge guard						
Sugar cane bagasse	10,195,166	768,090,444	75,339	42,946,610	6.5	6.2
As per IBGE, 2016	9,681,815	733,824,496	75,794		(−)4.1	

Sl. no.	Name of fibers	Botanical name of fibers
1.	Giant cane or reed/wild cane	*Arundo donax*
2.	Bamboo	*Bambusoideae*
3.	Banana	*Musa Sapientum*
4.	Brazil nut	*Bertholletia excelsa*
5.	Buriti	*Mauritia flexuosa*
6.	*Pampas* grass	*Capim-dos-pampas (Cortaderia selloana)*
7.	Coir	*Cocus nucifera*
8.	Cotton	*Gosspium*
9.	Curauá	*Ananas erectifolius*
10.	Jacitara	*Desmoncus polyacanthos*
11.	Jute	*Corchorus Capsularis*
12.	Malva	*Malva sylvestris*
13.	Piassava	*Attalea funifera*
14.	Pineapple	*Ananas comoscus*
15.	Mesquite (shrub or small tree)	*Prosopis juliflora*

(continued)

Table 1 (continued)

Sl. no.	Name of fibers	Botanical name of fibers
16.	Ramie	*Boehmeria,*
17.	Sisal	*Agave sisalana*
18.	Sponge Gourd	*Luffa cylindrica*
19.	Sugarcane bagasse	*Saccharum officinarum*
20.	Taquara-lixa	*Merostachys skvortzovii* Sendulsky
21.	Various types of straw	
22.	Various types of wood	

Fig. 1 **a** Photographs of some Brazilian Plants used for fiber extraction: (i) Arundo donax; (ii) Bamboo; (iii) Buriti palm tree; (iv) Capim-dos-pampas (Cortaderia selloana) plant; (v) Curaua plantation (*Ananas erectifolius*); (vi) Piassava; (vii) Prosopis juliflora; (viii) Ramie plant; (ix) Taquara-lixa (*Merostachys skvortzovii* Sendulsky); (x) Brazil nut (*Bertholletia excelsa*) tree; (xi) Brazil nut shell and bur fiber. Except for (x) all others Reproduced from Internet websites; (x) Reproduced from Inamura et al. (2011) with the kind permission of the publishers. **b** Photographs of some fibers obtained from Brazilian plants: (i) Buriti fiber; (ii) Curauá; (iii) Piassava; (iv) Ramie fiber; (v) taquara-lixa. Reproduced from (i) Monteiro et al. 2011; (ii) Fererira et al. (2010); (iii) Piassava Nascimento et al. (2012); (iv) Monteiro et al. Unpublished); (v) from Neto (2014); (vi) Arundo fibers isolated from the stem Fiore et al. (2014); (vii) Pine fibers, 5x; (viii) Eucalyptus fibers, 5x, both showing varying lengths of respective fibers from Naldony et al. (2016); (ix) Taquara-lixa fiber Neto (2014). Reproduced with the kind permission of the authors and Publishers

From the above, it becomes evident that these renewable resources are available abundantly in Brazil in quite good amounts. Similar data is also published for other countries of the world with updates every year.

According to IBGE, in Brazil, with better climatic conditions for the agricultural production in 2013, the total harvested area during this year was 527, 55, 763 ha producing total 64 crops producing about 188, 052, 443 tons of products yielding a value of R$ 122, 456 015 (IBGE 2013). This presented 20.4 variations in the value of the production in relation to that of the previous year.

2.2 About Fibers

Figure 2 shows the classification of various lignocellulosic fibers of Brazil.

In the following pages, all the Brazilian fibers will be briefly discussed:

2.2.1 *Arundo Donax* (Giant Cane or Reed/Wild Cane)

This plant is grown in about 48 countries and is called as "arundo grass" or "bamboo reed" or cane or "giant reed" in English and "cana- do-reino" or "cana-do-brejo" or "capim-plumoso" in Portuguese (Brazil). It is tall (about 6 m) and robust grass having wide (5 cm) and long (30–61 cm) leaves (http://issg.org/database/species/ecology.asp?si=112&fr=1&sts=sss&lang=EN). It mostly occurs in agricultural areas, coastland, desert, natural forests and grows near the water tables. Due to its appearance, it is grown as an ornamental species. It is used in making reeds for woodwind instruments and also as medicine. Fibers from *Arundo donax L.*,

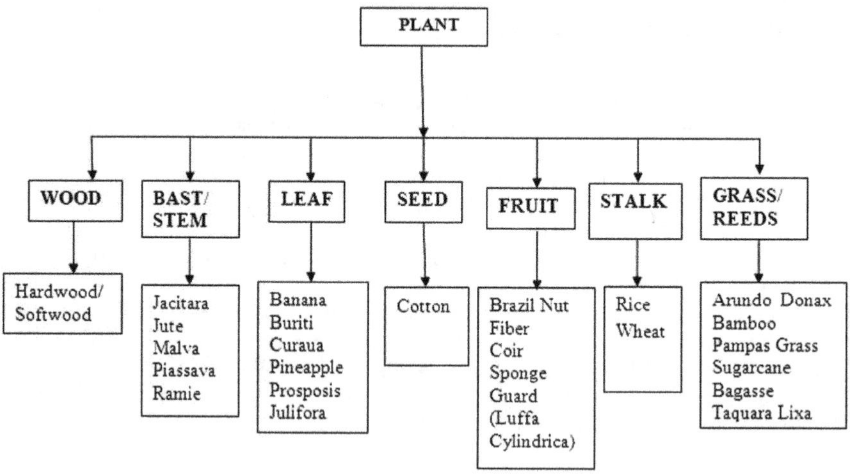

Fig. 2 Classification of lignocellulosic fibers of Brazil

extracted from the outer part of the stem of the plant are abundant in the Mediterranean area and grow naturally in Sicily, representing an invasive and aggressive species.

2.2.2 Bamboo

According to Jessica McMathis of Pennsylvania, USA, "bamboo is Mother Nature's magic material." This is an abundant plant material that grows in most developing countries; it is sustainable and strong, and can replace steel in places where it is available in plenty (McMathis 2014).

There are about 80 varieties of bamboo grown in Brazil particularly in sub-tropical and temperate zones with João Santos Industrial group (108 km^2) in the Northeast of the country being the largest commercial cultivator (Filho and Bahr 2004).

Taquara-lixa (*Merostachys skvortzovii* Sendulsky) belongs to the family of bamboo with about 155 taquaras species and bamboos (Liebsch and Reginato 2009; Neto 2014). This is mostly grown in Rio Grande do Sul (Brazil) to a height of about 7 m and diameter of about 4–6 cm in with hollow stem having life varying from 3 to 120 years (Liebsch and Reginato 2009). Like other varieties of bamboos, this has a variety of uses including construction sector starting from beams and light sleepers, light fences, in making baskets and as pipes (indigenous musical instrument of Northeast Brazil) (https://pt.wikipedia.org/wiki/Taquara).

2.2.3 Banana Fibers

Brazil ranks third in the production of banana and its fibers (Satyanarayana and Wypych 2007; Satyanarayana et al. 2009b; Homma 2001) and therefore it is not surprising that this fiber has been extensively studied by the researchers here. These fibers are obtained from the pseudostem (24% of this) using decorticators with 1–2% of fiber yield on dry weight basis. According to IBGE, annually 7.1 million tons of fibers are produced in the country (IBGE 2013). The dimensions of these stems are about 3–9 m long and 200–370 mm in diameter. In view of their existence in different layers with different colors, these exhibit different properties. Although large quantity of these stems are available in the country, only 2% is utilized annually for artistic works compared to about 10% in other banana-producing countries both by the artisans, as energy source and in other industrial applications (Satyanarayana et al. 2007; Kozlowski et al. 2004).

2.2.4 Brazil Nut Fibers

Brazil nut fruit (*Bertholletia excelsa)* is from the tree which mostly available in Latin America, mostly in Amazon basin River in Brazil and partly in Bolivia and

Peru (Inamura et al. 2011; Santos et al. 2010). This tree is large with a possible height of about 50 m and a trunk diameter of about 10–12 m. Each tree produces 30–1000 burs depending on the age of trees, which range from 200 to 500 years. The nut is characteristically a spherical capsule having bur and shell fibers on it. Considering the increased interest in the use of lignocellulosic fibers characterization of these fibers has been made in respect of their chemical properties including composition, thermal properties, and morphology (Inamura et al. 2011).

2.2.5 Buriti Fibers

These fibers are obtained from leaves of buriti (*Mauritia flexuosa* Mart.) or moriche palm, which is "mostly grown in Amazon Forest and Cerrado biomes and throughout South America" (Cittani and Baruque-Ramos 2016; www.ispn.org.br). *This tree is most common and abundant palm tree in the country with its grouping under about 2000 genera represented by about 1500 species* (Henderson et al. 1995; Lorenzi et al. 2004). It is reported that rural population of many states in Brazil depend on this tree known as "Tree of Life" as it is a major source of income to them because of full use of every part of this tree (Cittani and Baruque-Ramos 2016). In fact, its fruits, being nutritious, are used for juices, oil extracted from the fruit of this palm is reported to have medicinal value and used by traditional people, while other parts of the palm are used in creams, shampoos, etc. (www.ispn.org.br; http://www.dsenyo.com/blogs/dsenyo/14106133-buriti-a-gift-from-the-tree-of-life#). Its stem is used to make furniture as it is resistant to degradation. The fibers also called as buriti "linen or silk" are obtained from the leaves in their early days (Cittani and Baruque-Ramos 2016; www.ispn.org.br). In general, while these fibers are used in handicrafts (bags, carpets, tablecloths, and toys), there is good scope for their use in design and industry.

Good quality fibers are extracted from the leaves (from the petiole or leafstalk) immediately after harvesting using small and sharp knives after carefully pulling a membrane from the leaves (Barbosa et al. 2012; http://www.dsenyo.com/blogs/dsenyo/14106133-buriti-a-gift-from-the-tree-of-life#). In order to make the fibers strong, they are boiled in water. Then the fibers are dried in sun followed by making them into different strands based on the length of the fibers. Further processing of fibers depends on their final usage; for example, for crochet work, fibers are twisted, for making handbags they are woven, etc.

2.2.6 Capim-dos-pampas (*Cortaderia Selloana*) *Pampas* Grass Fibers

These fibers are obtained from the evergreen grass, which is an ornamental garden (flowering) plant with razor-sharp leaves and silvery white or silvery pink plumes. It is perennial and native of some of the South American countries including Brazil, while it was introduced to some of the European countries, North America and Australia (https://en.wikipedia.org/wiki/Cortaderia_selloana; http://www.florese folhagens.com.br/capim-dos-pampas-cortaderia-selloana/). It has different nomenclature depending on its country of origin. This grass having numerous stems of about 1.5–3 m height, is cultivated mainly for its leaf fiber, which is used for making paper, while its plumes picked and dried immediately when they come out, are used in flower arrangements. These leaves are tender with about 1–2 m long and about 1 cm wide and having different colors (blue–green or silver–gray). Fibers are extracted from these leaves.

2.2.7 Coconut Fiber or Coir Fiber

Coconut is grown in most parts of Brazil, but is a typical species of northern part of Brazil with yield per tree being the highest among the coconut-growing countries. It is grown in about 300, 000 ha with the annual production of about 2.8 million tons of coconut (IBGE 2013).

Its fiber or "coir" as it is also called is obtained from husks of coconuts (mesocarp of the fruit), which constitutes about 25% of the fruit—coconut. Typical to this country, green husks are mostly used to extract fibers from mostly immature coconuts (plucked after 6–8 months), which are collected from various places such as the beaches under a social program [Poverty and Environment in the Amazon (POEMA)], where the nuts are used for drinking its tender water. The extraction process is by mechanical means and it is reported that 8 tons of fibers are obtained from about 1.2 million nuts per month (Satyanarayana et al. 2009b). It may be noted that properties of these fibers from green coconuts would be different from those extracted from mature coconuts (plucked after about 12 months) as will be seen later in the chapter. In fact, this is one of the fibers which has high demand in the country particularly by the automotive industry and thus provides income to many families in many parts of the country.

2.2.8 Cotton

This fiber, one of the textile fibers (wool, ramie, silk, and jute), is extracted from plants of the genus Gossypium (Valadares 2014). The world's main producers of this fiber are China, India, USA, Pakistan with 26, 20, 14, and 9%, respectively, while Brazil contributes only 6%, with other countries accounting for the balance 25% of total production during 2012 (Valadares 2014). Brazil having the largest area planted with cotton and achieves the highest productivity in the country, this fiber constitutes about 46 and 93% of the natural and chemical fiber production of the world and the natural fiber production of Brazil, respectively (Filho et al. 2004; Valadares 2014). Although earlier most of the fiber produced is used for domestic consumption mainly for textile purposes (Satyanarayana et al. 2007; Valadares 2014), in view of large quantities of production of this fiber in the country in recent times, Brazil has been one of the most important cotton exporters almost since the 80s. During 2016, cotton was cultivated in 975,046 ha and 3,378,502 tons of cotton was produced. Some of the states in the country that produced cotton were Roraima, Tocantins, Maranhão, Piauí, Ceará, Rio Grande do Norte, Paraíba, Pernambuco, Alagoas, Bahia, Minas Gerais, São Paulo, Mato Grosso, Mato Grosso do Sul, and Goiás.

2.2.9 Curauá Fibers

Curauá plants are grown in about 100 ha (IBGE 2013). This plant belongs to the pineapple family and is a hygrophilous species from the Amazon region (http://www.revistaagroamazonia.com.br/13-fibras.htm). The leaf of this plant is about 1–1.5 m long and 4–5 cm wide with about 5 mm thickness, yielding ca. 8% fiber on dry weight basis. In view of its high strength properties (as will be seen later in the chapter) and hence most competitive among country's traditional fibers, there is greater demand for this fiber particularly from the automotive industry in the country, because of which, the area in which this plant is cultivated has been increased substantially in recent years

2.2.10 Jute Fibers

This plant was introduced to Brazil between 50s and 80s and the harvested area is about 396 ha and the amount of fibers produced is about 490 tons per annum during 2013 (IGBE 2013). The fibers from this plant belong to the stem fibers variety and the production of these fibers has dwindled over the years.

2.2.11 Malva Fibers

Herbaceous Malva plants belong to about 25–30 species of annual or biennial or perennial nature (https://en.wikipedia.org/wiki/Malva). These plants are cultivated mostly in the meadows of rivers with the largest producer being the Amazon region in the country (http://www.conab.gov.br/download/cas/semanais/Semana07a11022005/Juta%20-2007a11-02-05.pdf). The plants are harvested in an area of about 6612 ha (IBGE 2013). The leaves of this plant are not only used for fibers, but also in the traditional Austrian medicine (as tea or as baths for treatment of disorders of the skin, gastrointestinal tract, and respiratory tract) (https://en.wikipedia.org/wiki/Malva). The fibers belonging to bast or stem variety extracted (Process similar to that used for jute fibers) from plants and about 9502 tons per annum are produced (IBGE 2013).

2.2.12 Piaçava (Piassava) Fibers

Piaçava plant is a special plant of the country, being the type of a palm tree and grown mostly in Atlantic rainforest region of the country. In view of their low costs involved in their planting, maintenance, and exploitation, these plants have become very attractive in the country. The fibers also known as "piaçaba" or "piassava" fibers obtained from the leaf stalks of the palm tree belong to the stem fiber variety and are manually extracted with most production ($\sim$90%) coming from the state of Bahia (Satyanarayana et al. 2007, 2009b). The yield of about 8–10 kg of fiber per year from one plant is reported with the plant being productive for about 20 years. Most of the long fibers thus produced are exported to mainly to Europe and USA, while the shorter varieties are used in the country itself.

2.2.13 Pineapple Fibers

Brazil is the third largest producer of these fibers after Thailand and Philippines with its northeast part being the major contributor (Satyanarayana et al. 2007). The total harvested area in the country is about 655,887 ha (IBGE 2013). The fiber is extracted using decorticators (called "periquita") from the leaves, which weigh about 65 g and have dimensions of length, width, and thickness of about 1–1.5 m, 4–5 cm and 5 mm, respectively. The fiber yield is about 8% per leaf and the most competitive among the traditional Brazilian fibers.

2.2.14 Prosopis Juliflora

This plant belonging to Fabaceae family is a shrub or small tree and native of South America besides Mexico and Caribbean (https://en.wikipedia.org/wiki/Prosopis_juliflora). The height of the plant is about 12 m with its trunk having a diameter of about 1.2 m. While the stem of the plant is used as alternate to wood for both fuel and construction (Sirmah et al. 2008), its leaves with 12–20 leaflets with 20–30 cm long are used for extracting fibers (https://en.wikipedia.org/wiki/Prosopis_juliflora).

2.2.15 Ramie Fibers

Ramie plants are cultivated in an area of about 150 ha. Fibers from this plant are obtained from their stems. Brazil was third largest producer of this fiber in the world in the 90s, while it is reported that the annual production of the fibers is about 495 tons (IBGE 2013).

2.2.16 Sisal Fibers

These plants are cultivated in an area of about 243,759 ha (Harvested area: 176,739 ha) (IBGE 2013). There are different varieties in agave family with *Agave sisalana* variety being grown in Brazil with production being largely in small scale. It is reported that in recent years, Brazil is the world's leading producer of sisal, generating about 56% of the global total (http://www.ipsnews.net/2006/10/ecobreves-brazil-plant-fibers-to-replace-asbestos/). This plant (*Agave sisalana* variety) is an important crop in the country, since it provides employment to more than one million people mostly in northeastern part of Brazil (Satyanarayana et al. 2007). This being a leaf fiber with each leaf containing about 700–1400 long fibers of sufficient length produced by using a decorticator (called "Periquita"). Annual production of this fiber is about 150,584 tons (IBGE 2013).

2.2.17 Sponge Gourd (*Luffa Cylindrica*) Fibers

Sponge gourd (*Luffa Cylindrica*) plant, perennial and trailing herb belonging to a generic group of eight species known as "vegetable sponges" is another special fiber of Brazil, being the largest producer of this fiber besides Africa (Satyanarayana et al. 2007, 2009b). These fibers belong to fruit fiber variety and are in the form of mats with multi-direction arrangement in the fruit showing entangling around a beehive-like structure of the nucleus. These fibers when used as reinforcements in polymers are reported to change the failure mode to controlled type from abrupt type (Almeida et al. 2000).

2.2.18 Sugarcane Bagasse

Brazil is one of the leading growers of sugarcane, being second after Cuba (Filho and Bahr 2004). Bagasse being 30% of sugarcane is produced after squeezing the juice from it and has high potential in view of its good properties, both physical and mechanical (Sartori et al. 2001).

2.2.19 Other Lignocellulosic Fibers

There are many other types of lignocellulosic fibers available in Brazil such as varieties of wood, straws of rice and wheat, rice husk ash, various types of weeds, etc. All these can be used as fillers in composites (polymer and ceramic/cement matrices) and other products such as active carbon (Costa et al. 2000; Satyanarayana et al. 2004). However, except for rice straw and wood fibers, no scientific data is available in respect of others in this category and hence brief mention is made on some of these here.

Prosopis juliflora bark have been used for fuelwood, fodder and can grow in poor soils, tolerate high temperatures (48 °C), and a great amount of rain (150–750 mm). Its high amount of cellulose (61.65%) and low density makes it a potential reinforcement for polymeric matrices to prepare composites (Saravanakumar et al. 2013).

Yet another lignocellulosic fiber of Brazil, which has not received much attention, is Amazonian vegetable fiber called jacitara (*Desmoncus polyacanthos* Mart.). This native from Amazon basin region of Negro River, Brazil is another very common palm species (Arecaceae family) (Fonseca et al. 2013). These fibers are extracted from the stem of jacitara palm and widely used for preparing handicraft utensils and instruments by traditional Amazonian communities and thus play an important cultural role for these communities. It is widely commercialized in South America. From the point of view of morphology and quality of these fibers, they are similar to the Asian rattans (*Calamus* spp.), which are used widely in furniture industry due to their quality. However, except for only one study, there seems to be no other published report on the technical and scientific information about these fibers, which would help in exploring new applications for this species of Amazon (Fonseca et al. 2013).

2.2.20 Textile Fibers

Textile fibers are fibers that differ in their characteristics from those of normal fibers in that they can be made into yarn or fabric by spinning, knitting, twisting, or braiding (http://textilelearner.blogspot.in/2012/02/what-is-textile-fiber-types-of-textile.html; http://textilelearner.blogspot.in/2011/07/textile-fiber-properties-of-textile.html; http://textilefashionstudy.com/textile-fiber-general-properties-of-textile-fiber/). Thus textile is referred as the filaments, yarns (both being one dimensional) and as woven fabrics (two or three dimensional), which can be natural (cotton, jute, pineapple, etc.) and manmade (synthetic) variety (Nurul Fazita et al. 2016).

They should be uniform, fine (structural property) having luster, long (min. 5 mm or 100 times its diameter or width), flexible with good strength and durable. In addition, they should not be soluble in water or any organic solvent (http://textilelearner.blogspot.in/2011/07/textile-fiber-properties-of-textile.html; https://textileapex.blogspot.in/2015/11/properties-of-textile-fibre.html). Such modifications not only improve their wettability with the polymer matrices, but also reduce the moisture absorption and sometimes impart special properties and ease of processing (Satyanarayana et al. 2009b).

2.3 *Surface Treatments/Modification of Fibers*

Hydrophilic characteristics along with their higher polarity of lignocellulosic fibers pose a problem in the adhesion with polymeric or cement matrices (due to hydrophobicity). This results in fiber pull-outs leading to low mechanical properties. This problem can be overcome by giving surface treatments through chemical modification of lignocellulosic fibers or the use of coatings to these fibers or by using compatibilizers. This not only helps in providing good adhesion between the fibers and the matrix through improved wettability of the fibers (this transfers the applied load to the fibers through shear stresses at the interface) but also makes the composite unsusceptible to moisture. Besides this, sometimes the surface modifications of lignocellulosic fibers impart special properties and ease of processing (Satyanarayana et al. 2009a, b). While the surface modification enhances the strength properties, the coating of fibers enhances hydrophobicity and degradation resistance when exposed to environment leading to their enhanced durability (Abdelmouleh et al. 2004; Almeida et al. 2000; Amico et al. 2005; Barretto et al. 2010; Bisanda 2000; Bledzki and Gassan 1999; Boynard and Almedia 2003; Horeau et al. 2004; John and Venkata Naidu 2004; Mani and Satyanarayana 1990; Mohanty et al. 2000, 2001; Prasad et al. 1983; Ramos 2003;. Ramos et al. 2005; Santos et al. 2006; Satyanarayana 2004; Tanobe et al. 2005; Tomczak et al. 2006; Trindade et al. 2004).

It is relevant at this juncture to note that no surface modification is required for the fibers in their use for the preparation of biodegradable composites. But attempts have also been made by some researchers working with such composites to use these treatments with a view to improve further adhesion and to impart special features (Satyanarayana et al. 2009a, b). These features include fiber incorporation up to 60–85%, reduction in moisture sensitivity and durability of the composites and their nonstructural but high-grade applications.

Beside this, processing temperatures also play an important role in the good performance (properties) of composites and hence this should be properly selected (Satyanarayana et al. 2007, 2009b). In fact, surface treatment aspects of lignocellulosic fibers have been reviewed from time to time (Bledzki and Gassan 1999; Mohanty et al. 2000, 2001, 2002; Nurul Fazita et al. 2016). It seems mostly chemical methods (alkali treatment or mercerization, treating with methacrylamide aqueous solutions) (Nascimento et al. 2012; Simão et al. 2016; Tanobe et al. 2005) and physical methods (subjecting to radiation, steam explosion) (Barbosa et al. 2012; Ricardo et al. 2011) have been used by the Brazilian researchers. There seems to be only one study reported on the matrix treatment (annealing) by Brazilian researchers (Lomeli Ramirez et al. 2011).

2.4 *Justification for This Chapter and Objectives*

From the foregoing, it becomes clear that various lignocellulosic materials are available in Brazil and some in large quantities. Their prices could be either low or comparable with international prices. Considering the international scenario about the characteristics and possible and potential uses of some these fibers, one can assume these lignocellulosic materials of Brazil have a good future not only in the composite technology, but also in various other sectors. Thus, if proper attention is paid to utilize these to their potential, one can safely assume that many of Brazilian lignocellulosic fibers would meet the socioeconomic parameters of the country. For this purpose, it is necessary to study, understand their characteristics, and applications in various sectors. Accordingly, in the following sections, these are discussed based on the published reports on the above-mentioned items. For example, dimensions of various fibers including those of wood fibers, their chemical, physical, and mechanical properties, surface modifications of these fibers and their utilization in applications particularly in composites are reported. Some of the glaring observations from these reports include: (i) limited availability of both jute and ramie although they are good as reinforcements in polymers; (ii) very limited work on malva fiber-based composites is done in the country; (iii) no report of such work on the use of bamboo fibers, straws of wheat and rice in composites; (iv) although viability of preparing hybrid composites using cotton fibers is established in the country, availability of only small amounts of these fibers in view of most of the available cotton fibers is used for textile purpose, etc. All these point to the fact that the trends are available in Brazil for the utilization of these fibers both short-term [synthesis and characterization of composites based on the fibers] and long-term [alternatives for synthetic fibers and as possible substitutes for wood] objectives (Satyanarayana et al. 2005, 2007). More details of the attempts made so far in the utilization of lignocellulosic fibers in composites will be presented in a later section (Sect. 5.1).

Considering the importance of these fibers for Brazil and the scattered data on their availability and properties, an attempt is made in this chapter to bring out consolidated data on the sources and availability of all the useful Brazilian lignocellulosic fibers, organizing the available information on morphology, properties (chemical, physical, mechanical thermal, electrical, environmental, and moisture absorption), and methodologies used for the same and current uses. In addition, attempts made to prepare polymeric composites using these fibers and use of these as well as those of fibers themselves in arts will be discussed. It is relevant here to state that since this book would contain many specific chapters on various aspects of different types of lignocellulosic fibers available in various countries of the world and possibly their composites [thermoplastic, thermoset, "green" and nanocomposites], only brief discussions on the composites developed in Brazil using its fibers will be presented. This will include fabrication, morphology, properties, and product development as well as their applications in various sectors. The specific use of lignocellulosic fibers of Brazil in arts, an area not generally given due recognition in many books dealing with these fibers is presented in this chapter, mostly based on the work of two of the present authors. Finally, some future directions to be followed for further development of research on these fibers including suggestions for better utilization, which may help in the evaluation of their so far unknown properties, will be presented as part of concluding remarks. This is aimed at motivating the readers and interested researchers to explore new avenues for these fibers to underline the future potentials of these natural resources, which is expected to contribute toward national development. Besides, it is hoped that the presentation of the work on lignocellulosic fibers of Brazil attempted here will be applicable to fibers from other parts of the world, which would help to open up new areas of their utilization in a scientific manner enabling new value-added application for these agro-industrial byproducts, which may otherwise go waste.

3 Characterization of Plant Fibers

Recognizing the facts that (i) globally lignocellulosic fibers have been increasingly used in the development of composites keeping the environmental concerns and (ii) a number of lignocellulosic fibers are available in the country, but are not properly utilized thus requiring serious attention for their characterization, proper use, and thus value addition, Brazilian researchers have undertaken characterization of the available fibers. Of course, some of the fibers have been characterized as part of the development of composite materials in the country (Satyanarayana et al. 2007, 2009a, b), but not much attention has been paid for the systematic and complete characterization of all the available fibers except for a few. For example, there are studies of some fibers (coir, curauá, and piassava) dealing with the determination of tensile properties as functions of the dimensions of fibers and strain rate (Satyanarayana et al. 2005; Tomczak et al. 2007a, b). In the following section, methods of characterization used in the country by various researchers and results thereon will be presented.

3.1 Methods of Characterization (Techniques and Equipment Used)

Once the quantity of plant fibers available is known, their complete characterization becomes essential for their increased utilization. This includes physical properties, chemical aspects, thermal and mechanical properties, as well as the morphology of various plant fibers along with their fractographs. It is also essential that fiber surfaces have to be modified when they are used in the preparation of composites. This is because of two reasons, viz., in view of these fibers being hygroscopic they should be protected from moisture absorption on exposure to the environment. Second, to improve adhesion between the fibers and the matrix used in view of their higher polarity and hydrophilic nature. It is reported that sometimes, special properties are imparted by such treatments along with ease of processing of composites (Satyanarayana et al. 2009a, b). For this purpose, different techniques have been used by the researchers both in Brazil and elsewhere. These include physical (exposure to different types of rays or plasma and steam explosion), chemical (mercerization, acetylation, silanation, selective oxidation, etc.), physico-chemical (solvent extraction), mechanical (rolling, swaging, etc).

Researchers in Brazil have used both qualitative and quantitative methods to characterize the plant fibers for various properties mentioned above in the as-received condition as functions of species, age, location, etc., and on surface modification of these fibers. Similarly, even characterization of their composites has been made. For this purpose, various techniques and equipment based on standard methods of testing such as ASTM, APTCP M-11/77, and TAPPI T13 m standards have been used. These are discussed in the following pages:

3.1.1 Physical Properties

Physical properties include dimensions (length and diameter) of fibers, their density, helical or micro-fibrillar angle (the angle subtended by the helically wound cellulose microfibrils with the fiber axis), crystallinity index (measure of crystalline part), moisture absorption, electrical resistivity, etc. These are briefly mentioned.

Dimensions of Fibers

Dimensional measurements and morphology studies of fibers have been carried out using both optical and scanning electron microscopes. Dimension includes length and diameters of fibers. While the length of the fibers can be easily and accurately measured using measuring scale as well as using a projection or a stereo microscope, reliable diameter measurement by a contact procedure with any kind of metallic device (caliper or micrometer) poses problem due to the relatively soft nature of most of these fibers (Monteiro et al. 2011a, b). This is due to holding of these soft lignocellulosic fibers in between the grips of the metallic device whereby natural squeezing of the fiber occurs by the application of pressure by the operator. This may result in deforming the surface of the fiber and thus reducing the diameter. Therefore, many researchers have used a reliable way to measure lignocellulosic fiber diameter using a profile projector. In this noncontact technique, amplified image or shadow of the fiber is obtained by the light beam whereby measurements of one hundredth of a millimeter of precision can be made through a mobile-graduated scale (Monteiro et al. 2011a, b). Since lignocellulosic or plant fibers vary in their dimensions, measurements are done at least at 10 distinct positions along its length. In each of these, both the larger and the smaller diameters can be recorded to overcome the problem of differing fiber cross section (neither circular nor uniform along the length). Using these values one can obtain the effective mean diameters for each fiber. Alternately, one can also use the statistical method from measurements of dimensions made on a random lot of 100 fibers using the histogram. Some researchers have used scanning electron microscope to get both the longitudinal and cross-sectional features of the fibers using gold-coated fibers and appropriate voltage and current ratings (normally, 15 kV and 20 mA).

On the other hand, to measure the dimensions of cells or "ultimate" fibers, these have to be obtained using the as-received fibers. Normally, the chemical method is used to extract these cells or "ultimate" fibers, wherein the fibers are softened by heating the soaked fibers in a mixture (1:1) of acetic acid and hydrogen peroxide in a hot air oven at 60 °C (Lomeli Ramirez 2011). Resulting "ultimate" fibers are washed and colored (saffron) following the standard technique normally used with plant fibers. Dimensions of these "ultimate" fibers are then measured using an appropriate optical microscope with appropriate lenses (magnification) fitted with a digital camera, coupled with suitable software.

Density of Fibers

Density is an important property, which is helpful in determining the characteristics of a material. Normally researchers working with the plant fibers follow one of the following methods. They are usual Archimedes Principle, the methods used by Foelkel et al., and Brasil (Lomeli Ramirez 2011) using a pycnometer or density gradient method, which is as per the standard ABNT NBR 11931, which is similar to ASTM D 1505 and ASTM D 792. The latter case is generally used for polymeric materials.

In the case of Archimedes's principle first weight of dried fiber (m_{af}) is determined in air using a suitable weighing scale having good accuracy. Then, the fibers are immersed in ethyl alcohol to obtain its apparent weight (m_{bf}) and taking the density of ethyl alcohol (ρ_x) at 25 °C as 789 kg m^{-3}, the following equation is used to obtain the density of the fibers:

$$\rho_f = m_{af} * \rho_x / (m_{af} - m_{bf}) \tag{1}$$

On the other hand, following the standards mentioned above, the density of the fibers is estimated as follows: First, immerse a portion of the sample in a beaker containing a mixture of miscible solvents (chloroform having a density of 1484 kg m^{-3} and methylene chloride having a density of 1330 kg m^{-3}) of different densities. Then, measure the volume of each solvent used and fill the beaker until the fibers stay floating at the center of the liquid mass (density of the fiber is equal to that of the liquid mass). Then, the density of the fiber is calculated with the known volumes and density values of the solvents.

Micro-fibrillar Angle and Crystallinity of Fibers

Due to the interrelation between the mechanical and thermal properties of lignocellulosic fibers with their crystallinity, the potential of lignocellulosic fibers as reinforcement in polymer matrices can be determined by the degree of crystallinity of these fibers (Oliveira and Marques 2015). Both crystallinity and microfibrillar angle of lignocellulosic fibers are determined using X-ray diffraction (XRD) method using a diffractometer with monochromatic Cu $K\alpha$ radiation (k = 1.5418 Å), at selected operating conditions (say, 40 keV, 20 mA, and slits of

1°). For this, the fiber sample was cut into a suitable size (say, 0.5–1 mm), to make a sample with a known volume (say, 100 mm^3), with a view to obtaining the best random distribution from the fibers to estimate the crystallinity index and the microfibrillar angle. Some researchers have used a bunch of fibers in view of the beam width used is 1 mm, while in some cases, with a view to avoiding any interference in the results to be obtained, the fiber samples was covered using a polyester film such as Mylar of Dupont (Satyanarayana et al. 2014).

In the method followed using Buschle–Diller and Zeronian (Buschlediller and Zeronian 1992), the following equation was used to calculate the crystallinity index:

$$Icr = 1 - I_{\min} I_{\max} \tag{2}$$

where $I_{\min}$ is the intensity at the minimum of the crystalline peak (18° < 2θ < 19°) and $I_{\max}$ is the intensity at its maximum (22° < 2θ < 23°).

The other method using area under the curve in diffractogram, the following equation is used:

$$Icr(\%) = [I(2\theta)/\text{Total area}] \times 100, \tag{3}$$

where $I(2\theta)$ is the area of the diffraction peak at the angle (2θ) associated with the crystalline region of the fiber.

On the other hand, micro-fibrillar angle (θ) can be determined by single crystal method (Laue transmission) using appropriate conditions such as radiation with suitable filters, voltage and current conditions, the orientation of the fiber sample to the incident X-ray beam and sample-film distance. Then, using a microdensitometer, the intensity distribution along the diffraction arcs, microfibrillar angle (θ) is calculated following the Preston method (Preston 1978). In recent times, modern techniques such as charge-coupled devices are used, which can collect diffraction data at large solid angles.

However, the micro-fibrillar angles of most of the Brazilian fibers are not determined experimentally as mentioned above, but some researchers working with these fibers have used the following empirical/regression equations correlating Young's modulus (*Y*) and micro-fibrillar angle or an unified relationship between tensile properties with cellulose content (*C*) and structural parameters such as cell dimensions (*L*/*D*) and micro-fibrillar angle (θ) (Caraschi and Leão 2000) reported by others elsewhere (Almedia et al. 2005; Horeau et al. 2004; Satyanarayana et al. 2007; Tomczak et al. 2007a, b; Trindade et al. 2004). The equations used are as follows:

(a) Empirical equation relating Young's modulus and micro-fibrillar angle found applicable mainly for bast and seed fibers (Treloar 1977):

$$Y_c = Y_f \times \cos 2\theta, \tag{4}$$

where Yc and Y_f are the axial Young's modulus values of fiber and cellulose and θ is the microfibrillar angle;

(b) Another empirical equation relating Young's modulus, cellulose content and micro-fibrillar angle of the fibers found applicable to all fibers (Mc Laughlin 1980):

$$Y_f = W_c Y_c \cos 2\theta + W_{nc} Y_{nc}, \tag{5}$$

where Y_f is the effective modulus, W_c is the weight fraction of the crystalline region, Y_c is the modulus of the crystalline region, W_{nc} is the fraction of the noncrystalline region, and Y_{nc} is the modulus of the noncrystalline region;

(c) Yet another empirical equation used is the one correlating tensile properties with cellulose content (C) and structural parameters such as cell dimensions (L/D) and micro-fibrillar angle (θ) (Caraschi and Leão 2000):

$$\text{Pth} = \text{Pobs} = KC_{a1}\,(L/D)_{a2}\theta_{a3} + A, \tag{6}$$

where P denotes the value of one of the properties mentioned above while K and A being constants. With this, a number of equations can be obtained depending on the number of fibers considered. Solutions for these have been found with a_1, a_2, and a_3 taking suitable values for the relation to holding good. Then, some regression equations can be derived for tensile strength and % elongation, as explained by Ramos et al. (2005). The above correlation can be used either to calculate microfibrillar angle by knowing other parameters shown in the equation or knowing any two of the structural components of the fibers, one can calculate the strength properties of fibers for which such data are not available at present.

It is interesting to note that Brazilian researchers have calculated the crystallinity index of plant fibers following different methods, such as area method (Tomczak et al. 2007a, b; Tanobe et al. 2005), or intensity method following Buschle–Diller and Zeronian (Guimarães et al. 2009).

Moisture Absorption of Fibers

Moisture content of lignocellulosic fiber is important from the point of view of their stability in various conditions. There are standards for this determination such as TAPPI T212 om-02. Determination of this is as follows:

Fiber samples of known weight were taken in a previously weighed crucible and then thoroughly washed using distilled water. The crucible was kept in a hot air oven for 1 h at some suitable temperature (60–70 °C) followed by cooling after taking it out of the oven and then weighed. This procedure should be repeated till constant weight is obtained. Then, moisture content is calculated using the equation (Satyanarayana et al. 2013):

$$M(\%) = \left(W_f - W_o\right)/W_o \times 100, \tag{7}$$

where M is the moisture content and W_f and W_o were the weights of the fibers in wet and dry conditions.

Some Brazilian researchers have used another method as briefly mentioned below: In a previously weighed beaker about 2 g of samples are taken (m_s) and heated to 105 °C for 24 h in a hot air oven. The samples are then taken out from the oven and kept in desiccators. Then, the final mass of these samples was measured (m_u). Moisture content (M) of the fibers is then calculated using the equation

$$M = m_s/m_u. \tag{8}$$

To avoid any discrepancy in the measurement, the experiment is carried out in triplicate.

Electrical Properties

By understanding the electrical properties of lignocellulosic fibers, it would be possible to find suitable applications (insulating materials for special applications, for example, bushings, gaskets, spacer panels, etc.) for these fibers in any electric sector. These properties include insulation resistance, dielectric strength, which indicates dielectric constant, current leakages at certain voltages and stability of these fibers under the application of electric fields. To determine electrical properties (resistivity and dielectric strength) the fibers should be measured at room temperature and relative humidity ($\sim$65%) and also at other temperature and humidity after the fibers were subjected to dry heat at temperature higher than the boiling water (say, about 110 °C) to remove all the moisture in the fibers.

While the electrical resistance of the fibers suitable equipment such as million megaohmmeter may be used and the fibers are gripped between the terminals of the equipment, dielectric strength of the fibers can be measured using insulation breakdown tester.

Then, the volume resistivity (ρ) of the fibers having an area of cross section A, length l and can be calculated using the following equation:

$$\rho = RA/1 \tag{9}$$

where R is a constant. It may be noted that since lignocellulosic fibers do not have uniform dimensions and hence the area, the density of these fibers can be computed using the following equation:

$$R/1 \times m/1 \times 1/d \tag{10}$$

where m is the weight of fibers used and d is the density.

It may be relevant to note here that these properties of lignocellulosic fibers might not have been measured globally except for one study each in India and Brazil (Barretto et al. 2010; Kulkarni et al. 1981). The former study has suggested that most of the lignocellulosic fibers exhibit high electrical resistance and dielectric strengths indicating their possible substitution of wood in insulating applications. The latter study has reported the effect of chemical treatment of banana fibers of Brazil on various properties including dielectric permittivity and dielectric loss. The authors observed that concentration of alkaline solution dictated the dielectric behavior (permittivity and the loss factor) of the fiber, while the electrical modulus (observed by the dielectric relaxation) was dependent on the temperature of measurement. This could be related to the water molecules present in the fiber (hydrophilic groups of the anhydroglucose). Based on the observed dielectric properties, the authors contended that the values of permittivity and the loss factor were reasonable and therefore the fibers could be utilized as an electronic device along with other materials to do a composite phase.

Flammability

Flammability of lignocellulosic fibers, an important property particularly if the fibers or their products are exported, can be overcome by treating these materials by commercial flame retardant agents such as ammonium phosphate, zinc chloride, boric oxide, etc., or expensive method of fixation of organo-phosphate compounds. However, one of the drawbacks of using these materials is a reduction in tensile properties of the fibers.

There are certain standard tests such as vertical burning tests (ASTM-E-69-50) to determine the flammability of inflammable materials such as fibers. A laboratory level an inexpensive method using urea-diammonium hydrogen orthophosphate with a coating of phenol formaldehyde has been found to impart fire retardancy to coir fibers and their products (Venugopal et al. 1990). Reduction of tensile properties of treated fibers and their products has been reported. However, no report on the flammability of Brazilian fibers seems to be reported till now.

3.1.2 Chemical Characterization

It should be noted that understanding of chemical aspects (chemical composition of the plant fibers such as essential oils, cellulose, hemicelluloses, lignin, and sugars), thermal behavior, and predominant functional groups at the fiber surface, etc.) any material including a lignocellulosic fiber is essential and relevant for finding their uses. In the case of lignocellulosic fibers, this is particularly important for their use as reinforcements in several types of natural and synthetic polymeric matrices. Some of the techniques used for the above include: Fourier transform infrared spectroscopy (FTIR), Raman spectroscopy [for determining the chemical composition], thermogravimetric analysis (TGA) or derivative thermogravimetric analysis (DTG), differential thermal analysis (DTA), differential scanning calorimetry (DSC), and dynamic mechanical thermal analysis (DMA) [for understanding the chemical changes such as oxidation, degradation, and glass transition temperatures, dynamic mechanical properties]. Some of the above are also complimented by XRD and microscopy techniques such as electron spectroscopy for chemical analysis (ESCA) also known as X-ray photoelectron spectroscopy (XPS), X-ray fluorescence (XRF), energy-dispersive X-ray spectroscopy (EDS). Details of these techniques can be seen in any textbook on Materials Science and hence these are not given here.

Information about chemical constituents of fibers would help in understanding the chemical compatibility between the polymer matrix and the lignocellulosic fiber, since interface plays an important role in dictating good adhesion and consequently good mechanical properties of their composites. However, one should note that the complete chemical characterization of a specific fiber may be difficult considering the fact that the composition of lignocellulosic fibers is affected by a number of factors (nature, collection place, age, genetic variety, species, type of soil, growth conditions, maturity of the coconut at harvesting time, method of chemical analysis followed as well as the experimental conditions, etc.) (Asasutjarit et al. 2007; Bessadok et al. 2008; Guimarães et al. 2009; Rowell et al. 2000; Satyanarayana et al. 1986; Satyanarayana and Wypych 2007; Sen and Jagannatha Reddy 2011; Van Dam et al. 2006).

Chemical composition of lignocellulosic fibers can be determined following some standards such as Association of Agricultural Chemist (AOAC 2012), TAPPI standards (Moisture content: TAPPI T212 om-02; Ash content: TAPPI T211om-02 standards; Soluble products in cold/hot water: TAPPI T207 cm-99; Solubility in 1% NaOH solution: TAPPI T212 om-02; Extraction of fats, greases, some resins, and gums using a mixture of ethanol/toluene and extraction of tannins, gums, sugars, starches, and colored materials using hot water: TAPPI T204 cm-97 and TAPPI T264 cm-97; Soluble and insoluble lignin: TAPPI T222 om-02 standard and cellulose fractions (*a*, *b*, and *c*): TAPPI T203 cm-99 standards) or ASTM standards. These standards give a detailed methodology for the determination of each of the constituents of the lignocellulosic fibers, viz., cellulose, hemicelluloses, lignin, volatile components—soluble and insoluble components, moisture content, wax, etc. Keeping in view the space limitation, no details on the procedure followed in each of these standards is not given here. However, it is suggested to interested readers to refer to any of these standards to get the complete procedure for the determination of chemical constituents of the fibers.

Spectroscopic methods such as Fourier transform infrared (FTIR) spectroscopy technique, Raman spectroscopy, etc., have been used for studying the chemical structure of the fibers, where one gets the spectra of the fibers. This measurement is carried out with distilled water cleaned fiber samples. The samples are then dried and cut into small pieces and pressed into a disk shape along with a standard such as KBr sample (in a suitable ratio, say 1:3) used for calibration and suitable wavenumber range (say, 4000–400 cm^{-1}) to obtain the spectrum, which is then compared with that obtained for pure KBr sample. An instrument having a suitable resolution (say, 2 cm^{-1}) is used and required numbers of scans are accumulated.

3.1.3 Thermal Characterization

As mentioned earlier, thermal characterization of fibers is important particularly to understand the thermal stability of the fibers while using them for many applications. Besides, such a characterization throws light on the stability of various constituents of the fibers, chemical changes such as oxidation and glass transition temperatures and dynamic mechanical properties. Equipment used for this characterization include differential thermal analyzer (DTA), thermogravimetric analyzer, differential scanning calorimeter, and dynamic mechanical thermal analyzer. Normally, measurements with DTA or TGA are done under chosen operating conditions (heating rate, say in the range of 40–800 °C), and two atmospheres, viz., nitrogen and air or oxygen both could be at selected flow rate say, 50 mL/min and at a suitable pressure, say 10^4 Pa. Finally, the samples are maintained at 800 °C for some more time, say 10 min. Nitrogen being inert atmosphere, prevents combustion, and allows the degradation of components of the fibers to take place one by one. This leads to the possibility to identify the degradation regions of the fiber components. On the other hand, with air/oxygen atmospheres, both the reactions occur at the same time similar to oxidative atmosphere, because it is not possible to separate the different degradation processes of the fiber components such as hemicelluloses, cellulose, and lignin, since all the reactions would be very complex and they would be overlapped in the temperature ranges of these reactions (Tomczak et al. 2007a, b).

Differential scanning analysis (DSC) is performed in a suitable instrument with a selected range of temperatures, (say 40–550 °C) at a chosen heating rate (say 10 ° C/min) in a selected atmosphere at a chosen pressure (say 10^4 Pa).

Finally, dynamic mechanical analysis (DMA), which is also known as dynamic mechanical spectroscopy) is another thermal analysis technique used to study and characterize materials for their static properties (change in properties under tensile, bend or shear stress conditions and structure with a change in temperature) including lignocellulosic fibers following suitable ASTM standard. By this technique, it is possible to study the viscoelastic behavior of materials such as polymers and lignocellulosic materials. The parameters measured by this technique include glass transition temperature, storage modulus (characterizes the elastic response of a material), loss modulus (characterizes the viscous response of a material), and phase angle or dissipation factor (tan δ-helps in predicting the life expectancy of materials). In this technique, a sinusoidal stress is applied and the resulting strain is measured as the temperature is varied.

3.1.4 Mechanical Characterization

This includes tensile properties, flexural and impact properties as well as fracture analysis as applicable to lignocellulosic fibers and their composites. Most of these are done following available standards such as ASTM standards (such as ASTM D 3822) particularly for composites using ASTM D638 M standard. For all these determinations, first as-received fibers should be cleaned with freshwater and dried at a suitable temperature (say, 60–70 °C) and time (say, 24 h) in a hot air oven to remove any absorbed moisture in the fiber or impurity on the surface of the fibers. Selection of temperature for drying the fibers is normally based on the TGA carried out on the fibers at which the absorbed moisture of hydrophilic fibers begins to be released and no degradation occurs (Elzubair et al. 2007). Generally, tensile testing of fibers of a chosen gauge length is carried out at room temperature and relative humidity using low load cell (Say, 0.5–5.0 kN) and crosshead speed (5–300 mm/min). Pneumatic grips are used to hold the fiber samples with appropriate pressure to avoid any slippage without damaging the fiber. Normally, appropriate gauge length (200 mm) is chosen, but some researchers have chosen it due to the following reason: In order to make reasonably accurate failure analysis by applying the Weibull analysis to the obtained tensile properties, small samples are chosen wherein the Weibull model based on the failure of a chain, in which the weakest link controls rupture and seems to be well adapted to describe a set of tensile test results carried out at one gauge length (Monteiro et al. 2011a, b).

In the tensile testing, stress–strain curves of fibers are recorded (as a function of fiber diameter or at a given diameter), which are used to calculate tensile properties (Young's modulus–YM, ultimate tensile strength–UTS and % elongation) of the fiber (Monteiro et al. 2008; Tomczak et al. 2007a, b). YM is calculated from the straight line part of the curve, while strength is calculated using the values of the force (F) and the area of the curve. Percentage elongation is calculated using the standard method with monitoring the extension through the displacement observed without the use of the extensometer. Tests are carried out with a reasonably large number of fibers to take care of the large variation in their dimensions.

Flexural, Charpy, and Izod impact tests of composites have been carried out following mostly ASTM D–790 and ASTM D256 have been followed (Monteiro et al. 2013). Notches of 2.54 mm deep with 45° angle and radius of curvature of 0.25 mm at the bottom of the notch were made using a German steel mill as per DIN 847 standards (Din 847 2005).

3.1.5 Theoretical Considerations for Dimensional Analysis and Prediction of Properties

In order to understand the variation in properties of fibers and also to make a prediction of these properties based on its constituents, statistical methods, and mathematical models have been used by various researchers. The statistical method used is Weibull analysis for the understanding the observed strength properties of fibers, while one of the mathematical models used for the prediction of properties of fibers is the artificial neural networks (ANN). On the other hand, in the case of lignocellulosic fiber incorporated composites, factorial design of experiments, and statistical method of analysis of variance (ANOVA) can be used to arrive at proper processing parameters and to analyze the obtained mechanical properties of prepared lignocellulosic fiber-containing polymer composite laminates.

Tensile strength is one of the important properties of fibers (synthetic or natural) required for their use in composite development. This is particularly important in the case of lignocellulosic fiber, since they are not uniform in their dimensions and normally possess inherent flaws/defects, which affect the tensile properties. For example, tensile strength of sisal is reported to vary from 280 to 700 MPa with or without a standard deviation of the sampling (Amico et al. 2005). In view of nonuniformity of dimensions, although average values of tensile strength are obtained from a number of samples, they do not represent the actual values in view of the large divergence in the dimensions of fibers. Thus, the inability of such simplified analysis is evident and hence this approach poses a problem. Therefore, researchers have been resorting to statistical approaches including Weibull analysis to obtain representative values. However, it should be noted that large number of samples are required for the conventional statistical approaches to get correct strength values (Abernethy 2006; Klaine 2009; Monteiro et al. 2011a, b, 2014a) while with minimum data, even modeling using binomial and Poisson distributions would not give precise estimations. Further, Weibull analysis has an advantage among the statistical methods as it requires less number of samples (data) to get the meaningful value of strength of fibers along with reasonably accurate failure analysis and failure forecasts. Hence, this has been used as one of the methods to evaluate the obtained strength properties. In addition to the above-mentioned merits of this analysis, other attributes include (i) relation between inherent (due to drastic change in the cross section of the fibers, sequence of growth of cells) and created (resulted by the fiber processing methods—by extraction or while twisting, etc.) defects of the fibers and the strength of the fibers, (ii) determination of flaw size distributions, flaws (Satyanarayana and Wypych 2007; Satyanarayana et al. 2009b). The quantity of these defects depends on both diameter and length of the fibers, both of which in turn dictates strength properties.

Weibull analysis can be made either by two parameter analysis with simple numerical calculations or three parameter analysis needing laborious computer analysis. In the former case, the Weibull probability density [$f(t)$] and cumulative distribution functions [$F(t)$] are considered. These are given by Eqs. 11 and 12, respectively:

$$f(t) = \left(\frac{\alpha}{\beta^{\alpha}}\right) t^{\alpha-1} e^{(-t/\beta)^{\alpha}} \tag{11}$$

$$F(t) = 1 - e^{(-t/\beta)^{\alpha}} \tag{12}$$

where α and β are the shape and scale of the distribution, respectively. It may be noted that α being a dimensionless positive number, it would give the shape of distribution to be close to normal distribution for values has in the range $3 \leq \alpha$ 4. On the other hand, β, the "characteristic life" has its value to be approximately 63.2% with the same unit as t. These two parameters can be estimated by two methods. These are "maximum likelihood technique" and a probability plot.

In the former case, solution for two nonlinear equations is needed, while in the second method comprehensive reviews of the features regarding this distribution are considered (Amico et al. 2005). With the aid of an analytic test, it is possible to test whether a set of data follows Weibull distribution or not. For example, if the concentration of surface treating is low (<5%), more number of tests will have to be carried out to confirm that any treatment actually decreases fiber tensile strength or not, while confidence level >90% suggests a decrease in strength due to the treatment. Thus, this methodology can be applied to surface-treated fibers whereby the severity of the treatment can be assessed (Amico et al. 2005; Monteiro et al. 2011a, b, 2014a; Tanobe et al. 2005).

The simpler methodology followed to estimate the defects, also called "weak links", is by correlating the strength (σ) of the fiber with its length (L), which gives a linear relationship. Variation of strength with varying diameter and length of fibers can be understood by this relationship. A reported number of defects determined by the above methodology in the case of banana, coir, pineapple, and sisal fibers are 1376, 2320, 4461, and 2368 respectively (Satyanarayana et al. 2011).

Another theoretical method used is the ANN, particularly to predict the tensile properties of lignocellulosic fibers.

In the case of the mathematical modeling technique— "artificial neural network" (ANN), which is mostly used in many areas of Materials Science, can be considered as an extension of the statistical approach to develop a comparative study of the behavior of different natural fibers. This helps in determining the properties of required materials through proper selection of constituents and processes, particularly for analyzing the complex materials having different phases such as alloys and composites.

It uses mathematical equations interconnecting various groups of nodes (properties) similar to the vast network of neurons in a brain from the output of one neuron (characteristic of the material) to the input of another to arrive at the desired material property (https://en.wikipedia.org/wiki/Artificial_neural_network dated 13 Dec. 2016).

Initial evaluation of the good performance of ANN in Materials research has been studied by looking at its architecture, training the functions followed by training the process and finally evaluation method (Tomczak 2010). Once the network knows the problem based on the training it has undergone, new data should be included to verify the solution. Then, process of creating ANN for the properties of materials can be divided into the following stages: (i) Collection and analysis of data followed by preprocessing; (ii) training the neural network; (iii) testing of network and evaluation of its performance and finally (iv) Application of the network for simulation.

The ANNs have been found to be very useful in recent times in various engineering applications particularly to model the mechanical behavior of materials (Al-Haik et al. 2006; Bezerra et al. 2007; El Kadi 2006; Sha and Edwards 2007; Tomczak 2010). This is because of their high parallel structure, which helps to arrive at an accurate analytical solution in systems having many multivariables for which it is very difficult to obtain it (El Kadi and Al-Assaf 2007).

Probably considered to be the first attempt reported to apply this method of statistical analysis to the dimensional characterization and tensile properties of some lignocellulosic fibers such as flax, jute, abaca, and sisal was by Piponi et al. (2008). They used semi-parametric algorithms based on the Fiori and co-workers reported neural network literature (Fiori 2001; Fiori et al. 2001) to estimate the true distribution functions. For this purpose, a distribution of the geometrical properties and the mechanical behavior of the lignocellulosic fibers were used.

Apart from the above, there seems to be only one study carried out on the use of ANN related to lignocellulosic fibers from Brazil (Tomczak 2010) using 13 neurons in 3 hidden layers have been used. In this study, the researcher has used one of the strength properties (Strain at break, Tensile strength, and Young's modulus) represented by a neuron as an output, while one of physico-chemical characteristics (microfibrillar angle, cellulose content, crystallinity, and fiber diameter and test conditions (strain rate and gauge length) of the fibers as inputs. Also, two of the strength properties were used as inputs while the third one was simulated. One new ANN was defined for each of the simulated properties with the same structure as the others, keeping different inputs and outputs. When the correlation between the experimental data and the simulated data is better than 90% for any property, the model is considered to be the best fit.

3.1.6 Textile Properties

As is the case with other fibers, textile fibers are also characterized by 3 basic characteristics, viz., physical (length-15–150 mm, fineness: l/d = 1000, density or denier: number of grams per 9000 m of filaments, cross-sectional shape, crimp, etc.), chemical (reaction to dyes, moisture absorption capability, and resistant toward light, bacteria and even temperature, etc, and nonsoluble in water or any organic solvent) and mechanical properties (tenacity—rupture force per unit size (as well as denier) where the size is expressed as mass per unit length, elasticity, stiffness, etc.). These properties are essential to meet the requirements of textile aspects such as for spinning, weaving, etc (Franck 2005; Satyanarayana et al. 2009b; Wallenberger and Weston 2004). Two other important properties of textile fibers are the refractive index (important for the observation of fibers for fine details without misleading artifacts) and blending. The former helps in the visibility of all colorless and transparent objects, while the latter helps in reducing the cost along with in getting improved properties or to produce fabrics for special applications.

3.1.7 Surface Treatments of Lignocellulosic Fibers

Surface treatments of lignocellulosic fibers can be broadly classified under physical, chemical (Oliveira and Marques 2015), physico-chemical, and mechanical methods (El-Sabbagh 2014; Le Moigne et al. 2014; Pereira et al. 2015; Saha et al. 2016; Satyanarayana et al. 2009a, b). In the following, not only a brief of all the methods are presented, but also advantages and disadvantages of these treatments are mentioned by various researchers. These can be seen some of the reviews including the recent one (Pereira et al. 2015; Saha et al. 2016; Satyanarayana et al. 2009a, b).

Use of electric discharge using different rays or plasma and corona treatment, steam explosion constitutes physical methods. These methods are the most used techniques for surface oxidation activation of lignocellulosic fibers wherein the surface energy of the fibers are changed without any change in chemical composition, structure and surface properties, while tensile properties show improvement.

Before proceeding further about the surface modification aspects of lignocellulosic fibers, it may be necessary to briefly mention about the steam explosion because this method seems to have attracted the attention of a number of Brazilian researchers dealing with lignocellulosic materials and not much is reported on the effect of the steam explosion of plant fibers in many publications. Besides, this technique is reported to be the most successful options for fractionating of wood along with enhancing susceptibility of cellulose to enzymatic attack (Satyanarayana et al. 2009b). This process being a high pressure steaming, steps involved are (i) heating lignocellulosic materials at high temperatures and pressures and (ii) mechanical disruption of the pretreated material by violent discharge into a collecting tank (explosion). This method has been applied to many lignocellulosic materials or biomass materials, sisal and sugarcane bagasse fiber, *Miscanthus Sinensis* wood and wood flour (see References in Satyanarayana et al. 2009b]). Another advantage of this method is zero emissions model for agro-materials industry.

Acetylation, alkali treatment or mercerization, silane treatment, silane treatment, and peroxide treatment with various coupling agents, delignification (extraction using reagents such as alcohol or benzene reagents and with NaOH followed by drying at room temperature) and others constitute chemical methods. Selective oxidation of some units of lignin with furfuryl alcohol is another new chemical method attempted by some researchers in recent times (Satyanarayana et al. 2009a, b). In the case of chemical methods, an important role is played by concentration, time and temperature of these treatments to provide a quality modification to the fiber surface, which in turn help improvement of mechanical and other properties.

The structure (composition) and surfaces (morphology) of fibers are normally affected by the chemical methods involving solvents due to the extraction of some of the constituents such as pectin, lignin, and hemicelluloses, which in turn affect the properties of the fibers as reported in many of published reports mentioned above. On the other hand, change from hydrophilic fiber surfaces to hydrophobic, improvement in the dispersion of fibers, change in fibers due to the extraction of some of its constituents (pectin, lignin, and hemicelluloses) besides the changes in morphology are observed with chemical methods. In view of these, mostly improvement in tensile properties and moisture absorption of the composites result along with lowering the viscosity of the composite mixture and reduction in strength of fibers (alkali). Similarly, some treatments such as silane treatment and peroxide treatment with various coupling agents are reported to improve fiber strength, fiber fitness and fiber-matrix adhesion in natural fiber composites (Pereira et al. 2015; Saha et al. 2016; Satyanarayana et al. 2009a, b). Besides, high-quality fibers are reported to have been produced by the common alkali treatment. It is also reported that of weight loss and improved tensile strength are observed by the use of commercial oxidative bleaching agents such as alkaline calcium or sodium hypochlorite and hydrogen peroxide (Pereira et al. 2015). Action of the bleaching agent or alkaline reagent on the noncellulosic constituents of fibers such as hemicelluloses and lignin are reported to be the reason for the weight losses.

On the other hand, solvent extraction constitutes physicochemical methods, rolling, swaging, splitting etc., attempted by the researchers all over the world including Brazilian researchers constitute mechanical methods for the surface modification of natural fibers. While mechanical methods do not seem to have been employed in Brazil, steam explosion seems to have attracted their attention (Satyanarayana et al. 2009b).

Even the physicochemical methods improve strength properties and moisture absorption of composites while the mechanical methods improve bonding and tensile properties (Pereira et al. 2015; Saha et al. 2016; Satyanarayana et al. 2009b).

Effects of surface treatment of various lignocellulosic fibers of Brazil have been studied using different experimental techniques, which included FTIR, XPS/ESCA, XRD, TGA, and SEM, has been studied (Barretto et al. 2010; Br´ıgida et al. 2010; Nascimento et al. 2012; de Campos et al. 2013; Monteiro et al. 2008, 2011a, b, 2013; Ricardo et al. 2011; Simão et al. 2016; Tanobe et al. 2005; Vilay et al. 2008). Some of the results of microscopic studies are presented in the next section.

3.1.8 Structural Aspects (Morphology and Fractography)

Macro photographs of lignocellulosic fibers are obtained using both optical and scanning electron microscope in both longitudinal and cross-sectional views of fibers to understand the nature. For the optical microscopic study, first steel mold is filled with a mixture of polyester resin with 2% catalyst and 2% initiator to which a bunch of fibers are inserted vertically until the resin is set. Then it is cured followed by grinding successively using silicon carbide papers of grit size ranging from 220 to 600 mesh sizes. Then, the specimens are polished with a selvyt cloth mounted on disc polisher using a diamond paste of different sizes. Thus, obtained polished specimens are observed in reflected light using a suitable optical microscope.

On the other hand, for scanning electron microscopic observation of fiber samples or fracture surfaces of tensile-tested fibers are attached to a metallic support (brass stud) and coated with gold or platinum to make the fibers conducting. They were then observed with SEM at 15 keV and current of about 0.65 A.

4 Observations of Structure and Properties

4.1 *Physical Characterization*

Physical characterization includes determination of dimensions of both the microfibers and microfibrillar ones, density, microfibrillar angle, crystallinity, moisture absorption, electrical properties, flammability of fibers. Techniques used for the above-mentioned properties have been presented in the previous section. Here in results obtained of all these determinations are discussed. Table 2 lists available properties with respect to most of the Brazilian lignocellulosic fibers. Available properties of similar fibers reported globally are also listed for comparison.

4.1.1 Dimensions of Fibers

Figure 3 presents histograms showing the statistical distribution of length and diameters of some of the Brazilian fibers. It can be seen distribution is different for fibers obtained from different parts of the plants. Values of these reported by various researchers for different fibers are listed in Table 2.

4.1.2 Density of Fibers

These measured for various fibers by different researchers are listed in Table 2. As can be seen most of these lie between 1000 and 1500 kg m^{-3} compared to more than 2000 kg m^{-3} for synthetic fibers. This property is important from the point of view in understanding that specific strength properties of these fibers are almost equal or better than those of synthetic fibers thus suggesting the substitution of these fibers by the lignocellulosic fibers in composites used for different applications.

4.1.3 Microfibrillar Angle and Crystallinity of Fibers

Figure 4 show the XRD patterns of various fibers, which are used to calculate the crystallinity index of the fibers. These values are also listed in Table 2. It can be seen from the figures that the shape of XRD patterns depends on the fibers obtained from different parts of plants. They also indicate that peak of a crystalline part in each fiber is almost the same, while the amount of crystalline part in each of the fibers is different for each type of the fiber.

Moreover, the microfibril angle of arundo fibers (i.e., 7.37°) is also found to be comparable to the ones of other natural fibers as jute (8.1°), flax (5°), hemp (6.2°), *prosopis juliflora* (10.64°), and banana (11–12°), respectively (Kulkarni et al. 1983; Saravanakumar et al. 2013).

Table 2 Some physical properties of Lignocellulosic fibers of Brazil and others

Fiber	Density (kg m^{-3})	Technical length (mm)	Dia (μm)	Cell length (mm)	Diameter (μm)	Crystallinity (%)	Micro-fibrillar angle (θ)	References
Amazonian jacitara palm (*Desmoncus polycanthos* Mart.)	–	1.3–4.4	10.1–24.7	–	–	–	13–15	Fonesca et al. (2013)
Bagasse (sugarcane)	450–492	–	–	0.8–2.8	–	47–48	–	Simão et al. (2016), Neto (2014)
Bamboo Taquara-lixa in natura	880–1210		100–200				59.0	Liebsch et al. (2009), Neto et al. (2014), Monteiro et al. (2011a)
Banana	670–1500	60	150			39		Satyanarayana et al. (2007), Monteiro et al. (2011a), Neto (2014)
Cissus quadrangularis (CQSF)/root fibers	1220/1510							Indran et al. (2014, 2015)
Jute	1300–1450	60	18–20/20–30	2.3	25–30	50	7–9 17.1*	Satyanarayana et al. (2005), (2009a), Neto (2014) Monteiro et al. (2011a, b)
Piassava	1100–1450							Monteiro et al. (2011a)
Ramie	1500	120	40–50	154.0		–	7.5–12	Satyanarayana et al., (2007), Monteiro et al. (2011a, b)
Buriti	630–1120	2000				8.5		Monteiro et al. (2011a, b), Cittani and Baruque-Ramos (2016)
Pineapple	1440	20–80	1.56/4.5	10			8–15	Satyanarayana et al. (2007)
Sisal	1260–1500	3	18	1–8	8–41	72.2	20	Satyanarayana et al. (2007), Neto (2014), Monteiro et al. (2011a, b)

(continued)

Table 2 (continued)

Fiber	Density (kg m^{-3})	Technical length (mm)	Dia (μm)	Cell length (mm)	Diameter (μm)	Crystallinity (%)	Micro-fibrillar angle (θ)	References
Curauá,	570–920		97		7–10 &	75.6 66.3	18.8*	Satyanarayana et al. (2007), Neto (2014) Monteiro et al. (2011a)
Coir	1390–1520	10–150	40–400	0.75	12–14	43–44	47.8–51	Satyanarayana et al. (2007), Neto (2014)
Cotton	1510	35	19		10–45		33–34	Satyanarayana et al. (2007)
Sponge Guarc (*Luffa Cylindrica*)	820				25–60	59.1		Satyanarayana et al. (2007), Neto (2014)

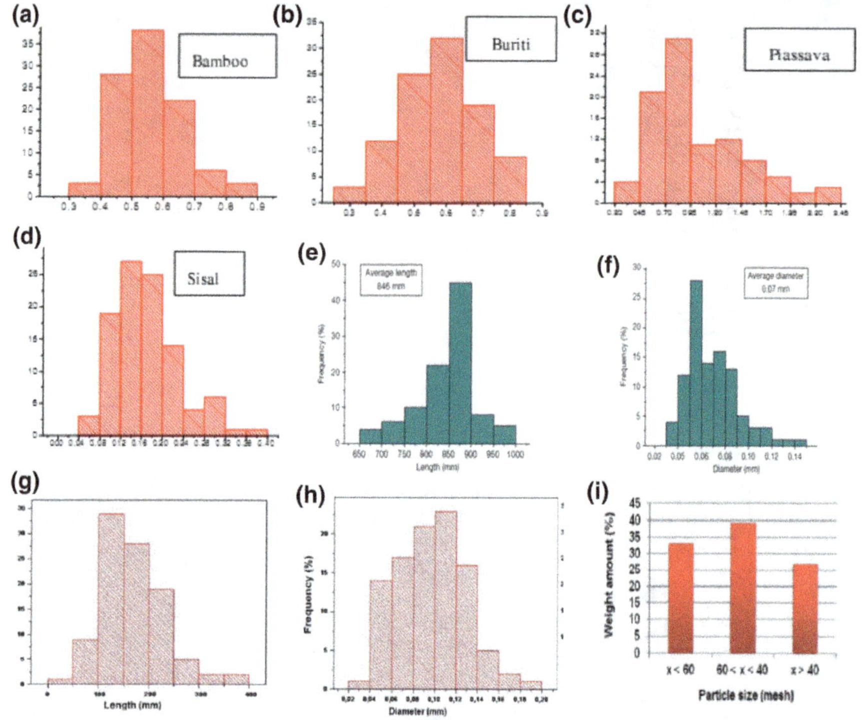

Fig. 3 Histograms showing statistical distribution: **a** Bamboo; **b** Buriti fiber diameter Monteiro et al. (2011a, b); **c** Piassava; **d** Sisal; **e** Curauá length and **f** Diameters Monteiro et al. (2013); **g** Ramie fiber length and **h** Ramie fiber diameter Monteiro et al. (Unpublished); **i** saw dust particles Santos et al. (2015). Reproduced with the kind permission of the Authors and Publishers

Some Brazilian researchers have also determined the crystallinity of surface-modified fibers. For example, Oliveira and Marques (2015) have measured crystallinity index of malva fibers with and without surface treatment by various concentrations (5–20%) of NaOH. Crystallinity of the surface-treated fibers was found to increase from 58.5% for untreated fiber to the value were 64.2 for 5% NaOH-treated fibers for 1 h and to the maximum of 70.6% for fibers treated with 7% NaOH for the same time. However, the values varied between 60.3 to 63.2% for treatments with 10–20% NaOH. These results suggested that it was the concentration of NaOH was the predominant factor for the increase in the crystallinity and not the time for which they were treated contrary to the earlier observation by Reddy et al. (2013). A change from cellulose I to cellulose II was observed with the latter being thermodynamically more stable crystalline polymorphic structure. The authors also observed NaOH treatment removed amorphous extractables in the fibers, increased thermal degradation resistance of the fibers as well as their composites with polyhydroxy butyrate (PHB).

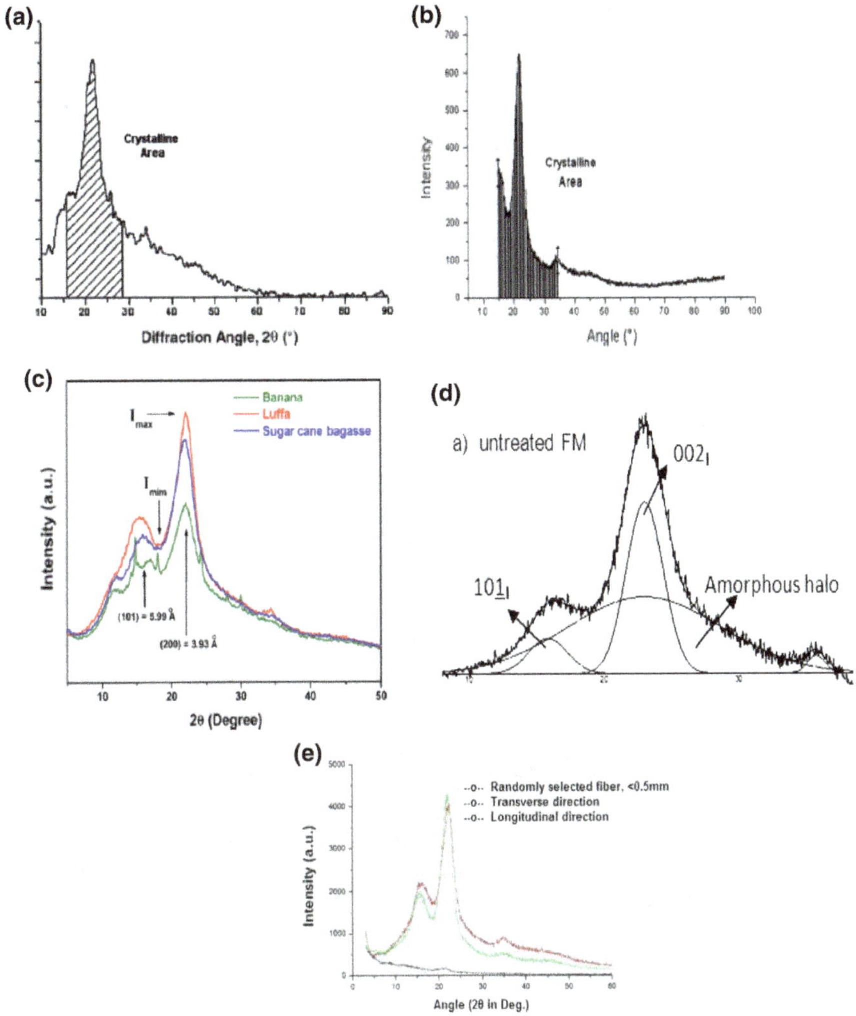

Fig. 4 X-ray diffraction patterns of some Brazilian Fibers: **a** Coir fiber; **b** Curauá fiber; **c** Banana, sponge guard, and sugarcane Bagasse fibers; **d** Malva; **e**; **f** Taquara-Lixa Diffractograms superimposed, of the samples in longitudinal and transverse direction. Reproduced **a** Tomczak et al. (2007a); **b** Tomczak et al. (2007b); **c** Guimarães et al. (2009); **d** Oliveira and Marques (2015); **e** Neto (2014). With the kind permission of the Publishers

4.1.4 Moisture Absorption of Fibers

Values reported for some of the lignocellulosic fibers of Brazil are listed in Table 2. This is an important property to understand the stability of the products made of various fibers particularly on exposure to the environment and also during the processing of composites.

4.1.5 Electrical Properties of Fibers

There seems to be only one study on this aspect in respect of Brazilian fibers (only banana fibers) with measurement of dielectric properties as a function of temperature (Barreto et al. 2010). For this purpose, banana fibers were taken in the form of disks (as capacitors) with and without alkali treatment and using an impedance analyzer AC dielectric behavior was detected at room temperature (25 °C). Relative permittivity [both the real (ϵ') and imaginary (ϵ'') parts] as well as the dielectric loss factor ($tg\delta = \epsilon''/\epsilon'$) were measured in the region of 10 Hz to 700 kHz on the disk samples after removing them from the vacuum environment. For the measurement of these properties as a function of temperature (30–90 °C), the authors have used an oven. Further, the dependence of the interfacial polarization effect (generates electric charge accumulation around the fiber) on the frequency and temperature, the following equation was used:

$$M = 1/[\varepsilon' - j\varepsilon''] = M' + jM'' \tag{13}$$

where M is the electric modulus, M' and M'' are the real and imaginary parts of modulus. The value of dielectric permittivity (ϵ') was found to increase from 5.49 for the as-received fibers to a maximum value of 12.59 for fibers treated with 1% NaOH (max. concentration) at 10 Hz. Similarly, the values of dielectric loss ($tg\delta$) also increased (approximately between 10^{-1} and 10^{-2}) with chemical treatment used in the fibers. These results suggested a dependence of both the dielectric permittivity and the loss factor on the concentration of the alkaline solution used. These results have been attributed to the partial removal of the amorphous components, which possess ϵ' values lower than that of cellulose crystalline phase. On the other hand, a variation of these two properties (dielectric permittivity and loss factor) with frequency was found to be minimum, almost remaining constant after the frequency value of 1 kHz. The above results indicated the possibility of use of these fibers in a composite phase in conjunction with other materials as an electronic device.

In the case of electric modulus, the dielectric relaxation was found to depend on the temperature. This was attributed to the presence of water molecules of anhydroglucose.

4.2 *Chemical Characterization*

Table 3 compares chemical composition of various Brazilian lignocellulosic fibers along with available values of similar fibers reported globally. Depending on the source of cellulose, the length of different types of chains may be found (1–4 linked ß-d-anhydroglucopyranose units) together with the hemicelluloses, a combination of 5-and 6-ring carbon ring sugars (polysaccharides, except pectin). Lignin, a highly branched polymer acts as the matrix material along with hemicelluloses

Table 3 **a** Chemical composition of lignocellulosic fibers of Brazil and others, **b** comparison of FTIR spectral data obtained in some of the Brazilian lignocellulosic fibers

Fiber	Hollow cellulose/cellulose %	Insoluble lignin (%)	Ashes (%)	Total extractives (%)	References
Amazonian jacitara palm (*Desmoncus polycanthos* Mart.)	66.9 (85.3)	14.7	1.8 (Minerals)	11.6	Fonesca et al. (2013)
Arundo donax	64.09	17.2	1.9		Fiore et al. (2014)
Bamboo	33–45	20–25	–		Satyanarayana et al. (2007)
Banana	50.92	17.44	4.14 ± 0.92	6.2	Guimarães et al. (2009)
Buriti	65–71	21–27	–	5.4–6.0	Cittani and Baruque-Ramos (2016)
Capim-dos-pampas	–	–	–	–	–
Cissusquadrangularis (CQSF)/root fibers	82.73/77.17	11.27/10.45	–		Indran et al. (2014), Indran and Raj (2015)
Coir	43.4–53	38.3–40.77	–	–	Satyanarayana et al. (2007)
Cotton	90	–	–	–	Satyanarayana et al. (2007)
Curauá	83.1	7.6	3.1	6.2	Tomczak et al. (2007b)
Jute	60	15.9–11.8	1.0	–	Satyanarayana et al. (2007)
Jute	59–71	11.8–12.9	0.7		Satyanarayana et al. (2007)
Piassava	31.6	48.4	–	–	Satyanarayana et al. (2007)
Pineapple	83	12	0.1–1		Satyanarayana et al. (2007)
Prosopis juliflora	77.79	17.11	5.2		Saravana Kumar et al. (2013)
Ramie	80–85	0.5			Satyanarayana et al. (2007)
Sisal	74–75.2	7.6–7.98	0.14–0.87	–	Satyanarayana et al. (2007)
Sponge Guard (*Luffa cylindrica*)	82.4	11.2	0.4	3.2	Tanobe et al. (2005), Satyanarayana et al. (2007)
Sugarcane Bagasse	71.30	23.33	2.75 ± 0.19		Guimaraes et al. (2009)
Taquara-lixa	65.07	27.47	0.78 ± 0.02	6.67	Neto (2014)

(continued)

Table 3 (continued)

Fiber	Hollow cellulose/cellulose %	Insoluble lignin (%)	Ashes (%)	Total extractives (%)	References
Rice straw	51–70	12–16	15–20		Satyanarayana et al. (2007)
Pine sawdust	64.06	28.98	0.25	6.71	Trombetta (2010)
Silk-floss	81.6	15.1	0.62	0.92	Annunciado et al. (2005)

Fibers	Hydroxyl band cm^{-1}	Hydrogen-bonded stretching cm^{-1}	C=O stretching cm^{-1}	C=C stretching cm^{-1}	C–H stretching cm^{-1}	References
Açai	3350	2920–2846	1729			Gehlen (2014)
Banana	3430 3600–3100	−3400	1058 1725	1635 1500–1600	2920 2910	Barreto et al. (2010), Guimaraes et al. (2009)
Banana	3600–3100	3600	1654	1026–1244	2913	Becker et al. (2013)
Coir (green)	3500–3200	2900–2750	1740			Lomeli Ramirez (2011)
Curauá	3330	2920–2853	1736			Gehlen (2014)
Sisal	3330	2875	1735			Mochnaz (2003)
Taquara-lixa	3390	2903	1738			Neto (2014)

embedding the cellulose microfibrils constituting the lignocellulosic fibers as natural composites (Thakur et al. 2014).

It can be seen from the Table 3a that chemical composition of lignocellulosic fibers varies for fibers obtained from different parts of plants. Further, it also varies for different regions/localities. In the case of coir fiber, it can also be seen that chemical composition differs from "green" (younger ones) variety to brown variety (matured coconut). While curauá and sisal fibers of Brazil posses the lowest lignin content (7.6%) among the lignocellulosic fibers, coir fibers show higher amount (35.4%) which seems to be lower than that of fibers from other regions of the world. It can also be seen that fibers such as taquara-lixa, pine sawdust, and sugarcane bagasse possess intermediate values of lignin (27.47, 28.98 and 23.33%, respectively). Curauá, silk-floss and sisal fibers show higher amounts of holocellulose (cellulose + hemicelluloses) resulting in their high tensile strength values.

There are some of the other fibers whose complete data on their chemical composition is not available. For example, fibers of silk-floss have limited data of their properties. This possesses very low water intake (2.5–5%) and high hydrophobicity (97.6%) when compared to other lignocellulosic fibers as sisal (27–31%), pine sawdust (21–27), sponge gourd (50–51%), and coir fibers (42–45%) (Annunciado et al. 2005). This fiber-floss is reported to be not only an excellent sorbent material for crude oil, with sorption capacity of 85 g oil/g sorbent, but also, has good potential as reinforcing fiber in polymeric matrices, especially in highly hydrophobic polyolefins (Annunciado et al. 2005).

Other fibers having limited reported characteristics include cortaderia selloana ("capim dos pampas") (Bonetti et al. 2005) and silk-floss (similar to kapok fibers) (Neto et al. Unpublished). These fibers are soft due to their low lignin content (Bonetti et al. 2005; Guimarães et al. 2009; Lomeli Ramírez et al. 2014; Neto et al. 2015; Sydenstricker et al. 2003; Tomczak et al. 2007a; Trombetta 2010).

Fibers from *Arundo donax L.*, extracted from the outer part of the stem of the plant are abundant in the Mediterranean area and grow naturally in Sicily, representing an invasive and aggressive species. Its high tensile strength (248 MPa) and Young Modulus (9.4 GPa) has recently drawn the attention of researchers as a potential reinforcement in polymer composites (Fiore et al. 2014).

Another new natural cellulosic fiber in terms of use in composites is the *Cissus quadrangularis* stem fibers (CQSF) that present high modulus (56–234 GPa), density of 1.22 g/cm^3, high content of cellulose (82.73%), low content of lignin (11.27%), and moisture of 6.6%. Also, *Cissus quadrangularis* root fibers (CQRF) present high modulus (68–203 GPa) and tensile strength (1857–5330 MPa), the density of 1.51 g/cm^3, high content of cellulose (77.17%), low content of lignin (10.45%) and moisture of 7.3%. CQRF is common in the tropical regions of India and its root and stem may be used for medical applications and for composites (Indran and Raj 2015; Indran et al. 2014).

Characterization of lignocellulosic fibers for some infrared characteristic transmittance bands of their constituents (Cellulose, hemicelluloses, lignin, water content, etc) is normally carried out by IR spectroscopy. In fact, FTIR spectroscopy more often used for the chemical characterization of the lignocellulosic fibers because of its simplicity to identify functional groups and molecules that are present in the fibers.

Figure 5 shows Fourier Transform Infrared (FTIR) spectra of some of the Brazilian lignocellulosic fibers obtained by various researchers.

As can be seen from the figure, four types of bands occur in almost all the lignocellulosic fibers, peaks occurring at similar frequencies typical of most of the lignocellulosic fibers indicating a complex of lignin and cellulosic materials (cellulose and hemicelluloses) in the basic structure. The small variations in the peaks obtained in the FTIR spectra reported for similar fibers from different regions may be due to the change in the testing methods used and also on the origin of the fiber. These bands represent mostly absorption process, stretching, etc. These are bands at 3500–3100, 2900–2300, 1700–1800, 1500–1600, 1328–1374, and 1000–1200 cm^{-1}. Based on the occurrence of these bands (absorption, stretching, etc) it is

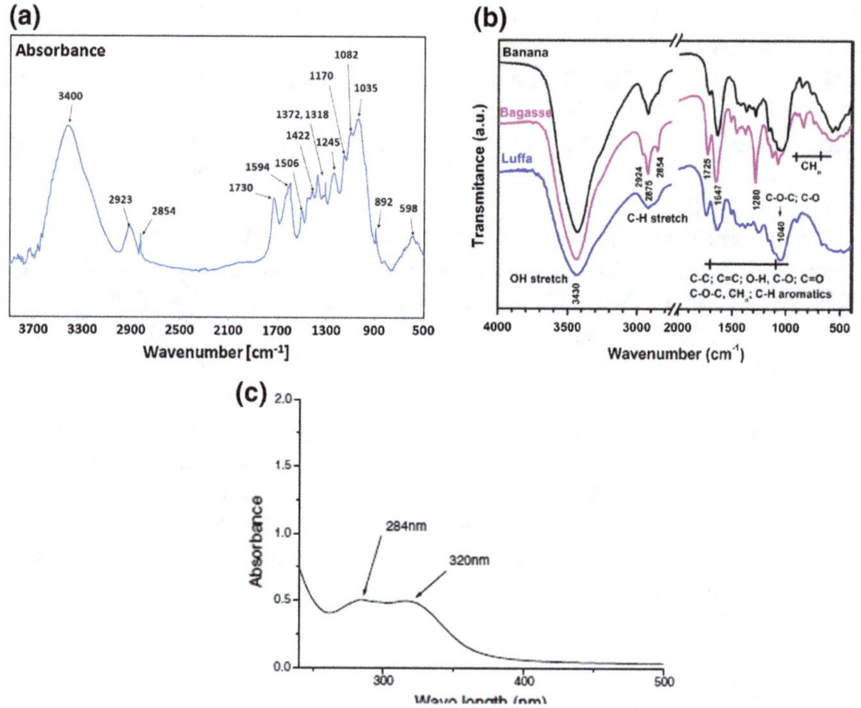

Fig. 5 Fourier transform infrared spectra of Brazilian **a** Arundo fibers **b** Banana, Sugarcane Bagasse and Sponge Guard fibers; **c** UV visible spectrum of Brazilian Curauá Fibers. Reproduced from Refs. **a** Friore et al. (2014), **b** Guimarães et al. (2009); **c** Tomczak et al. (2007b); with the kind permission of the Publishers

possible to identify various components of the fibers. This is illustrated in Table 3b. Thus, the spectral lines in the range of 3450–3000 cm^{-1}, which is characteristic of stretching vibrations of OH group, bonded to cellulose/carbons 2, 3, and 6 carbons of the glucose (Tomczak et al. 2007a, b, Van Dam et al. 2006). In fact, the above broadband may also indicate the presence of moisture in the some of the fibers. Similarly, the peaks in the range of 2900–2360 cm^{-1} are due to –C=C– stretching, the peaks in the range of 1500–1620 cm^{-1} to the medium –C=C– stretching and the ones in range of 1700–1800 cm^{-1} is unconjugated C=O stretching, related to carboxylic ester or acetyl groups (hemicelluloses), the one in the range of 1200–1400 cm^{-1} is due to C–H rock or absorption of angular strain of CH (cellulose and lignin) (Brigida et al. 2010; Mahato et al. 2013), which is similar in various lignocellulosic fibers such as coir (Tomczak et al. 2007a, b), okra (*Abelmoschusesculentus*) (De Rosa et al. 2010), vegetable fibers (Poletto et al. 2014), banana, sugarcane bagasse and *luffa cylindrica* (Guimarães et al. 2009). In some cases such as coir from a green coconut, some additional bands have also been observed such as a peak at 1080 cm^{-1} and band at 730 cm^{-1}. While the former has been attributed to =C–H bends, the latter to aromatic H in the lignin (Oliviera and Marques 2015).

4.3 *Thermal Characterization*

Figures 6 shows DTA/TGA/DSC curves obtained for some of the Brazilian lignocellulosic fibers. These are similar to those obtained for similar fibers elsewhere. From these, degradation temperatures of various constituents of the fibers have been arrived at. From the figures, a three-stage process is evident in the thermal degradation of lignocellulosic fibers with 250–300 °C presenting the first stage, 300–400 °C the second and around 450 °C, the third stage. Researchers working with lignocellulosic fibers have attributed the first stage to the low molecular weight components such as hemicelluloses, the second associated with degradation of cellulose whereas the third stage to breaking down of lignin (Nurul Fazita et al. 2016). It should be noted that one of the crucial aspects in the development of lignocellulosic fiber-incorporated polymer composites is the thermal degradation of these fibers, which would certainly pose limitation for the use of some thermoplastics. Further, porous polymer products can be expected due to the volatiles produced during the processing of such composites due to the degradation occurring at temperatures above 200 °C. Such products would exhibit lower densities and inferior mechanical properties.

The data that can be obtained from the above figures is shown in Table 4 for 3 Brazilian fibers.

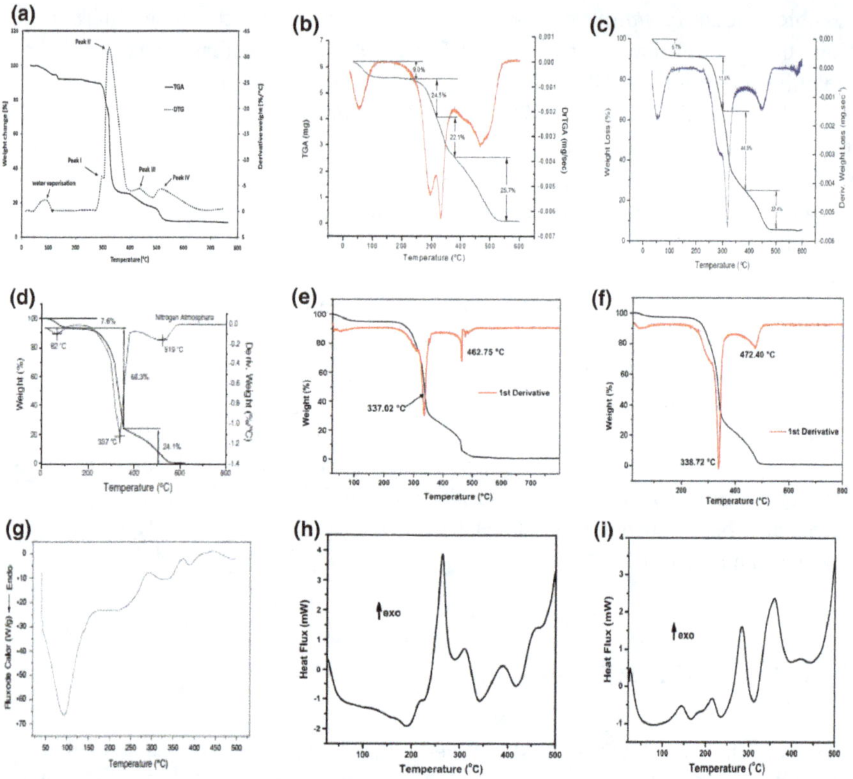

Fig. 6 DTA/TGA/DSC curves of some of the Brazilian lignocellulosic fibers: TG/DTG curves **a** Arundo fibers Fiore et al. (2014); **b** Brazilian nutshell fiber in air Inamura et al. (2011); **c** Brazilian nut bur fiber in air Inamura et al. (2011); **d** Curauá Fiber in N_2 Atmosphere Tomczak et al. (2007b); **e** Sugarcane Bagasse fibers in air Guimarães et al. (2009); **f** Sponge Guard fibers in air Guimarães et al. (2009); DSC Curves: **g** Açaı fiber Martins et al. (2008); **h** Banana fibers Guimarães et al. (2009); **i** Sponge Gourd fibers Guimarães et al. (2009) (Reproduced with kind permission of the Publishers)

4.4 Mechanical Properties (Tensile, Flexural, and Impact Properties)

One of the major mechanical properties studied in respect of Brazilian fibers is their tensile properties. In some cases, these properties are studied as a function of diameter and length of the fibers and also as a function of strain rate. For this purpose first stress–strain curves of the fibers are obtained for from which Young's modulus (YM), ultimate tensile strength (UTS) and percentage elongation are normally derived. Figure 7 shows stress–strain curves thus obtained in respect of curauá fibers while Table 5 lists tensile properties derived from such curves for various Brazilian fibers by different researchers.

Table 4 TGA data derived from the curves of Taquara-lixa, açaí and curauá some fibers (Reproduced from Gehlen 2014)

Fiber	Mass loss (%)	Peak (oC)	Observations
Taquara-lixa (*Merostachysskvortzoviisendulsky)*	7.1 (45 °C)	360	Due to the high lignin content, 27.5%, the fiber is stable up to 250 °C
	67.9 (380 °C)		
	22.3 (380–800 °C)		
	2.7 (800 °C)		Residue
Açai (*Euterpe edulis*)	8.4 (100 °C)		Loss of water
	30.0 (257 °C)		Starts the degradation of hemicelluloses
	31.0 (338 °C)		Final degradation of lignin and cellulose
	5.4 (650 °C)		Residue
Curauá (*Ananaserectofolius*)	5.9 (80 °C))		Loss of water
	65.8 (290 °C))		Degradation of nonstructural lignocellulosic materials
	7.4 (650 °C)		Residue

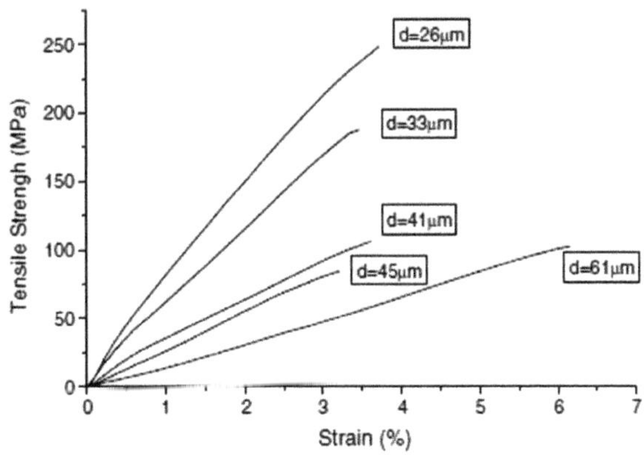

Fig. 7 Typical stress–strain curves for different diameters of Curaua fibers of Brazil Tomczak et al. (2007b). Reproduced with kind permission of the Publishers

Table 5 Tensile properties and flexural modulus of some lignocellulosic fibers of Brazil and others

Fiber	Young´s Modulus (GPa)	Ultimate tensile strength (MPa)	Elongation at break (%)	FM/FS (MPa)	References
Amazonian jacitara palm (*Desmoncus polycanthos* Mart.)	0.76–1.87	38.3–74.4	–	–	Fonesca et al. (2013)
Arundo donax	9.4	248	–	–	Fiore et al. (2014)
Banana	27–32	700–800	–	2–5	Satyanarayana et al. (2007), Monteiro et al. (2011a)
Bamboo	18–55	391–800 106–204		391–713	Osório et al. (2011), Monteiro et al. (2011a)
Buriti		129–254			Monteiro et al. (2011a)
Capim-dos-pampas					
Cissusquadrangularis (CQSF)/roots	56–234/ 68–203	1857–5330			Indran et al. (2014), Indran and Raj (2015)
Coir	2.5–4.5	95–174	13.7–41	50–275	Satyanarayana et al. (2007), Tomczak et al. (2007a)
Green coir	From Lupita paper				
Cotton	12	400		0.03–0.10	Satyanarayana et al. (2007)
Curauá	30–80@	1250–3000@	4.5–6@	87–310	Satyanarayana et al. (2007), Monteiro et al. (2011a) Tomczak et al. (2007b)
		117–3000	1.3		
	10.5	439–495 (MOR)	1.3–3.2		
	27.1	117 (MOR)/ 495			
Jute		465	0.7	0.3–0.5	Satyanarayana et al. (2007), Monteiro et al. (2011a)
	13–27	393–800			
Malva					
Piassava	1.07–4.59 5-6	134.58–142.9 109–1750	6.4–21.9		100; Monteiro et al. (2011a)
Pineapple	82	180	3.2		Satyanarayana et al. (2007)
Ramie	44 61–128	500–870 400–1620	1.2	–	Satyanarayana et al. (2007), Monteiro et al. (2011a)
Sisal	19.00 9–28	324–329/ 577 287–913	2–2.5	324.2 ± 173.9	Satyanarayana et al. (2007), Sydenstricker et al. (2003), Monteiro et al. (2011a)
Sponge guard					
Sugarcane Bagasse	27.1	222	1.1		Satyanarayana et al. (2007)

FM Flexural modulus; *FS* Flexural strength

4.5 Surface Treatments of Lignocellulosic Fibers

Banana fibers have been subjected to alkaline treatment and characterized for the crystalline fraction, dielectric behavior, metal removal (governed by solution pH) and biodegradation (Barretto et al. 2010). Results indicated this fiber is a low-cost alternative for metal removal in aqueous industry effluents. Thus, for regions with low resources, the biosorbents are an alternative to diminish the impact of pollution caused by local industries, besides being a biodegradable product.

Effect of surface treatment by alkali and methacrylamide to sponge gourd fibers using different techniques, which included FTIR, XPS/ESCA, X-Ray, TGA, and SEM, has been studied Tanobe et al. (2005). The results revealed no change in the fiber integrity along with improvement in surface area and thermal stability along with lowering of chemical constituents of the fibers with NaOH treatment. On the other hand, methacrylamide affected the integrity of the fibers besides changed in the chemical constituents of the fiber.

Alkali treatment of sugarcane bagasse fibers revealed enhancement in chemical constituents of the fiber along with the modification of the fiber surface, but without any significant change in the thermal resistance of the fibers (Simão et al. 2016). Modification of fiber surface was reflected in the improved impact strength of its composites with polypropylene matrix by almost 100%. FTIR studies of these fibers showed increased peak intensity at certain frequencies and reduction in carbonyl absorption regions suggesting extraction of lignin and other extractives leading to higher hydrophilic nature of the treated fibers.

On the other hand, similar treatment of sisal and bagasse fibers resulted in improvement of surface of the fibers, which was reflected in the good dispersion of the fibers in the matrix along with enhanced interaction between the fibers and the matrix resulting in improved strength properties (de Campos et al. 2013; Vilay et al. 2008).

Steam explosion and alkali treatments have been given to sugarcane bagasse fibers (Ricardo et al. 2011). Removal of acid-insoluble lignin by almost 60% was observed. On the other hand, steam explosion (physical treatment) of these fibers increased the amount of lignin along with decreasing the hemicellulose content. As a result of these, enhancement of properties (density, thermal stability, water absorption and thermo-mechanical properties) was observed in their composites with polyester suggesting improved chemical modifications, particularly with a steam explosion.

An interesting study was undertaken by the University of Parana′ West, Brazil, to evaluate the durability of bamboo for using it as reinforcement of concrete using bamboo splints and aimed at establishing some design procedures which guarantee the durability and the safety of structural member (Lima et al. 2008). Tensile properties of bamboo were measured with and without their exposure to wet and dry conditions in tap water and $CaOH_2$ solution up to 60 cycles. No change in these properties of the fibers was observed even after 60 cycles of exposure indicating that neither exposure to $CaOH_2$ solution nor tap water has any effect on these properties.

4.6 *Theoretical Aspects (Weibull Analysis and Artificial Neural Network)*

4.6.1 Weibull Analysis

Although many research groups in Brazil have been working on lignocellulosic fibers and their composites, only a few groups involving the present authors of this chapter seems to have reported on the Weibull analysis related to lignocellulosic fibers with majority of publications coming out from only one group (Amico et al. 2001a, b, 2005; Bevitori et al. 2010; Costa et al. 2010, 2011; Inacio et al. 2010; Margem et al. 2010; Monteiro et al. 2011a, b, 2014a; Nascimento et al. 2010; Portela et al. 2010; Santafe et al. 2009; Tanobe et al. 2005).

Recognizing the fact that lignocellulosic fibers exhibit large variation in their strength properties, and therefore averaging of the obtained strength values and use of standard deviation may not be proper, Amico and co-workers used Weibull analysis in their studies on sisal fibers to understand the effect of surface treatment on tensile strength of fibers (Amico et al. 2001a, b, 2005). They used two methods, viz., the maximum likelihood technique (requiring the solution of two nonlinear equations), and the analytical method. Details of their work can be seen in the references given here. Briefly, it is as follows:

They used two approaches in the second method; the first one gave the low coefficient of variation indicating a poor Weibull distribution ruling out the necessity for the use of diameter in their statistical analysis for sisal fiber. In the second approach called "Weibull distribution of strength" histograms were used with mean single fiber diameter, for all the fibers (0.195 mm for the 100 fibers) and the actual fiber diameter along with the Weibull parameters obtained by the maximum likelihood method. Obtained results were tested using a plot of tensile strength values with both single fiber diameter of the actual fiber diameter with logarithm fitting. Smaller scatter was observed in the corrected tensile strength values as per this fitting suggesting appropriateness of the model (Weibull probability) used.

On the other hand, correlation between tensile strength and diameter of eight lignocellulosic fibers (sisal, ramie, curauá, jute, bamboo, coir, piassava, and buriti) of Brazil origin has been studied through statistical evaluation using the Weibull analysis (Bevitori et al. 2010; Costa et al. 2010, 2011; Inacio et al. 2010; Margem et al. 2010; Monteiro et al. 2010, 2011a, b, 2014a; Nascimento et al. 2010, 2011; Portela et al. 2010; Santafe et al. 2009).

For this, Weibull statistical method was applied for the tensile strength values obtained for each of the diameter intervals of all the eight fibers shown in Fig. 3 with the arbitrarily chosen number of diameter intervals based on the range of the diameter of each fiber studied.

Based on the specific Weibull distribution with related deviation, the analysis of obtained strength values was carried out using a computer program using 4 parameters, viz., characteristic strength (q), Weibull modulus (b), statistical precision or adjustment coefficient (R^2) and the average strength, σ_{Av}. Prepared Weibull plots of Arundo fibers and those of one diameter for all 8 chosen fibers of one diameter are shown in Fig. 8A, B) respectively, while plots of the Weibull average tensile strength as a function of the equivalent diameter of corresponding statistical intervals of the chosen fibers are shown in Fig. 8C.

It may be noted that Fig. 8A showing the Weibull distributions for (a) tensile strength, and (b) Young's modulus of Arundo fibers has been calculated through an equation and the model used by the authors provides a good fitting of the data (Fiore et al. 2014). This figure also shows the Weibull shape and scale parameters for both the tensile properties with the shape parameter indicating the reliability of the data. No Weibull distribution for % elongation at break is shown although it would have been complete in respect of tensile properties of this fiber. However, the Weibull scale and shape parameters for % elongation are reported to be 3.24% and 5.30, respectively.

On the other hand, Fig. 8B corresponds to double log plot of the reliability *versus* the location parameter and the obtained graphs are unimodal showing a straight line fitting the points within the same diameter interval. Thus, this suggests that similar strength behavior of fibers tested belong to a unique group for each diameter interval. Further, it also becomes evident that there exists an inverse relationship between the tensile strength (σ) and the fiber diameter (d) in the case of all the tested fibers, which may be hyperbolic type [$\sigma = A/d - B$, where, A-in MPa. mm and B-in MPa are constants also called as hyperbolic coefficients having different value for each fiber]. It has been found that values of A vary from 13 MPa. mm for coir to 260 MPa.mm for Piassava fiber and those of B vary from −349 MPa for piassava to +389 MPa for ramie fiber (Monteiro et al. 2011a, b).

The hyperbolic relation between the tensile strength and the diameter of fibers indicates that smaller the diameter of the chosen fiber, higher is its tensile strength as in the case of sisal, ramie, curauá, and piassava fibers [see Fig. 8C-(e, f and h) respectively], which show about 1000 MPa as the tensile strength of these fibers. Other interesting points to be noted from the Fig. 8C is a constant value (shown by horizontal line) going through the limits of error bars as in the case of bamboo, coir, and buriti fibers [see Fig. 8C-(e, f, and h) respectively].

It is also reported that (i) this type of hyperbolic behavior may also be applicable to other fibers of Brazil as well as of other countries and (ii) the role played by the defects and flaws in the fiber could be responsible for this type of relationship between the fiber tensile strength and its diameter (Monteiro et al. 2011a, b, 2014a). The latter has been supported by the fractographic studies of these fibers as will discussed later in the chapter [see sec.—Microstructural and fractographic studies].

Considering that low diameter fibers exhibit higher tensile strength values comparable to metallic alloys, Monteiro and his co-workers have suggested that such lignocellulosic fibers should be recognized as a class of engineering materials. With this, they have made an attempt to locate these fibers suitably in tensile

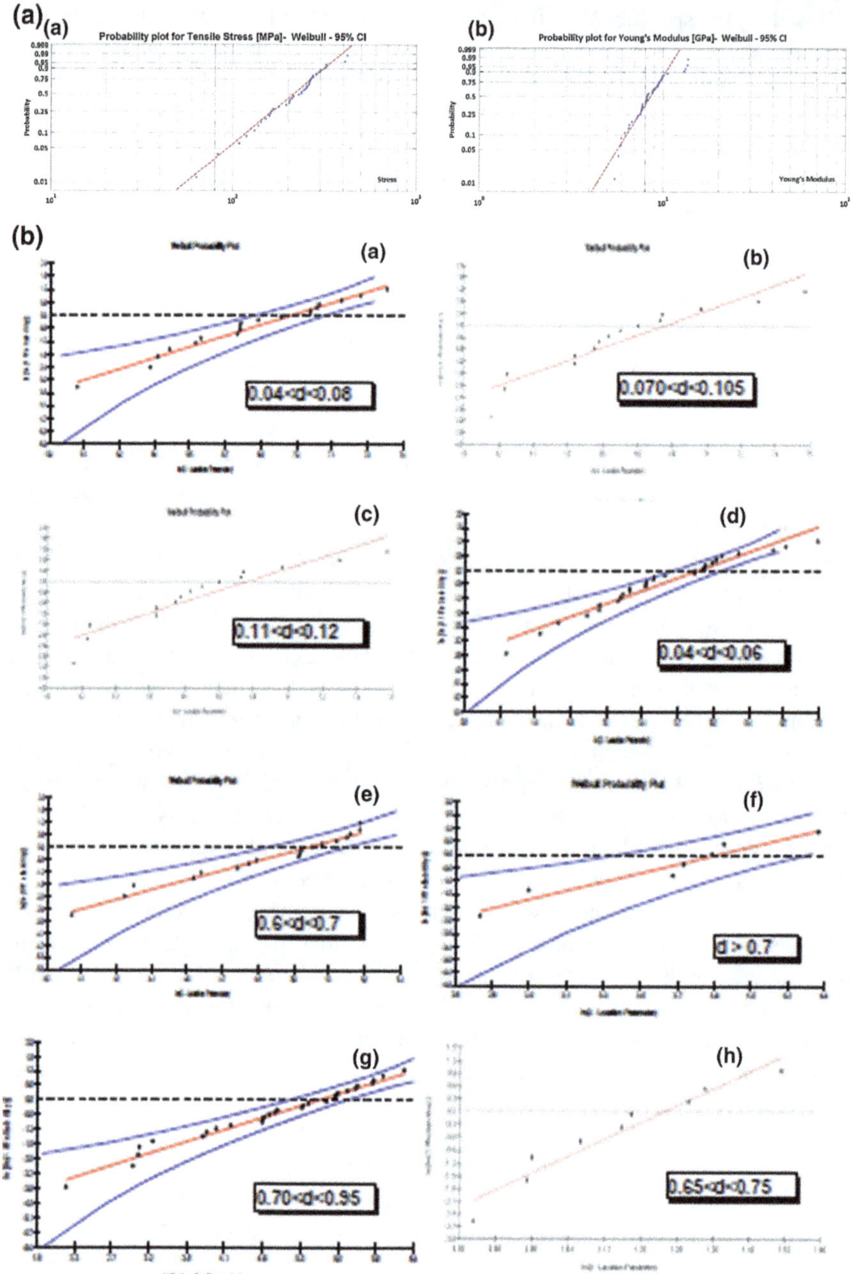
(a)
(a)
Probability plot for Tensile Stress [MPa]- Weibull - 95% CI
Probability
Stress
(b)
Probability plot for Young's Modulus [GPa]- Weibull - 95% CI
Probability
Young's Modulus
(b)
(a)
0.04<d<0.08
(b)
0.070<d<0.105
(c)
0.11<d<0.12
(d)
0.04<d<0.06
(e)
0.6<d<0.7
(f)
d > 0.7
(g)
0.70<d<0.95
(h)
0.65<d<0.75

◂**Fig. 8** **A** Weibull distribution of Arundo fibers for **a** Tensile stress **b** Young's modulus. Reproduced from Fiore et al. (2014) with the kind permission of the Publishers. **B** Weibull graphs for the eight lignocellulosic fibers: **a** Bamboo, **b** Buriti, **c** Coir, **d** Curauá, **e** Jute, **f** Piassava, **g** Ramie and **h** Sisal. Reproduced from Monteiro et al. (2011b) with the kind permission of the Publishers. **C** Weibull average strength as a function of the mean diameter associated with intervals in Fig. 3 for the eight fibers considered: **a** Bamboo, **b** Buriti, **c** Coir, **d** Curauá, **e** Jute, **f** Piassava, **g** Ramie, and **h** Sisal. Reproduced from Monteiro et al. (2011b) with the kind permission of the Publishers. **D** Ashby's density versus strength chart. Reproduced from Moneiro et al. (2011b) with the kind permission of the Publishers

strength versus density chart in a specified field in a proper Ashby's property chart (Monteiro et al. 2011b). This is shown in Fig. 8D. As can be seen, the field for lignocellulosic fibers is of ellipse shape, which includes some of the lignocellulosic materials having density values between 0.45 and 1.60 g/cm^3 and strength values between 95 and 3000 MPa based on the observed values of density and tensile properties of the fibers listed in Tables 2 and 5.

It can also be seen from the figure that location of lignocellulosic fibers in the chart is at the same level as that of a common type of engineering materials (alloys and composites). Furthermore, based on the density values of the fibers, these fibers would be more preferable from the minimum weight design in the two modes of loading, [viz., $\sigma^{2/3}/\rho$ and $\sigma^{1/2}/\rho$], in the development of new products particularly in automobile applications. In support of this, it may be relevant to point out that Monteiro and coworkers have prepared some polymeric composites with low diameter fibers, which exhibited high strength values (Costa et al. 2011). The foregoing also suggests the need for the development of a methodology to separate low diameter fibers from the lot obtained from the supplier. In fact, these authors have developed such a method wherein differences in densities are considered based on the use of suitable fluids while determining the density by Archimedes's principle (Monteiro et al. 2011b).

4.6.2 Artificial Neural Network (ANN)

Coir and curauá fibers of Brazil origin have been characterized by some physical, chemical, thermal, and strength properties with the latter as a function of different test conditions (fiber dimensions and speed of testing) (Tomczak 2010). In addition, application of ANN in strength properties of these two lignocellulosic fibers has been considered, which is the only pioneering study carried out on this aspect to the best knowledge of the authors. In this proposed computational model, strength properties [YM, UTS, and % elongation—outputs] are studied as a function of cellulose content, test speed (strain rate), and dimensions (gauge length and diameter)[inputs] of the fibers. The ANN scheme employed for the simulation of YM is shown in Fig. 9.

Some of the experimental and simulated values of YM, UTS, and % elongation of curauá fibers are listed in Table 6a–c.

(c)

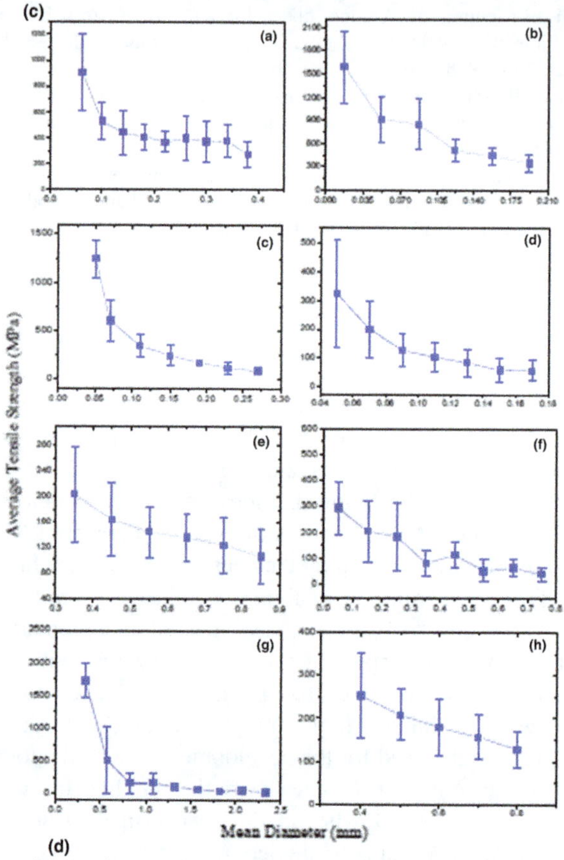

(d)

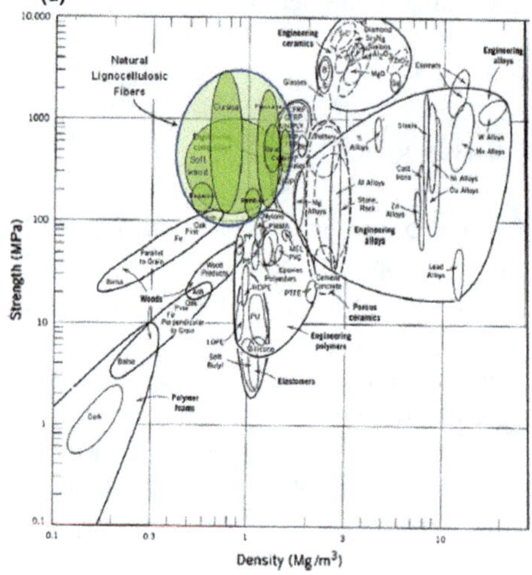

Fig. 8 (continued)

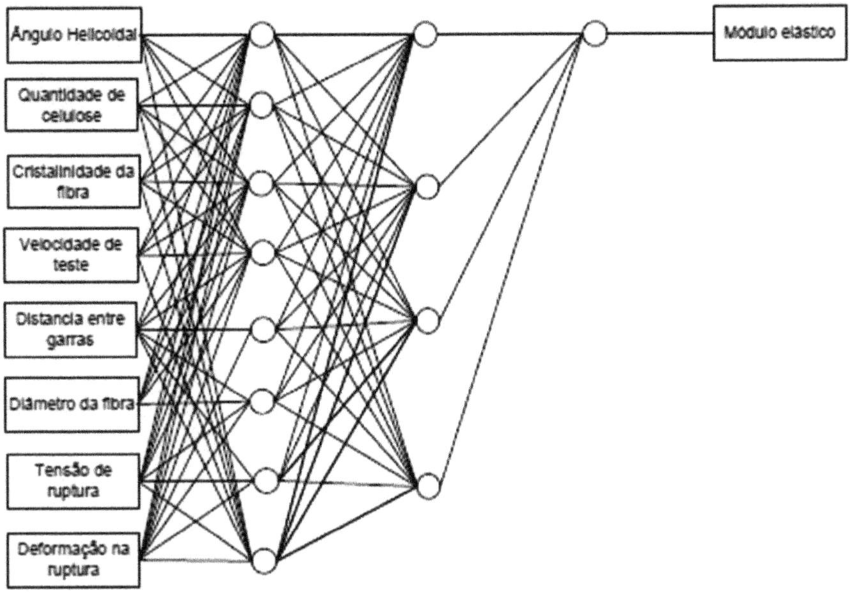

Fig. 9 ANN scheme employed for the simulation of the YM of the fibers

The correlation between the data on the influence of diameter of the fiber, gauge length, and test speed (strain rate) on experimental and simulated values of YM of curauá fibers was found to be 91.25, 99.55, and 94.79% respectively.

The aim of such a study is to enable the researchers to use these fibers from any region in any applications such as in composites as the expected results would take care of variations in the origin of fibers and their related properties. Thus, ANN may help in finding the potential of lignocellulosic fibers of any origin. For the modeling study by ANN, suitable software such as Pythia has been used. Experimental data of strength properties obtained were randomly divided into training data and test data, while initial experimental data was divided into two sets, one (70%) for network training and the other (30%) for verification. Helical angles, cellulose content, crystallinity, diameter of the fibers along with the test speed and gauge length of the fiber were considered as input signals. For each of these inputs, strength properties were considered as outputs or responses. It may be noted that any of the tensile property was defined as output when it was not used as an initial signal. The efficiency of each model was assessed by degree of correlation between observed data and simulated data by drawing graphs between these two data and calculating the correlation coefficient using a suitable equation. Thus three models were proposed, one for each of the evaluated properties.

Thus, this study has revealed feasibility of modeling the strength properties of lignocellulosic fibers using their physical, chemical, and structural characteristics. Probably, other input parameters such as climate and soil type can also be used to find their influence on ANN for modeling the tensile properties of lignocellulosic

Table 6 **a** Comparison of experimental and simulated elastic modulus values of Curauá fibers, **b** comparison of experimental and simulated ultimate tensile strength values of Curauá fibers, **c** comparison of experimental and simulated % elongation values of Curauá fibers

Speed (mm, min^{-1})	Distance between grips (mm)	Young's modulus experimental (MPa)	Young's modulus simulated (MPa)
5	5	26.39	26.12
	10	43.91	42.76
	15	44.32	44.49
	20	46.25	43.96
	25	51.89	50.00
10	20	47.99	44.59
20	20	47.00	46.31
50	20	53.69	51.64

Speed (mm, min^{-1})	Distance between grips (mm)	Ultimate tensile strength experimental (MPa)	Ultimate tensile strength simulated (MPa)
5	5	2213.00	2185.23
	10	2314.00	2236.81
	15	1950.05	1806.16
	20	1784.59	1661.35
	25	1730.61	1608.25
10	20	1898.04	1829.12
20	20	1921.82	1788.80
50	20	2174.67	2077.92

Speed (mm, min^{-1})	Distance between grips (mm)	Experimental elongation (%)	Simulated elongation (%)
5	5	10.20	10.71
	10	6.14	6.08
	15	4.83	4.89
	20	4.57	4.59
	25	3.74	3.89
10	20	4.70	4.79
20	20	4.52	4.49
50	20	4.49	4.53

fibers from anywhere. This would enable the expansion of utilization of these fibers in various applications without any limitation of origin of these fibers.

4.7 *Structural Aspects*

4.7.1 Morphology and Fractography

Figure 10 illustrates the typical cross sections of some of the Brazilian lignocellulosic fibers. It can be seen from the figure that depending on the source of fibers (stem/leaf/fruit etc.) distinct shapes and sizes exist for the transverse sections of each type of lignocellulosic fiber despite all these fibers have cellulose and lignin as their major constituents. In most cases, these are approximately elliptical with small eccentricity. In fact, when visually observed, apparent round fibers may display microscopically quite distinct eccentric cross-section shapes as can be seen in Fig. 10d–f. Moreover, even in a single fiber, significant changes in cross-section dimensions along the axial direction are common features of all lignocellulosic fiber as is evident in these figures. Therefore, as mentioned earlier, a careful measurement of the transversal dimensions has to be conducted to permit a reliable estimation of the diameter mean value, which will allow an exact determination of the area for the correct evaluation of the tensile strength.

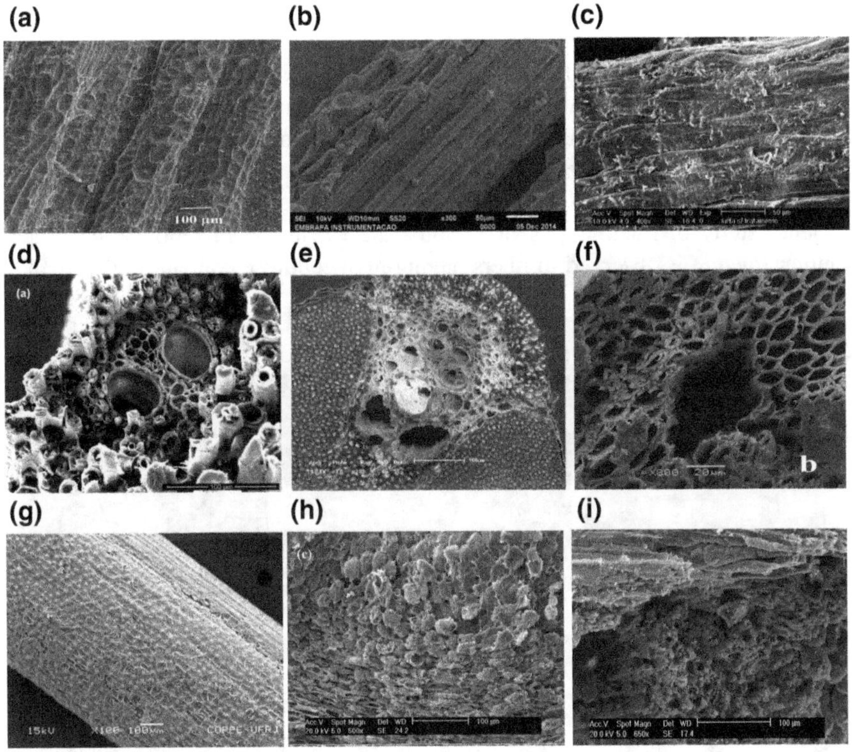

Fig. 10 Photographs of cross section and longitudinal section of some of the Brazilian fibers: longitudinal sections of **a** Acai Fiber; **b** Bagasse; **c** *Luffa cylindrica*; **d** Cross section of Arundo fibers; **e** transverse cross section of Piassava fiber with a number of lacunas at the end of the cross section and number of cells of different sizes; **f** cross section of green coir fiber; **g** longitudinal section of the piassava fiber showing the surface of the fiber; **h** cross section Brazilian nutshell fiber; (i) cross-section Brazilian nut fur fiber. Reproduced with the kind permission of authors and Publishers **a** from Martins et al. (2008); **b** Simão et al. (2016); **c** Tanobe et al. (2005); **d** Friore et al. (2014 **e**, **f** Nascimento et al. (2012); **f** longitudinal section of this fiber; **g** Cruz et al. (2015); **h** and **i** Inamura et al. (2011)

It is interesting to note that shell fiber present different morphology, depending on the fiber layers while the bur fiber presents irregular fibers, in different directions, like polymers cross-linking. This is reported to hinder fracture of the fiber.

Further, it is assumed that all types of lignocellulosic fibers consist of a number of continuous cylindrical cells with different arrangements and cell wall thickness. These cells are cemented by the intercellular substances. This substance is isotropic, non-lignocellulosic and ligneous in nature. As can be seen from Fig. 10d–f, there is a cavity termed "lacuna" in all types of fibers, whose position and dimensions differ for each type of fiber. Similarly, there is a cavity at the center of each cell, called "lumen", whose dimensions also differ from fiber to fiber as is evident from these figures.

On the other hand, longitudinal sections of the fibers show surface structures, which also differ for each type of fiber. It is also pertinent to note here that the tensile strength fiber is dictated by the dimensions of the individual cells (l/*d* ratio), which in turn differ for each type of fiber.

Further, in view of the differences in the structures of these fibers having different shapes, sizes, and arrangements of cells, lacuna and lumen, as well as thickness of cell walls, fracture mode of these fibers also differ. Figure 11a–f shows fractographs of various types of fibers revealing differences in their fracture mode with some showing intercellular or intracellular or mixed modes of fracture. Based on detail study of fractographs of various fiber, it has been reported that intergranular fracture is observed in fibers having high cellulose content, high length to

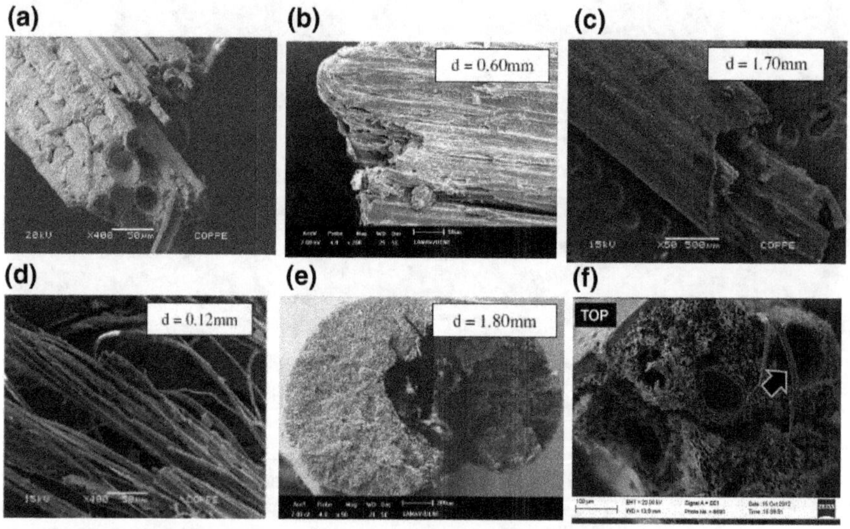

Fig. 11 Scanning electron microscopy (SEM) photographs of the fracture surfaces: **a** Bamboo—transverse section of giant bamboo showing the fracture tip of a thicker bamboo fiber; **b** Thick Buriti fiber; **c** Thick Coir fiber; **d** Thick Curauá fiber; **e** thick Piassava fiber; **f** jacitara fiber (in all height positions of the plant) after mechanical test. Arrows indicate the presence of unbroken fibers. Reproduced from **a** Cruz et al. (2015); **b–e** Monteiro et al. (2011a, b); **f** Fonseca et al. (2013)

diameter ratio (l/*d*) of cells or microfibrils and low microfibrillar angle exhibiting high tensile strength and modulus with low % elongation. On the other hand, the intercellular fracture is observed in fibers having low cellulose content, high microfibrillar angle, low l/*d* ratio exhibiting low strength and modulus values, but longer % elongation (Satyanarayana et al. 1986).

4.7.2 Results of Surface Treatments

Figure 12a–i shows the photomicrographs depicting the effect of surface treatments on some of the Brazilian lignocellulosic fibers in terms of morphology and fractographs. In the case of Piassava fiber, alkali treatment has led to the removal of protrusions and waxy layers, which are normally on the surface of the fibers (Nascimento et al. 2012). It was also observed the degree of removal depended on the concentration of the alkali and time of treatment. This is illustrated in Fig. 12,

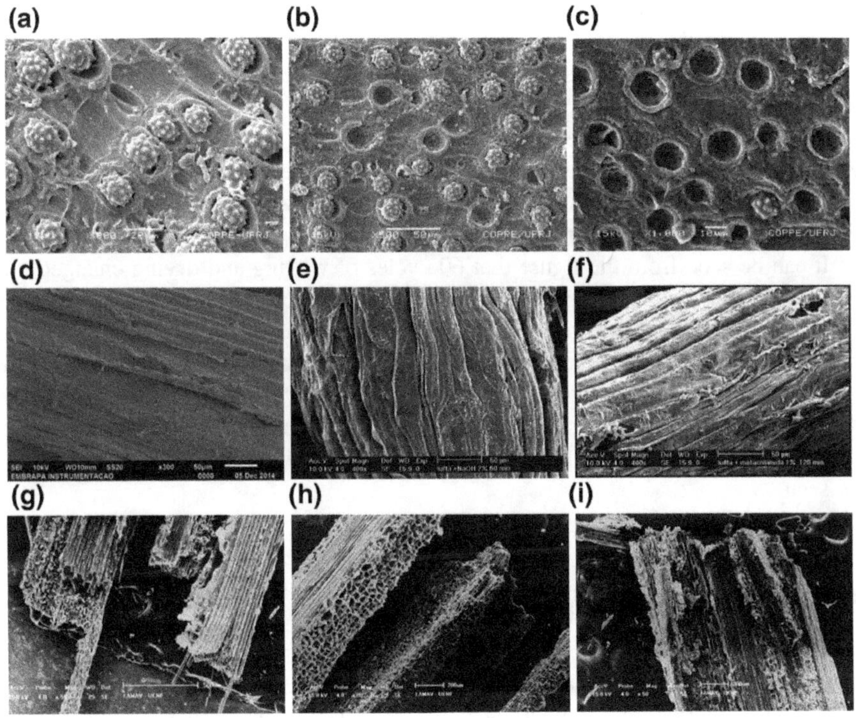

Fig. 12 SEM micrographs of surface treated lignocellulosic fibers of Brazil: **a**–**c** Piassava fibers untreated fiber, 0.1 and 10% alkali treated fibers for 1 min. and 60 min respectively; **d** surface of alkali treated bagasse fiber; **e** and **f** Longitudinal surface of luffa cylindrical fiber after alkali treatment and methacrylamide treatment; **g**–**i** Tensile ruptured Buriti fibers subjected to electron beam irradiation doses of **i** 50, (ii) 250 and (iii) 500 kGy (50X). Reproduced **a**–**c** from Ref. Nascimento et al. (2012); **d** From Simão et al. (2016); **e** and **f** Tanobe et al. (2005); **g**–**i** from Barbosa et al. (2012); with the kind permission of the Publishers

which shows photomicrographs of the surface of piassava fiber in untreated as seen in Fig. 12a and alkali-treated conditions with Fig. 12b showing surface morphology for 0.1% NaOH for 1 min treatment and Fig. 12c showing that for 10% NaOH for 60 min. It can be observed that with the former treatment resulting in partial removal of the protrusions while the latter one almost complete removal of protrusions from the surface of the fiber. In fact, it was found that the former gave better adhesion, while the latter poor adhesion. Figure 12d shows the surface of alkali treated bagasse fiber revealing the effect of alkali on the fiber. On the other hand, Fig. 12e, f shows a superficial layer of extractives on the untreated fiber and the surface of the treated Luffa cylindrica fiber, respectively. It can be seen that the latter does not show any superficial layer, which has been attributed to the removal of hemicelluloses and other extractives from the fiber (Tanobe et al. 2005).

Electron beam irradiation (Physical treatment) of buriti fibers extracted from the petiole of the palm tree revealed enhancement in tensile strength with a low dosage of 50 kGy, while it decreased with higher dosages (250 and 500 kGy). These observations have been supported by the fractographic studies of the fibers (Fig. 12g–i, which revealed a different type of fracture with lower doze treated fibers showing the plastic type of fracture (Fig. 12g), while higher doses treated ones showing the brittle type of fracture along with degradation of the surface of the fiber (Fig. 12e, f).

Figure 13a–d shows the morphology of bamboo splints with cement paste exposed to tap water and $Ca(OH)_2$ solution by alternate wet and dry conditions for 60 cycles as observed in SEM (Lima et al. 2008). It may be noted that the surface of the fibers was cleaned before taking photographs.

It can be seen from the figure that 60 cycles of wetting and drying changed the surface of bamboo although the figures show the cement paste on the splint embedded in concrete prism and a layer of $Ca(OH)_2$ deposition in splint aged in a solution of calcium hydroxide. However, there is no observable deterioration of the bamboo fiber after aging except for some changes of the fiber surface close to the splint surface. Neither deposition of cement products nor crystals of $Ca(OH)_2$ could be seen within the veins of the fiber, while completely empty vascular system could be seen.

5 Applications of Lignocellulosic Fibers

Considering many lignocellulosic fibers are coarse and strong they have been used in both conventional and new application areas. The conventional uses include as brushes, twine, ropes, string, yarn and which can also be woven into carpets, mats, and mattings, in various handicrafts, in papermaking, textiles (buffing cloth), as reinforcements in plastics and rubber, in biogas generation (waste or short fibers) (http://www.fao.org/economic/futurefibres/fibres/sisal/en/-22 Oct. 2016). Almost all the fibers-producing countries have utilized their fibers in these applications.

Fig. 13 Scanning electron micrographs of bamboo splint after aging: **a** splint surface with cement paste, **b** splint surface with deposition of $Ca(OH)_2$, **c** bamboo fibre aspect after being tested in tap water, **d** bamboo fibre aspect after being tested in calcium hydroxide solution, **e** nodal region after aging in tap water and **f** vascular system after aging in solution of calcium hydroxide From: Lima et al. (2008)

Of the new applications, nanomaterials, composites, and materials in arts are three important applications, which are being followed in Brazil also. For this purpose, the country has been supporting research and development activities through many of its funding agencies such as National Council for Scientific and Technological Research (CNPq) with a view to open areas for value addition and employment generation, besides working toward biodegradation and renewal of the ecological cycle. One aspect of such endeavor is a presentation of the results of these activities by Brazilian composite researchers constituting about 12% of the 235 in 2007, which might have grown greater than this in recent times. Materials and Metallurgical Engineering Groups in the country in both national and international conferences conducted in Brazil on natural resources dealing with lignocellulosic fibers and their composites. These groups work in different aspects of lignocellulosic fibers. Furthermore, recognizing the merits and demerits of these

Table 7 Merit list for the use of plant fibers incorporated composites (Satyanarayana et al. 2007, 2009; Nurul Fazita et al. 2016)

Merits	Demerits
Abundant availability, renewable resource and biodegradable	Heterogeneous quality
Non-brittle fracture on impact	Moisture sensitivity
Same performance for lower weight	Low durability
Stronger (25–30%) for the same weight	Lower mechanical properties especially impact resistance
Low cost—less than the base resin	Low thermal stability
Fully and easily recyclable	Poor fire resistance
Reduced molding cycle time—up to 30%	Poor fiber-matrix adhesion
Non-abrasive to machinery	Price fluctuation by harvest results or agricultural politics
Natural appearance and non toxic	
Low thermal expansion coefficient	
Good sound abatement capability	
Better energy management characteristics	
More shatter resistant	
Low mold shrinkage	
Easily colored	
High flex modulus—up to 5X base resin	
High tensile modulus—up to 5X base resin	
High notched impact—up to 2X base resin	
Lower processing energy requirements	
Meets minimum recycle content requirements including possibility for thermal recycling	

fibers shown in Table 7, the country follows some of the international trends in the search of potentials of lignocellulosic fiber covering various sectors such as automotive sector with commercial production and use of some parts and also in building sector (Satyanarayana et al. 2009b). In addition, there has been a trend in some of the countries in adopting better cultivation methods including genetic engineering and treatments to get quality fibers having uniform properties suitable for different applications (Anandjiwala et al. 2007). In the following pages, some details of work in the areas of composites and fine arts are discussed.

5.1 Nanomaterials

Recognizing that lignocellulosic materials are attractive due to their abundance, renewability, and low cost and existence of some limitations for industrial uptake of these materials in structural and other applications, reduction in the size of these materials below the normal microlevel could be one option to overcome the problem. Reduction of size of lignocellulosic fibers could offer improved behavior and enhanced properties. Besides, there has been increasing interest in the development of functional materials. Thus, an advantage in behavior and functionality exhibited by the nano-sized materials has led to the increasing attempts to the development and use of nanomaterials along with the need to find new sources to produce them. Many attempts have been reported to obtain nanomaterials (nanocellulose, nanolignin, and nanosilica) from lignocellulosic materials including wood (Alemdar and Sain 2008; Anandjiwala et al. 2007; Carneiro et al. 2015; Carpenter et al. 2015; Kalia 2016; Kalia et al. 2011; Pilla 2011; Thakur 2013, 2014; Thakur et al. 2013, 2014; Thomas et al. 2013) although most of nanofibers from lignocellulosic materials are concerned with nanocellulose. Surface modification and functionalization of nanocellulose from various sources including natural fibers can lead to various nanomorphologies which have the potential for application in many areas of engineering, medicine, etc. All these are possible through nanotechnology.

Nanotechnology may be defined as the manipulation of materials measuring 10 nm or less in at least one dimension. New materials and products may be developed when it is fully understood how materials and their interfaces are constructed at nano- and atomic scales (Kamel 2007).

It is reported that greater expectations have been placed on various aspects of nano-related things (Science of nanomaterials, nanotechnology including nanomanufacturing, etc.) by different persons be they are academicians, investors, governments, and industrial sectors (Serrano et al. 2009; Tuuadaij and Nuntiya 2008). This is because both nanomaterials and processes to obtain them (Nanotechnology) exhibit unique characteristics with nanomaterials exhibiting enhanced properties and performance, while the technology having the capability to fabricate new structures at the atomic scale. Therefore, it is not surprising to know attainment of both industrial and economic reality of these nano-related materials with estimated worldwide annual sources of naturally occurring nanoparticles to be the lowest from biomass (~ 1.8 million tons) (Gaffet 2011). Therefore, it is logical to explore the use and application of nanotechnology-based methods as a promising approach for efficient utilization of the lignocellulosic resources. In the following pages, brief mention of preparation of nanomaterials (nanocellulose, nanolignin, and nanosilica) from lignocellulosic materials, their characteristics and applications is made:

Different nomenclatures such as "crystallites", "“nanocrystals", "whiskers", "nanofibrils", and "nanofibers" have been given to these nanosized materials with one dimension in the nanometer range. These have been attracting scientists toward their isolation, characterization and possible applications (Klemm et al. 2011). These are called "nanocellulose", "nanolignin" and "nanosilica" derived from cellulose, lignin, and silica respectively.

Cellulose nanofibrils represent a unique material due to their low cost when compared to other types of new materials such as carbon nanotubes that are stronger (nanofibrils have about 25% of the strength of carbon nano tubes), but more expensive and not biodegradable (Kamel 2007; Yang et al. 2013). Cellulose nanocrystals and nanofibers may confer a large reinforcing effect on nanocomposites (Neto et al. 2012). Considering the abundant source of cellulose of plants and lignocellulosic residues, these materials may be prepared using an almost infinite number of plants. However, plants with a high amount of cellulose are preferred to prepared cellulosic nanofibers and nanocrystals.

It is known that well-developed agricultural, mining, manufacturing, and service sectors exist in Brazil, which is reported to be the South America's leading economic power being the world's seventh largest economy in 2011. In the area of nanotechnology it has not lagged behind with other developed/developing countries promoting nano-related research and its applications through a number of committed nanoscience (http://www.azonano.com/article.aspx? ArticleID=3589). These include 3 universities and 4 initiatives and networks focusing on nanotechnology (http://www.nanowerk.com/nanotechnology/Nanotechnology_Companies_Research_and_Degree_Programs_in_Brazil.php).

In Brazil, nanofibers of silk-floss, jacitara, and capim dos Pampas are being developed by some research groups, while nanomaterials have been produced with curauá, sugarcane bagasse, and other fibers including wood (Alves et al. 2012; Lunz et al. 2012; Kumode 2013; Neto et al. 2012). In the case of nanosilica, several silica processing industries have started producing nanosilica particles considering the progress of nanotechnology and increase in demand for these new materials (Carneiro et al. 2015). Accordingly, this chapter presents preparation methods of cellulose, lignin, and silica in nano form, their characterization, and applications.

Processing methods of various nanomaterials from lignocellulosic materials are shown in Table 8. Obtained nanomaterials from lignocellulosic materials have been characterized by their physical properties (dimensions, density, electrical thermal, etc.), mechanical properties (tensile strength, modulus, elongation) and other properties using various techniques (Satyanarayana et al. 2017). On the other hand, surface morphology and dimensions as well as their shape have been determined generally using optical (OM)/scanning and transmission electron microscopy (SEM and TEM) and atomic force microscopy (AFM). Dynamic light scattering has also used in the characterization of nanomaterials. Some of the nanomaterials obtained are shown in Fig. 14.

Some of the potential and existing applications of these nanomaterials are listed in Table 9.

Nanocomposites may be classified by nanofiller dimensionality, that is to say, one-dimensional (nanofiber), two-dimensional (nanolayer), and three-dimensional (interpenetrating network) systems (Kamel 2007).

5.2 *Composites of Lignocellulosic Fibers of Brazil*

It is reported that the southeast or the northeast of the country accounts for about 90% of these researchers (Satyanarayana et al. 2005). These researchers have used

Table 8 List of processing methods of various nanomaterials from lignocellulosic materials (adopted from Satyanarayana et al. 2017)

Type of nanomaterial	Processing method and characteristics
Nanocellulose MFC CNC, NFC	Mechanical-shearing and grinding/steam explosion/high pressure homogenization/chemical- acid hydrolysis, alkaline treatment and bleaching/combined chemical and mechanical treatments/ultrasonication/ electro-spinning/ Enzymatic Dia: 5–60 nm; length: several μms Dia: 5–70 nm; length: 100–250 nm (from plant celluloses)
Nanolignin	[a]Precipitation method/chemical modification/electrospinning followed by surface modification/freeze drying followed by thermal stabilization and carbonization/supercritical anti-solvent technology/chemo-mechanical methods/nano Lignin by self-assembly/Lignin nanocontainers by mini-emulsion method/template mediated synthesis
Nano silica	Acid treatment → either mechanical treatment or thermal combustion under controlled conditions/ activation techniques/combined chemical and thermal treatment or ball milling → oxidation/grinding → chemical treatment/enzymatic crystallization of amorphous nano silica

MFC Micro-fibrillated cellulose, Nanofibrils, microfibrils, nanofibrillated cellulose
CNC Cellulose nanocrystals, crystallites, whiskers, rod-like cellulose microcrystals
NFC Nanofibrillated cellulose
[a]Not known how many of these methods are applicable to lignocellulosic materials

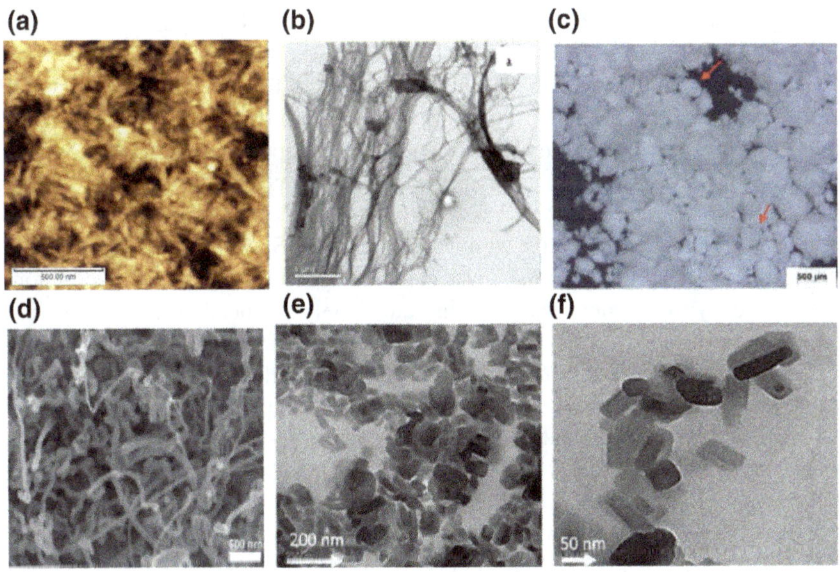

Fig. 14 **a** AFM image of cellulose whiskers extracted from Kraft pulp of Eucalyptus urograndis Neto et al. (2012); **b** TEM photograph of nanocellulose of Balsa wood Kumode (2013); **c** SEM of silica nanoparticle obtained from *Equisetum arvenses* L showing agglomerates (Carneiro et al. 2015); **d** High magnification SEM image of catalyzed material using sugarcane bagasse after pyrolysis at 1000 °C Alves et al. (2012); **e** and **f** TEM micrographs of nanoparticles of *Luffa cylindrica* fiber at two magnifications of 200 and 50 nm respectively Anupama et al. (2017)

Table 9 Applications of nanomaterials obtained from plant-based materials

Material	Applications	Reference
Nanocellulose	• Biomedical applications → Artificial skin and cartilage, medical implants, tissue engineering, drug delivery, wound-healing, vessel substitutes, cardiovascular applications, etc. • Dielectric materials → in oil-filled high-power transformers and cables • In composite manufacturing for various applications • Advanced functional materials → Aerogels, composites for light emitting devices/electronic applications, oil/Water absorbing gel materials, water purification systems • As templates for the chemical synthesis of inorganic materials	
Nano lignin	• In the development of composites/biobased polymers: as thermoplastic polymer composites, rubber-based lignin composites, lignin-reinforced biodegradable composites, foam-based composites, besides many others indicating the vast expansion of application for nanolignin • In drug delivery • As UV barrier and antibacterial agent → in finishing process in the textile industries for functional textiles by coating the fabric surface to overcome the problem of color change	
Nano silica	• In composites • In encapsulation of living cells	

both synthetic and natural polymers for the development of composites as has been done elsewhere globally. In the case of lignocellulosic fibers, they have been used primarily for substituting synthetic fibers such as glass fibers with the possibility of addressing environmental concerns along with supporting global sustainability and diversification of nonfood areas enabling stable economy through agro-industry. Besides, Brazilian researchers working with plant fibers have observed success in the use of lignocellulosic fibers-based composites in various areas such as automotive, building, and various engineering sectors (Satyanarayana et al. 2007, 2009b). Many such attempts have also reported development of both mono composites and hybrid composites involving either only lignocellulosic fibers or in a combination of these with synthetic fibers (Satyanarayana et al. 2009a, b).

One of the nonconventional applications of lignocellulosic fibers has been in using them as fillers or as reinforcements in the preparation of composites with both polymers and ceramic/cement matrices. In the case of polymers both synthetic and natural types have been used. Figure 15 illustrates progress in the development of polymer-based composites during 2000–2014.

The studies carried out on the above (polymer and ceramic/cement based) composites using Brazilian lignocellulosic fibers till the year 2011 have been already published by two of the authors (Dr. KGS/Dr. THDS) in journals, Conference Proceedings and in some books (Ramos et al. 2005; Satyanarayana 2005, 2006; Satyanarayana and Wypych 2007; Satyanarayana et al. 2004, 2005,

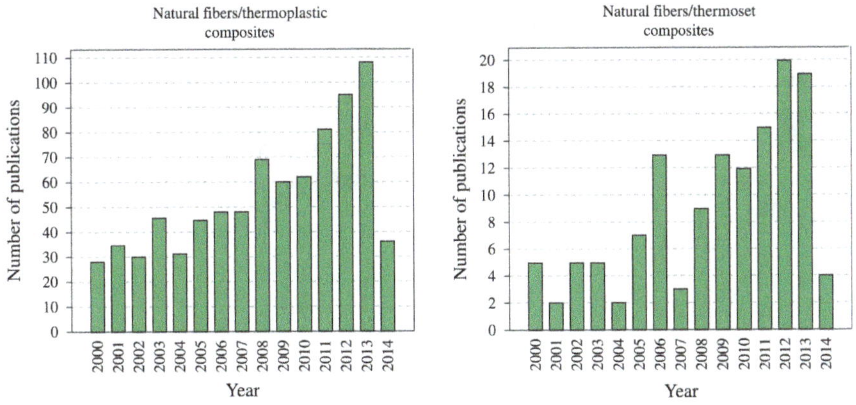

Fig. 15 Number of publications about natural fibers composites with thermoplastic and thermoset matrices in the last years from Web of Science database. Reproduced from Pereira et al. (2015) with the kind permission of Publishers

2009a, b; Tomczak et al. 2006). Some of the recent ones are mentioned in the following pages. Besides, some patents have also been taken by one the authors of this chapter on the subject matter (Guimaraes et al. 2006a, b).

In view of the above, only publications related to use of lignocellulosic fibers of Brazil origin including wood/sawdust in composites from 2012 till to date will be summarized in this chapter. Thus, preparation methods of composites and their characterization in respect of morphology and properties as well as product developments will be discussed in this chapter considering the fact this book would contain specific chapters on various types of composites [thermoplastic, thermoset, "green" and nanocomposites]. Interested readers for the publications before 2012 may see the references mentioned above.

5.2.1 Polymer-Based Composites

One important aspect of use of lignocellulosic materials in composites, particularly in polymer matrices, is their compatibility with the matrix materials to be used. This problem arises due to the weak fiber/matrix adhesion, which may result in lower mechanical performance of the composite even if a high strength fiber is incorporated. Of course, one of the reasons for this is the hydrophilic nature of the fibers and the hydrophobic nature of the polymers resulting in weak bond between the two. One of the methods used to assess is by measuring interfacial shear strength (IFSS) at the fiber/matrix interface. There are several methods available for this including theoretical modeling and estimation by inferred critical length values from mechanical tests (Monteiro et al. 2011a, b). The experimental methods include (i) single fiber pull out test, (ii) critical length pull-out curve, (iii) single fiber fragmentation test, (iv) micro-bond test, and (v) micro-compression tests. All these employ almost similar experimental conditions. Of these, very common test used is the single-fiber pull-out test, wherein a partially embedded single fiber is pulled from the block of an intended matrix material. The pull-out tests give values of shear strength of the interface and energies of debonding and pull-out (Satyanarayana and Wypych 2007). More details on this can be seen in this reference. One of the recent publications gives some details of polyester-malva fiber composites (Margem et al. 2016).

Processing Aspects of Polymer-Based Composites

The matrix materials used include both thermoset and thermoplastics in synthetic polymers as well as elastomers. In fact, Brazilian researchers in composites have also used recycled plastics, besides the use of natural or bio-polymers. Use of the latter is based on the consideration of environmental concerns, which are easily biodegradable along with better fiber-matrix adhesion. Table 10a shows the list of polymers used while Table 10b lists some of the matrix-fiber-processing method used by various groups of composite researchers of Brazil. As can be seen from the Table 10b processing of composites of lignocellulosic materials with polymers (synthetic and natural) has been by different researchers in Brazil, which depended on the available facilities with them and suitable to the systems studied. More details of each system can be seen in the references mentioned therein. However, there are hardly any reports on hybrid composites involving lignocellulosic fibers of Brazil in recent years.

Some of the Brazilian researchers have used factorial design of experiments (2^3) considering various process parameters such as composition of constituents (wt/wt), applied load (ton) and processing temperature (°C) (Vanderleia 2009) or size of reinforcement, amounts of plasticizer (For e.g., glycerin), if used, and pressure/temperature time (Kumode 2013). This methodology helps not only in minimizing the number of experiments to be carried out and arriving at optimum processing conditions to get the best properties in the final composites, but also to understand the interaction effects of the processing conditions. For this purpose, suitable software such as Minitab-16 can be used to model the effect of process parameters and the desirability approach can be used to optimize the response parameters, which are normally the strength properties.

Structure-Properties Aspects of Polymer-Based Composites

Table 11a, b lists some of the observed properties of a few polymer (synthetic and natural)-based composites containing lignocellulosic fibers of Brazil. Of course, it is not exhaustive, but includes mainly reported by the authors' group and a few of other research groups in Brazil. This is mainly due to the length of the chapter, which is already big enough. Interested readers are advised to more details by seeing the given references as well as other Brazilian researchers' publications.

Table 10 a List of polymers used in the preparation of composites, **b** matrix-reinforcement-processing method

Synthetic polymer thermoset/ thermoplastics	Natural/bio-polymer	Elastomers	References
Thermoplastics: Polypropylene, polyethylene, high-density polyethylene, high impact polystyrene; polyvinyl chloride **Thermoset**: Polyester, polyurethane, epoxy, phenolic, vinylester	**Thermoset** Biodegradable aliphatic polyester, starches (cassava, corn, castor seed cake), starch/EVOH, Soya protein; Polyhydroxy butyrate	Natural rubber	Pereira et al. (2015)

Matrix material	Reinforcement	Processing method used	References
Epoxy	Piassava fibers	Press-molded	Nascimento et al. (2012)
Epoxy	Buriti	Press-molded	Barbosa et al. (2014a, b)
Epoxy	Jute	Press-molded	Monteiro et al. (2014d)
Epoxy	Sisal	Press-molded	Monteiro et al. (2014g, h)
High-density polyethylene (recycled HDPE)	Coconut, eucalyptus and pine	Extrusion	Naldony et al. (2015)
Polyester	Bamboo	Press-molded	Monteiro et al. (2014e, f)
Polyester	Banana	Press-molded	Monteiro et al. (2014k)
Polyester	Buriti	Press-molded	Monteiro et al. (2014b, c)
Polyester	Curauá	Press-molded	Ferreira et al. (2010), Monteiro et al. (2013)
Polyester	Malva	Press-molded	Monteiro et al. (2014i)
Polyester	Ramie	Press-molded	Monteiro et al. (2014j)
Polyester	Sugarcane bagasse	Compression molding	Ricardo et al. (2011)
Polyester	Sponge guard (*Luffa cylindrica*)	Compression molding	Tanobe et al. (2014)
High-density biopolyethylene	Curauá	Thermo-molding/ extrusion/injection molding	Castro et al. (2012)
Polyethylene	Jute fabric	Press-molded	Lima et al. (2012)
Polypropylene	Pine saw dust (*Pinus Elliotti*)	Extrusion/compression molding	Santos et al. (2014, 2015)
Cassava starch	Green coir	Thermo-molding	Lomelí Ramírez et al. (2010, 2011, 2014)
Corn starch	Green coir	Thermo-molding	Lomelí Ramírez et al. (2010, 2011, 2014)
Corn starch	Banana and sugarcane bagasse	Thermo-molding	Guimarães et al. (2010)
Castor oil cake	Sugarcane bagasse	Thermo-molding	Vanderleia (2009)
Castor oil cake	–	Thermo forming	Kumode et al. (2013)
Castor oil cake	Banana and sugarcane bagasse	Thermo-molding	Guimarães et al. (2014, 2016)
Castor oil cake	Nanocellulose of balsa tree	Thermo-forming	Kumode et al. (2016)

Table 11 **a** List of composite systems with synthetic polymers studied, their tensile and flexural properties, **b** list of composite systems with natural polymers studied, their tensile and flexural properties

Composite system	Young's modulus (MPa)	Ultimate tensile strength (MPa)	Elongation (%)	Flexural modulus (MPa/GPa)	Flexural strength (MPa)	References
PP-75 sugarcane bagasse						Simão et al. (2016)
PPF	–	–	–	1.1 GPa	18.7	
PPFN	–	–	–	0.3 GPa	10.6	
PPFS	–	–	–	0.6 GPa	25.1	
PPFNS	–	–	–	0.5 GPa	21.4	
PP-0.5–1% MAH/ 20-35 saw dust Uncoated Coated with MAPP Coated with MAPP + 0.5–1% DCP Coated with MAPP + 1% BPO	1.09–1.50 1.48–1.90 0.72–0.82 1.86–2.16	15.25–19.45 21.63–23.00 23.37–26.28 19.72–25.18	2.46–3.39 2.64–4.57 3.75–5.21 1.95–4.47	– – –	36.99–47.57 43.87–51.68 46.37–50.83 42.71–53.38	Lucas et al. (2015)
Polyester-10–50 vol. % Curauá	10.4 GPa	190	–	~8000	~160	Oliveira et al. (2012)
Glycerol-dicarboxylic acid/2-10 Piassava	1098–2247	2.2–3.2	49–63	–	–	Miranda et al. (2012)
Recycled PE/5-40 jute fabric (new)	–	–	–	0.78–2.06	25.2–41.9	Lima et al. (2012)
Recycled PE/5-40 jute fabric (used)				0.74–2.03	17.5–28.7	

Composite system	Young's modulus (MPa)	Ultimate tensile strength (MPa)	Elongation (%)	Flexural modulus (MPa/GPa)	Flexural strength (MPa)	References
PHBV-20 Curauá	3.7	28.5	–	–	–	Beltrami et al. (2014)
COC/RG-25–35 banana (treated)	418–596	0.65–1.36	1.47–1.78	–	–	Guimaraes et al. (2014)
25–35 sugarcane bagasse	303–462	0.85–1.32	1.43–1.57	–	–	
CS-G/20-35 banana (manual)	71–97	3.56–4	10.63–14.96			Guimaraes et al. (2010)
25 banana (automatic control)	484 ± 151 80.88 ± 8.81	4.34 ± 2.30 1.73±0.16	1.20 ± 0.22 4.50 ± 0.25			
CS/RG -30 banana (manual)	162 ± 19	4.15 ± 0.2	10.30 ± 1.31			
25 bagasse (automatic control)	484.04 ± 150.86	4.34 ± 2.30	1.20 ± 0.22			

(continued)

Table 11 (continued)

Composite system	Young's modulus (MPa)	Ultimate tensile strength (MPa)	Elongation (%)	Flexural modulus (MPa/GPa)	Flexural strength (MPa)	References
CS/RG-25 bagasse (manual)	468 ± 59.23	3.85 ± 0.35	2.75 ± 0.46			
CaS-G/5-30 green coir-UT	10–11	3				Lomelí Ramírez et al. (2011)
CaS-G/5-30 green coir-T	~176	373.5				
Natural rubber/5–10 Açaí fiber						Martin et al. (2008)
GT	3.4–4.5	10–13	295–307			
IAN-873	4.3–5.2	13–15	329–332			
RRIM-600	3.9–5.7	12–16	313–345			

GT Gondag Tapen; IAN-873: North Institute of Agronomy; RRIM-600: Rubber Research institute of Malyasia; PPF: 25PP+Fiber; PPFN: 25PP+NaOH; PPFS: PP+7SEBS-g-MA; PPFNS: 18 PP+NaOH+7 SEBS-g-MA; HDBPE: High-density biopolyethylene; LHPB: Hydoxyl-terminated polybutadiene; FPIN1: HDPE + Compatibilizer MAPP + pine fiber powder; FPIN2: HDPE + pine fiber powder without compatibilizer; FEUC1: HDPE +compatibilizer MAPP + eucalyptus fiber powder; FOC1: HDPE + compatibilizer MAPP + coconut fiber powder; *MAH* maleic anhydride; *PP* polypropylene; *DCP* dicumyl peroxide; *BPO* Benzoyl peroxide

CS Corn starch; *CaS* Cassava starch; *COC* Castor oil cake; *G* Pure glycerin; *RG* Recovered glycerol; *PHBV* Polyhydroxybutyrate-co-hydrxyvalerate

One of the methods followed by many of the researchers working on the lignocellulosic fibers incorporated polymer/ceramic-based composites to understand the observed strength properties is the study of their fractographs. Some of these are mentioned below:

Figure 16a–l shows SEM photographs of fracture surfaces observed in some of the composite systems. These are composites of HDBPE-10 wt% curauá fibers (Fig. 16a), extruded HDBPE- 5% LHPB- 10% curauá fibers (Fig. 16b), dicarboxylic acid- glycerol -10% piassava fibers (Fig. 16c), corn starch (Amidex-3001)–30% crude glycerin–25% banana fiber (Fig. 16d) (composite showing fracture surface with some fibers lying parallel to the crack surface, a pulled-out fiber and fiber fracture, pull-outs, and some voids), composites of HDPE-MAPP matrix-containing pine fiber and eucalyptus fibers (Fig. 16e, f) [in both voids are shown by green arrows and fiber pull-outs by red arrows], wood fiber/sawdust fibers of different sizes (Fig. 16g), polyester- 42.6 vol.% short-*Luffa* fibers (Fig. 16h) and castor oil cake-30% recovered glycerol (RG) -25% banana fiber (Fig. 16i), [showing pulled-out fiber with plasticizer coating and nonstructured starch showing small polymorph granules], PP-75% sugarcane fibers(Fig. 16j), thermoplastic starch-30% green coconut fibers (Fig. 14k) and recycled polyethlene-30 wt% used jute fabric (Fig. 16l).

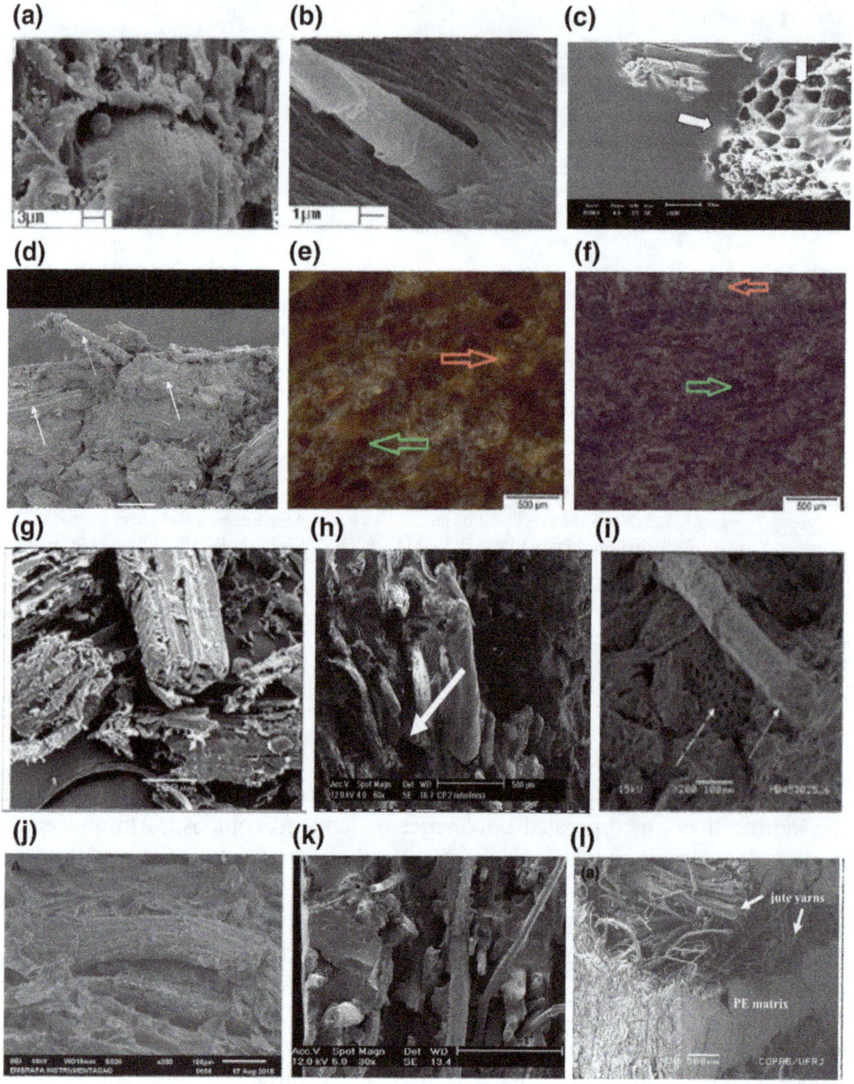

Fig. 16 SEM photographs of fracture surfaces observed in some of the composite systems. **a** HDBPE-10 wt% Curaua composites; **b** HDBPE/5% LHPB/10% fiber extruded.; **c** dicarboxylic acid- glycerol -10% piassava fiber(X600); **d** 45% starch (Amidex-3001)–30% crude glycerin–25% banana fiber composite showing fracture surface with some fibers lying parallel to the crack surface, a pulled-out fiber and fiber fracture, pull-outs and some voids; **e** HDPE-MAPP-Pine fiber composite; **f** HDPE-MAPP-eucalyptus fiber composite-in both voids are shown by green arrows and fiber pull-outs by red arrows; **g** wood fiber/sawdust fibers of different sizes (200); **h** polyester-42.6 vol.% short-*Luffa* fibers; (i) 70 castor oil cake—30 recovered glycerol (RG) and its composite with banana fiber showing pulled-out fiber with plasticizer coating and nonstructured starch showing small polymorph granules; **j** PP-75% sugarcane Fibers; **k** Thermoplastic starch-30% green coconut fibers; **l** PE-30 wt% jute fabric. Reproduced from: **a** and **b** Castro et al. (2012); **c** Miranda et al. (2012); **d** Guimarães et al. 2010; **e** and **f** Nalondy et al. (2015); **g** Santos et al. (2015); **h** Tanobe et al. (2014); **i** Guimarães et al. (2016); **j** Simão et al. (2016); **k** Lomeli Ramirez et al. (2011); **l** Lima et al. (2011a, b)

Analysis of the above fractographs reveals the following:

SEM studies were carried out on the fracture samples of composites of high-density biopolyethylene (HDBPE) obtained from ethylene derived from sugarcane ethanol with and without hydroxyl-terminated polybutadiene (LHPB) compatibilizer reinforced with curauá fiber prepared by two different processes (internal mixed-thermoforming and extruded using twin-screw extruder followed by injection molding). Highly homogeneous fiber distribution along with improved adhesion at the interface was observed in extruded injection-molded samples (Fig. 16b) resulting in higher flexural and impact strengths compared to those obtained by internal mixed-thermo-processing. The latter samples showed detachment of the fiber from the matrix (Fig. 16a), while the former confirms the compatibilizing action of LHPB.

Dicarboxylic acids-glycerol incorporated with 2–10% untreated and NaOH-treated piassava fiber composites prepared by molding and tensile tested showed improved adhesion between NaOH fiber and matrix due to the alkali treatment as revealed by the fracture surface (Fig. 16c) (Miranda et al. 2012). This resulted in the improved mechanical behavior of the matrix. This study also revealed the fiber content was more effective than the surface modification suggesting the potential of this material in the industrial marketplace

Fractographic studies of tensile tested green composites of cornstarch plasticized with commercial glycerin and reinforced with 20–25% banana and sugarcane bagasse fibers revealed their morphology was dictated by the properties of the fibers. Figure 16d shows fractograph of a composite containing 25% banana fibers revealing rough fracture surface with some cracks, fiber pull-outs and fibers lying parallel to the crack surface. These SEM studies helped in understanding observed improved tensile properties in these composites over those of the matrix.

Optical microscopy studies of flexural and compressive strengths tested recycled high-density polyethylene and three lignocellulosic fiber powders (Eucalyptus, pine, and coconut fiber) processed with and without compatibilizer by extrusion supported observed best results of these two properties of composites with eucalyptus and pine. Figure 16e, f show the fractographs of composites containing compatabilizer (MAPP) with pine fiber and eucalyptus fiber, respectively revealing better dispersion of fibers in the recycled HDPE+MAPP matrix. Also, fewer voids (shown by green arrows) and pull-outs (shown by red arrows) are evident from the fractographs of eucalyptus fiber-containing composites than those of pine fiber-containing composites. This suggests better fiber/matrix adhesion in these two composites, resulting in higher flexural and compressive strength values probably due to the use of a compatibilizer, which promotes better.

On the other hand, fractographic studies of PP-wood fiber/sawdust composites processed by extrusion revealed (Fig. 16g) the usual cellular nature of the fiber in the cross section revealing thick cell walls with the large, but irregularly shaped lumen, which is uncommon to many lignocellulosic fibers (Santos et al. 2015).

Studying the effect of chemical treatments on sponge gourd (*Luffa cylindrica*) fibers of Brazil in their composites with polyester resin) observed poor fiber/matrix adhesion (existence of gaps between the fiber and matrix), presence of voids or cavities (evidence for fiber debonding) with sizes in the range of 40–50 μm (darker areas in SEM micrographs—shown by the dotted arrow) and more fiber pull-outs in the composites (Fig. 16h). These observations have supported the observed low mechanical strength and brittleness of composites, particularly for composites with and without NaOH treatments.

Fractographic studies of a tensile-tested green composite of castor oil cake-recovered glycerol (plasticizer)-25% banana fiber showed rough fractured surface, fiber pull-outs lying parallel to the fracture surface along with nonstructured starch having small polymorph granules (Fig. 16i). Coating of fibers with plasticizer is also evident. These SEM studies supported observed significant improvement in all the tensile properties of the matrix by the fiber incorporation. A similar observation was made in composites containing sugarcane bagasse fibers.

SEM studies of fractured composite of PP with highly filled (75%) sugarcane bagasse fiber with and without alkali treatment and with and without coupling agent (SEBS-g-MA) prepared using mixer-torque rheometer showed the following: (i) improved of surface topography of the surface-treated fiber underlining that the alkali treatment modified the fiber surface; improved the interfacial adhesion between the fibers and the matrix PP due to the addition of coupling agent (Fig. 16j) (Simão et al. 2016).

Another study of green composite dealing with cassava starch plasticized by glycerin and incorporated with Brazilian green coir fibers prepared by thermo-molding reported fractographic studies made using SEM to understand the observed strength properties (Lomeli Ramirez et al. 2011). The authors observed that composite containing 30% fiber showing irregular fracture surfaces with cracks and pores in the matrix (Fig. 16k). Besides, pull-out of the coir fibers from the matrix after the test along with many holes are also evident in the figure suggesting not so good interfacial adhesion in these composites compared to those they observed with the composites containing both treated matrix and fibers.

An interesting fractographic study of tested for impact strength (both Charpy and Izod) of composites of recycled polyethylene containing used jute fabrics as well as continuous and aligned lignocellulosic fibers prepared by thermal compression molding with different percentages up to 40 wt% of jute fabric with oriented yarns has revealed fractures occurring predominantly by delamination (separation of jute fabric from the PE matrix) longitudinally to the impact direction (Lima et al. 2011a, b). These can be seen in Fig. 16l, which is a fractograph of PE-30% jute fabric composite and are attributed to the propagation of cracks following the plan of the fabrics caused by the low shear stress of the fabric/matrix interface. This figure also reveals high participation of the fibers in the rupture of the composite thus underlining the resistance offered to crack propagation due to the contribution of jute fabric.

Besides, the above there are some more interesting studies published recently on composites using bamboo incorporated into epoxy (Cruz et al. 2015), LDPE with and without glycerol and PE-g-MA (Delgadoa et al. 2012), UF and PF (Arruda et al. 2011), Poly (hydroxybutyrate-cohydroxyvalerate) with curauá fibers (Beltrami et al. 2014), recycled HDPE-micro fibers of sugarcane bagasse for possible use as masonry bricks (Cestari et al. 2016), etc. One of these presents results of an epoxy resin reinforced with giant bamboo fiber with ceramic tile backed by a plate at the front used as multilayered armor subjected to ballistic tests with high-velocity ammunition. While the ceramic layer spalled the projectile, the giant bamboo fiber composite helped in dissipating the remaining energy indicating the composite acted as an efficient barrier for the fragments originated from the ceramic brittle rupture. This composite seems to have many advantages such as lightweight and low cost compared to the conventional aramid fabric (Cruz et al. 2015). Another bamboo composite with HDPE matrix is reported to be a potential food packaging material as it exhibited an interesting antioxidant activity (Delgadoa et al. 2012).

5.2.2 Ceramic/Cement-Based Composites

Another important composite application of lignocellulosic fibers is ceramic/cement-based composites though this area is less researched compared to that of polymer-based composites. Since ceramic/cement-based materials are very brittle in nature, application of composites technology is very relevant. In fact, asbestos fibers have been used since a long time in these matrices. Considering the ill effects of asbestos and thus banned its use in many counties along with the advent of the use of lignocellulosic fibers in polymer composites, use of lignocellulosic fibers has been extended to ceramic/cement matrices. While many publications till 2009 have been already reported elsewhere (Satyanarayanan et al. 2009b), a couple of examples are presented below for completion purpose.

After the successful feasibility studies on the development of Brazilian giant bamboo fiber (*Guadua magna*)-reinforced polymer composites for possible application of particle boards (Arruda et al. 2011), attempts were made to develop cement and gypsum-based composites using particles of the same bamboo variety (Araújo et al. 2011). However, recognizing the curing of cement by the addition of bamboo fibers and thus on the strength development in the resultant product, the authors used $CaCl_2$ as an additive to accelerate the curing process in view of its efficiency and its cost. With the addition of $CaCl_2$, density of both the boards did not vary much (1180–1240 kg m^{-3}), While increases in the values of moisture content (7.92–10.5%), swelling thickness (0.92–1.8%), internal bonding (from 0.21 to 0.31 N/mm^2), modulus of rupture (7.36–7.91 N/mm^2), and modulus of elasticity (3000 N/mm^2) were observed. But there was a decrease in water absorption (19.32–15.50 for 2 h dipping).

Another study undertaken at the University of Parana′ West, Brazil, has reported the durability of bamboo splints reinforced in concrete wherein attempt was to establish some design procedures to guarantee the safety aspect also of the structural member made of this composite (Lima et al. 2008). Results of durability test revealed that tensile properties (YM: 23.75–24.99 GPa and UTS: 277.19–277.26 MPa) of the composites remained almost constant even after 60 cycles (each cycle of exposure to tap water and Ca $(OH)_2$ lasting for 24 h) and up to almost

6 months of aging. SEM studies of these samples taken on samples after cleaning the surface shown in Fig. 17a–f revealed that surface of the bamboo splint changed after 60 cycles of wetting and drying. However, one can see the presence of cement paste on the splint embedded in the concrete prism and a layer of $Ca(OH)_2$ deposition in a splint. Neither there is any deposition of cement product nor the crystals of $Ca(OH)_2$ in the veins of the fiber or the deterioration of bamboo fiber can be seen in the figure except for those very near to surface of the splint.

Fig. 17 SEM of Bamboo splint after aging: **a** splint surface with cement paste, **b** splint surface with deposition of $Ca(OH)_2$, **c** bamboo fiber aspect after being tested in tap water, **d** bamboo fiber aspect after being tested in calcium hydroxide solution, **e** nodal region after aging in tap water, and **f** vascular system after aging in solution of calcium hydroxide. Reproduced from Lima et al. (2008) with the kind permission of the Publishers

5.3 Fiber Used as a Material in Fine Arts

It is well known that professionals who work with lignocellulosic fibers in the technical areas identify their origins, followed by thorough characterization of these fibers. Based on these they may suggest applications in various fields such as reinforcing them in other materials such as polymers and ceramics. The latest trend of the applications of lignocellulosic fibers is the nanotechnology, which in turn finds applications in several areas. Considering the availability of a large variety of lignocellulosic fibers, vegetable nanofibers, and the possibility of developing nanocomposites of these, recent developments in the recent past years is focused on their use in various fields including arts.

Artists often develop a personal relationship with their work. They search for materials and techniques that somehow help them to express feelings, emotions, or, that may be seen as a silent way of communication. When artists use lignocellulosic fibers in their work, values of mechanical, thermal, optical or electrical properties do not matter. The search for specific materials is related to texture, color, sensation, perception, appearance or technique. In this context, this section describes the work of Pamela Bowman, a fine artist who enjoys working with lignocellulosic fibers in artistic installations. Why a specific material has been chosen, among so many others is an almost impossible question to be answered. However, engineers and artists may have in common curiosity, doubts, and enquiries that drive them ahead and contribute to building our world. Therefore, a true dialogue between art and technology is still a challenge that may benefit both areas, providing new techniques and materials for arts and beauty for technology.

As a fine artist Pam Bowman (one of the authors of this chapter) often uses fiber as a primary material in her work. She does this for several reasons; (1) It is a material reference to domesticity and relates to the concepts she wishes to portray, (2) the historical context of the use of fiber also relate to the concepts she wishes to portray, (3) she is attracted to its appearance and its tactile nature, more than she is attracted to other more traditional sculptural materials such as steel or stone, and (4) before she was a fine artist she was a craftsman with experience in sewing, weaving, basket making, and other fiber arts.

Pam Bowman's primary art medium is "Installation". This is a contemporary art form that is conceptual in nature. She creates large works that fill a space, are usually site-specific, and use a variety of materials and processes that reinforce the concept of the work.

Bowman endeavors to express the value of the human experience through installations that incorporate natural materials and traditional, labor-intensive processes. She is particularly interested in exploring how our repetitive tasks and efforts compound, interact, and contribute to a greater sense of purpose.

Materiality is the foundation for Pam Bowman's installations. Natural materials and fibers relate to the earth, the lives of men and women throughout time, and to the work they need to accomplish. String, cordage, and rope are particularly meaningful as symbols of tools helpful to perform work and as signifiers of accomplishment.

The repetitive and labor-intensive processes Bowman uses to manipulate materials reflect the tasks of living and the steady, continual efforts of life. Many of these processes relate to traditional fiber techniques such as weaving, braiding, and quilting. These techniques are reminiscent of ancient repetitive work necessary for home and life. Their historical associations with work and meticulous constructive processes provide a powerful metaphor for the human experience.

Through small, sustained effort over time, routines, and relationships contribute to the learning and progress of an individual. The rhythms, routines, and rituals of life intrigue Bowman. Daily routines and interpersonal relationships are repetitive. They are interlinking tasks we perform that create patterns and bonds in our lives. Over time, these small efforts join, connect, and converge creating a whole that is ultimately stronger than the individual—much like the accumulation of fibers in a rope. Experiences, like fibers, do not idly lineup but pull, cover, tug, and jockey for position in a complex, are shared the journey.

Through natural, relatable materials, and laborious, repetitive processes, Pam Bowman's installations give voice to the mundane and ordinary. They reveal with the magnitude and grace the extraordinary results of simply living.

5.3.1 Fiber as a Material Reference and Historical Reference to Domesticity

Throughout time people have used string or string-like materials to accomplish many tasks. Fiber has been used since ancient times for making objects that are useful for daily living. Traditional fiber techniques such as sewing, weaving, basketry, braiding, and rope making use fibrous materials that have been made into thread, cordage or yarn. The gathering of the materials was careful and labor intensive, and the building techniques repetitive. Hands performed the same movements over and over, developing skill and rhythm. Few people today weave their own fabric or make their own ropes, but we still use a string as a tool to help us accomplish many different tasks.

Many of the tasks that are traditionally thought of as women's tasks are repetitive in nature. Cooking, laundry, and cleaning are all tasks that are performed over and over. While the division of labor varies within cultures and within families, there are some tasks that have traditionally belonged to women throughout history and within most cultures. Bowman was interested in why the labor of home and family has been structured this way.

The research of Judith Brown explains the division between men's and women's activities. She states:

Nowhere in the world is the rearing of children primarily the responsibility of men, and in only a few societies are women exempted from participation in subsistence activities. Societies are able to draw on womanpower because their subsistence activities are compatible with simultaneous child watching. Such activities have the following characteristics: they do not require rapt concentration and are relatively dull and repetitive; they are easily interruptible and easily resumed once interrupted; they do not place the child in potential danger, and they do not require the participant to range very far from home (Barber 1991).

This would explain the pervasiveness of work that is traditionally seen as "women's work", such as making clothing, preparing food, and childcare.

Making clothing and preparing food fit the model that Brown has determined are appropriate tasks while caring for children. Before industrialization, making clothing included spinning, weaving, and sewing. Barber has shown that the textile arts have been dominated by women for 20,000 years (see *Women's Work: the first 20,000 years, by E.J.W. Barber.)* "Domestic spinning, weaving, fiber preparation, etc. have been found the world around to be ideal for women's chores: they are not dangerous to the children, they can be done at home, and they are repetitious and simple enough to be interrupted and resumed easily around the frequent little crises of child raising" (Barber 1995). In today's industrialized world, some of the reasons for this division of labor are no longer valid. Nevertheless, women have a long history with repetitive tasks and with fiber.

Figure 18 shows the *Webwork* references of the historical domestic tasks of weaving, braiding, and rope making.

Fig. 18 Webwork, cotton string, cotton rope, loom, 2015. Reproduced courtesy of the artist

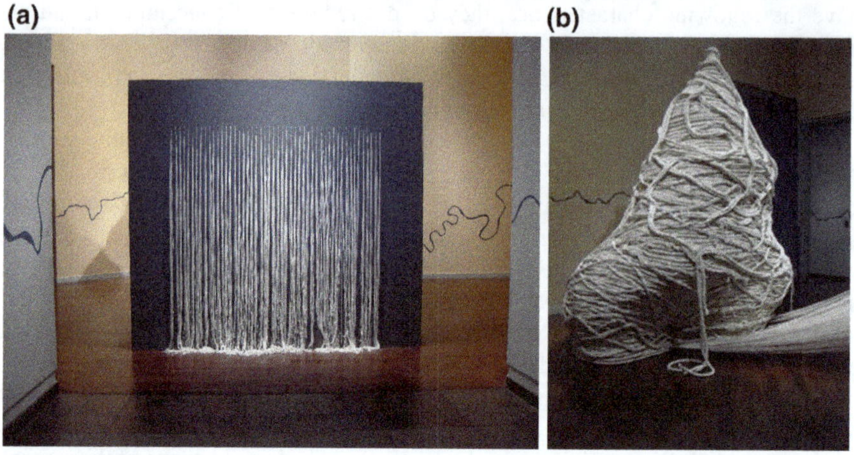

Fig. 19 **a** Wall hanging: Cotton rope, cotton string, wood, steel, paint, 2013. Reproduced courtesy of the artist. **b** Cotton rope. Reproduced courtesy of the artist

With *Becoming* the viewer first approaches a wall with cascading strings (see Fig. 19a). As they walk around the wall they see the same strings go through the wall and into a large form made of coiled rope, appearing as if the strings take this the form shown in Fig. 19b. Each string represents many repetitive acts of service performed over a lifetime.

As they accumulate, that lifetime of service becomes something of magnitude and significance. Repetitive works can be domestic in nature or other types of service such as going to a job daily, church and community service, or time spent in interpersonal relationships.

Traditional fiber techniques that are referenced in this work are weaving, coiled basketry, sewing, and rope making.

5.3.2 Choosing Fiber for Its Aesthetic Appearance and Tactile Nature

Every artist has materials of preference. Often artists will choose a certain medium because they enjoy the tasks associated with that medium. Those who work with steel typically enjoy welding for example. Artists also make choices based on their appreciation of the appearance or the feel of that medium—the beauty and feel of stone for example. Pam Bowman loves the look and feel of fiber and enjoys working with it. With fiber, she can portray expressions of beauty and goodness. The following discussion explains her interpretation of that expression and its importance.

The principles of morality, beauty, goodness, acceptance, and unity combined together form an elegant and graceful whole. Such a harmonious whole has concinnity, which can be defined as (1) Harmony in the arrangement of parts with respect to a whole, (2) Studied elegance and facility in the style of expression, and (3) An instance of harmonious arrangement or studied elegance and facility.

Bowman sees a connection between aesthetics and morality, beauty, and goodness. Such terms are difficult and complex and deserve definition. Beauty has been discussed at length by artists, critics, and theorists. Within some contemporary circles, it has been dismissed as no longer relevant. For example, Wittgenstein said, "'Beautiful' is an odd word to talk about because it's hardly ever used" (Proudfoot 1996). Within the bounds of this paper, beauty is that which gives visual pleasure and exalts the mind and spirit. Too often beauty is thought of only in terms of formal elements such as line, form, and composition. Over time a canon of acceptability has developed that helps us agree on such elements. Certainly, art can meet the formal criteria for elements and principles of design and have an appearance that is pleasing to the eye, yet have disparaging content. Dave Hickey elaborates on this type of situation in a discussion of Robert Mapplethorpe's work. Hickey feels it is the discontinuity between the formal beauty and the subject matter that makes the work so charged. In other cases, works are visually appealing but contain no content at all. They fall into the realm of pretty, but do not move the viewer. Bowman feels art must promote positive feelings in the viewer in order to be truly beautiful. Beauty has consistency in form and content, and it is connected with goodness and unity.

Goodness is probity, virtue, excellence, and light. It feeds your soul, enlarges you, and ennobles you. Goodness gives hope, helps you move forward, and can make it possible for you to look outward to others. Goodness is connected to reverence, benevolence, compassion, and wholeness. In contrast, darkness drains your energy, makes you lose interest in what is around you, and turns your thoughts inward to your fears, inadequacies, and aspects of you that are incomplete. Darkness makes you feel that your path forward is hopeless. Goodness and virtue have been studied for centuries in the field of ethics, also called moral philosophy, which "involves systematizing, defending, and recommending concepts of right and wrong behavior." [Fieser]The fact that this has been an area of study for so long, and continues to be a viable part of academia attests to the validity and importance of morality and goodness in society.

These principles of morality, beauty, and goodness build upon each other and create an elegant harmony. This concinnity of ideals that influence Bowman's work would not be complete, however, without including acceptance, and unity. Acceptance can mean contentment, peace of mind, resignation, and reconciliation. Acceptance within a relationship entails respect, patience, forgiveness, and unconditional love. In relationships between parent and child, and between person and God, there is also an element of submission. If family relationships have acceptance and unity, they can get closer to the ideal of "home".

Pam Bowman describes such a home as a place where you have feelings of comfort, familiarity, and security. You feel free to be yourself and express your ideas because it is safe to do so. Family members do not control or manipulate each other, and their communication is open and thoughtful. It is a place where you feel valued and needed. You like to be at home because you feel happy there, and your

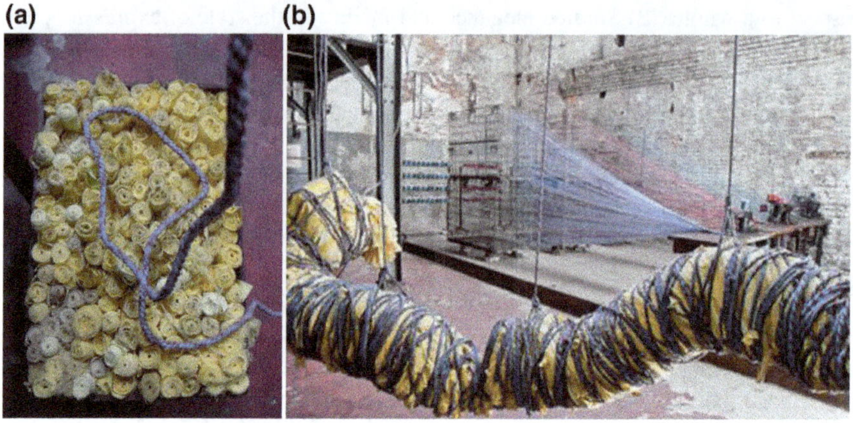

Fig. 20 **a** *Seamless (Detail)*, Dyed Cotton, cotton warp thread, handmade rope, historical artifacts (2012); **b** *Seamless*, Dyed cotton, cotton warp thread, handmade rope, historical artifacts (2012). Reproduced courtesy of the artist. (*Note* The title of the artwork (in this case, seamless), is in italics, and the materials used are not. Also, Fig. 20b has been altered in the ratio of its dimensions. It is "squished" to fit this format. The appearance of the images is important and should not be altered from the original. They can be made larger or smaller, but the ratio cannot be changed.)

happiness enables you to reach out to others instead of being self-centered. The ideal home has a feeling of sacredness within it. The possibility of this type of "home" is something that influences the work of Pam Bowman. It is a harmonious whole, a home of concinnity.

Bowman feels that she can best portray concepts of such a home and the values of morality, beauty, and goodness when she uses fiber as her primary material. She also likes to use ceramics, wood, steel, and other materials, but fiber is usually the primary material (see Fig. 20a, b).

5.3.3 Taking the Craft of Fiber Arts into the Realm of Fine Art

Pam Bowman has always been an object maker. Before she was an artist she was a craftsperson. She developed an expertise with fiber arts, particularly needlework and basketry. As she becomes more involved in art, she became more aware of the hierarchal relationship between art and craft. Bowman likes to use fiber arts and furniture in her installations. Both speak of domesticity, but one is thought of as typically female while the other is thought of as typically male. Both are given low status in a hierarchal relationship to painting and sculpture. During the modern era, crafts such as needlework and weaving were viewed as insignificant compared to fine art.

Feminist theory has helped to redefine such boundaries. "In its broadest terms, we would define the impact of feminism on art history as an adjustment of historical perspective".

Broude and Garrard say, "But the issue of folk arts and the crafts and their association with women also poses a larger question for art historical revision. The creation of hierarchies of value in the arts, in which the so-called fine arts of painting, sculpture, and architecture are held to be more significant than the crafts and decorative arts, has been the work both of artists, who since the Renaissance have controlled the art academies and defined their values, and of subsequent historians, who have selected out and emphasized the achievements of those in positions of cultural authority (or those in self-conscious rebellion against that authority). The resulting devaluation of what are called the 'minor arts' has led to the exclusion from our histories of such non-fine art activities as weaving, needlework, and quilt making. But in the broadest historical sense, when the small trickle of "high art" activity that has occurred in a few centuries at our own end of the historical spectrum are measured against the endless millennia in which weaving and pot making were the world's principal forms of art-making, one may conclude that it is not the crafts and traditional arts, but the fine arts, that are history's aberration (Broude and Garrard 1979).

Contemporary art practice has opened the way for artists to use a variety of common and mundane materials, yet artists still struggle at times with issues of hierarchy. Bowman has been successful at using fiber and referencing fiber arts within a fine art setting (see Fig. 20a–c). It may be noted that Fig. 20c has been again "squished"—the ratio of it has been altered (Fig. 21).

In conclusion, as a fine artist, Pam Bowman usually prefers to use fiber as a primary material in her work. By doing this, she is able to reference domesticity as well as allude to the historical meanings of fiber and their uses in many cultures. She also uses fiber because she is able to best express her feelings of morality, beauty, and goodness when she builds installations that use fiber. Finally, Bowman is able to build upon a lifetime of experience when she uses skills related to fiber

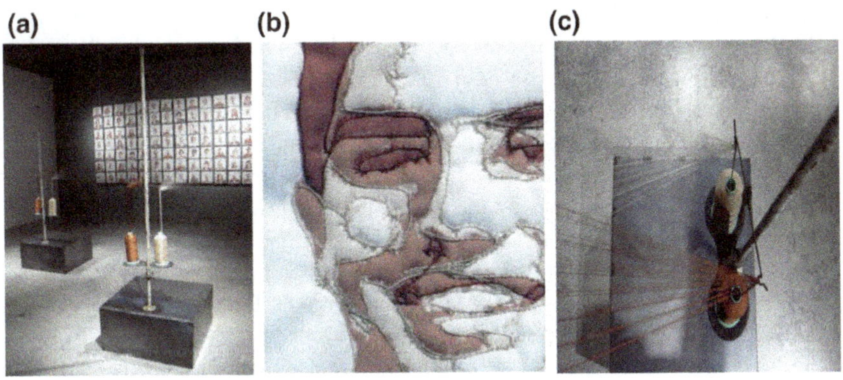

Fig. 21 **a** *Aggregation,* Cotton fabric and batting, thread, found objects, steel (2015); **b** *Aggregation* (*detail*), Cotton fabric and batting, thread, found objects, steel, (2015); **c** *Aggregation* (*Detail*), Cotton fabric and batting, thread, found objects, steel, (2015). Reproduced courtesy of the artist

arts and crafts. She is able to elevate them to the realm of fine art as she portrays the human experience, with its routines, relationships, and progressions with magnitude and grace.

5.3.4 Lignocellulosic Fiber Composites in Arts

While technologists address environmental issues in many ways, artists may add a significant amount of representativeness to it with their work by exploring the meanings of technologies for both the innovations and inventors as well as for society as a whole (Gangadharan 2009). This intersection between art and technology seems like an oxymoron, but it could be the new frontier for those who work with creative technology (Harvey 2011). The idea of connecting art to technology may also be enriched by thermoplastic texts. It is interesting to note that polymer science and technology, being an extraordinary field, has been contributing immensely to enhance the quality of life of people by offering them the possibility of developing new products and improved processes that bring comfort, safety, and lower cost goods (Flores-Sahagun et al. 2017). However, despite its many developments in polymerization techniques, new polymers, Ziegler–Natta catalysts, molding processes, etc., professionals in the field always deal with very technical issues.

One of the early uses of lignocellulosic fiber-based composites is based on thermoplastic molding techniques, wherein attempts have been made in developing a dialogue between art and technology by cognition in both fields. This would eventually lead not only to enhance potentials of thermoplastic molding, but also result in the effects, beauty, and textures of sculptures (Flores-Sahagun et al. 2017). The first reported work on the use of polymer-based composites in arts and sculptures seems to be of viscous Aspen wood fiber/polypropylene composites in 1998 using Haake Rheomix 600 extruder and a small silicone rubber mold of a feminine figure named Organic Beauty, whose original sculpture was made using molds having distinctive and flawless geometric shapes as found in many Egyptian arts. This is shown in Fig. 22. Because of random viscous flow of the polymer used, there was no repetition of the flow of polymer and thus uniqueness was observed in the resultant sculptures. Observed defects in the sculptures by the technical person, if any, found in the sculptures were termed as the effects by the artist embraced thus underlining the different views by the two different types of persons.

Recently, one of the authors of this chapter has made the Organic Beauty on major scale using Pinus wood fiber-polypropylene composites in the form of viscous filaments to fill a bigger mold of the same Organic Beauty sculpture. For this purpose, initially they conducted some laboratory experiments using a Braskem (H503) PP or low-density polyethylene (LDPE) having melt flow indices (MFI) of 3.2 g/10 min as matrix, maleated polypropylene (MAPP) (F4, MFI equal to 53.8 g/10 min. and F6, MFI 108.1 g/10 min) with 5 wt% pine wood powder. These composite mixtures were prepared using a bowl before feeding to the extruder. This was earlier prepared and characterized by Santos et al. (2015). It was observed that

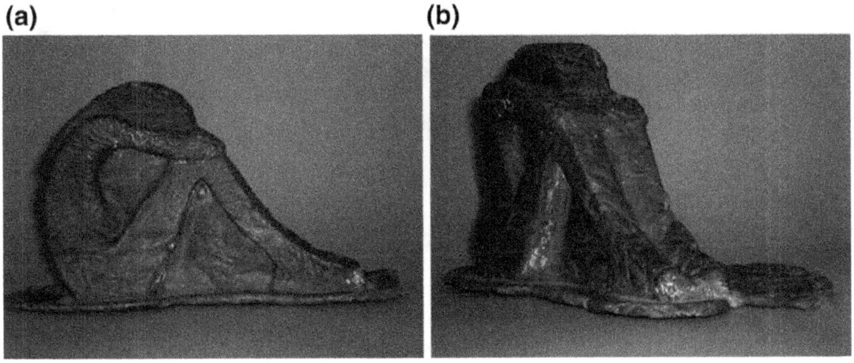

Fig. 22 Original composite sculptures of the 1998 Organic beauty: **a** Lateral image **b** Frontal image. Reproduced from Flores-Sahagun et al. (2017) with the kind permission of the Publishers

the resulting filaments with compatibilizer F6 MAPP were more opaque and difficult to compact into the mold resulting in a sandy texture, while the filaments with compatibilizer F4, MFI filling approximately two-thirds of the mold using a LDPE was very sticky within the extruder temperatures used having a sandy texture.

Flores-Sahagun et al. (2017) have observed the identification of the best composite composition and processing conditions were some of the challenges in their work. It was found that a polymer melt flow index (MFI) was very important, while the added amount of fiber dictated the flexibility of filaments as revealed by enough flexibility of composite mixture containing up to 10 wt% of fibers to manipulate. They also reported the composition of the composite dictated the color of the product with the addition of fiber turning to darker from the initial brown material. Also, the drying temperature of fibers and composites may affect the material color. Another factor that had its influence was the temperature with the observation of degradation of the product (darker product) at higher temperatures (100 °C and above for about 8 h), while no change in color of the product was observed when processed at 60 °C even for 24 h. However, considering the importance of the work underlining change of color of filaments and flexibility for the preparing sculptures and not for improving the strength properties, the researchers used mainly 5 wt% pine wood powder and drying at the highest temperature of 120 °C for 8 h resulting in darker brown product with apparently no reduction in mechanical strength. It was inferred that to obtain a special effect long vegetable fibers might be used in sculptures whereby instead of a chemical bonding in the composite possibility a mechanical anchoring is used. Accordingly, seven different vegetable fibers [Pampas grass (*Cortaderia selloana*), silk-floss, curauá (*Ananas erectifolius*), coir fibers, sponge gourd (Luffa cylindrical) and sisal (*Agave sisalana*)] were used with processing condition of 170 °C temperature in a extruder at 25 rpm and a PP of MFI equal to 80 g/10 min. to evaluate the appearance of some Brazilian fibers in these sculptures. As mentioned earlier their characteristics are known (Annunciado

Fig. 23 Sculptures of **a** Horse; **b** Rhinoceros; **c** Hippopotamus. Reproduced from Flores-Sahagun et al. (2017) with the kind permission of the Publishers

et al. 2005; Tomczak et al. 2007a, b; Tanobe et al. 2005; Sydentstricker et al. 2003; Naldony et al. 2015).

However, Sydentstricker et al. (2015) used finally MAPP, (F5) (Santos et al. 2015) having a MFI of 27.8 g/10 min. as a compatibilizer. The PP/MAPP blend was then used to coat the dried Pinus wood fibers for its use in preparing the composite. This procedure was followed by molding other sculptures as well (Flores-Sahagun et al. 2017). They have also explored the possibility of making the finished sculptures without using any mold, but using only the filaments. In both cases they used a rotatory base. It is reported that with the help of a sculptor for artistic manipulation to create the products, these researchers produced a horse, a hippopotamus, a rhinoceros, and a delicate ballerina. These are shown in Fig. 23, which had 5 wt% reinforcements, and were found to be unique. With these, they have identified technical aspects that would affect the final composite products. It was concluded that the molding using viscous filaments may add value to recycled products.

6 Concluding Remarks

From the foregoing, it becomes evident that Brazil has abundant availability of one of the renewable resources for lignocellulosic fibers. Data on the cultivated harvested land of these lignocellulosic materials, annual production of various types of fibers produced in the country, methods of their extraction are mentioned. Attempts have been made to collate the data on the physical, chemical, thermal, and mechanical properties as well the structural aspects of various lignocellulosic fibers of the country. Besides, methods/techniques to obtain values of these properties are mentioned. Also, two of the theoretical aspects (Weibull Analysis and artificial neural network) used in the characterizing of these fibers have been presented. While presenting the applications of these fibers, main attention has been focused on two unique ones; viz., in composites and fine arts. In the case of application of these fibers in composites, development of composite materials with the polymer and ceramic matrices, fabrication, structure, properties and product development of such composites are presented to underline many benefits to the society in general. This may be the first attempt to present these particularly on fine arts. Such studies would not only generate a good wealth of data on their structure and properties, but also help to open up new avenues for exploiting their full potential. These include opening up new uses for the fibers thus minimizing their non-utilization and thus going to waste, which would address the environmental safety along with generation of employment particularly in rural sectors. It should be noted that all these would contribute to the sustainable world order. Some of the perspectives that would enhance their utilization include bringing in improvements in the processing of these fibers and their composites leading to better properties. These, would in turn, present a bright future for the above-said applications.

Acknowledgements At the outset, the authors express their sincere thanks to Prof. Susheel Kalia, Editor of this book for inviting us to contribute this chapter. The authors place on record and appreciate the kind permission given by some of the authors (who have given permission to use their figures), M/s. Elsevier Inc Publishers, Springer, Wiley Publishers, SAGE Publications to reproduce some of the figures from their publications free of charges.

The authors wish to thank the National Council for Scientific and Technological Research (CNPq, process 310272/2014-9) for the fellowship given. Funding the authors also wish to thank the National Council for Scientific and Technological Research (CNPq, process 446263/2014-1) for the financial support given.

References

Abdelmouleh M, Boufi S, Belgacem MN, Durante AP, Salha B, Gandini A (2004) Modification of cellulosic fibers with functionalised silanes: development of surface properties. Intern J Adhes Adhes 24:43–54

Abdul Khalil HPS, Bhat AH, Ireana Yusra AF (2012) Green composites from sustainable cellulose nanofibrils: a review. Carbohydr Polym 87:963–979

Abernethy RB (Ed. & Publisher) (2006) The new Weibull handbook, 5th edn., Reliability and statistical analysis for predicting life, safety, supportability, risk, cost and warranty claims, 5th edn. The New Weibull Handbook, Florida, USA. ISBN 0-9653062-3-2

Alemdar A, Sain M (2008) Isolation and characterization of nanofibers from agricultural residues —Wheat straw and soy hulls. Bioresour Technol 99:1664–1671

Al-Haik MS, Hussaini MY, Garmestani H (2006) Prediction of nonlinear viscoelástico behaviour of polymeric composites using an artificial neural network. Int J Plast 22:1367–1392

Almeida JMR, Boynard CA, Monteiro SN (2000) Effect of chemical treatments on the surface morphology of sponge gourd (Luffa cylindrica) fibers. In ISNaPol 2000:27–31

Almeida JMR, Acquino RCMP, Monteiro SN (2005) Tensile mechanical properties, morphological aspects and chemical characterization of piassava (Attalea Funifera) fibers. Compos-Part A 37:1473–1479

Alves JO, Tenório JAA, Zhuo C, Levendis YA (2012) Characterization of nanomaterials produced from sugarcane Bagasse. J Mater Res Technol 1(1):31–34

Amico SC, Costa THS, Silva PS (2001a) Tensile strength of sisal fibers I: the Weibull distribution. In: Proceedings of 11th international molecular colloquium and 6th Brazil polymers congress, Gramado, pp 1537–1540

Amico SC, Costa THS, Monachnacz S (2001b) Tensile strength of sisal fibers II: influence of chemical treatment from the point of view of a Weibull distribution. In: Proceedings of 11th international molecular colloquium and 6th Brazil polymers congress, Gramado, pp 1541–1544

Amico SC, Sydenstricker THD, Silva PSCP (2005) Evaluation of the influence of chemical treatment on the tensile strength of sisal fibres by a Weibull distribution analysis. Met Mater Proces 17(3–4):233–242

Anandjiwala R, Hunter L, Kozlowski R, Zaikov G (eds) (2007) Nova Publishers Inc., New York (USA); ISBN No: 978-1-60021-559-9 (hardcover)

Annunciado TR, Sydenstricker THD, Amico SC (2005) Experimental investigation of various vegetable fibers as sorbent materials for oil spills. Mar Pollut Bull 50:1340–1346

AOAC Official Methods of Analysis-AOAC international, 19th edn. 2012

Araújo PC, Arruda LM, Del Menezzi CHS, Teixeira DE, Souza MR (2011) Lignocellulosic composites from Brazilian giamt bamabo (*Guadua mana*). Part 2: properties of cement and gypsum bonded particleboarda. Maderas Ciencia Tecnol 13(3):297–306

Arruda LM, Del Menezzi CHS, Teixeira DE, De Araújo PC (2011) Lignocellulosic composites from Brazilian giant bamboo (*Guadua magna*) part 1: properties of resin bonded particleboards. Maderas Ciencia Tecnol 13(1):49–58

Asasutjarit C, Hirunlabh J, Khedari J, Charoenvai S et al (2007) Development of coconut coir-based lightweight cement board. Constr Build Mater 21(2):277–288

ASTM D 256 84, Standard test method for impact resistance of plastic and electrical insulating materials. USA: ASTM; 1984. pp 81–102; (b) ASTM D 790—Flexural strength testing of plastics

Barber EJW (1991) Prehistoric textiles: the development of cloth in the neolithic and bronze ages with special reference to the Aegean. Princeton University Press, Princeton (USA)

Barber EJW (1995) Women's work: the first 20,000 Years—Women, cloth, and society in early times paperback. W.W. Norton & Co., New York. ISBN: 0-393-31348-3

Barbosa AP, Costa LL, Portela TGR, Moura EA, Del Mastro NL, Satyanarayana KG, Monteiro SN (2012) Effect of electron beam irradiation on the mechanical properties of buriti fiber. Materia Revista 17(4):1135–1143

Barbosa AP, Picanço OM, Giulio RA, Margem FM, Monteiro SN (2014a) Charpy impact test of epoxy matrix composites reinforced with Buriti fibers. Mater Sci Forum 775–776:296–301

Barbosa AP, Picanço OM, Giulio RA, Margem FM, Monteiro SN (2014b) Manufacturing of epoxy composites incorporated with Buriti fibers and evaluation of thermogravimetric behavior. Mater Sci Forum (Online) 805:460–465

Barreto ACH, Costa MM, Sombra ASB, Rosa DS, Nascimento RF, Mazzetto SE, Fechine PBA (2010) Chemically modified banana fiber: structure, dielectrical properties and biodegradability. J Polym Environ 18:523–531. doi:10.1007/s10924-010-0216-x

Beltrami LVR, Bandeira JAV, Scienza LC, Zattera AJ (2014) Biodegradable composites: morphological, chemical, thermal, and mechanical properties of composites of poly (hydroxybutyrate-cohydroxyvalerate) with curaua fibers after exposure to simulated soil. J Appl Polym Sci.
DOI 10.1002/APP.40712 (8 pages)

Bessadok A, Marais S, Roudesli S, Lixon C, Metayer M (2008) Influence of chemical modifications on water-sorption and mechanical properties of Agave fibres. Compos A 39 (1):29–45

Bevitori AB, Silva ILA, Monteiro SN et al (2010) Weibull analysis of the tensile strength variation with diameter for jute fibers, In: Biomaterials symposium, first TMS-ABM international materials congress, Rio de Janeiro, Brazil, pp 1–10

Bezerra EM, Ancelotti AC, Pardini LC, Rocco JAFF, Iha K, Ribeiro CHC (2007) Artificial neural networks applied to epoxy composites reinforced with carbon and E-glass fibers: analysis of the shear mechanical properties. Mater Sci Eng A 464:177–185

Bisanda EST (2000) The effect of alkali treatment on the adhesion characteristics of sisal fibers. Appl Compos Mater 7:331–339

Bledzki AK, Gassan J (1999) Composites reinforced with cellulose based fibers. Prog Polym Sci 24:221–274

Bonetti TF, Sydenstricker THD, Amico SC (2005) Avaliação da *Cortaderia selloana* (Capim-dos-Pampas) na forma de fibras e mantas como material sorvente para o setor do petróleo. Biomassa Energia 2(4):311–317

Boynard CA, Almedia JRM (2003) Aspects of alkali treatment of sponge gourd (*luffa cylindrica*) —polyester matrix composite. J Appl Polym Sci 87(12):1927–1932

Br'ıgida A, Calado V, Gonc¸alves L, Coelho M (2010) Effect of chemical treatments on properties of green coconut fiber. Carbohydr Polym 79(4):832–838

Buschlediller G, Zeronian SH (1992) Enhancing the reactivity and strength of cotton fibers. J Appl Polym Sci 45(6):967–979

Caraschi JC, Leão AL (2000) Mechanical properties of curauá fiber reinforced polypropylene composites, in ISNaPol, pp 450–453, (CD-ROM)

Carneiro ME (2012) Obtaining nano silica from *Equisetum arvenses* L and their use in the modification of veneers of schi*zolobium parahyba var. amazonicum* (Huber ex Ducke) wood. Ph.D. thesis, Federal University of Parana, Curitiba, Brazil (in Portuguese)

Carneiro ME, Magalhães WLE, Muñiz GIB, Nisgoski S, Satyanarayana KG (2015) Preparation and characterization of nano silica from *Equisetum arvenses*. J Bioprocess Biotech 5:205. doi:10.4172/2155-9821.1000205

Carpenter AW, De Lannoy CF, Wiesner MR (2015) Cellulose nanomaterials in water treatment technologies. Environ Sci Technol 49(9):5277–5287

Carrijo OA, Liz RS, Makishima N (2002) Fibra da casca do coco verde como substrato agrícola. Hortic Brasil Brasília-DF 20(4):533–535

Cestari SP, Mendes LC, Altstädt V, Lopes LMA (2016) Upcycling polymers and natural fibers waste- properties of a potential building material. Recycling 2016(1):205–218

Cittani IM, Baruque-Ramos J (2016) Brazilian Buriti fiber. In: Fangueiro R, Rana S (eds) Natural fibres: advances in science and technology towards industrial applications-from science to market. Springer USA, 456 pages. ISBN: 978-94-017-7513-7

Costa HM, Visconte LY, Nunes RR, Furtado CR (2000) Rice husk ash—a new resource of silica for natural rubber compounds. In: International symposium ISNaPol, pp 37–42

Costa LL, Loiola RL, Monteiro SN (2010) Tensile strength of bamboo fibers: Weibull analysis to characterize the diameter dependence. In: Biomaterials symposium, first TMS-ABM international materials congress, Rio de Janeiro, Brazil, pp 1–10

Costa LL, Monteiro SN, Loiola RL (2011) Title of the paper? Proceedings of characterization of minerals, metals and materials symposium–TMS conference, San Diego, pp 1–8

Costalonga AM, Pessin RL, Helena GF, Jonas A, Garcez AAR, Monteiro SN (2014) Banana aqueous extract as a potential addition to clay ceramics. Mater Sci Forum (online) 798–799:246–250

Cruz RB, Lima Jr EP, Monteiro SN, Louro LHL (2015) Giant Bamboo fiber reinforced epoxy composite in multilayered ballistic armor. Mat Res 18(supl 2):70–75. http://dx.doi.org/10.1590/1516-1439.347514

de Campos A, Tonoli GH, Marconcini JM, Mattoso LH, Klamczynski A, Gregorski KS et al (2013) Composite reinforced with treated sisal fibers: property, biodegradation and water-absorption. J Polym Environ 21(1):1–7. http:// dx.doi.org/10.1007/s10924-012-0512-8

De Rosa IM, Kenny JM, Puglia D, Santulli C, Sarasini F (2010) Morphological, thermal and mechanical characterization of okra (*abelmoschus esculentus*) fibers as potential reinforcement in polymer composites. Compos Sci Technol 70(1):116–122

Delgado PS, Lana SLB, Ayres E, Patrício POS, Oréfice RL (2012) The potential of bamboo in the design of polymer composites. Mater Res 15(4):639–644

Dias DSAMF, Rosane T, Scarpini CV, Monteiro SN, Fontes VCM, Fernando V, Matheus M (2014a) Influence of firing temperature on the behavior of clay ceramics incorporated with elephant grass ash. Mater Sci Forum (online) 798–799:526–531

Dias DSAMF, Albernaz SK, Scarpini CV, Monteiro SN, Fontes VCM (2014b) Characterization of elephant grass ash as a potential clay ceramic addition. Mater Sci Forum (online) 805:585–590

Dicker MPM, Duckworth PF, Baker AB, Francois G, Hazzard MK, Weaver PM (2014) Green composites: a review of material attributes and complementary applications. Compos Part A Appl Sci Manuf 56:280–289

DIN Standard-847 (2005) DIN standards committee tools and clamping devices

El Kadi HA (2006) Modeling the mechanical behavior of fiber-reinforced polymeric composite materials using artificial neural networks—a review. Compos Struct 73:1–23

El Kadi HA, Al-Assaf Y (2007) The use of neural networks in the prediction of the fatigue life of different composite materials. In: Kageyama K, Ishikawa T, Takeda N, Hojo M, Sugimoto S, Ogasawara T (eds) Proceedings of 'A giant step towards environmental awareness: from green composites to aerospace', 16th international conference on composite materials (ICCM-16), 7–13 Jul, 2007, Kyoto Japan, organized by Japan Society for Composite Materials. 9 pages

El-Sabbagh A (2014) Effect of coupling agent on natural fibre in natural fibre/polypropylene composites on mechanical and thermal behaviour. Compos Part B Eng 57:126–135

Elzubair A, Bonelli CMC, Rademaker H, Suarez JCM, Mano EB (2007) Morphological, structural, thermal and mechanical characterization of piassava fibers. J Nat Fibers 4(2):13–31

Ferriera AS, Monteiro SN, Lopes FPD et al (2009) Curaua fiber mechanical properties evaluation by the Weibull analysis (in Portuguese). In: 64th international congress of the Brazilian Association for metallurgy and materials, Belo Horizonte, Brazil, pp 1–12

Ferreira AS, Lopes FPD, Monteiro SN, Satyanarayana KG (2010) Charpy impact resistance of alkali treated Curaua reinforced polyester Composites. Revista Matéria 15(2):132–139. http://www.materia.coppe.ufrj.br/sarra/artigos/artigo11206

Fieser J (1995) Ethics, *The internet Encyclopedia of Philosophy.* www.iep.utm.edu/ethics, p 1

Filho PA, Bahr O (2004) Biomass resources for energy in North-Eastern Brazil. Appl Energy 77 (1):51–67

Filho NOL, Ladchumananandasivam R, Veri´ssimo SA, Medeiros JI (2004) The importance of cotton fibre in the Brazilian market and its perspectives. In: Conference on textile engineering (SINTEX II), September 2004-Paper No. TTN-A-015.pdf

Fiore V, Scalici T, Valenza A (2014) Characterization of a new natural fiber from *Arundo donax* L. as potential reinforcement of polymer composites. Carbohydr Polym 106:77–83

Fiori S (2001) Probability density function learning by unsupervised neurons. Int J Neural Syst 11 (5):399–417. doi http://dx.doi.org/10.1142/S0129065701000898

Fiori S, Bucciarelli P (2001) Probability density estimation using adaptive activation function neurons. Neural Process Lett 13, 31. doi:10.1023/A:1009635129159

Flores-Sahagun THS, Veroneze SMA, Carneiro NR, Paredes RSC (2015) Processos de moldagem para obtenção de peças plásticas poliméricas revestidas ou não com metais a partir de filamentos poliméricos, BR1020150076959 (deposition), assigned to Universidade Federal do Paraná and Universidade Federal do Rio de Janeiro

Flores-Sahagun THS, Carneiro NR, Flores-Sahagun DL (2017) Thermoplastic polymeric composites and polymers: their potentialities in a dialogue between art and technology. In: Thakur VK, Thakur MK, Kessler MR (eds) Handbook of composites from renewable materials, vol 2. Scrivener Publishing LLC, pp 263–286

Fonseca AS, Mori FA, Tonoli GHD, Savastano H Jr (2013) Properties of an Amazon vegetable fiber as a potential reinforcing material. Ind Crop Prod 47:43–50

Franck RR (ed) (2005) Bast and other plant fibers. Woodhead Publishing Ltd. & CRC Press, Cambridge England (UK), 432 pages

Gaffet E (2011) Nanomaterials: a review of the definitions, applications, health effects. How to implement secure development. Compt Rend Phys 12(7):648–658

Ganadharan SP (2009) Mail art: networking without technology. New Media Soc 11(1–2):279

Gehlen LR (2014) Effect of use of lignocellulosic fibers (açai and curauá fibers) in unsaturated polyester resin matrix. Dissertation submitted as partial fulfillment to get Master's degree in Engineering and Materials Science, Post-Graduate program in Materials Science & Engineering, Federal University of Parana, Curitiba (PR-Brazil). 104 pages

Guimarães JL, Satyanarayana KG, Wypych F, Ramos LP (2006a) Thermally molded direct-rolled plastic composites is based on starch with glycerol hydrate reinforced with fibers of vegetable or mineral and use of degradable polymer materials. International patent, BR200602428-A, 2006

Guimarães JL, Satyanarayana KG, Wypych F, Ramos LP, Cursino ACT (2006b) Method for thermal molding for deriving and extracting vegetable oil from castor bean cake for using these materials in packaging industry, involves using plastic with glycerol derivative. International patent, BR200605742-A, 2006

Guimarães JL, Frollini E, da Silva CG, Wypych F, Satyanarayana KG (2009) Characterization of banana, sugarcane bagasse and sponge gourd fibers of Brazil. Indus Crops Prod 30:407–415

Guimarães JL, Cursino ACT, Saul CK, Sierrakowski MR, Ramos LP, Wypych F, Satyanarayana KG (2016a) Effect of processing conditions on tensile properties of green composites of castor oil cake starch-plant fibers. Guimarães et al. J Bioproces Biotech 2014, 5:1 (8 pages). http://dx.doi.org/10.4172/2155-9821.1000198

Guimarães JL, Cursino ACT, Saul CK, Sierrakowski MR, Ramos LP, Satyanarayana KG (2016b) Evaluation of castor oil cake starch and recovered glycerol and development of "Green" composites based on those with plant fibers. Materials 9, 76 (18 pages); doi:10.3390/ma9020076

Harvey KC (2011) Art + technology: the new frontier of creativity and innovation. ACM Siggraph Comp Graph 45(1):1

Henderson A, Galeano G, Bernal R (1995) Field guide to the palms of the Americas. Princeton University Press, New Jersey, 414 pages

Homma AKO (2001) O Desenvolvimento da agroindústria no estado do Pará, http://www.captu.embrapa.br/pup_outros/rev20011213_08.pdf. Accessed 10th Dec 2016

Horeau W, Trindade WG, Siegmund B, Castelan A, Frollini E (2004) Sugarcane bagasse and curauá lignins oxidised and reacted with furfuryl alcohol: characterization and stability. Polym Degrad Stab 86:567–576

https://en.wikipedia.org/wiki/Artificial_neural_network dated 13 Dec 2016

https://pt.wikipedia.org/wiki/Taquara dated 24th Nov 2016

http://www.floresefolhagens.com.br/capim-dos-pampas-cortaderia-selloana/- Accessed 24 Nov 2016

http://www.fao.org/economic/futurefibres/fibres/sisal/en/. 22 Oct 2016

https://en.wikipedia.org/wiki/Prosopis_juliflora. 24 Nov 2016

http://www.azonano.com/article.aspx?ArticleID=3589. Accessed 1st Feb 2017

http://www.faculty.ucr.edu/~legneref/botany/fibers.htm. Accessed 23 Oct 2016

https://en.wikipedia.org/wiki/Cortaderia_selloana. Accessed 24 Nov 2016

http://www.revistaagroamazonia.com.br/13-fibras.htm. Accessed February 2005

http://lifeofplant.blogspot.in/2011/02/plant-fibers.html-accessed on 24th Oct. 2016

http://textilelearner.blogspot.in/2011/07/textile-fiber-properties-of-textile.html. **textile** fashion-study.com/ **textile-fiber**- general-**properties**-of-**textile-fiber/**

http://textilelearner.blogspot.in/2011/07/textile-fiber-properties-of-textile.html

http://textilelearner.blogspot.in/2012/02/what-is-textile-fiber-types-of-textile.html

http://www.conab.gov.br/download/cas/semanais/Semana07a11022005/Juta%20-2007a11-02-05.pdf

http://www.dsenyo.com/blogs/dsenyo/14106133-buriti-a-gift-from-the-tree-of-life#. 23rd Nov 2016

http://www.ipsnews.net/2006/10/ecobreves-brazil-plant-fibers-to-replace-asbestos/. Accessed 22 Oct 2016

http://www.nanowerk.com/nanotechnology/Nanotechnology_Companies_Research_and_Degree_Programs_in_Brazil.php-Accessed on 1st Feb 2017

https://en.wikipedia.org/wiki/Malva→24 Nov. 2016

https://textileapex.blogspot.in/2015/11/properties-of-textile-fibre.html]

IBGE-Municipal agricultural production, temporary and permanent crops 2013. (2013) Prod Agric Munic, Rio de Janeiro, 40, 1–102

Inacio WP, Lopes FPD, Monteiro SN (2010) Tensile strength as a function of sisal fiber diameter through a Weibull analysis, In: Biomaterials symposium, first TMS-ABM international materials congress, Rio de Janeiro, Brazil, pp 1–10

Inamura PY, Kraide FK, Armelin MJA, Scapin MA, Moura EAB, del Mastro NL (2011) Characterization of Brazil nut fibers. https://www.ipen.br/biblioteca/2011/eventos/17481

Indran S, Raj RE (2015) Characterization of new natural cellulose fiber from *Cissus quadrangularis* stem. Carbohydr Polym 117:392–399

Indran S, Raj RE, Sreenivasan VS (2014) Characterization of new natural cellulosic fiber from *Cissus quadrangularis* root. Carbohydr Polym 110:423–429

John K, Venkata Naidu S (2004) Sisal fiber-glass fiber hybrid composites: the impact and compressive properties. J Reinf Plast Comp 23(12):1253–1258; 23(15), 1601–1605 and 23 (17), 1815–1819

Kalia S (ed) (2016) Biodegradable green composites. Wiley, Hoboken. ISBN: 9781118911099 (Print)

Kalia S, Kaith BS, Kaur I (eds) (2011) Cellulose fibers: bio- and nano-polymer composites: green chemistry and technology. Springer, New York ISBN: 978-3-642-17369-1 (Print)

Kamel S (2007) Nanotechnology and its applications in lignocellulosic composites, a mini review. eXPRESS Polym Lett 1(9):546–575

Klaine C (2009) Characteristic strength, Weibull modulus, and failure probability of fused silica galss. Optical Eng 48(11):113401 (pages: 1–10). doi: 10.1117/1.3265716

Klemm D, Heublein B, Fink HP, Bohn A (2005) Cellulose: fascinating biopolymer and sustainable raw *material*. Angew Chem Int Ed 44:3358–3393

Klemm D, Kramer F, Moritz S, Lindstrom T, Ankerfors M, Gray D, Dorris A (2011) Nanocelluloses: a new family of nature-based materials. Angewand Chem-Internat Edn 50 (24):5438–5466

Kozlowski R, Rawluk M, Barriga J (2004) State of the art-production, processing and applications of fibrous plants. In: Sivam RL, Araújo RC (eds) Proceedings of 2nd international conference on textile engineering (SINTEX—2004), Sept 7–11, 2004, Natal, in CDROM, Paper No. TIP-1-007

Kulkarni AG, Satyanarayana KG, Rohatgi PK (1981) Electrical resistivity and dielectrical strength of plant fibers. J Mater Sci 16:1719–1726

Kulkarni AG, Satyanarayana KG, Rohatgi PK (1983) Weibull analysis of strengths of coir fibres. Fibre Sci Technol 19(1):59–76

Kumode MMN (2013) Potential of the use of the castor oil cake (*Ricinus communis L.*) as matrix for the production of composites and nanocomposites [Potencial do Uso da a Torta de Mamona (*Ricinus communis* L.) Como Matriz Para Produção de Compósitos e Nanocompósitos]. Ph.D. thesis, Engineering and Materials Science. Federal University of Parana, Curitiba (PR-Brazil), (in Portuguese)

Kumode MMN, Magalhães WLE, Satyanarayana KG, Lengowski EC (2013) Evaluation of physical and mechanical properties of nanocomposites of castor oil cake and glycerol with nanocellulose: In: Proceedings of Embrapa Instrumentação, São Carlos (Brazil), 10–13 June 2013, pp 241–243

La Mantia FP, Morreale M (2011) Green composites: a brief review. Compos Part A Appl Sci Manuf 42:579–588

Le Corre D, Bras J, Dufresne A (2009) Nanoparticles for eco-efficient packaging: influence of botanic. In: Jimenez A (ed) 2nd international conference on biodegradable polymers and sustainable composites (BIOPOL 2009), Alicante, Spain (6 pages)

Le Moigne N, Longerey M, Taulemesse JM, Bénézet JC, Bergeret A (2014) Study of the interface in natural fibres reinforced poly(lactic acid) biocomposites modified by optimized organosilane treatments. Ind Crop Prod 52:481–494

Liebsch D, Reginato MR (2009) Florescimento e frutificação de Merostachys skvortzovii Sendulsky (taquara-lixa) no estado do Paraná. IHERINGIA, Sér. Bot., Porto Alegre 64(1):53–56, jan./jun. 2009 (Available at https://www.researchgate.net/publication/286889771. 24 Nov 2016)

Lima HC Jr, Willrich FL, Barbosa NP, Rosa MA, Cunha BS (2008) Durability analysis of bamboo as concrete reinforcement. Mater Struct 41:981–989

Lima AC, Monteiro SN, Satyanarayana KG (2011a) Recycled polyethylene composites reinforced with jute fabric from Sackcloth: part II-impact strength evaluation. J Polym Environ 19:957–965

Lima AC, Monteiro SN, Satyanarayana KG (2011b) Recycled polyethylene composites reinforced with jute fabric from Sackcloth: part i-preparation and preliminary assessment. J Polym Environ 20:245–253

Lomelí Ramírez MG (2011) Development of biocomposites of thermoplasticized starch reinforced with green coconut fiber. Ph.D. dissertation, Federal University of Parana, Curitiba (PR), Brazil, 168 pages

Lomeli-Ramírez MGL, Satyanarayana KG, Manríquez-González R, Seitsu IS, Graciela IBM, Sydenstricker THF (2014) Bio-composites of cassava starch-green coconut fiber: part II—structure and properties. Carbohydr Polym 102:576–583

Lomelí-Ramírez MG, Satyanarayana KG, Seitsu I, Graciela IBM, Tanobe VOA, Sydenstricker THF (2011) Study of the properties of biocomposites: part I—Cassava starch-green coir fibers of Brazil. Carbohydr Polym 86(4):1712–1722

Lorenzi H, Souza HM, Madeiros-Costa JT, Cerqueira LSC, Ferreira E, Costa JTM (2004) Palmeiras Brasileiras e Exóticas Cultivadas. São Paulo, Plantarum, p 242

Lunz JN, Cordeiro SB, Mota JCF, Marques MFV (2012) Statistical experimental design for obtaining nanocellulose from curaua fiber. Macromol Symp 2012(319):99–107

Mahato DN, Mathur BK, Bhattacherjee S (2013) DSC and IR methods for determination of accessibility of cellulosic coir fibre and thermal degradation under mercerization. Indian J Fibre Text Res 38(1):96–100

Mani P, Satyanarayana KG (1990) Effect of surface treatment of lignocellulosic fibres on their debonding stress. J Adhes Sci Technol 4(1):17–24

Margem FM, Neto BJ, Monteiro SN (2010) Ramie fibers mechanical properties evaluation by the Weibull analysis (in Portuguese). In: Brazilian congress on materials science and engineering. Campos do Jordão, Brazil, pp 1–10

Margem JI, Vinicius AG, Margem MF, Gomes Dias RC, Fabio OB, Monteiro SN (2016) Pullout tests behavior of polyester matrix reinforced with malva fiber. Mater Sci Forum (online) 869:371–376

Martins MA, Pessoa JDC, Gonc¸alves PS, Souza FI, Mattoso LHC (2008) Thermal and mechanical properties of the açaı′ fiber/natural rubber composites. J Mater Sci 43:6531–6538. doi 10.1007/s10853-008-2842-4

McLaughlin EC, Tait EC (1980) Fracture mechanism of plant fibers. J Mater Sci 5:89–95

McMathis J (2014) Could a bamboo fiber composite replace steel reinforcements in concrete? American Ceramic Society. Post Published on 22nd Aug 2014

Miranda CS, Fiuza RP, Oliveira J, Carvalho RF, Guimarães DH, Jose NM (2012) Thermal, mechanical and morphological properties of composites developed from glycerol and dicarboxylic acids reinforced with piassava fiber. Macromol Symp 319:74–82

Mochnaz S (2003) Preparation and characterization of sisal (*Agave Sisalana*) fibers for use in polymeric composites. Dissertation submitted as partial fulfillment to get Master's degree in Engineering and Materials Science, Post-Graduate program in Materials Science & Engineering, Federal University of Parana, Curitiba (PR-Brazil). 79 pages

Mohanty AK, Misra M, Hinrichsen H (2000) Biofibres, biodegradable polymers and biocomposites: an overview. Macromol Mater Eng 276(3–4), 1–24

Mohanty AK, Misra M, Drzal LT (2001) Surface modifications of natural fibers and performance of the resulting biocomposites: an overview. Comp Interfaces 8(5):313–343

Mohanty AK, Misra M, Drzal LT (2002) Sustainable biocomposites from renewable resources, opportunities and challenges. J Polym Environ 10:19–26

Monteiro SN (2009) Properties and structure of *Attalea funifera* piassava fibers for composite reinforcement—a critical discussion. J Nat Fibers 6:191–203

Monteiro SN, Margem FM, Lima ES, Bevitori AB, Satyanarayana KG, Drelich J Reinforcement of polyester with renewable Ramie Fibers (Unpublished)

Monteiro SN, Terrones LAH, Almeida JRM (2008) Mechanical performance of coir fiber/ polyester composites. Polym Test 27:591–595

Monteiro SN, Lopes FPD, Barbosa AP, Bevitori AB, Silva ILA, Costa LL (2011a) Natural lignocellulosic fibers as engineering materials—an overview. Met Mater Trans A 42: 2963–2974

Monteiro SN, Satyanarayana KG, Ferreira A, Nascimento D, Lopes F, Silva I, Bevitori A, Inacio A, Bravo Neto J, Portela T (2011b) Selection of high strength natural fibers. Rev Mater 15(4):488–505

Monteiro SN, Satyanarayana KG, Margem FM, Ferreira A, Nascimento DCO, Santafe Jr HPG, Lopes FPD (2011c) Interfacial shear strength in lignocellulosic fibers incorporated polymeric composites. In: Susheel Kalia BS, Kaith, Kaur I (eds) Cellulose fibers: bio- and nano-polymer composites, green Chemistry and Technology, Springer, New York. Chapter 9, pp 241–262

Monteiro SN, Felipe Perisse DL, Nascimento DCO, Ferreira AS, Satyanarayana KG (2013) Processing and properties of continuous and aligned curauá fibers incorporated polyester composites. J Mater Res Technol 2(1):2–9

Monteiro SN, Margem FM, Santafé HPG Jr, Martins LBS, Oliveira MP (2014a) Correlation between the diameter and the density of coir fiber using the Weibull statistical methodology. Mater Sci Forum 775–776:266–271

Monteiro SN, Margem FM, Giulio RA, Leite LR, Oliveira MP (2014b) Tensile strength of polyester composites reinforced with thinner buriti fibers. Mater Sci Forum (online) 775–776:183–188

Monteiro SN, Margem FM, Oliveira MP, Giulio RA (2014c) Impact tests with polyester matrix reinforced with buriti fibers. Mater Sci Forum (online) 775–776:330–335

Monteiro SN, Margem FM, Leao SI, Bevitori A, Oliveira MP (2014d) Bending mechanical behavior of epoxy matrix reinforced with jute fiber. Mater Sci Forum (online) 775–776:314–318

Monteiro SN, Margem FM, Barboza DSML, Liete LR, Oliveira MP (2014e) Dynamic-mechanical analysis of polyester composites reinforced with giant bamboo (*Dendrocalmus giganteus*) Fiber. Mater Sci Forum (online) 775–776:302–307

Monteiro SN, Margem FM, Barboza DSML, Liete LR, Oliveira MP (2014f) Tensile strength of polyester matrix composites reinforced with giant bamboo (*Dendrocalmus giganteus*) fibers. Mater Sci Forum (online) 775–776:308–313

Monteiro SN, Margem FM, Wellington PI, Arthur CP, Oliveira MP (2014g) Tensile properties of epoxy composites reinforced with continuous sisal fibers. Mater Sci Forum (online) 775–776:284–289

Monteiro SN, Margem FM, Arthur CP, Noan S, Oliveira MP (2014h) Charpy impact tests in epoxy matrix composites reinforced with continuous sisal fiber. Mater Sci Forum (online) 775–776:290–295

Monteiro SN, Margem FM, Margem JI, De Souza MLB, Goncalves OC, Oliveira MP (2014i) Dynamic-mechanical behavior of malva fiber reinforced polyester matrix composites. Mater Sci Forum (online) 775–776:278–283

Monteiro SN, Margem FM, Bevitori A, Leao SI, Oliveira MP (2014j) Thermal characterization of polyester composites reinforced with ramie fibers. Mater Sci Forum (online), 775–776, 272–277

Monteiro SN, Margem FM, Salgado DAF, Leite LR, Oliveira MP (2014k) Izod impact tests in polyester matrix composites reinforced with banana fibers. Mater Sci Forum (online), 775–776, 261–265

Naldony P, Flores-Sahagun THS, Satyanarayana KG (2016) Effect of the type of fiber (coconut, eucalyptus, or pine) and compatibilizer on the properties of extruded composites of recycled high density polyethylene. J Compos Mater 50(1):45–56

Narayan R (2006) Biobased and biodegradable polymer materials: principles, concepts and technology exemplars. World polymer congress and 41st international symposium on macromolecules, MACRO-2006, Rio de Janeiro, Brasil 16–21 July, CD-ROM, paper No. 2103

Nascimento DCO, Motta LC, Monteiro SN (2010) Weibull analysis of tensile tested piassava fibers with different diameters. In: Characterization of minerals, metals and materials—TMS 2010, Seattle, USA, pp 1–8

Nascimento DCON, Ferreira AS, Monteiro SN, Aquino RCMP, Satyanarayana KG (2012) Studies on the characterization of piassava fibers and their epoxy composites. Composites: Part A 43:353–362. doi:10.1016/j.compositesa. 2011.12. 004

Neto CB (2014) Development of epoxy resin and taquara-lixa (Merostachysskvortzoviisendulsky) fibers composites for structural applications. Dissertation submitted for Doctoral degree in Engineering and Materials Science, Post-Graduate program in Materials Science & Engineering, Federal University of Parana, Curitiba (PR-Brazil), 214 pages

Neto WPF, Silve´rio HA, Vieira JG, Alves HCS, Pasquini S, Nascimento de Assunca˜o RM, Dantas NO (2012) Preparation and characterization of nanocomposites of carboxymethyl cellulose reinforced with cellulose nanocrystals. Macromol Symp 319:93–98

Neto CB, Sydenetricker THF, de Andrade AS, Mazzaro I (2015) Caracterização de fibras de taqura-lixa (*Merostachysskvortzovii*Sendulsky) Utilizáveis em compósitos poliméricos. Anais do 13° Congresso Brasileiro de Polímeros, 18 a 22 de outubro de 2015, Natal, RN

Neto BC, Flores-Sahagun THS, Andrade AS, Mazzaro I (2016) Characterization of taquara-lixa (Merostachys skvortzovii Sendulsky) fibers for use in polymer composites (Unpublished)

Nurul Fazita MR, Krishnan J, Bhattacharyya D, Mohammad Hafiz MK, Saruabh CK, Hazwan Hussain M, Abdul Khalil HPS (2016) Green composites made of bamboo fabric and poly (Lactic) acid for packaging applications—a review. Materials 9(6):435 (29 pages); doi:10. 3390/ma9060435

Oliveira PF, Marques MFV (2015) Chemical treatment of natural malva fibers and preparation of green composites with poly(3-hydroxybutyrate). Chem Chem Techmol 9(2):211–222

Osorio L, Trujillo E, Van Vuure AW, Verpoes I (2011) Morphological aspects and mechanical properties of single bamboo fibers and flexural characterization of bamboo/epoxy composites. J Reinf Plast Compos 30(5):396–408

Peponi L, Biagiotti J, Torre L, Kenny JM, Mondragòn I (2008) Statistical analysis of the mechanical properties of natural fibers and their composite materials. I. Natural fibers Poly Compos 29(3):313–320

Pereira BM, Nascimento DOC, Soares MP Jr, Fontes VCM, Monteiro SN, Jonas A (2014) Effect of banana fiber in the properties of clayey ceramic. Mater Sci Forum (online) 798–799:229–234

Pereira PHP, Rosa MF, Cioffi MOH, Benini KCCC, Milanese AC, Voorwald HJC, Mulinari DR (2015) Vegetal fibers in polymeric composites: a review. Polímeros 25(1):9–22

Pilla S (ed) (2011) Handbook of bioplastics and biocomposites engineering applications. Scrivener Publishing/Wiley, USA, ISBN: 978 0-470-62607-8

Poletto M, Ornaghi HL Jr, Zattera AJ (2014) Native cellulose: structure, characterization and thermal properties. Mater 7(9):6105–6119

Portela TGR, Costa LL, Lopes FPD et al (2010) Characterization of fibers from different parts of the buriti palm tree. Characterization of minerals, metals and materials—TMS 2010. Seattle, USA, pp 1–7

Prasad SV, Pavithran C, Rohatgi PK (1983) Alkali treatment of coir fibers for coirpolyester composites. J Mater Sci 18:1443–1454

Preston RD (1978) Physical biology of plant cell walls. Chapman & Hill, London

Proudfoot M (1996) Aesthetics, an encyclopedia of philosophy. In: Parkinson GHR (ed) Florence, KY, Routledge, 851

Ramos LP (2003) The chemistry involved in the pretreatment of lignocellulosic materials. Quimica Nova, 863–871 (in Portuguese)

Ramos LP, Silva TA, Martins LF, Satyanarayana KG (2005) Conversion of lignocellulosics to fuels, chemicals and environmentally-friendly materials: Met. Mater Processes 17(3–4):299–318

Rangan A, Manchiganti MV, Rajendran TM, Satyanarayana KG, Menon R (2017) Novel method for the preparation of Lignin-1 rich nanoparticles from lignocellulosic fibers. (Being accepted by Ind. Crops Prod)

Reddy OK, Uma Maheswari C, Shukla M, Song JI, Varada Rajulu A (2013) Tensile and structural characterization of alkali treted Borassus fruit fine fibers. Compos Part B 44(1):433–438

Ricardo JB, Satyanarayana KG, Wypych F, Ramos LP (2011) The effect of steam explosion on the production of sugarcane bagasse/polyester composites. Compos A 42:364–370. doi:10.1016/j.compositesa.2010.12.009

Rowell RM, Han JS, Rowell JS (2000) Characterization and factors effecting fiber properties. In: Frollini E, Leão AL, Mattoso LHC (eds) Natural polymers and agrofibers composites. Pub. Embrapa Instrumentação Agropecurária, São Carlos, Brasil, pp 115–134

Saha P, Chowdhury S, Roy D, Adhikari B, Kim JK, Thomas S (2016) A brief review on the chemical modifications of lignocellulosic fibers for durable engineering composites. Poly Bull 73:587–620

Santafe HPG Jr, Monteiro SN, Costa LL et al (2009) Weibull distribution as an instrument of statistical analysis for coir fiber tensile tests (in Portuguese), In: 64rd international congress of the Brazilian association for metallurgy and materials, Belo Horizonte, Brazil, pp 1–12

Santos AM, Sydenstricker THD, Amico SC (2006) The study of hybrid polypropylene/ glass and coir fiber composites for engineering applications, ibid, paper No.1623

Santos OV, Lopes AS, Azevedo GO, Santos AC (2010) Processing of Brazil-nut flour: characterization, thermal and morphological analysis. Ciênc Tecnol Aliment, Campinas, 30 (Supl.1):264–269

Santos LP, Flores-Sahagun THS, Satyanarayana KG (2015) Effect of processing parameters on the properties of polypropylene–sawdust composites. J Compos Mater 49(30):3727–3740

Santos LP, Trombetta E, Flores-Sahagun THS, Satyanarayana KG (2016) Effect of domestic compatabilizer on the performance of polypropylene-sawdust composites. J Compos Mater 50 (10):1353–1365

Saravanakumar SS, Kumaravel A, Nagarajan T, Sudhakar P, Baskaran R (2013) Characterization of a novel natural cellulosic fiber from *Prosopis juliflora* bark. Carbohydr Polym 92:1928–1933

Sartori MMP, Florentino HO, Basta C, Lea˜o AL (2001) Determination of optimum quantity of crop residues for energy in sugarcane crop using linear programming in variety selection and planning strategy. Energy 26:1031–1040

Satyanarayana KG (2004) Steam explosion- a boon for value addition to renewable resources. Metal News 22:35–40

Satyanarayana, KG, Wypych F (2007) Characterization of natural fibers. In: Fakirov S, Bhattacharyya D (eds) Engineering biopolymers: homopolymers, blends and composites, vol 1. Auckland, New Zealand, Munich: Hanser Publishers, pp 3–47. ISBN No: 978-3-446-40591-2

Satyanarayana KG (2015) Recent developments in 'Green' composites based on plant fibers-preparation, structure property studies. J Bioprocess Biotech 5(2):206 (12pages) doi:10.4172/2155-9821.1000206

Satyanarayana, KG, Anupama R, Prasad VS, Magalhães WLE (2017) Preparation, characterization, and applications of nanomaterials (Cellulose, Lignin, and Silica) from renewable (lignocellulosic) resources. In: Vijay Kumar Thakur (ed) Book: handbook of composite from renewable materials. Wiley-Scrivener, USA, 66 pages (in print)

Satyanarayana KG, Sukumaran K, Mukherjee PS, Pillai SGK (1986) Materials science of some lignocellulosic fibers. Metallography 19:389–400

Satyanarayana KG, Ramos LP, Wypych F (2004) Development of new materials based on agro and industrial wastes towards ecofriendly society. In: Ghosh TK, Chakrabarti T, Tripathi G

(eds) Biotechnology in energy management, vol. II. New Delhi: APH Publishing Corporation, pp 583–624
Satyanarayana KG, Wypych F, Guimarães JL, Amico CS, Sydenstricker THD, Ramos LP (2005) Studies on natural fibers of Brazil and green composites. Met Mater Proc 17(3–4):183–194
Satyanarayana KG, Guimarães JL, Wypych F (2007) Studies on lignocellulosic fibers of Brazil. Part I—source, production, morphology. Prop Appl Compos Part A 38:1694–1709
Satyanarayana KG, Gregorio AC, Wypych F (2009a) Biodegradable composites based on lignocellulosic fibers-an overview. Prog Polym Sci 34:982–1021
Satyanarayana KG, Ramos LP, Wypych F (2009b) Comparative study of Brazilian natural fibers and their composites with others. In: Thomas S, Pothan LA (eds) Natural fiber reinforced polymer composites. Chapter 18. Publishers: Old City Publishing, Inc., USA, pp 473–522. ISBN No: 978-2-914610-99-5
Satyanarayana KG, Monteiro SN, Felipe Perisse DL, Frederico MM, Helvio Pessanha GS Jr, da Costa LL (2011) Dimensional analysis and surface morphology as selective criteria of lignocellulosic fibers as reinforcement in polymeric matrices. In: Susheel Kalia BS, Kaith, Kaur I (eds) Cellulose fibers: bio- and nano-polymer composites, green chemistry and technology. Springer, New York (USA). Chapter 8, pp 215–240
Satyanarayana KG, Sydentstricker THF, Lucas PDS, Juliana DS, Mazzaro I, Mikowski A (2013) Characterization of blue agave bagasse fibers of Mexico. Compos A 45:153–161
Satyanarayana KG, Sydentstricker THF, Irineu M, Sukumaran K, Pillai SGK, Ravikumar KK (2014) Characterization of *Aechmea magadalenae* leaf fiber yarns. J Biobased Mater Bioener 8:184–191. doi:10.1166/jbmb.2014.1415
Sen T, Jagannatha Reddy HN (2011) Application of sisal, bamboo, coir and jute natural composites in structural upgradation. Int J Innov Manage Technol 2(3):186–191
Serrano E, Rus G, García-Martínez J (2009) Nanotechnology for sustainable energy. Renew Sustain Ener Rev 13:2373–2384
Sha W, Edwards KL (2007) The use of neural networks in materials science based research. Mater Des 28:1747–1754
Simão JA, Vitor Brait C, Marconcini JM, Mattoso LHC, Barsberg ST, Sanadi AR (2016) Effect of fiber treatment condition and coupling agent on the mechanical and thermal properties in highly filled composites of sugarcane Bagasse fiber/PP. Mater Res 19(4):746–751. doi:10.1590/1980-5373-MR-2015-0609
Sirmah P, Muisu F, Mburu S, Dumarçay P, Gérardin P (2008) Evaluation of *Prosopis juliflora* properties as an alternative to wood shortage in Kenya. Bois et Fortêts Des Tropiqes No 298 (4):25–35
Spinacé MAS, Lambert CS, Formoselli KKG et al (2009) Characterization of lignocellulosic curaua fibers. Carbohyd Polym 77:47–53
Sydenstricker THD, Mochnaz S, Amico SC (2003) Pull-out and other evaluations in sisal-reinforced polyester biocomposites. Polym Test 22:375–380
Tanobe VOA, Munaro M, Syndenstricker THD, Amico SC (2005) A comprehensive chemically treated Brazilian sponge gourds (*luffa cylindrica*). Polym Test 24:474–482
Tanobe VOA, Flores-Sahagun THS, Amico SC, Muniz GIB, Satyanarayana KG (2014) Sponge Gourd (*Luffa Cylindrica*) reinforced polyester composites: preparation and properties. Def Sci J 64(3):273–280. doi:10.14429/dsj.64.7327
Thakur VK (ed) (2013) Green composites from natural resources. ISBN: 978-1-466-57069-6
Thakur VK (ed) (2014) Lignocellulosic polymer composites: processing, characterization, and properties. ISBN: 978-1-118-77357-4
Thakur VK, Thakur MK, Kessler MR (eds) (2013) Handbook of composites from renewable materials, structure and chemistry. Scrivener Publishing/Wiley, USA. ISBN: 978-1-119-22362-7
Thakur VK, Thakur MK, Gupta RK (2014) Review: raw natural fiber-based polymer composites. Int J Polym Anal Charact 19:256–271
Thomas S, Visakh PM, Mathew AP (eds) (2013) Advances in natural polymers: composites and nanocomposites. Springer, Berlin Heidenberg. ISBN:978-3-642-20939-0

Tomczak F (2010) Studies on the structure and properties of Brazilian coconut and curauá fibers (Estudos sobre a estrutura e propriedades de fibras de coco e curauá do Brasil), Post graduation Program in Materials Engineering and Science (PIPE), Federal University of Parana, Curitiba-PR, Brazil. 140 pages (in Portuguese)

Tomczak F, Syndenstricker THD, Satyanarayana KG, Gondak MO (2006) The influence of time in the alkali treatment of coir fibers in PP-coir composites in world polymer congress and 41st international symposium on macromolecules—MACRO 2006, 11–16 July 2006, Rio de Janeiro, Brazil, [CD-ROM], paper No. 1011

Tomczak F, Sydenstricker THD, Satyanarayana KG (2007a) Studies on lignocellulosic fibers of Brazil. Part II: morphology and properties of Brazilian coconut fibers. Compos Part A 38(7):1710–1721

Tomczak F, Sydenstricker THD, Satyanarayana KG (2007b) Studies on lignocellulosic fibers of Brazil: Part III—morphology and properties of Brazilian curaua fibers. Compos A 38:2227–2236

Treloar LRG (1977) Physics of textiles. Phys Today 30(12):23–30

Trindade WG, Horeau W, Ruggiero IAT, Frollini E, Castelan A (2004) Phenolic thermoset matrix reinforced with sugarcane bagasse fibers: attempt to develop a new fiber surface chemical modification involving formation of quinines followed by reaction with furfuryl alcohol. Macromolec Mater Eng 289:728–736

Trombetta E (2010) Utilization of natural fiber (saw dust) of pine reinforced in polypropylene composites for automotive components. Master thesis (in Portuguese), University of Paraná (UFPR), Curitiba, Brazil

Tuuadaij N, Nuntiya A (2008) Preparation of nano silica powder from rice husk ash by precipitation method. Chiang Mai J Sci 35:206–211

Valadares LF (2014) Biobased fibers and materials in Brazil. Chem Biol Technol Agri 1:16 (6 pages)

Van Dam JEG, Van Den Oever MJA, van den Teunissen W et al (2006) Process for production of high density/high performance binder less boards from whole coconut husk Part 2: coconut husk morphology, composition and properties. Indust Crops Prod 24:96–104

Vanderleia A (2009) Obtaining of thermolded composites through the use of castor oil cake plastified with glycerol, derivative of the transesterefication of oils anf fats. Masters' thesis, Federal University of Parana (UFPR), Curitiba (PR-Brasil). (In Portuguese) Obtenção de compósitos termomoldados a partir da torta de mamona plastificada com glicerol, derivado do processo de transesterificação de óleos e gorduras. Dissertação (Mestrado)—UFPR, Curitiba, Paraná, Brazi

Venugopal B, Pillai CKS, Satyanarayana KG (1990) Studies on fire retardance of needle-felt coir fibers. Res Ind 35:108–110

Vilay V, Mariatti M, Taib MR, Todo M (2008) Effect of fiber surface treatment and fiber loading on the properties of bagasse fiber–reinforced unsaturated polyester composites. Compos Sci Technol 68(3–4):631–638. http://dx.doi.org/10.1016/j.compscitech. 2007.10.005

Wallenberger FT, Weston N (eds) (2004) Natural fibers, plastics and composites. Kluwer Academic Publishers, New York

www.ispn.org.br. Accessed 23-Nov 2016

Yang HS, Kiziltas A, Gradner DJ (2013) Thermal analysis and crystallinity study of cellulose nanofibril-filled polypropylene composites. J Therm Anal Calorim 113:673–682

Retting Process as a Pretreatment of Natural Fibers for the Development of Polymer Composites

L. Sisti, G. Totaro, M. Vannini and A. Celli

Abstract The development of high-performance materials made from natural resources is increasing worldwide. Within this framework, natural fiber reinforced polymeric composites now experience great expansion and applications in many fields, ranging from the automotive to the construction sector. The great challenge in producing composites containing natural fibers and with controlled features is connected to the great variation in properties and characteristics of fibers. The quality of the natural fibers is largely determined by the efficiency of the treatment process and can dramatically influence the properties of the final composites. The overall fiber extraction processes, applied to vegetable fibers, is called retting and consists in the separation of fiber bundles from the cuticularized epidermis and the woody core cells. Today, many efforts are being made to optimize the retting methods in terms of fiber quality production, reduction of environmental issues and production costs. This chapter aims to provide a classification and an overview of the retting procedures that have been developed during years and are applied to extract mainly bast fibers.

Keywords Plant fibers · Retting · Polymer composites

Contents

L. Sisti (✉) · G. Totaro · M. Vannini · A. Celli
Dipartimento di Ingegneria Civile, Chimica, Ambientale e dei Materiali,
Università di Bologna, Via Terracini 28, 40131 Bologna, Italy
e-mail: laura.sisti@unibo.it

S. Kalia (ed.), *Lignocellulosic Composite Materials*, Springer Series on Polymer and Composite Materials, https://doi.org/10.1007/978-3-319-68696-7_2

1 Introduction

Increasing environmental concerns and depletion of petroleum resources are driving energy and chemical industries to move from petrochemical to renewable resources. As a major renewable resource, lignocellulosic fibers derived from structural plant tissues will play an important role in this transition. Fiber crops are the commodities with the longest tradition among technical and nonfood agricultural products. Actually, bio-fibers have gained popularity especially for the development of sustainable materials, thanks to their great potential for use in composite materials for applications in packaging, automotive, and other industries (Dicker et al. 2014; Frollini et al. 2013, 2015; Miller 2013), especially as reinforcing fibers.

Up to now man-made fibers, such as glass, aramid, and carbon fibers were the predominant reinforcing fibers for polymers, but they feature environmental problems both during the production and disposal processes. Natural fibers, instead, have many remarkable advantages over the synthetic ones, such as lightweight, low cost, and biodegradability. However, at present, the offer of natural fibers does not fulfill industrial demand. In fact, the intensive cultivation of some fiber crops, such as hemp, flax, kenaf, and jute has increased, while stalk processing, for the extraction of fibers, relies on the traditional methods. Natural fibers from fiber crops, as a commodity, are facing competition on two sides: synthetic fibers on the consumer's side, and more remunerative crops on the grower's side. In order to face such dual challenges, the production of fiber crops must adopt a strategy based on the agricultural and industrial research and development and market promotion of both traditional and diversified fiber crops products (van Dam and Bos 2004). The improvements in fiber processing techniques play a significant role in obtaining fibers with enhanced standardized quality, capable to match the high performances required in composite materials but also in what concerns control and a lower environmental impact in terms of energy and water waste and consumption. The present essay will provide an overview of the traditional and new emerging fiber extraction processing techniques, which are known as retting or degumming and which involve the extraction of fibers or fiber bundles from harvested stems.

Moreover, the effect of these treatments on the performance of the resulting natural fibers in the composite application is addressed.

2 Plant Fibers as Renewable Resources for Polymer Composites

The species producing fibers of commercial relevance are mainly jute, flax, ramie, hemp, and kenaf, because they are much stronger than other fibers; they are also referred to, usually, as bast fibers. Plant fibers are classified by the part of the plant from which they are obtained, such as leaf, seed, fruit, stem, and bast. As its name implies, bast fibers are obtained from the outer layer of bast surrounding the plant stem. Since the bast fiber task is to act as reinforcement within the plant and to provide stability, they supply a good reinforcement in composite materials. Leaf fibers are reported to impart improved toughness in composites and the insertion of seed or fruit fibers gives elastomeric type properties to the component under consideration. Figure 1 lists the classification of plant fibers and Table 1 reports the main bast fibers along with their cultivation area and production capacities.

Bast fibers develop as bundles of cells embedded within other tissues in the stem. The process of separation and extraction of fibers from the nonfibrous tissues and woody part of the stem through separation, dissolution, and decomposition of pectins, gums, and other mucilaginous substances is called retting (Pallesen 1996; Dasgupta et al. 1976). The quality of the fiber is largely determined by the efficiency of the retting process, which relies on the breakdown of pectic materials through which fibers are liberated. Knowing the physical and chemical structure of plants is really crucial in designing fiber processing since these factors affect the

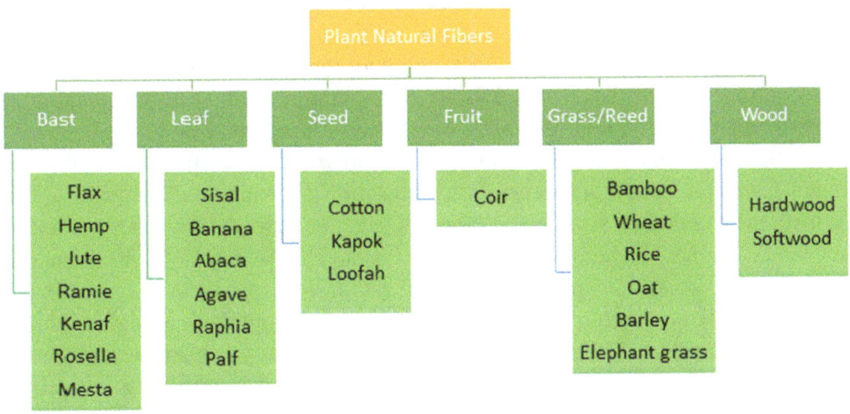

Fig. 1 Classification of plant natural fibers according to which part of the plant they are obtained from

Table 1 Cultivation area and production capacity of commercial important bast fibers and of the principal leaf, seed, fruit, and grass fibers (Yan et al. 2014; Faruk et al. 2012)

Fiber		Botanical name	Main producer countries	Production capacities per year (10^6 t)
Bast	Jute	*Corchorus capsularis, Corchorus olitorius*	India Bangladesh	2.5
	Kenaf	*Hibiscus cannabinus*	China India Thailand	0.45
	Flax[a]	*Linus usatissimum*	China Europe	0.5–1.5
	Ramie	*Boehmeria nivea*	China	0.15
	Hemp	*Cannabis sativa* L.	China Europe	0.10
Leaf	Sisal	*Agave sisalana*	Brazil China Tanzania	0.30
Seed	Cotton	*Gossypium* sp.	China USA India Pakistan	25
Fruit	Coconut	*Cocus nucifera*	India Sri Lanka	0.45
Grass	Bamboo	*Bambusa vulguaris*	China	30

[a]The real production of flax is understimated because it does not take into consideration the production of Canada

quality of the final fibers. In general, bast fibers consist of cellulose, hemicellulose, and lignin in various proportions. Other compounds, such as pectin, wax, minerals, and water-soluble compounds, are present in addition to the main components. Pectin serves as a glue to hold fibers together in bundles and bundles to nonfibrous tissues. Hemicelluloses, pectin, and lignin act like a matrix, whereas cellulose acts as reinforcement to the matrix contributing to the strength of the fiber (Thomas et al. 2011). In Fig. 2 major bast fibers and the most common leaf, seed, fruit, and grass fibers are shown while their chemical composition is reported in Table 2.

The big disadvantage of natural fibers, as compared to synthetic ones, is that they do not have the same consistency in composition and therefore in quality. This inconsistency is due to a variety of reasons, such as climate, crop variety, retting process, and processing equipment used for fibers (Thomsen et al. 2006). In natural fibers, climatic conditions play an important role in fiber production. For example, low temperature and high relative humidity during the growing season contribute to fineness and length of fibers. The role of retting in obtaining high-quality standardized fibers is crucial and research and development are heading toward the industrialization of treatment processes.

Fig. 2 Common bast fibers

Table 2 Chemical composition of major bast fibers and of the most common leaf, seed, fruit, and grass fibers (Gurunathan et al. 2015; Faruk et al. 2012; Wang et al. 2010)

	Fiber	Cellulose (wt%)	Hemicellulose (wt%)	Lignin (wt%)	Pectin (wt%)	Water soluble (wt %)	Wax (wt%)
Bast	Hemp	70.2–74.4	17.9–22.4	3.7–5.7	0.9	2.1	0.8
	Jute	61–71.5	13.6–20.4	12–13	0.2	1.2	0.5
	Flax	71–78	18.6–20.6	2.2	2.3	3.9–10.5	1.7
	Kenaf	45–57	21.5–23	8–13	3–5	–	–
	Ramie	68.6–76.2	13.1–16.7	0.6–0.7	1.9	6.1	0.3
Leaf	Sisal	67–78	10–14	8–11	10	1.3	2
Seed	Cotton	85–90	5.70	0.7–1.6	0–1	1.0	0.6
Fruit	Coconut	36–43	0.15–0.25	41–45	3–4	5.2–16.0	–
Grass	Bamboo	26–73.8	12.5–30	10.2–31	0.4	3.2	–

3 The Retting Techniques

The retting process, also known as degumming, involves the extraction of fiber bundles from the harvested stem. To date, several retting methods are applied; the most traditional, still widely used approaches, i.e., water retting and dew retting (Tamburini et al. 2004) are based on the microbiological retting. Other approaches

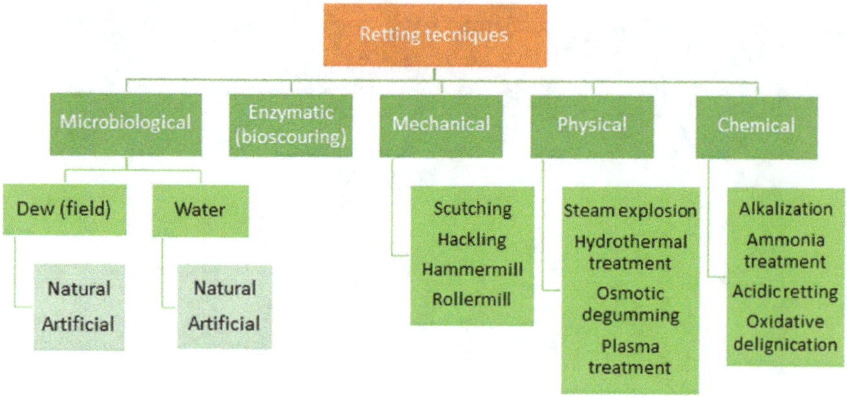

Fig. 3 Retting techniques

involve mechanical, physical, chemical, and enzymatic retting. The latter is very promising but not yet practiced on an industrial scale. Figure 3 shows a rationalization of all the retting treatments.

3.1 Microbiological Retting

Microbiological retting is a traditional and highly widespread retting method. Two different types of microbiological retting are mainly adopted: dew and water retting. Both of them are carried out by pectin enzymes secreted by indigenous microflora.

3.1.1 Dew Retting

In dew retting, also called field retting, harvested plants are thinly spread out for 2–10 weeks in fields (Fig. 4). During this period, microorganisms, mainly filamentous fungi or aerobic bacteria present in soil and on plants, attack noncellulosic cell types, removing pectins, and hemicelluloses from parenchyma cells and the middle lamellae, without attacking cellulose fibers. In this process, the colonizing fungi possess a high level of pectinase activity and the capacity to penetrate the cuticular surface of the stem: thus, fiber bundles come out separated into smaller bundles and individual fibers. Several fungal species and bacteria have been isolated from dew retted plants: *Cladosporium* sp., *Penicillium* sp., *Aspergillus* and *Rhodotorula* sp. (Ribeiro et al. 2015; Fogarty et al. 1972; Ahmed and Akhter 2001). During flax dew retting, other fungi were isolated, such as *Cladosporium herbarum*, *Epicoccum nigrum*, *Alternaria alternate*, *Fusarium* sp., *Aureobasidium pullulans*, *Phoma* sp., *Mucor* sp., *Rhizomucor pusillus*, and *Rhizopus oryzae* (Akin et al. 1998; Henriksson et al. 1997; Sharma 1986; Xiao et al. 2008;

Fig. 4 Dew retting (Flax composites 2014)

Molina et al. 2001; Booth et al. 2004). For more lignified fibers different enzymes are needed and often a mechanical step is added to separate the fibers (Ribeiro et al. 2015).

Currently, dew retting is the most used process for the industrial production of bast fibers, mainly flax and jute, because of its low cost (Bacci et al. 2010). Unfortunately, the method is limited to geographic regions, where the weather is suitable for fungi proliferation. Moreover, often low and inconsistent fiber quality is produced as compared to other methods, such as water retting. Risks of under retting and over retting are also reported: they may cause difficulties in separation or weaken the fiber (Jankauskiene et al. 2015). For example, cellulotyc enzymes secreted by the microbiota can damage the fibers if exposition lasts too long. Therefore, it is necessary to monitor the retting process to ensure the quality of the fibers. Occupation of lands for several weeks during retting and the presence of a product contaminated with soil and fungi are other drawbacks affecting this treatment.

As far as artificial dew retting is concerned, during the last few years, the employment of fungi in a more controlled environment for the extraction of natural fibers is being investigated. The controlled parameters involve the type of fungi, temperature, and period of treatment, in order to provide retting treatments with lower costs, higher efficacy, and environmentally friendly alternatives (Pickering et al. 2007a).

Among fungi, the white rot fungus (basidiomycetes) has been shown to be the only one able to degrade noncellulosic compounds from natural fibers, thus improving the mechanical properties of the resulting natural fiber reinforced composites (Pickering et al. 2007a).

Another recent study reports an enzymating enrichment of the dew retting process through the use of customized enzymatic blends (Texazym® SER). Such enzymes, developed by INOTEX, are sprayed on the field before pulling out or within the first 3 days of dew retting. It was found out that this method can increase flax long fiber yields by more than 40%. In this case, the enzymes in combination with mild mechanical treatments can replace aggressive and energy-intensive processing like Laroche cottonization (Antonov et al. 2007).

3.1.2 Water Retting

In water retting, quite widespread 50 years ago, straws are soaked in freshwater, today in large tanks, while in the past rivers or ponds were used (Fig. 5). During this treatment, that is carried out for a period of 7–14 days on most bast fiber crop straws, water penetrates into the central stalk portion, by breaking the outermost layer, and thus provoking an increased absorption of moisture and the development of a pectinolytical bacterial community (Donaphy et al. 1990). The duration of the treatment depends on the water type and temperature and on any bacterial inoculum (Bismark et al. 2005). The first stage of the process consists in the growing of aerobic microorganisms which consume most of the dissolved oxygen, ultimately creating an environment favorable for the growth of anaerobs. In the aerobic phase microorganisms belonging to the genus *Bacillus*, such as *B. subtilis*, *B. polymyxa*, *B. mesentericus,* and *B. macerans* have been found to be active (Munshii and Cathoo 2008; Ali 1958; Kapoor et al. 2001; Tamburini et al. 2003) while in the anaerobic phase different microbiota, of the genus *Clostridium*, such as *C. aceto-butylicum*, *C. felsineum*, *C. aurantibutyricum,* and *C. tertium* have been isolated from retting water (Munshii and Cathoo 2008; Tamburini et al. 2003; Zheng et al. 2001; Di Candilo et al. 2000; Donaghy et al. 1990).

Water retting generally produces fibers with a higher quality than those produced by dew retting (Amaducci and Gusovious 2010; van Sumere 1992), but the water

Fig. 5 Water retting of bast fibers (**a**) in river (Frontier Culture Museum of Virginia 2012), (**b**) in tank (Retting the flax 2014)

retting process impacts the environment due to the consumption and contamination of large amounts of water (van Dam and Bos 2004) and energy (Van der Werf and Turunen 2008). With freshwater resources becoming increasingly scarce, an alternative or improvement in water retting will have to be foremost in dealing with water scarcity and pollution reduction. Enzyme retting, for example, which will be discussed in the next paragraph, has been evaluated as a replacement for current retting methods. Moreover, Zhang et al. (2008a, b) have explored the possibility of substituting freshwater with seawater, which is a natural resource considered to be inexhaustible and abundant. They demonstrated that seawater retting treatments yield good retting results and good pectinolytic strains belonging to *S. maltophilia* species and *O. antrophi* species. Artificial water retting, employing warm water and bacterial inoculum, has also been used to produce homogeneous and clean high-quality fibers in 3–5 days (Bismark et al. 2005; Sisti et al. 2016).

3.2 *Enzymatic Retting*

A modification of water retting is the enzymatic treatment, also called bioscouring, where degrading enzymes are directly added to tank water or in a bioreactor (Ouajai and Shanks 2005). This technique has been demonstrated to be a promising replacement for traditional retting methods in terms of time-saving, ecology friendliness, and convenient characteristics. The duration of enzymatic retting ranges from 8 to 24 h. The high energy input and nonreusability of enzymes are the main concerns, which affect the cost-effectiveness of the process (Tahir et al. 2011).

Pectinases are the main enzymes employed for retting, in order to free the fibers from other tissues. Today, with the advancement of biotechnological tools, pectinases or pectinolytic enzymes are of significant importance thanks to their all embracing applications, such as in fruit juice extraction and its clarification, scouring of cotton, degumming of plant fibers, wastewater treatment, vegetable oil extraction, tea and coffee fermentation, bleaching of paper, in poultry feed additives, and in the alcoholic beverages and food industries (Jayani et al. 2005). Pectinolytic enzymes are a heterogeneous group of related enzymes that hydrolize the pectic substances, mostly present in plants. They are widely distributed in higher plants and microorganisms (Mohnen 2008), since they help in cell wall extention and in softening some plant tissues during maturation and storage. They also help maintain ecological balance by causing decomposition and recycling of waste plant materials (Ridley et al. 2001). Plant pathogenity and spoilage of fruits and vegetables by rotting are some other major manifestations of pectinolytic enzymes. Pectinolytic enzymes, which are classified according to the main mechanism catalyzing the respective reactions, gather pectin esterase, polygalacturonase, and pectin lyase. Evans et al. and Zhang et al. reported that polygalacturonase activity plays a fundamental role during the retting process (Evans et al. 2002; Zhang et al. 2000), but also pectate lyase has been shown to be potentially important for retting bast plants (Akin et al. 2007; Bruhlmann et al. 2000).

Polygalacturonase and pectate lyase are both depolymerizing enzymes for pectin. Polygalacturonases are the pectinolytic enzymes that catalyze the hydrolytic cleavage of the polygalacturonic acid chain (Jayani et al. 2005), while pectate lyase carries out a non-hydrolitic breakdown of pectates and pectinases by a transelimination split of the pectic polymer (Sakai et al. 1993).

Pectinases can be produced from bacteria (Dosanjh and Hoondal 1996; Kapoor et al. 2001), yeast (Blanco et al. 1999), fungi (Singh et al. 1999), and actinomycetes (Beg et al. 2000). Most commercial preparations of pectinases are produced from fungal sources (Henriksson et al. 1997). *Aspergillus niger* is the most commonly used fungal species in the industrial production of pectinolytic enzymes (Jayani et al. 2005). Among bacteria sources, certain genera namely *Bacillus*, *Lactobacillus*, *Pediococcus*, and *Leuconostoc* are known to be effective during the fermentation process to produce pectinolytic enzymes (Kouhoundè et al. 2014).

Enzyme retting via the pectinases is capable of producing consistent high strength renewable fibers with variable fineness values for use in novel resins (Foulk et al. 2011). Some of the major commercialized enzymes are shown below.

Viscozyme® L, produced by Novozymes, is a multienzymatic solution containing a wide range of carbohydrases, including arabase, cellulose, β-glucanase, hemicellulose, and xylase. It has been tested by different authors in flax fiber extraction (Bacci et al. 2010; Foulk et al. 2008; Akin et al. 2003; Adamsen et al. 2002; Evans et al. 2002).

Pectinex® Ultra SP-L, produced by Novozymes, is a highly active pectolytic enzyme obtained by a selected strain of *Aspergillus aculeatus* that contains pectolytic and a range of hemicellulolytic activities that can disintegrate plant cell walls (Bacci et al. 2010).

Scourzyme® L has been developed by Novozymes for the bioscouring technique; it contains an alkaline pectate lyase, which degrades the pectin from the primary cell wall of fibers without degrading the fiber itself (Oujai and Shanks 2005).

Flaxzyme®, a commercial enzyme mixture from Novo Nordisk, consists of pectinases, hemicellulases, and cellulases and was developed specifically for enzyme retting (Akin et al. 1999).

BioPrep® 3000 L was developed by Novozymes as the first commercially available alkaline pectate lyase. It was isolated and produced for its unique ability to degrade the pectin layer between the waxy cuticle and cellulosic fabric cotton (Durden et al. 2001; Etters et al. 2001).

Texazyme® BFE was developed by Inotex for elementarization of bast fibers through the degradation of pectin layers. It consists of a multicomponent product without cellulase activity (Foulk et al. 2008).

For each enzyme, specific conditions are identified for employment in retting, since the activity can change dramatically with pH, temperature, and enzyme concentration. Moreover, chelators and surfactants are usually employed in formulations to improve activity. For example, the role of Ca(II) chelators, such as ethylenediaminetetracetic acid (EDTA), is well known (Akin et al. 2004) to improve retting, particularly, in removing the epidermal/cuticle material from the fiber and fiber bundles. In fact, EDTA acts by destabilizing bridges between Ca(II) and polygalacturonic acid, thus leading to the disruption of tissues.

Foulk et al. (2008) reported that, with a specific knowledge of the composition of the enzyme mixture, enzyme retting could be used to tailor fibers/fiber bundles with particular properties, such as strength and fineness, and for specific applications. Strength, for example, which is a major concern in many applications, is preserved by retting with relatively pure pectinase, either pectate lyase, or polygalacturonase. However, a mixed enzyme preparation containing cellulase could be used for advantageous applications where the fibers will be shortened, such as for paper/pulp or injection molding. In fact, it has been found that enzyme formulations like Viscozyme, containing cellulase as a component, can weaken the bast fibers, since the nodes of the fibers are particularly sensitive to the attack by this enzyme (Foulk et al. 2008). The final application, therefore determines the retting formulation.

3.3 Mechanical Retting

The mechanical extraction of fibers consists of various steps, as developed since ancient times, mainly to recover hemp and flax fibers (Fig. 6). Today, this treatment is a completely automated process but steps have not changed (Fig. 7): a first separation is carried out by **breaking**, that is the stalks are passed between fluted rollers to crush and break the woody core (shive) into short pieces (hurds); the remaining fibers and hurds are subjected to **scutching** (traditionally performed with boards and hammers), by which the fiber bundles are gripped between rubber belts or chains and carried past revolving drums with projecting bars that beat the fiber bundles, separating the hurds, and broken short fibers (tow) from the remaining long fibers. Finally, in the **hackling** (realized in the past by pulling the fibers through a set of pins) thick fibers are divided by passing the long fibers through a series of combs of increasing fineness to clean and align the long fibers and separate the remaining tow and debris. Interestingly, the modern mills maintain the integrity of the long fibers, by disentangling and aligning the fibers, without destroying length. Another process currently used to mechanically separate the fibers is called **decortication** and can be performed by **hammermilling** or **rollermilling** (Deyholos and Potter 2014). In the first case (Fig. 8), single or multiple concurrent drums rotating with hammers projecting transversely from the drum surface beat the straw until the separated hurd/shive and fiber particles can pass freely through some meshes placed inside the machine. In the second case, (Fig. 9), long cylindrical corrugated rollers are assembled in such orientations as to crack the straw stalks while producing minimal damage to the fiber. The two processes differ in the *pro* and *con*: if the hammermilling is characterized by higher throughput capability, the rollermilling gives much greater length control, producing even very long fibers and will better preserve the integrity of the fibers, without damages or entanglements.

The choice of the preferable mechanical retting depends on the type of the fiber, its final application, type of ensuing treatments. Therefore, with these multiple

Fig. 6 Traditional retting of hemp fibers: scutching (**a** and **b** frames) and hackling (**c** and **d** frames) (Weald and Downland Open Air Museum 2016)

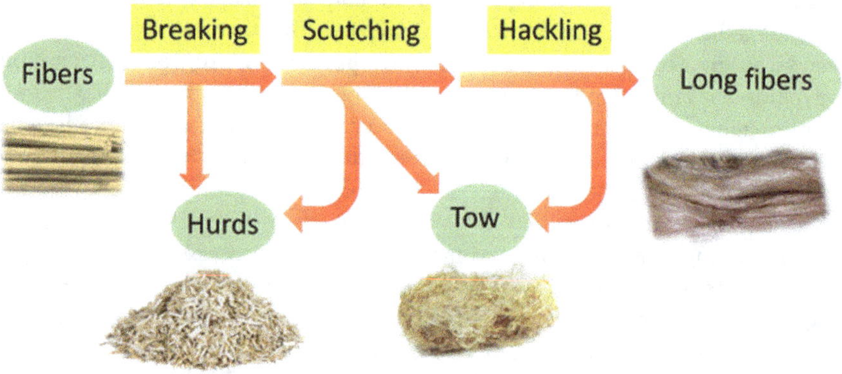

Fig. 7 Mechanical processing of fibers

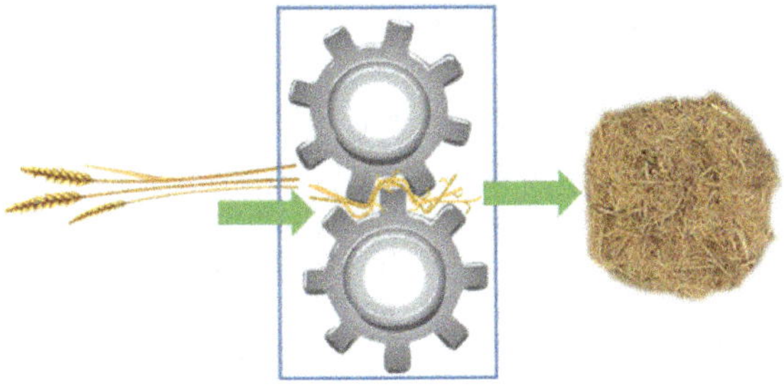

Fig. 8 Automated decortication by hammermilling

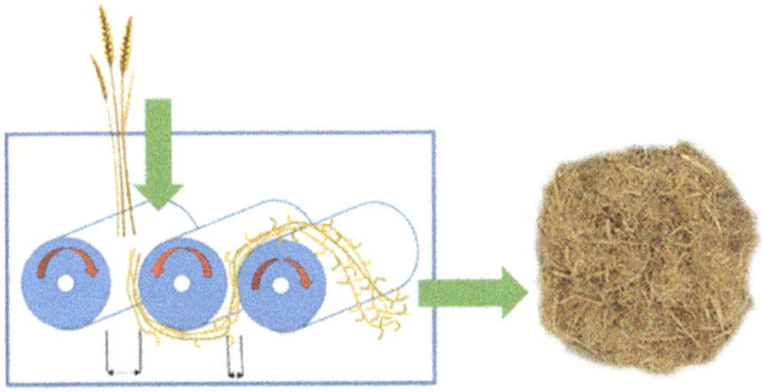

Fig. 9 Automated decortication by rollermilling

variables in mind, it is continuously under investigation. However, it is well known that mechanical retting tends to disrupt the fiber cell wall structures, leading to dislocations, kink bands, or nodes, which negatively affect the tensile mechanical properties (Davies and Bruce 1998; Baley 2002, 2004) and could even impair the composite performances (Huges et al. 2000; Hänninen et al. 2012). Although on the one hand, it was established that the defect extent of the mechanical retting is strongly related to the previous retting treatments (Van de Weyenberg et al. 2003), on the other hand, some researchers found that the fibers can be decorticated without significant damage and can be produced with high quality, even when no pretreatments are carried out (Hobson et al. 2001; Gañán et al. 2008; Bacci et al. 2010; Kengkhetkit and Amornsakchai 2012).

These findings are important and noteworthy: indeed, the elimination of pre-decortication enzymatic or microbiologic treatments should prove advantageous in terms of market costs, as it would reduce the variability of fiber quality, the uncertainties of production, and the amount of time necessary for the field to be available for the next crop. Moreover, new decortication methods can be developed in order to obtain better quality fibers and to reduce production costs by decreasing the times and increasing the treated biomass volumes.

3.4 Physical Retting

Among the physical treatments of fibers, the processes using electromagnetic radiation, high temperature, and/or pressure can be considered.

Steam explosion is an autohydrolysis process involving the use of saturated steam at high pressure followed by a sudden decompression, which causes the substantial breakdown of the lignocellulosic structure, the hydrolysis of the hemicellulose fraction, the depolymerization of the lignin components, and the defibrillation. High decompression rates lead to improved fiber freeness but shorter fiber length. During the process, high temperature softens the material and mechanical action during the high-pressure discharge results in fiber separation: the partially depolymerized lignin becomes more or less soluble in various organic solvents, such as alcohols, acetone, and in alkaline solutions, whereas the cellulose, much more resistant to hydrolysis than pectinic and hemicellulosic polysaccharides, retains its structure. The steam explosion treatment is a fast and well-controlled process, with a low cost and very flexible treatment parameters and is well adapted for the processing of various fibers, including those not previously retted (Zhang et al. 2008a, b). It is successfully applied on banana fibers (Sheng et al. 2014), semi-retted hemp fibers (Thomsen et al. 2006; Garcia-Jaldon et al. 1998), coir fibers (Abraham et al. 2013), flax (Kessler et al. 1998), and wheat straw (Zhang et al. 2008a, b). The steam explosion can be considered as a pretreatment that facilitates the ensuing retting processes (Zhang et al. 2008a, b). It can be carried out downstream, after alkali treatments, bleaching, and sometimes acid hydrolysis (Abraham et al. 2013; Sheng et al. 2014) in order to completely degrade the hemicellulose and lignin fractions. Usually, steam explosion is combined with an alkaline pre-soaking to favor the cleavage of lignin-hemicelluloses bonds. The reaction results in the increased solubility of the lignin alkaline solvent and in an enhanced water solubilization of hemicellulose (Sheng et al. 2014; Thomsen et al. 2006; Garcia-Jaldon et al. 1998; Abraham et al. 2013).

Another interesting physical treatment to extract the fibers is based on the **hydrothermal method**: the lignins and hemicellulose are degraded by using water at elevated pressure and temperature. This approach has been applied on hemp fibers by Thomsen (Thomsen et al. 2006) and on flax fibers by Stamboulis (Stamboulis et al. 2000), who carried out the so-called Duralin treatment, a sort of curing process at different times and temperatures, that leads to an easy separation of the fibers from the stem by a simple breaking and scutching operation. The fibers obtained via such procedures are fiber bundles rather than individual fibers and they exhibit improved moisture resistance and a somewhat higher and more uniform strength.

Recently, a Polish group (Konczewicz and Wojtysiak 2015; Kozlowski et al. 2013) has developed a new **osmotic degumming** of flax fibers. The degumming mechanism is based on the diffusive penetration of water inside the stem, where the long bundles of cellulosic fibers are clustered in slivers with polysaccharides, mostly pectins. The pectins, which are highly absorbent, increase their volume several times, which results in a considerable increase of hydrostatic pressure inside the stem and leads to pressing the epidermis. As the peripheral tension is stronger than the longitudinal one, cracks of the epidermis occur lengthwise, without breaking and shortening the fibers. Since the pectins become diluted and solved (together with other bast substances) in water, the technological liquid is subjected to proper filtration, which also serves to recover pectins for further use in the cosmetic industry. This osmotic method produces fibers characterized by good tenacity, divisibility, and soft touch. It is equally as efficient as the warm water retting method and could be applied to other bast fibers by simply changing the degumming parameters (temperature, flow velocity, and process duration).

To specifically modify the surface of fibers in order to improve their compatibility with polymeric matrix, the **plasma treatment** is an effective physical method, which can be performed at both atmospheric and high pressure under the flow of different types of gas (usually oxygen or argon).

Depending on the material to be treated, plasma flow can cause ablation, cross-linking or surface activation. Ablation consists in the removal of organic residues as well as surface layers at a molecular level. Cross-linking occurs as a result of the interaction between two or more radicals leading to the formation of covalent links while surface activation increases the surface energy as a result of the generation of polar groups on the reinforcement surface (Kafi et al. 2011). Exposure times, pressures, and discharge power are the variables that must be carefully considered to achieve the best results in terms of the surface modification. This kind of treatment is widely used for common natural fibers as flax, cane, coir, and bamboo (Bozaci et al. 2013; Scalici et al. 2016; Praveen et al. 2016; Xu et al. 2006) because, unlike chemical treatments, it is a simple nonpolluting process that can be considered as dry and clean.

The physical treatments are surely not yet completely developed and only few papers describe their use. However, they are characterized by high quickness, easy scalability, and process flexibility, that make this kind of retting noteworthy of further investigation.

3.5 Chemical Retting

With respect to water retting or dew retting, chemical processes are sometimes preferable since they produce fibers characterized by high-constant quality, regardless of weather conditions, usually in shorter times. Numerous chemical treatments can be performed on the fibers depending on their type, the ensuing retting process to be applied, their final applications. The most used chemical process is alkalization, a treatment aimed at removing hemicelluloses: it is usually carried out with sodium hydroxide, added as an aqueous solution at a variable concentration in the range 1–25% by weight. Considering the coir fibers, the NaOH effect is sometimes ambiguous and the reported results are controversial: indeed, if the alkali treatment seems to increase the elongation at break and the surface roughness (Silva et al. 2000) while improving the ultimate tensile strength, the initial modulus, the electrical properties, and the thermal stability (Mahato et al. 1993, 1995), at the same time it decreases the fiber tensile strength with increasing NaOH concentration, demonstrating that the alkalization could induce damages on fiber (Gu 2009). In order to minimize the fiber deterioration, a combination of sodium hydroxide, sodium carbonate, and sodium sulfide and short soaking treatments (only 2 h) have been successfully used on raw coconut fibers (Basu et al. 2015), achieving a reduction in diameter, linear density, and flexural rigidity of fibers.

Similar considerations also apply to kenaf: the alkali retting produces stronger, more flexible and less brittle fibers (Parikh et al. 2002a; Amel et al. 2013) but the base concentration influences the fiber morphology (Ramesh et al. 2015) because the high alkaline medium increases the microvoid volume fractions, leading to the reduction of the tensile strength of the fibers (Kawahara et al. 2005). Milder conditions prevent the degradation of kenaf fibers: the use of a weak base as sodium sulfite leads to cleaner and brighter fibers, smoother, and softer materials, having the higher tensile strength (Umoru et al. 2014). The same reagents have been tested on hemp fibers (Hurren et al. 2002) and it turns out that the combination of NaOH and Na_2SO_3 produces fibers with the whitest color and the finer diameter but not the best spinnability (obtained after acidic treatment with HCl). The need to select weak alkali treatments even on flax fibers has been highlighted by Van de Weyenberg who demonstrates that soaking in a diluted NaOH solution for only 45 s enables better adhesion between flax fibers and epoxy matrix, improving the mechanical properties of the resulting materials (Van de Weyenberg et al. 2006).

To completely remove the lignins and most hemicelluloses, aqueous ammonia treatments are an interesting alternative to alkali retting. Ammonia pretreatment techniques include the ammonia recycle percolation (ARP) and the soaking in aqueous ammonia (SAA). With ARP the biomass is pretreated with aqueous ammonia in a flow-through column reactor; after the reaction, the solid fraction of the biomass is separated whereas the liquid fraction is sent into a steam-heated evaporator in order to separate lignin, sugars, and ammonia, that will be further recycled. This pretreatment was first proposed by Iyer (Iyer et al. 1996) who tried the alkali hydrolysis on herbaceous biomass, as corn cobs and stover mixture and switchgrass, obtaining high efficiency of delignification but preserving the cellulose chains, after only 1 h or less of reaction at 170 °C (Kim et al. 2003). On the other hand, even at low temperatures, the SAA efficiently removes the lignin in the raw material by minimizing the interaction with hemicellulose and leading to an increment of surface area and pore size. The SAA has been applied on barley hulls (Kim et al. 2008) and the scientists obtained 66% of lignin solubilization when treating biomass with 15% aqueous ammonia at 75 °C during 48 h.

A further method to remove the lignin is by oxidative degradation that is commonly used when the target is the complete degradation and consumption of the lignocellulosic mass in order to recover the residues for the chemicals industry or for fuel production. In these cases, degradation is realized using peroxides and the raw biomasses are really varied: wheat straw (Klinke et al. 2002; Bjerre et al. 1996), mulberry (Cong and Dong 2007), poplar wood (Chang and Holtzapple

2000). Interestingly, the pretreatment with hydrogen peroxide of cane bagasse greatly enhances the fiber susceptibility to further enzymatic hydrolysis (Azzam 1989): the reported results show that about 50% of lignin and most of the hemicellulose content of cane bagasse have been solubilized by 2% alkaline hydrogen peroxide within 8 h.

A similar bleaching treatment on kenaf fibers (Shi et al. 2011) completely removes the hemicellulose (the lignin was already eliminated by alkali retting) within only 1 h, retaining the fiber diameter and increasing its crystallinity. Recently, a thorough investigation on the better oxidizing agent, among H_2O_2, NaOCl, NaClO/NaOH, for delignification of green coconut fibers has been described (Brígida et al. 2010). The latter study reports that hydrogen peroxide is more effective to eliminate waxes and fatty acids from the fiber surface but the treatment with NaOCl/NaOH is the most efficient in hemicellulose removal and, consequently, in cellulose exposition (which determines the inherent hydrophilicity and is useful for further functionalization).

A different method to prepare the fibers for ensuing treatments is based on the acidic retting: in such case, the risk to degrade the cellulosic fraction is quite high and the management of the processing waste is likewise complex. Therefore, such methodology is not popular and only a few papers describe this treatment. The use of concentrated strong acids, such as H_2SO_4 and HCl to treat lignocellulosic materials is widely reported when the main target is the complete cellulose hydrolysis (Sun and Cheng 2002). More frequently, strong acids were used in diluted solutions in processes exploiting less severe conditions (Shi et al. 2011a) but usually leading to sugar recovery (i.e., cellulose degradation) (Chen et al. 2007). A fascinating alternative to inorganic acids consists in the organic acids, like maleic and fumaric acids, which can be used for a diluted acid pretreatment (Kootstra et al. 2009). All acids, tested on the retting of wheat straw, drive to sugar formation.

Certainly, chemical treatments are an effective alternative to microbial dew retting which suffers from climatic risks leading to substantial harvest losses. Indeed, chemical retting is not affected by weather variability and can retain the fiber quality. However, difficulties in waste-management and the moderate risk to degrade the fibers currently make such treatments less attractive than in the past.

4 Retted Fibers in Polymer Composites Application

The use of natural fibers for composites is an attractive field from an environmental and sustainable perspective. In particular bast fibers, given their high cellulose and low lignin content, are particularly suited to composite applications and are the most promising replacement for glass fibers in composites (Deyholos and Potter 2014).

Some European countries, where a great environmental consciousness exists, have already introduced the use of lignocellulosic fibers in polymer composites for automotive applications and a great amount of products have been exported to the

US and other countries. Such materials are mostly characterized by nondegradable polymeric matrices, such as polyester, poly(propylene) (PP), epoxy, and phenol formaldehyde (PF). However, recent EU directives establish that all new vehicles must use 95% of recyclable materials in order to mitigate environmental impacts, and this boosted the development of commercially viable biodegradable composites. Therefore, bio-based lignocellulosic composites have a steadily growing market, with a growth projection in North America from US$ 155 million in 2000 to US$ 1.38 billion by 2025 (Satyanarayana et al. 2009).

Natural vegetable fibers can be used to reinforce thermosetting resins (e.g., unsaturated polyesters, epoxy, etc.), thermoplastics (e.g., polypropylene, polyethylene, etc.), and natural polymers (e.g., polylactic acid, polyhydroxybutirric acid, etc.). The most commonly used polymers in natural fibers composites matrices, and an overview of their properties are schematically shown in Tables 3 and 4.

It is well known that lignocellulosic fibers have been used in the field of composites since historical times, and one of their specific applications in aircraft took place as early as the 1940s. In the last decades, reasons for their use have increased,

Table 3 Most commonly used polymeric matrices for composites with natural fibers

Matrix material		
Thermosetting plastics	Thermoplastics	Rubber and natural polymers
Phenolic	Polypropylene	India-rubber
Epoxy	Polyamide	Modifies starch
Polyester	Polyethylene	Polylactic acid
Polyimide	Polystyrene	Cellulose esters
Polyurethane	Polyvinyl chloride	Polyhydroxybutirric acid

Table 4 Overview of the properties of some polymers commonly used in composite applications (Dicker et al. 2014)

Polymer type		Density (kg/m^3)	Glass transition T (°C)	Melt T (°C)	Young's modulus (GPa)	Tensile strength (MPa)	Elongation (%)
Bio	PLA	1210–1250	45–58	150–162	0.35–3.83	21–60	2.5–6
	PHB	1180–1260	4–15	168–182	3.5–4	24–40	5–8
	Starch	1000–1390	–	110–115	0.125–0.85	5–6	31–44
Synthetic	Epoxies	1110–1400	66.9–167	–	2.35–3.08	45–89.6	2–10
	Polyester	1040–1400	147–207	–	2.07–4.41	41.4–89.6	2–2.6
	PP	890–910	0.9–1.55	–	0.9–1.55	28–41	100–600

because of (i) the abundance of sources throughout the world, which represent value-added opportunities for agricultural industries and the possibility to create rural jobs, (ii) the possibility of reducing the dependence on petroleum products, (iii) the availability of complete data concerning fiber structure and properties. Other reasons for their increased use are related to the reduced cost, weight, and energy content in comparison with synthetic fibers (such as glass, kevlar, and carbon) along with comparable tensile strength values, as well as a lower abrasiveness for tooling and the absence of airborne particles, thus reducing respiratory problems for workers (Satyanarayana et al. 2009; Frollini et al. 2013). More in detail, lignocellulosic fibers present unique attributes such as: non-brittle fracture on impact, high flex and tensile modulus, high notched impact, low-thermal expansion coefficient, good sound abatement capability, natural appearance, and the fact that they can be easily colored (useful for textiles applications), full recyclability (Satyanarayana et al. 2009). In Table 5, some properties of the most employed lignocellulosic fibers are reported and compared with those of some man-made fibers. Such properties result from their structure and chemical composition.

Several factors limit the use of bast fibers in composites. The first problem is the lack of homogeneity. As with many natural products, there is large variation in fiber characteristics and composition according to sources (even in fibers grown in the same field and season). Quality production efficiency depends on the environmental conditions, on the cultivar and on growing, harvesting, and primary processing conditions. This lack of homogeneity greatly complicates optimization and standardization of manufacturing processes, where high reliability and stability are required (Deyholos and Potter 2014).

Table 5 Typical properties of some fibers commonly used in composite applications (Dicker et al. 2014; Callister 2007)

Fiber		Density (kg/m^3)	Tensile strength (MPa)	Young's modulus (GPa)	Elongation at break (%)	Moisture absorption (%)
Synthetic	Carbon HS	1800–1840	4400–4800	225–260	0	–
	E-glass	2550–2600	1900–2050	72–85	1.8–4.8	–
	Kevlar 49™	1444	3600–4100	131	2.8	–
Bast	Flax	1420–1520	750–940	75–90	1.2–1.8	8–12
	Hemp	1470–1520	550–920	55–70	1.4–1.7	6.2–12
	Jute	1440–1520	400–860	35–60	1.7–2	12.5–13.7
	Ramie	1450–1550	500–680	38–44	2–2.2	7.5–17
	Kenaf	1435–1500	195–666	60–66	1.3–5.5	–
Fruit	Coconut	1150–1220	135–240	4–6	15–35	8
Seed	Cotton	1520–1560	350–800	7–12	5–12	7.8–8.5
Leaf	Sisal	1400–1450	550–790	10–25	4–6	10–22
Grass	Bamboo	600–1100	140–800	11–32	2.5–3.7	–

The retting process, which represents the focus of this chapter, is a successful pretreatment for the improvement of matrix/fiber adhesion and the development of high-grade composites. An optimal retting process allows an optimum retting degree, which means that the elementary fibers are released well from the technical fibers and separated from each other by less non-fiber tissue attached on the surface. The performances of the materials increase accordingly, due to the increase of cellulose portions left on the fiber surface, and also by the higher fiber aspect ratio (length/diameter) due to the extent of the separation of elementary fibers. When the retting degree is low, impurities (pectin, lignin, hemicelluloses, etc.) between the fibers can cause stress concentrations in composites and lead to an early fracture (Hu et al. 2012a, b). The development of lignocellulosic fibers-based biocomposites requires the selection of an appropriate biopolymer matrix, suitable surface treatments of the fibers, along with low-cost but high-speed fabrication techniques. Many attempts were made to apply and adapt conventional polymer processing methods, such as compression molding and extrusion followed by injection molding, to the vegetal fiber composite preparation. However, some important aspects of composite manufacturing should be taken into account: preservation of mechanical properties of the lignocellulosic fibers by minimizing attrition and thermal degradation, attaining a high degree of fiber dispersion with control of fiber orientation and ensuring good wettability, despite of incorporation of a high volume fraction of fibers. In general, natural fiber reinforced composites can be manufactured via most production techniques, without using high temperatures (Fig. 10).

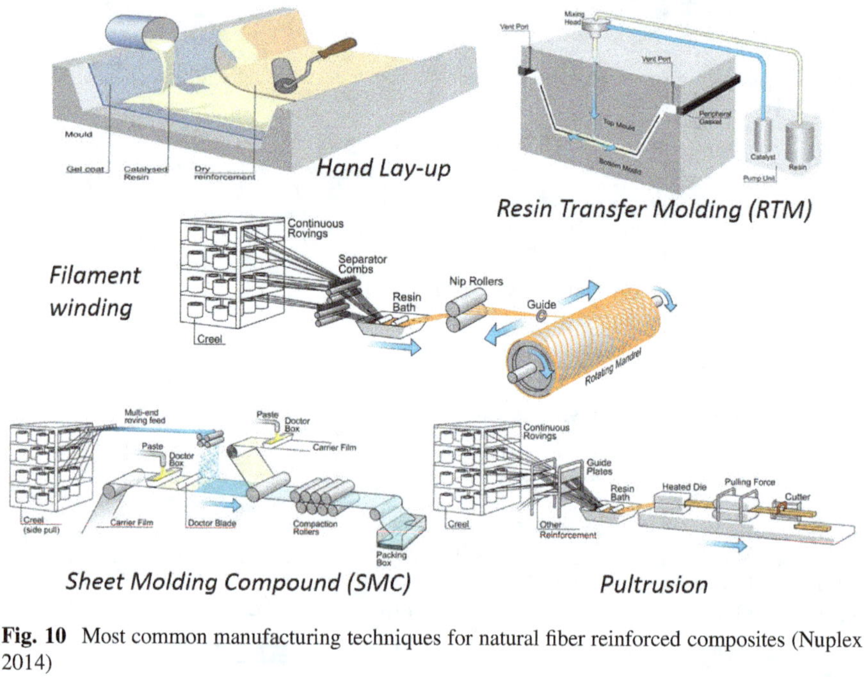

Fig. 10 Most common manufacturing techniques for natural fiber reinforced composites (Nuplex 2014)

Press molding seems to be the best preferred technique (Satyanarayana et al. 2009) since extrusion or melt mixing do not allow to take advantage of fiber length.

A wide range of biodegradable products has been produced using biopolymers containing lignocellulosic fibers for different applications, ranging from automotive vehicles including trucks, construction, and insulation panels, to special textiles (geotextiles and nonwoven textiles). Other identified uses for these materials include bathtubs, archery bows, golf clubs, boat hulls, maintenance-free roofing panels, and longer lasting and better-looking lightweight components, such as cosmetic packaging, tableware, and furniture. Furthermore, thanks to their load-bearing potential, the use of natural fiber-based composites has spread to various sectors, including aircraft, grain and fruit storage, and footwear. Thanks to the identification of all such new applications, a remarkable growth of the market for these new biocomposites is expected in a next future (Satyanarayana et al. 2009).

4.1 Flax

Flax (*Linum usitatissimum* L.) is an important agricultural crop. It is one of the first natural fibers to be extracted, spun, and woven into textiles and now widely used. Flax fibers are produced from the stems of flax bast plants. Like cotton, flax fiber is rich in cellulose, but its cellulose structure is more crystalline, making it stronger, crisper, and stiffer to handle, and more easily wrinkled.

The mechanical properties of flax fibers are reported to be highly dependent on the stem section used to obtain them. Fibers with a different chemical composition (e.g., in terms of cellulose, pectin, and lignin) resulting in different characteristics (e.g., depectinization efficiency and selectivity) are obtained from different stem sections. Fibers from the middle section exhibit higher strength than the fibers from the bottom section (Liu et al. 2015a). Lower fiber grades are used as reinforcement and filler in composites for automotive interior substrates and furniture (Yan et al. 2014).

Concerning the flax retting, Sharma et al. (1999) compare dew-, water-, and enzymatic retted fibers in terms of fiber fineness, strength, ash, and weight-losses, that are the key parameters for determining fiber quality. They are strictly connected to the proportion of residual pectins, hemicelluloses, lignin, and lipids present in the retted fiber, coming from differences in the activities of the polysaccharide-degrading enzymes involved.

In particular, in dew retting, pectinases, and hemicellulases released by the fungal colonists ret the flax slowly in 5–8 weeks. In contrast, water or enzymatic retting can be performed rapidly in 3–7 days, and seem to result in a superior retted material. Most of the pectin from the fiber can be hydrolyzed to improve fiber fineness, but only a limited proportion (nearly 10%) of hemicellulose can be removed without lowering the fiber strength. Removal of residual hemicelluloses bound with lignin will reduce fiber strength. This study, although based on a limited sample size, revealed the superiority of water retted fibers compared to the others.

As far as the enzymatic treatment of flax is concerned, Foulk compared some pectinase-rich commercial enzyme products and demonstrated that Texazym® BFE and Bioprep® 3000 L, as well as Viscozyme®, confer higher tenacity to the fibers, but indicate a potential advantage for Bioprep® 3000 L, in view of its monocomponent nature, commercial availability, price, and ability to ret flax, in combination with ethylenediaminetetraacetic acid (EDTA) at high pH (Foulk et al. 2008).

Good flax fiber quality with a minimum loss in fiber yield was also obtained by steam explosion treatment of flax following dew retting (Kessler et al. 1998).

On the other hand, Van Der Velde et al. test the tensile properties of scutched and hackled long flax fibers on individual fibers and on bundles and notice no significant influence of the retting degree on tensile strength (Van Der Velde and Baetens 2001). Pillin et al. also validate such conclusion in a study about the mechanical properties of oleaginous flax fibers as a function of variety, culture year, dew retting degree and agronomic factors. Despite the interesting mechanical properties obtained, they are not affected by the retting degree. The same conclusion is obtained with agronomic factors, such as seeding rate and plant height (Pillin et al. 2011).

In any case, numerous Authors discuss about the great importance of the retting to achieve high-performance materials.

Some representative examples of retted flax composites are discussed below.

Two kinds of retted Canadian linseed flax fibers, dew, and enzymatic retted fibers, are used in PP composites. The aspect ratio of the enzymatic retted fibers is much higher than that of dew retted as the elementary fibers are better separated. Indeed, the elementary fibers in dew retting remain tightly bundled into technical fiber wrappings with more noncellulose portions. This result induces a lower retting degree and is responsible for a lower thermal stability of dew retted fiber. Moreover, a better retting degree and fewer damages on enzymatic retted flax fibers endowed better tensile properties, thus resulting in a better reinforcement in composites (Hu et al. 2012a, b).

Enzymatic and dew retted short flax fiber composites with recycled high-density polyethylene (HDPE) were prepared and compared with similar products made with wood pulp, glass, and carbon fibers. The composites, with a 30 wt% of fiber loading, were easily blended with a Brabender mixer, because the flax fibers length was 0.21–0.40 mm. A commercial multienzyme product, Viscozyme® L, was combined in varying amounts with a commercial chelator product (Mayoquest® 200). Enzyme and chelator concentrations were denoted: low (0.05% and 5 mm), middle (0.1% and 10 mm), and high (0.3% and 25 mm). Dew retted fiber composites resulted in both lower strength and percent elongation, confirming that enzymatic retting is more efficient in producing fibers suitable for composite preparation. Moreover, enzyme and chelator concentrations did not much affect the final composites properties. Flax fibers improve the elastic modulus with respect to wood pulp/HDPE composites and HDPE alone (Foulk et al. 2004).

Enzymatic retted flax fibers and polylactic acid (PLA) were blended with a twin-screw extruder. The flax fiber content was 30 and 40 wt%. Triacetin was added as a plasticizer, in order to improve the impact properties, because of the brittle nature of PLA. The material properties have been studied and compared to those of PLA and

PP/flax fiber composite. Preliminary results reveal that the thermal properties of PLA improve with the addition of flax fibers. Moreover, the composite strength is enhanced with respect to that of PP/flax fiber composite, even if a poor interfacial adhesion is observed by morphological characterization (Oksman et al. 2003).

Short flax fibers/PP composites have been prepared by Martin et al., who assess the influence of the degree of dew retting of flax on the properties of composites. The length of the fibers was 3 mm, therefore, it was possible to prepare the composites by extrusion, with a 20% of volume fraction. The matrix was grafted with 4 wt% of maleic anhydride, in order to improve the compatibility. Water sorption studies indicate a higher water uptake for fibers with a low retting degree. The tensile properties of single fibers increased with the degree of retting in terms of both Young's modulus and strength at break. A high degree of retting enables easier splitting of the fiber bundles during fiber/matrix extrusion and injection molding, resulting in smaller fiber bundle diameters and a higher aspect ratio. Moreover, a skin/core effect is highlighted: defects are mostly located in the core of samples because the fibers are oriented parallel to the flow mostly in skin areas, resulting in better tensile properties (Martin et al. 2013).

Bodros et al. prepared natural fiber reinforced thermoplastic composites using a film-stacking technique. The materials were manufactured with a 20–30% of volume fraction of dew retted flax fibers 10 mm long, and various thermoplastics, such as PLA, poly(L-lactide) (PLLA), poly(3-hydroxybutyrate) (PHB), poly-ε-caprolactone (PCL), thermoplastic starch (MaterBi® Z), poly(butylene succinate) (PBS), and poly(butylene adipate-*co*-terephtalate) (PBAT). The tensile properties are compared to those of PP/flax composites. Preliminary results show that the tensile properties are improved by increasing the fiber volume fraction. The tensile strength and Young's modulus of PLLA and PLA flax composites prove to be higher than those of similar PP/flax fiber composites. The tensile strength and modulus of flax fiber/PLLA composite are very close to those of glass fiber polyester composites (Bodros et al. 2007).

Chemical treatments are often associated with other types of retting. Baley et al. prepared flax/unsaturated polyester resin composites using dew retted fibers (33 mm long), further subjected to sodium hydroxyde/acetic anhydride or formic acid-based treatments. This classical treatment induces a general increase of the flax fiber/polyester adhesion. The microbond test of a unique microdrop of the matrix on flax fibers allows the assessment of the interfacial adhesion, showing a good mechanical anchorage (Baley et al. 2006).

The chemical treatment with sodium hydroxide, sodium carbonate, and hydrogen peroxide, in addition to a microbiological retting was explored by Abdel-Halim et al. They prepared flax-reinforced epoxy composites by hot pressing from raw, scoured ($NaOH + Na_2CO_3$) and bleached ($NaOH + H_2O_2$) retted, and semi rotted flax fibers in order to investigate the effect of the retting degree as well as subsequent chemical treatments on the mechanical properties of the composites. The fibers were 1 cm long and were loaded in a high amount (almost 67 wt%). The chemical treatment improved the fiber fineness and the surface energy of the flax fibers, thus enhancing the mechanical properties of the composite. Such improvement follows the order: bleached fibers > scoured fibers > raw fibers. Among raw fibers, retted fiber composites had better mechanical properties than semi-retted ones. No improvement is obtained by the use of a titanate on modifying the fiber surface (Abdel-Halim et al. 2008).

4.2 Hemp

Hemp (*Cannabis sativa* L.) fiber plant is one of the world's oldest cultivated annual crops, traditionally grown for its long and strong bast fibers and seeds. In most western countries the cultivation of hemp was interrupted for decades (after 1980) as a result of competition with other feedstock, such as cotton and synthetic fibers, high labor costs, and the prohibition of cultivation due to the use of cannabis (*C. indica*) as a narcotic. The crop can grow in a wide range of geographic and climatic zones, and adjusts well to most regions of the world.

Hemp is used for a wide range of products, and has integrated many agro-industrial fields such as agriculture, textile, biocomposites, paper-making, automotive, constructions (as materials for building and insulation), bio-fuel, food, oil, cosmetics, personal care, household electrical appliances, packaging, and pharmaceutical industry (Salentijn et al. 2015).

Hemp hurds (also termed "shives"), i.e., the woody and lignified core tissues of the stems, are used as horse-bedding, pulping, and concreting. In addition to the traditional uses, novel applications for fiber hemp are being developed. The high cellulose content of hemp cell walls together with the [its??? Riferito a biomass] relatively high productivity make hemp biomass an interesting renewable feedstock for energy production, for the production of second generation bio-ethanol and as a reinforcement in "green composite" materials and concrete. It is estimated that the global market for hemp consists of more than 25,000 products (Salentijn et al. 2015). Hemp is more lignified than flax.

As reported for flax fibers, the effect of the stem section on the mechanical properties of hemp fibers is notable: fibers from the middle section of stems exhibit higher tensile strength and elongation than fibers from the bottom and the top. Compared to flax fibers, the most disadvantageous feature in hemp fibers is the presence of inferior secondary fibers (Liu et al. 2015a). In general, hemp bast fiber is organized in layers from the innermost xylem toward the surface consisting of cambium, secondary, and primary fibers, epidermis, and cuticle, as described by Liu et al. (2015b). Secondary fibers in hemp are primarily located at the bottom of the plant stem and are much shorter (approx. 2 mm long) and thinner (approx. 6 μm in diameter) than the primary fibers (approx. 20 mm long and 10–40 μm in diameter). The presence of secondary fibers may contribute to make the fibers in the bottom stem section poorer. Moreover, as already expressed, fibers from different sections show a different chemical composition, which may affect fiber extraction with biological methods, resulting in different responses (e.g., depectinization efficiency and selectivity) at different stem sections (Liu et al. 2015a).

The optimal growth stage for harvesting hemp fibers for use in composites is reported to be at the beginning of flowering when blooming begins (Liu et al. 2015b). Authors noticed that fibers harvested at an early blooming stage exhibit high tensile strength and strain, which decrease with plant maturity. Reduction in strength was related to the increase in the proportion of secondary fibers and decrease in cellulose deposition leading to inferior properties of fibers. The same Authors also noticed that extended retting dew (i.e., 70 days) had a detrimental effect on the mechanical performance of the fibers, presumably due to the accelerated degradation of cellulose by the action of microorganisms (Liu et al. 2015b).

Some representative examples of various types of retted hemp composites are discussed below.

Composites of PBS with retted and unretted hemp fibers were prepared by compression molding. The fiber content was 10, 20, and 30 wt%, while the fiber length was 3 cm. The efficiency of the retting process, carried out with warm inoculated water, was evaluated by analyzing the final performances of the materials. All the composites showed good mechanical properties, in particular, the composition of 20 wt% retted hemp fibers provided the best performances thanks to a stronger interface between fibers and matrix (Sisti et al. 2016).

Two low-cost retting methods were exploited for the preparation of short hemp fibers/PP composites. The fibers were retted in a bag in the first case, and in the presence of white rot fungi in a second experiment. The composites were prepared by extrusion, with 40 wt% of hemp fibers content (10 mm long). It was found that all the treatments increase the tensile strength of composites. In particular, composites prepared with white rot fungi *Schizophyllum commune* (*S. com*) have the highest tensile strength (Li et al. 2009).

Physical retting treatments have also been reported. Keller et al., for example, used hemp fibers with different length for the extrusion of poly(3-hydroxybutyrate-*co*-hydroxyvalerate) (PHBV) and copolyester amide (PEA) composites. The fibers, 8 mm long, were separated by a steam explosion process, while other fibers, 15 mm long, were degummed by biological processes. Composites with a fiber volume fraction of up to 42% could be achieved. The tensile strength and the Young modulus of PEA were improved by the reinforcement with 27% of fibers. In case of the brittle matrix PHBV instead, no improvement of the tensile strength was achieved, whereas its Young modulus was increased by short-fiber reinforcement, though this was at the expense of impact strength (Keller 2003).

PP/hemp fibers composite films with various amount of fibers (0–30 wt%) were prepared by melt blending and hot pressing. The fibers were chemically treated and then steam exploded. The mechanical properties of the materials resulted to be improved, in particular after a surface fiber treatment with polypropylene-maleic anhydride copolymer (Vignon et al. 1996). Similar materials have also been prepared by Pickering et al., who obtained the best mechanical results with the composite containing PP, 40 wt% of chemical retted hemp fibers and 3 wt% of a maleated polypropylene coupling agent (Pickering et al. 2007b).

Moreover, Bledzki et al. used alkali retted hemp fibers for PP and epoxy composites processed by the film-stacking technique in combination with the filament winding system (30–35 vol.% fiber content). Retted fibers and/or fibers treated with maleic anhydride-polypropylene copolymer were employed, obtaining good tensile strength (Bledzki et al. 2004).

Mwaikambo and Ansell observed an increase in the tensile properties of hemp fiber reinforced cashew nut shell liquid composites following alkalization of fibers. Nonwoven fiber mats and unidirectional fiber composites were manufactured by hand layup compression molding and the increase was observed for both types of composites (Mwaikambo and Ansell 2003).

The alkali treatment was successfully used by Mehta et al., who prepared hemp fiber mats/polyester resin composites by compression molding and the materials showed an increase in mechanical properties. The only exception was their impact strength which was found to decrease following alkalization treatment (Mehta et al. 2006).

Aziz and Ansell studied the mechanical properties of hemp and kenaf-fiber reinforced polyester composites, untreated and with alkali treatment. The composites have been prepared with short, long, and random mat fibers by hot pressing. The alkali treated fibers of both types of composites showed superior flexural strength and flexural modulus values compared to untreated fibers (Aziz and Ansell 2004).

4.3 Kenaf

Kenaf (*Hibiscus cannabinus* L.) belongs to the Malvaceae family, and is a fast growing, multipurpose crop with several harvested components: leaves and tender shoots are suitable for forage; seeds have an oil and protein composition similar to cotton seeds; the woody core can be a substitute for a number of different forest products; and the long bast fibers, traditionally used for cordage, are expanding into the field of composite materials (Ramesh et al. 2015). Kenaf grows in tropical and subtropical areas, 4–5 months after planting, with heights of 4–5 m and 25–35 mm in diameter (Umoru et al. 2014).

This plant features the highest carbon dioxide absorption among other plants (1 t of kenaf absorbs 1.5 t of atmospheric carbon dioxide). Kenaf fiber has a pale color because it contains less noncellulosic compounds than jute. Its fibers are coarse and quite brittle. It exhibits breaking strength similar to jute (Bledzki et al. 2015).

Kenaf bast fiber has been used for the production of fiber board and particle board, textiles, a fuel, and as a reinforcement material for composites (Amel et al. 2013). Applications of kenaf include newsprint, textiles, chemical sorbents, insulating, and noise-absorbing nonwoven materials for automobiles and structural applications and composite consumer products, such as laptops and cell phone cases (Ramesh et al. 2015).

Some representative examples of various types of retted kenaf composites are discussed below. The most common treatment seems to be the chemical retting.

Xia et al., for example, used kenaf fibers, 50.8 mm long, impregnated with aluminum hydroxide and alkali retted, to reinforce unsaturated polyester composites by the vacuum-assisted resin transfer molding process. The results show that the elastic modulus, tensile modulus, and tensile strength of the composites, as well as their water resistance, increase compared to those of composites made with untreated fibers. The presence of $Al(OH)_3$ proves to be advantageous in improving the interfacial compatibility between fibers and resin matrix (Xia et al. 2016).

Shi et al. prepared PP/kenaf composites by sheet molding. The fibers, 50.8 mm long, were alkali retted, impregnated with calcium carbonate and then compounded with PP films in the weight ratio of 50:50. First of all, the tensile strength of the individual fibers increases significantly (more than 20%) after the treatments and the compatibility fibers/matrix is improved by the inorganic particles. The tensile modulus and tensile strength of the composites, compared to those reinforced with untreated kenaf fibers, increase by 25.9 and 10.4%, respectively (Shi et al. 2011b).

Biocomposites based on the alkali retted kenaf fibers and natural polymer starch have been prepared by Song and Kim, by hot press. 20 wt% of fibers, 50 or 1 mm long, were used in composites. Various plasticizers, such as polyvinyl alcohol (PVA), polyethylene glycol (PEG), and glycerol (G), were added and the results show that the interfacial adhesion in the case of G and PVA is better, whereas in the case of PEG, a detachment of the fiber from starch was observed. Hence, the mechanical properties follow the general trend G > PVA > PEG (Song and Kim 2013).

Shi et al. applied progressive chemical treatments to kenaf fibers in order to prepare nano-scale to macro-scale cellulosic fibers from kenaf bast fibers. The chemical treatments included alkaline retting, bleaching, and acidic hydrolysis to obtain both pure-cellulose microfiber and cellulose nanowhisker (CNW). The chemical components of the different scale fibers were analyzed and CNWs were used for polyvinyl alcohol (PVA) composite reinforcement, prepared by film casting. The incorporation of 9 wt% of CNWs fiber (lengths of 100–1400 nm) in PVA composites increases the tensile strength by 46% with respect to the homopolymer (Shi et al. 2011a).

The successful use of chemically retted kenaf fibers in thermoformable composites with recycled polyester and off-quality polypropylene for automotive interiors has been reported by Parikh et al. (2002b).

Yang et al. prepared by compression molding a series of composites based on poly (hydroxybutyrate-*co*-valerate)/poly(butylene adipate-*co*-terephthalate) (80/20) and retted kenaf fibers. Two fiber lengths, 5 and 10 mm, and a hybrid 1:1 fiber mixture was utilized in composites. Both alkali and pectinase rettings were conducted on fibers. The alkali retted fibers showed synergistic benefits of the use of hybrid fiber lengths on modulus, that is to say, that the Young modulus of the composite containing the two length fiber mixture (5 + 10 mm) was higher respect to the samples with 5 mm and 10 mm alkali treated fibers. Such specific effect was not noticed with pectinase retted fiber samples, but in general, these composites outperformed the corresponding composites with alkali retted fibers (Yang et al. 2014).

Good mechanical properties have been obtained also by Bledzki et al. They prepared by extrusion a series of PP biocomposites with different types of fibers, obtaining the best performances with the sample containing 40 wt% of water retted kenaf fibers, 2 mm long, in comparison with abaca, jute, and wood microfibers composites (Bledzki et al. 2015).

Du et al. used mechanically retted short kenaf bast fibers, with different lengths (1.72–2.75–3.30 mm), to reinforce unsaturated polyester composites fabricated by compression molding. The effects of fiber loadings and aspect ratios on composite tensile properties were predicted using classical models in micromechanics. The results show that both composite tensile moduli and strengths increase consistently when increasing fiber loadings up to 75 vol.% (Du et al. 2010).

4.4 *Jute*

Jute is a long, soft, shiny fiber that can be spun into coarse, strong threads. Like cotton, it is cultivated just for its fiber. Jute fibers are extracted from the bark of the white jute plant (*Corchorus capsularis*) and from tossa jute (*C. olitorius*) by either biological or chemical retting process. It is also named "the Golden Fiber," because of its golden and silky shine. The length of a single fiber can range from 1 to 4 m. The structure of a jute fiber, consisting mostly of cellulose and lignin, has a polygonal section of various sizes, which results in an uneven thickness of fiber cell walls, and this, in turn causes variations in strength. Similarly to other lignocellulosic fibers, jute bast fiber is separated from the pith thanks to retting. In the case of water retting, cut jute stalks are placed in ponds for several weeks. Microbial action in the pond softens the jute fiber and weakens the bonds between the individual fibers and the pith. The fiber strands are then manually stripped from the jute stick and hung on racks to dry (Bledzki et al. 2015).

In hot and humid climates jute plants can be harvested in 4–6 months, therefore the suitable climate for growing jute is the monsoon season. Temperatures ranging from 20 to 40 °C and a relative humidity of 70–80% are favorable for successful cultivation. Jute requires a weekly rainfall of 5–8 cm with an extra amount during the sowing period. It is cultivated in the world and extensively grown in Bangladesh, China, India, Indonesia, and Brazil but the best quality comes from Bangladesh (Ahmed and Nizam 2008).

Jute fiber is a good insulator, has antistatic properties and moderate moisture retention. Due to a high lignin content (up to 20%), jute fibers are brittle, but strong and have a low extension to break (about 1.5%) (Bledzki et al. 2015). Due to its good spinning quality, it is a good textile fiber. It is suited for making jeans and other heavy-duty types of fabrics (Tahir et al. 2011). Jute fibers are used in many sectors of industry, like fashion, travel, luggage, furnishing, and in the production of carpets and other floor coverings, and, last but not least, as a reinforcement in biocomposites (Bledzki et al. 2015).

Some representative examples of various types of retted jute composites are discussed below.

Prasad et al. reported an example of composite made by chemical retted jute fibers and PP. More in detail, the materials have been prepared by injection molding with different fiber loading (0, 5, 10, 15, 20, 25 wt%) and fiber condition (untreated, NaOH treated at different concentrations followed by bleaching with H_2O_2).

The fiber length was 3 mm. The results showed that tensile strength increases with increase in the fiber loading up to 20%, and then there is a reduction. Regarding the alkali treatment, it seems that concentrations higher than 10% of NaOH, may cause strength deterioration. The tensile modulus of the composite with treated fibers has increased considerably when compared to plain PP and untreated fiber reinforced PP composite (Prasad et al. 2014).

Another example of chemical retted jute fibers has been reported by Graceraj et al. The Authors prepared composites based on a mixture of cashew nut shell liquid (CNSL) resin and polyester resin with alkali treated fibers, by hand layup technique. CNSL was used for the improvement of biocompatibility. The fibers were knitted in the mold, therefore, they were 250 mm long. Different amounts of CNSL (5–10–15%) and fiber volume fraction (5–10–15%) have been used. The fatigue strength could be controlled by the optimization of the composition parameters, that is almost 5–7% of CNLS, 8–24 h of NaOH treatment and 7.5–15% of volume fraction (Graceraj et al. 2016).

The water retting process was instead exploited by Gowda et al. The Authors evaluated the mechanical properties of jute-reinforced polyester composites made by hand layup. Microorganisms (mainly *bacillus bacteria*) decomposed the gums and softened the tissues in 5–30 days, depending on the temperatures (27 °C was the optimum) and the type of water used. The jute fibers contained a large numbers of short fibers and few long ones. The Authors tested the mechanical properties and concluded that the composites have better strengths than wood composites. Therefore, they could be suitable for indoor applications, such as shelves, partitions, wash basins, and table tops, and for outdoor uses, such as roofing, drainage pipes, automobile components, electrical fittings as well as larger items, such as light-weight fishing boats (Gowda et al. 1999).

Hassan studied the mechanical properties of phenol formaldehyde resin/jute composites. The fibers have been acetylated or steamed in a rotating autoclave and then sprayed with different levels of PF resin; 12 sheets were overlaid and pressed. The effect of the amount of resin (12–30%), on the flexural strength, tensile strength, water absorption, and thickness swelling of the composites was studied, as well as the effect of steaming and acetylation on the structure and thermal stability of jute fibers. Steaming resulted to be superior to acetylation in improving the dimensional stability. However, the Authors concluded that, due to the lower thermal stability of the steamed and acetylated jute as compared to untreated jute, the use of a resin having curing temperature <160 °C, for example, urea formaldehyde, is recommended, if high-mechanical properties along with high-dimensional stability are required (Hassan 2003).

Unfortunately, various authors do not specify exactly the type of retting in case of jute fibers. For example, Hu et al. used the film-stacking hot pressed method to prepare composites with polylactide film (0.3 mm thickness) and short jute retted fibers (10–15 mm length) in a volume fraction of 30, 40, and 50%. The Authors concluded that, when the volume fraction of fibers in the composite was 40%, the composite had better mechanical properties (Hu et al. 2007). Moreover, Hu et al. employed a more uniform fiber blending method to fabricate composites with retted jute fiber (100 mm) and short PLA fibers, reaching the 70% of fiber volume fraction. The fabrication process includes two steps: felt making and hot pressed molding. The composites showed very good formability and processability (Hu et al. 2012a, b).

Roe et al. used retted jute fibers and a polyester resin matrix to form uniaxially reinforced composites containing up to 60 vol.% of fibers. The composites were fabricated in a lossy mold comprising demountable steel plates forming interleaved troughs 12.7 mm wide and 40 mm deep and the fibers were cut exactly to the length of the mold. The tensile strength and Young's modulus, work of fracture determined by Charpy impact and interlaminar shear strength have been measured as a function of fiber volume fraction. The Authors conclude that the resin forms an intimate bond with jute fibers up to a volume fraction of 0.6, above which the quantity of resin was insufficient to wet fibers completely. Therefore, good quality composites with very acceptable specific properties were assessed (Roe and Ansell 1985).

Good mechanical performances have been achieved also in some mixed systems as reported, for example, by Ramnath et al., who prepared composites by hand layup technique with layers of abaca and retted jute fibers with different orientation. The materials were laminated on top and bottom with glass fibers, in order to improve the surface finish and strength (Ramnath et al. 2013). Sature et al. also reported about the good mechanical reinforcement given by jute/hemp epoxy composites laminated with woven glass mat on top and bottom by hand layup technique. Water retting was applied to jute fibers (Sature and Mache 2015).

Finally, examples of retted jute/epoxy composites made with the vacuum-assisted resin infiltration have been reported by Hossain et al. (2013), while a mixed system based on the hemp and jute/polyester composites has been reported by Hughes et al. (2002).

4.5 Other

In addition, the literature also reports examples of composites based on the other types of fibers, deriving from leaves (such as palm, abaca, sisal, etc.) or grasses (such as bamboo, elephant grass, etc.).

For example, sugar palm fibers, treated by a seawater retting process, were successfully used to reinforce epoxy composites. The results showed that the tensile and flexural strengths of the composites increased proportionally to the fiber treatment duration thanks to a good fiber-matrix bond (Leman et al. 2010).

Epoxy composites have been also reinforced with different fiber contents of royal palm fibers. The fibers were collected from a royal palm tree through the process of water retting and mechanical extraction, and further subjected to alkali treatment which resulted to be effective in improving the tensile and flexural properties while the impact strength decreased, with respect to the untreated fiber composite (Goud and Rao 2011).

Moreover, retted royal palm and bamboo fibers were used to reinforce a polyester resin matrix, tested for tensile, flexural, and dielectric properties and compared with similar sisal and banana fiber composites. It has been observed that the tensile and flexural properties increase in composites containing palm and bamboo fibers, as well as the dielectric strength in palm fiber composites, useful for electrical insulation applications (Rao et al. 2010).

Bamboo fibers, retted by the action of bacteria and moisture and then mechanically extracted, have been used as reinforcements in polyester resins for insulating materials. The results showed indeed that the thermal conductivity of composites decreased with the increase in fiber content (volume fraction 0.1–0.3). Therefore, this composite can be used in the building and automotive industries to save energy by reducing the rate of heat transfer (Mounika et al. 2012).

Polyester resin composites were also prepared with jowar. Experiments of tensile and flexural tests were carried out, and the samples were compared with sisal and bamboo composites. The fibers were extracted by water retting and manual process. The composites present high strength and rigidity, suitable for lightweight applications compared to conventional sisal and bamboo composites (Prasad and Rao 2011).

Sisal fibers extracted by field retting process were subjected to a chemical mercerization treatment and used for the production of PP composites, leading to better flexural and tensile properties, when compared to untreated and unreinforced PP matrix (Oladele and Agbabiaka 2015). Moreover, sisal fibers were used to reinforce unsaturated polyester resin composites. The fibers were subjected to both field and chemical retting (NaOH, KOH, H_2O_2, and ethanol). The tensile and hardness properties of the polyester composites turned out to be enhanced with KOH treatment (Oladele et al. 2014).

Other types of fibers investigated for polyester resin composites are wildcane grass fibers, that were water retted, in order to produce inexpensive materials with high toughness (Prasad et al. 2011). Elephant grass stalk fibers were incorporated in a polyester matrix in order to obtain materials with enhanced mechanical properties. The fibers were extracted using water retting or chemical treatment and the latter proved to be more promising (Rao et al. 2007). Chemically retted elephant grass fibers were also used to reinforce PLA composites, resulting in materials with higher tensile and impact strength when compared to similar lignocellulosic fibers/PLA composites (Gunti et al. 2016).

Cornhusk fibers instead were used in PP composites. The fibers were alkali and enzymatic (pulpzyme and cellulase) retted. The composites resulted in increased mechanical and sound absorption properties, therefore a possible application of these composites may be in structural components for automotive interiors such as headliner substrate (Huda and Yang 2008).

5 Conclusions and Future Prospectives

The main disadvantages of natural fibers in the reinforcement of composites are the poor compatibility between fiber and matrix and their relatively high moisture absorption. Therefore, natural fiber modifications trough extraction processes are necessary to improve their adhesion with different matrices. In recent years, the various retting techniques have experienced a sizable evolution and improvements that tend to minimize the lack of consistency in fiber qualities and the high levels of variability in fiber properties. As described above, the technology based on enzymes seems to be the most promising retting technique, due to its environmental friendliness, shorter processing times, and acceptable fiber quality. However, at present, it is not yet feasible and developed on a large scale, due to the high cost of the process.

References

Abdel-Halima ES, El-Rafie MH, Kohler R (2008) Surface characterization of differently pretreated flax fibers and their application in fiber-reinforced composites. Polym Plast Technol Eng 47:58–65

Abraham E, Deepa B, Pothen LA, Cintil J, Thomas S, John MJ, Anandjiwala R, Narine SS (2013) Environmental friendly method for the extraction of coir fibre and isolation of nanofiber. Carbohydr Polym 92:1477–1483

Adamsen APS, Akin DE, Rigsy LL (2002) Chelating agents and enzyme retting of flax. Text Res J 72:296–302

Ahmed Z, Akhter F (2001) Jute retting: an overview. J Biol Sci 1(7):685–688

Ahmed Z, Nizam SA (2008) Jute—microbiological and biochemical research. Plant Tissue Cult Biotechnol 18:197–220

Akin DE, Condon B, Sohn M, Foulk JA, Dodd RB, Rigsby LL (2007) Optimization for enzyme-retting of flax with pectate lyase. Ind Crops Prod 25:136–146

Akin DE, Henriksson G, Evans JD, Adamsen APS, Foulk JA, Dodd RB (2004) Progress in enzyme retting of flax. J Nat Fibers 1:21–47

Akin DE, Morrison WH III, Rigsby LL, Evans JD, Foulk JA (2003) Influence of water presoak on enzyme-retting of flax. Ind Crops Prod 17:149–159

Akin DE, Rigby LL, Perkins W (1999) Quality properties of flax fibers retted with enzymes. Text Res J 69:747–753

Akin DE, Rigsy LL, Henriksson G, Eriksson KL (1998) Structural effects on flax stems of three potential retting fungi. Text Res J 68:515–519

Ali MM (1958) Aerobic bacteria involved in the retting of jute. Appl Microbiol 6:87–89

Amaducci S, Gusovious HJ (2010) Hemp cultivation, extraction and processing. In: Mussig J (ed) Industrial application of natural fibres: properties and technical application. Wiley, UK

Amel BA, Paridah MT, Sudin R, Anwar UMK, Hussein AS (2013) Effect of fiber extraction methods on some properties of kenaf bast fiber. Ind Crops Prod 46:117–123

Antonov V, Marek J, Bjelkova M, Smirous P, Fisher H (2007) Easily available enzymes as natural retting agents. Biotechnol J 2:342–346

Aziz SH, Ansell MP (2004) Optimising the properties of green composites. In: Baillie CA (ed) green composites. Woodhead Publishing Limited, Cambridge

Azzam AM (1989) Pretreatment of cane bagasse with alkaline hydrogen peroxide for enzymatic hydrolysis of cellulose and ethanol fermentation. J Environ Sci Heal B 24:421–433

Bacci L, Di Lonardo S, Albanese L, Mastromei G, Perito B (2010) Effect of different extraction methods on fiber quality of nettle (*Urtica Dioica* L.). Text Res J 81:827–837

Baley C (2002) Analysis of the flax fibres tensile behaviour and analysis of the tensile stiffness increase. Compos Part A-Appl S 33:939–948

Baley C (2004) Influence of kink bands on the tensile strength of flax fibers. J Mater Sci 39: 331–334

Baley C, Busnel F, Grohens Y, Sire O (2006) Influence of chemical treatments on surface properties and adhesion of flax fibre–polyester resin. Compos Part A-Appl S 37:1626–1637

Basu G, Mishra L, Jose S, Samanta AK (2015) Accelerated retting cum softening of coconut fibre. Ind Crops Prod 77:66–73

Beg QK, Bushan B, Kappor M, Hoondal GS (2000) Production and characterization of thermostable xylanase and pectinase from *Streptomycetes* spp. QG-11-3. Res Bull Panjab Univ Sci 51:71–78

Bismark A, Mishra S, Lampke T (2005) Plant fibers as reinforcement for green composites. In: Mohanty AK, Misra M, Drzal LT (eds) Natural fibers, biopolymers, and biocomposites. CRC Press, Boca Raton

Bjerre AB, Olesen AB, Fernqvist T, Ploger A, Schmidt AS (1996) Pretreatment of wheat straw using combined wet oxidation and alkaline hydrolysis resulting in convertible cellulose and hemicellulose. Biotechnol Bioeng 49:568–577

Blanco P, Sieiro C, Villa TG (1999) Production of pectic enzymes in yeast. FEMS Microbiol Lett 175:1–9

Bledzki AK, Fink HP, Specht K (2004) Unidirectional hemp and flax EP- and PP-composites: influence of defined fiber treatments. J Appl Polym Sci 93:2150–2156

Bledzki AK, Franciszczak P, Osman Z, Elbadawi M (2015) Polypropylene biocomposites reinforced with softwood, abaca, jute, and kenaf fibers Ind Crops. Prod 70:91–99

Bodros E, Pillin I, Montrelay N, Baley C (2007) Could biopolymers reinforced by randomly scattered flax fibre be used in structural applications? Compos Sci Technol 67:462–470

Booth I, Goodman AM, Grishanov SA, Harwood RJ (2004) A mechanical investigation of the retting process in dew-retting hemp (*Cannabis sativa*). Ann Appl Biol 145:51–58

Bozaci E, Sever K, Sarikanat M, Seki Y, Demir A, Ozdogan E, Tavman I (2013) Effects of the atmospheric plasma treatments on surface and mechanical properties of flax fiber and adhesion between fiber–matrix for composite materials. Compos Part B-Eng 45:565–572

Brígida AIS, Calado VMA, Gonçalves LRB, Coelho MAZ (2010) Effect of chemical treatments on properties of green coconut fiber. Carbohydr Polym 79:832–838

Bruhlmann F, Leupin M, Erismann KH, Fiechter A (2000) Enzymatic degumming of ramie bast fibers. J Biotechnol 76:43–50

Callister WD Jr (2007) Composites. In: Callister WD Jr (ed) Materials science and engineering: an introduction, 7th edn. Wiley, New York

Chang VS, Holtzapple MT (2000) Fundamental factors affecting biomass enzymatic reactivity. Appl Biochem Biotechnol 84–86:5–37

Chen Y, Sharma-Shivappa RR, Keshwani D, Chen C (2007) Potential of agricultural residues and hay for bioethanol production. Appl Biochem Biotechnol 142:276–290

Cong R, Dong W (2007) Structure and property of mulberry fiber. Mod Appl Sci 1:14–17

Dasgupta PC, Sardar D, Majumdar AK (1976) Chemical retting of jute. Food Farming Agric 8:7–9

Davies DC, Bruce DM (1998) Effect of environmental relative humidity and damage on the tensile properties of flax and nettle fibers. Text Res J 68:623–629

Deyholos MK, Potter S (2014) Engineering bast fiber feedstocks for use in composite materials. Biocatal Agric Biotechnol 3:53–57

Di Candilo M, Ranalli P, Mastromei G, Polsinelli M, Bozzi C, Focher B (2000) Optimum conditions for microbial retting of hemp in tanks. In: Bioresource Hemp. Wolfsburg: www.nova-institut.de

Dicker MPM, Duckworth PF, Baker AB, Francois G, Hazzard MK, Weaver PM (2014) Green composites: a review of material attributes and complementary applications. Compos Part A-Appl S 56:280–289

Donaphy JA, Levette PN, Haylock RW (1990) Changes in microbial populations during anaerobic flax retting. J Appl Bacteriol 69:634–641

Dosanjh NS, Hoondal G (1996) Production of constitutive, thermostable, hyper active exo-pectinase from Bacillus GK-8S. Biotechnol Lett 18:1435–1438

Durden D, Etters JN, Sarkar AK, Henderson LA, Hill JE (2001) Advances in commercial biopreparation of cotton with alkaline pectinase. AATCC Rev 1:28–31

Du Y, Zhang J, Yu J, Lacy TE Jr, Xue Y, Toghiani H, Horstemeyer MF, Pittman CU Jr (2010) Kenaf bast fiber bundle-reinforced unsaturated polyester composites. IV: effects of fiber loadings and aspect ratios on composite tensile properties. Forest Prod J 60:582–591

Etters JN, Sarkar AK, Henderson LA, Liu J (2001) The influence of biopreparation of cotton with alkaline pectinase on dyeing properties. AATCC Rev 1:22–24

Evans JD, Akin DE, Foulk JA (2002) Flax-retting by galacturonase-containing enzyme mixtures and effects on fiber properties. J Biotechnol 97:223–231

Faruk O, Bledzki AK, Fink HP, Sain M (2012) Biocomposites reinforced with natural fibers: 2000–2010. Prog Polym Sci 37:1552–1596

Flax composites (2014) http://flaxcomposites.com

Fogarty M, Ward OP (1972) Pectic substances and pectolytic enzymes. Process Biochem 7:17–31

Foulk JA, Akin DE, Dodd RB (2008) Influence of pectinolytic enzymes on retting effectiveness and resultant fiber properties. BioResources 3:155–169

Foulk JA, Chao WY, Akin DE, Dodd RB, Layton PA (2004) Enzyme-retted flax fiber and recycled polyethylene composites. J Polym Environ 12:165–171

Foulk JA, Rho D, Alcock MM, Ulven CA, Huo S (2011) Modifications caused by enzyme-retting and their effect on composite performance. Adv Mater Sci Eng 9:121–129

Frollini E, Bartolucci N, Sisti L, Celli A (2013) Poly(butylene succinate) reinforced with different lignocellulosic fibers. Ind Crops Prod 45:160–169

Frollini E, Bartolucci N, Sisti L, Celli A (2015) Biocomposites based on poly(butylene succinate) and curaua: mechanical and morphological properties. Polym Test 45:168–173

Frontier Culture Museum of Virginia (2012) http://frontierculturemuseum.blogspot.it/2012/08/retting-flax.html

Gañán P, Zuluaga R, Restrepo A, Labidi J, Mondragon I (2008) Plantain fibre bundles isolated from Colombian agro-industrial residues. Bioresource Technol 99:486–491

Garcia-Jaldon C, Dupeyre D, Vignon MR (1998) Fibres from semi-retted hemp bundles by steam explosion treatment. Biomass Bioenergy 14:251–260

Goud G, Rao RN (2011) Effect of fibre content and alkali treatment on mechanical properties of Roystonea regia-reinforced epoxy partially biodegradable composites. Bull Mater Sci 34:1575–1581

Gowda TM, Naidu ACB, Chhaya R (1999) Some mechanical properties of untreated jute fabric-reinforced polyester composites. Compos Part A-Appl S 30:277–284

Graceraj PP, Venkatachalam G, Shankar AG, Kumar K (2016) Investigation on fatigue strength of the jute fiber reinforced hybrid polymer matrix composites. UPB Sci Bull Series D 78:185–196

Gu H (2009) Tensile behaviours of the coir fibre and related composites after NaOH treatment. Mater Des 30:3931–3934

Gunti R, Prasad AVR, Gupta AVSSKS (2016) Mechanical and degradation properties of natural Fiber reinforced PLA composites: jute, sisal, and elephant grass. Polym Composite doi:10.1002/pc

Gurunathan T, Mohanthy S, Nayak SK (2015) A review of the recent developments in biocomposites based on natural fibres and their application prespectives. Compos Part A-Appl S 77:1–25

Hänninen T, Thygesen A, Mehmood S, Madsen B, Hughes M (2012) Mechanical processing of bast fibres: the occurrence of damage and its effect on fibre structure. Ind Crops Prod 39:7–11

Hassan ML (2003) Recycling of jute textile in phenol formaldehyde–jute composites. J Appl Polym Sci 90:3588–3593

Henriksson G, Akin DE, Hanlin RT, Rodriguez C, Archibald DD, Rigsby LL, Eriksson KL (1997) Identification and retting efficiencies of fungi isolated from dew-retted flax in in the United States and Europe. Appl Environ Microbiol 63:3950–3956

Hobson RN, Hepworth DG, Bruce DM (2001) Quality of fibre separated from unretted hemp stems by decortication. J Agric Eng Res 78:153–158

Hossain MR, Islam MA, Vuurea AV, Verpoest I (2013) Effect of fiber orientation on the tensile properties of jute epoxy laminated composite. J Sci Res 5:43–54

Hu RH, Lim JK, Kim CI, Yoon HC (2007) Biodegradable composites based on polylactic acid (PLA) and China jute fiber. Key Eng Mater 353–358:1302–1305

Hu RH, Ma ZG, Zheng S, Li YN, Yang GH, Kim HK, Lim JK (2012a) A fabrication process of high volume fraction of jute fiber/polylactide composites for truck liner. Int J Precis Eng Man 13:1243–1246

Hu W, Ton-That MT, Denault J, Rho D, Yang J, Lau PCK (2012b) Comparison between dew-retted and enzyme-retted flax fibers as reinforcing material for composites. Polym Eng Sci 52:165–171

Huda S, Yang Y (2008) Chemically extracted cornhusk fibers as reinforcement in light-weight poly(propylene) composites. Macromol Mater Eng 293:235–243

Huges M, Sèbe G, Hague J, Hill C, Spear M, Mott L (2000) An investigation into the effects of micro-compressive defects on interphase behaviour in hemp-epoxy composites using half-fringe photoelasticity. Compos Interface 7:13–29

Hughes M, Hill CAS, Hague JRB (2002) The fracture toughness of bast fibre reinforced polyester composites part 1 evaluation and analysis. J Mater Sci 37:4669–4676

Hurren CJ, Wang X, Dennis HGS, Clarke AFK (2002) Evaluation of bast fibre retting systems on hemp. In: Proceedings of the 82nd Textile Institute World Conference, Cairo, Egypt

Iyer PV, Wu Z-W, Kim SB, Lee YY (1996) Ammonia recycled percolation process for pretreatment of herbaceous biomass. Appl Biochem Biotechnol 57–58:121–132

Jankauskiene Z, Butkute B, Gruzdeviene E, Ceseviciene J, Fernando AL (2015) Chemical composition and physical properties of dew- and water-retted hemp fibers. Ind Crops Prod 75:206–211

Jayani RS, Saxena S, Gupta R (2005) Microbial pectinolytic enzymes: a review. Process Biochem 40:2931–2944

Kafi AA, Magniez K, Fox BL (2011) A surface-property relationship of atmospheric plasma treated jute composites. Compos Sci Technol 71:1692–1698

Kapoor M, Beg QK, Bhushan B, Singh K, Dadich KS, Hoondal GS (2001) Application of alkaline and thermostable galacturonase from *Bacillus* sp. MG-cp-2 in degumming of ramie (*Boehmeria nivea*) and sunn hemp (*Crotolaria juncia*) bast fibers. Process Biochem 36:803–807

Kawahara Y, Tadokoro K, Endo R, Shioya M, Sugimura Y, Furusawa T (2005) Chemically retted kenaf fibers. Sen'I Gakkaishi 61:115–117

Keller A (2003) Compounding and mechanical properties of biodegradable hemp fibre composites. Compos Sci Technol 63:1307–1316

Kengkhetkit N, Amornsakchai T (2012) Utilisation of pineapple leaf waste for plastic reinforcement: 1. A novel extraction method for short pineapple leaf fiber. Ind Crops Prod 40:55–61

Kessler RW, Becker U, Kohler R, Goth B (1998) Steam explosion of flax—a superior technique for upgrading fibre value. Biomass Bioenergy 14:237–249

Kim TH, Kim JS, Sunwoo C, Lee YY (2003) Pretreatment of corn stover by aqueous ammonia. Bioresource Technol 90:39–47

Kim TH, Taylor F, Hicks KB (2008) Bioethanol production from barley hull using SAA (soaking in aqueous ammonia) pretreatment. Bioresource Technol 99:5694–5702

Klinke HB, Ahring BK, Schmidt AS, Thomsen AB (2002) Characterization of degradation products from alkaline wet oxidation of wheat straw. Bioresource Technol 82:15–26

Konczewicz W, Wojtysiak J (2015) The effect of physical factors on the process of physical-mechanical degumming of flax fibers. Text Res J 85:391–403

Kootstra AMJ, Beeftink HH, Scott EL, Sanders JPM (2009) Comparison of dilute mineral and organic acid pretreatment for enzymatic hydrolysis of wheat straw. Biochem Eng J 46:126–131

Kouhoundè SHS, Adeati K, Delvigne F, Savadogo A, Traore AS, Thonart P (2014) The use of Microrganisms of cassava retting for the production of pectinolytic enzymes. J Microbiol Biotech Food Sci 4:277–281

Kozlowski R, Konczewicz W, Wojtysiak J, Podsiedlik W (2013) Device for processing fibrous raw materials and method of fibrous plants processing. European Patent EP 2 242 876 B1

Leman Z, Sapuan SM, Ishak MR, Ahmad MMHM (2010) Pre-treatment by water retting to improve the interfacial bonding strength of sugar palm fiber reinforced epoxy composite. Polymers from Renewable Resources 1:35–45

Li Y, Pickering KL, Farrell RL (2009) Analysis of green hemp fibre reinforced composites using bag retting and white rot fungal treatments. Ind Crops Prod 29:420–426

Liu M, Fernando D, Meyer AS, Madsen B, Daniel G, Thygesen A (2015a) Characterization and biological depectinization of hemp fibers originating from different stem sections. Ind Crops Prod 76:880–891

Liu M, Fernando D, Daniel G, Madsen B, Meyer AS, Tutor Ale M, Thygesen A (2015b) Effect of harvest time and field retting duration on the chemical composition, morphology and mechanical properties of hemp fibers. Ind Crops Prod 69:29–39

Mahato DN, Mathur BK, Bhattacherjee S (1993) Effects of alkali treatment on electrical and spectral properties of coir fibre. J Mater Sci Lett 12:1350–1353

Mahato DN, Mathur BK, Bhattacherjee S (1995) Effects of alkali treatment on thermal stability and moisture retention of coir fibre. Indian J Fibre Text 20:202–205

Martin N, Mouret N, Davies P, Baley C (2013) Influence of the degree of retting of flax fibers on the tensile properties of single fibers and short fiber/polypropylene composites. Ind Crops Prod 49:755–767

Mehta G, Drzal LT, Mohanty AK, Misra M (2006) Effect of fiber surface treatment on the properties of biocomposites from nonwoven industrial hemp fiber mats and unsaturated polyester resin. J Appl Polym Sci 99:1055–1068

Miller SA (2013) Sustainable polymers: opportunities for the next decade. ACS Macro Lett 2: 551–554

Mohnen D (2008) Pectin structure and biosynthesis. Curr Opin Plant Biol 11:266–277

Molina SMG, Pelissari FA, Vitorello CBM (2001) Screening and genetic improvement of pectinolytic fungi for degumming of textile fibers. Braz J Microbiol 32:320–326

Mounika M, Ramaniah K, Prasad AVR, Rao KM, Reddy KHC (2012) Thermal conductivity characterization of bamboo fiber reinforced polyester composite. J Mater Environ Sci 3 (6):1109–1116

Munshii TK, Cathoo B (2008) Bacterial population structure of the jute retting environment. Microbial Ecol 56:270–282

Mwaikambo LY, Ansell MP (2003) Hemp fibre reinforced cashew nut shell liquid composites. Compos Sci Technol 63:1297–1305

Nuplex (2014) www.nuplex.com/composites/processes

Oksman K, Skrifvars M, Selin JF (2003) Natural fibres as reinforcement in polylactic acid (PLA) composites. Compos Sci Technol 63:1317–1324

Oladele IO, Agbabiaka OG (2015) Investigating the influence of mercerization treatment of sisal fiber on the mechanical properties of reinforced polypropylene composites and modeling of the properties. Fiber Polym 16:650–656

Oladele IO, Daramola OO, Fasooto S (2014) Effect of chemical treatment on the mechanical properties of sisal fibre reinforced polyester composites. Leonardo Electron J Pract Technol 24:1–12

Ouajai S, Shanks RA (2005) Morphology and structure of fiber after bioscouring. Macromol Biosci 5:124–134

Pallesen BE (1996) The quality of combine-harvested fibre flax for industrials purpuses depends on the degree of retting. Ind Crops Prod 5:65–78
Parikh DV, Calamari TA, Sawhney APS, Blanchard EJ, Screen FJ, Warnock M, Muller DH, Stryjewski DD (2002a) Improved chemical retting of kenaf fibers. Text Res J 72:618–624
Parikh DV, Calamari TA, Swahney APS, Blanchard EJ, Screen FJ (2002b) Thermoformable automotive composites containing kenaf and other cellulosic fibers. Text Res J 72:668–672
Pickering KL, Farrell RL, Lay MC (2007a) Interfacial modification of hemp fiber reinforced composites using fungal and alkali treatment. J Biobased Mater Bio 1:109–117
Pickering KL, Beckermann GW, Alam SN, Foreman NJ (2007b) Optimising industrial hemp fibre for composites. Compos Part A-Appl S 38:461–468
Pillin I, Kervoelen A, Bourmaud A, Goimard J, Montrelay N, Baley C (2011) Could oleaginous flax fibers be used as reinforcement for polymers? Ind Crops Prod 34:1556–1563
Prasad AVR, Rao KM (2011) Mechanical properties of natural fibre reinforced polyester composites: Jowar, sisal and bamboo. Mater Design 32:4658–4663
Prasad AVR, Rao KM, Gupta AVSSKS, Reddy BV (2011) A Study on flexural properties of wildcane grass fiber-reinforced polyester composites. J Mater Sci 46:2627–2634
Prasad GLSR, Kumar MVHS, Rajesh G (2014) Effect of fibre loading and successive alkali treatments on tensile properties of short jute fibre reinforced polypropylene composites. Int J Eng Sci 3:30–34
Praveen KM, Thomas S, Grohens Y, Mozetič M, Junkar I, Primc G, Gorjanc M (2016) Investigations of plasma induced effects on the surface properties of lignocellulosic natural coir fibres. Appl Surf Sci 368:146–156
Ramesh D, Ayre BG, Webber CL, D'Souza NA (2015) Dynamic mechanical analysis, surface chemistry and morphology of alkali and enzymatic retted kenaf fibers. Text Res J 85:2059–2070
Ramnath BV, Kokan SJ, Raja RN, Sathyanarayanan R, Elanchezhian C, Prasad AR, Manickavasagam VM (2013) Evaluation of mechanical properties of abaca–jute–glass fibre reinforced epoxy composite. Mater Design 51:357–366
Rao KMM, Prasad AVR, Babu MNVR, Rao KM, Gupta AVSSKS (2007) Tensile properties of elephant grass fiber reinforced polyester composites. J Mater Sci 42:3266–3272
Rao KMM, Rao KM, Prasad AVR (2010) Fabrication and testing of natural fibre composites: Vakka, sisal, bamboo and banana. Mater Design 31:508–513
Retting the flax (2014) https://blumenkinderheirlooms.wordpress.com/2014/08/30/retting-the-flax/
Ribeiro A, Pochart P, Day A, Mennuni S, Bono P, Baret JL, Spadoni JL, Mangin I (2015) Microbial diversity observed during hemp retting. Appl Microbiol Biothechnol 99:4471–4484
Ridley BL, O'Neill MA, Mohnen D (2001) Pectins: structure, biosynthesis, and oligogalacturonide-related signaling. Phytochemistry 57:929–967
Roe PJ, Ansell MP (1985) Jute-reinforced polyester composites. J Mater Sci 20:4015–4020
Sakai T, Sakamoto T, Hallaert J, Vandamme EJ (1993) Pectin, pectinaseand protopectinase: production, properties and applications. Adv Appl Microbiol 39:231–294
Salentijn EMJ, Zhang Q, Amaducci S, Yang M, Trindad LM (2015) New developments in fiber hemp (*Cannabis sativa* L.) breeding. Ind Crops Prod 68:32–41
Sature P, Mache A (2015) Mechanical characterization and water absorption studies on jute/hemp reinforced hybrid composites. Am J Mater Sci 5:133–139
Satyanarayana KG, Arizaga GGC, Wypych F (2009) Biodegradable composites based on lignocellulosic fibers-an overview. Progr Polym Sci 34:982–1021
Scalici T, Fiore V, Valenza A (2016) Effect of plasma treatment on the properties of *Arundo donax* L. leaf fibres and its bio-based epoxy composites: a preliminary study. Composites Part B-Eng 94:167–175
Sharma HSS (1986) An alternative method of flax retting during dry weather. Ann Appl Biol 109:605–611
Sharma HSS, Faughey G, Lyons G (1999) Comparison of physical, chemical, and thermal characteristics of water-, dew-, and enzyme-retted flax fibers. J Appl Polym Sci 74:139–143

Sheng W, Gao J, Jin Z, Dai H, Zheng L, Wang B (2014) Effect of steam explosion on degumming efficiency and physicochemical characteristics of banana fiber. J Appl Polym Sci 131:40598–44606

Shi J, Shi SQ, Barnes HM, Pittman CU Jr (2011a) A chemical process for preparing cellulosic fibers hierarchically from kenaf bast. BioRes 6:879–890

Shi J, Shi SQ, Barnes HM, Horstemeyer MF, Wang G (2011b) Kenaf bast fibers—part II: inorganic nanoparticle impregnation for polymer composites. Int J Polym Sci 2011:1–7

Silva GG, De Souza DA, Machado JC, Hourston DJ (2000) Mechanical and thermal characterization of native Brazilian coir fiber. J Appl Polym Sci 76:1197–1206

Singh SA, Ramakrishna M, Rao AGA (1999) Optimization of downstream processing parameters for the recovery of pectinase from the fermented broth of *Aspergillus Carbonarious*. Process Biochem 35:411–417

Sisti L, Totaro G, Vannini M, Fabbri P, Kalia S, Zatta A, Celli A (2016) Evaluation of the retting process as a pre-treatment of vegetable fibers for the preparation of high-performance polymer biocomposites. Ind Crops Prod 81:56–65

Song KH, Kim IS (2013) Effects of plasticizer on the mechanical properties of kenaf/starch bio-composites. Fiber Polym 14:2135–2140

Stamboulis A, Baillie CA, Garkhail SK, Van Melick HGH, Peijs T (2000) Environmental durability of flax fibres and their composites based on polypropylene matrix. Appl Compos Mater 7:273–294

Sun Y, Cheng J (2002) Hydrolysis of lignocellulosic materials for ethanol production: a review. Bioresource Technol 83:1–11

Tahir P, Ahmed AB, Saifulazry SOA, Ahmed Z (2011) Retting process of some bast plant fibres and its effect on fibre quality: a review. BioResources 6:5260–5281

Tamburini E, Gordillo León A, Perito B, Di Candilo M, Mastromei G (2004) Exploitation of bacterial pectinolytic strains for improvement of hemp water retting. Euphytica 140:47–54

Tamburini E, Leon AG, Perito B, Mastromei G (2003) Characterization of pectinolytic strains involved in the water retting process. Environ Microbiol 5:730–736

Thomas S, Paul SA, Pothan LA, Deepa B (2011) Natural fibres: structure, properties and applications. In: Kalia S, Kaith BS, Kaur I (eds) Cellulose fibers: bio- and nano-polymer composites. Springer, Berlin

Thomsen AB, Thygesen A, Bohn V, Vad Nielsen K, Pallesen B, Jørgensen MS (2006) Effect of chemical-physical pretreatment processes on hemp fibres for reinforcement of composites and for textiles. Ind Crops Prod 24:113–118

Umoru PE, Boryo DEA, Aliyu AO, Adeyemi OO (2014) Processing and evaluation of chemically treated kenaf bast (*Hibiscus cannabinus*). Int J Sci Technol Res 3:1–6

Van Dam JEG, Bos HL (2004) The environmental impact of fibre crops in industrial applications. Hintergrundpapier zu

Van de Velde K, Baetens E (2001) Thermal and mechanical properties of flax fibres as potential composite reinforcement. Macromol Mater Eng 286:342–349

Van de Weyenberg I, Ivens J, De Coster A, Kino B, Baetens E, Verpoest I (2003) Influence of processing and chemical treatment of flax fibres on their composites. Compos Sci Technol 63:1241–1246

Van de Weyenberg I, Truong TC, Vangrimde B, Verpoest I (2006) Improving the properties of UD flax fibre reinforced composites by applying an alkaline fibre treatment. Composi Part A-Appl S 37:1368–1376

Van der Werf HMG, Turunen L (2008) The environmental impacts of the production of hemp and flax textile yarns. Ind Crops Prod 27:1–10

Van Sumere CF (1992) Retting of flax with special reference to enzyme retting. In: Sharma HSS, Van Sumere CF (eds) The biology and processing of flax. M. Publications, Belfast

Vignon MR, Dupeyre D, Garcia-Jaldon C (1996) Morphological characterization of steam exploded hemp fibers and their utilization in polypropylene-based composites. Bioresource Technol 58:203–215

Wang YP, Wang G, Cheng HT (2010) Structure of bamboo fiber for textiles. Text Res J 80: 334–343

Weald & Downland Open Air Museum (2016) www.wealddown.co.uk

Xia C, Zhang S, Shi SQ, Cai L, Huang J (2016) Property enhancement of kenaf fiber reinforced composites by in situ aluminum hydroxide impregnation. Ind Crops Prod 79:131–136

Xiao Z, Wang S, Bergeron H, Zhang J, Lau PCK (2008) A flax retting endopolygalacturolase-encoding gene from *Rhizopus oryzae*. Antonie Van Leeuwenhoek 94:563–571

Xu X, Wang Y, Zhang X, Jing G, Yu D, Wang S (2006) Effects on surface properties of natural bamboo fibers treated with atmospheric pressure argon plasma. Surf Interface Anal 38: 1211–1217

Yan L, Chouw N, Jayaraman K (2014) Flax fibre and its composites—a review. Compos Part B-Eng 56:296–317

Yang B, Nar M, Visi DK, Allen M, Ayre B, Webb er CL III, Lu H, D'Souza NA (2014) Effects of chemical versus enzymatic processing of kenaf fibers on poly(hydroxybutyrate-co-valerate)/poly(butylene adipate-co-terephthalate) composite properties. Compos Part B-Eng 56:926–933

Zhang J, Henriksson G, Johanssson G (2000) Polygalacturonase is the key component in enzymatic retting of flax. J Biotechnol 81:85–89

Zhang L, Li D, Wang L, Wang T, Zhang L, Chen KD, Mao Z (2008a) Effect of steam explosion on biodegradation of lignin in wheat straw. Bioresource Technol 99:8512–8515

Zhang LL, Zhu RY, Chen JM, Feng XX (2008b) Seawater-retting treatment of hemp and characterization of bacterial strains involved in the retting process. Process Biochem 43: 1195–1201

Zheng L, Du Y, Zhang J (2001) Degumming of Ramie fibers by alkalophilic bacteria and their polysaccharide-degrading enzymes. Bioresource Technol 78:89–94

Pretreatments of Natural Fibers for Polymer Composite Materials

A. Orue, A. Eceiza and A. Arbelaiz

Abstract The use of lignocellulosic fibers as reinforcement in polymer composites is attracting interest due to their properties such as mechanical properties and environmental benefits. Nevertheless, the hydrophilic character of lignocellulosic fibers reduces the compatibility with the hydrophobic matrices resulting in composites with poor mechanical properties. Therefore, in order to reduce the hydrophilic character of fiber and improve the fiber/matrix adhesion, is necessary to modify the fiber surface morphology. In this chapter, different lignocellulosic fiber treatments and the effect of these treatments on fiber properties as well as on composite mechanical performance were discussed. Even though chemical treatments are the most widely used, physical and biological treatments are environmentally friendly and promising alternatives.

Keywords Lignocellulosic fiber · Treatments · Surface modification
Adhesion · Properties

Abbreviations

AC	Alternating current
CDT	Corona discharge treatment
DBD	Dielectric-barrier discharge
FTIR	Fourier Infrared Spectroscopy
HDPE	High-density polyethylene
IFSS	Interfacial shear strength
ILSS	Interlaminar shear strength
MA	Maleic anhydride
MAPE	Maleic anhydride grafted polyethylene
MAPP	Maleic anhydride grafted polypropylene
MDI	4,4′-diphenylmethane diisocyanate

A. Orue · A. Eceiza · A. Arbelaiz (✉)
Materials + Technologies Group, Chemical & Environmental Engineering Department, University of the Basque Country, Plaza Europa 1, 20018 Donostia-San Sebastián, Spain
e-mail: aitor.arbelaiz@ehu.eus

S. Kalia (ed.), *Lignocellulosic Composite Materials*, Springer Series on Polymer and Composite Materials, https://doi.org/10.1007/978-3-319-68696-7_3

PBS	Polybutylene succinate
PE	Polyethylene
PLA	Poly(lactic acid)
PP	Polypropylene
SEM	Scanning electron microscopy
SFE	Steam flash explosion
SFE-AT	Steam flash explosion with alkali solution
TAPPI	Technical Association of Pulp and Paper Industry
TCB	1,2,4-trichlorobenzene
TGA	Thermogravimetric analysis
VTMO	Vinyltrimethoxysilane
XPS	X-ray photoelectron spectroscopy
XRD	X-ray diffraction

Contents

1 Introduction

Synthetic fiber reinforced plastics have proven to meet the structural and durability requirements of components for various applications. Nevertheless, the growing global energy crisis and ecological risks have focused more research interests in natural fiber reinforced polymer composites due to their potential (Praveen et al. 2016). Lignocellulosic fibers can be used as reinforcement in polymeric matrix composites due to their characteristics. Among other properties, lignocellulosic fibers have acceptable specific properties, low density, biodegradability, low cost, and recyclability (Yu et al. 2014; Lu et al. 2014; Arbelaiz et al. 2005c).

The mechanical properties of polymeric composites are governed by the individual properties of the components and by the interface formed between the matrix and fiber. A critical factor in reinforced polymeric composites is the strength of the bond between the fiber and polymer matrix. The load is transmitted through the fiber-matrix interface and consequently, the optimal mechanical performance of fiber reinforced composite depends, among other variables, on the interfacial bond between fiber and polymer matrix (Praveen et al. 2016).

2 Fiber-Matrix Adhesion

The adhesion between fiber and polymeric matrix is sustained by the interfacial forces. These forces can be broadly divided into primary forces and secondary forces. Primary forces, also known as short-range forces, arise from chemical bonding. Chemical bonding compromises of interlinking between molecules of the substrate by covalent, ionic or metallic bonds. The secondary or intermolecular forces, such as Van der Waals forces, are caused from the physical attraction between two substrates (Kim and Pal 2010). There is no single theory of adhesion which can satisfactorily explain the mechanism of adhesion between lignocellulosic fiber and the polymeric matrix. Many adhesion mechanisms, such as mechanical interlocking, adsorption, and diffusion theory, and boundary layer theory are developed in order to explain lignocellulosic fiber/polymeric matrix adhesion (Kim and Pal 2010). In the literature, fibers are modified using different treatments in order to improve fiber/matrix adhesion. A brief discussion of different lignocellulosic fiber treatments and the effect of these treatments on fiber properties as well as on composite mechanical performance is given in this chapter.

3 Lignocellulosic Fiber

Natural fibers can be classified according to their source; plant, animal, or mineral (Saba et al. 2014). Plant fibers or lignocellulosic fibers are mainly composed of cellulose, hemicelluloses, and lignins. Table 1 shows the chemical composition of sisal, flax, and hemp lignocellulosic fibers determined by using the standards methods of Technical Association of Pulp and Paper Industry (TAPPI).

Cellulose, the main component of lignocellulosic fibers, is the most abundant natural organic compound which performs structure-forming functions in plant-cell wall (Daintith 2000; Allaby 1998). Cellulose is a semicrystalline polysaccharide consisting of a linear chain of hundreds to thousands of β-(1,4)-glycosidic bonds

Table 1 Chemical composition of sisal, hemp, and flax lignocellulosic fibers (Mondragon et al. 2014)

Fiber	Cellulose (%)	Lignins (%)	Hemicelluloses (%)	Others (%)
Sisal	60–65	7–9	10–15	11–23
Hemp	67–74	3–5	13–18	3–17
Flax	63–70	2–3	16–22	5–19

linked D-glucopyranose with the presence of a large amount of hydroxyl groups (Zhou et al. 2016). The repeating unit of cellulose is shown in Fig. 1.

Hemicelluloses have a heterogeneous monosaccharide composition mainly based on pentoses (D-xylose and L-arabinose), hexoses (D-glucose, D-galactose, D-mannose, L-rhamnose) and uronic acid units (D-galacturonic acid and D-glucuronic acid) (Egües 2013). Hemicelluloses are often referred to as matrix components and may be found in the middle lamellae that bind cell walls of fibers, in the primary wall regions and in the thicker, cellulose-rich, secondary layer of the plant-cell wall (Focher 1992). The heterogeneity of hemicelluloses extends to branching polymers, thus giving new dimensions and complexities within the cell wall (Focher 1992). However, the degree of polymerization of nature cellulose is 10–100 times higher than that of hemicellulose ones (Bledzki and Gassan 1999).

Lignins are phenolic complex, cross-linked polymer, comprising both aliphatic and aromatic constituents, that is found in plant-cell walls. Its function appears to be cement together and anchor cellulose fibers, to stiffen the cell wall and provide protection against microbial attacks, external agents, moisture, and weathering (Egües 2013). Structures of the three major precursors of lignin are shown in Fig. 2.

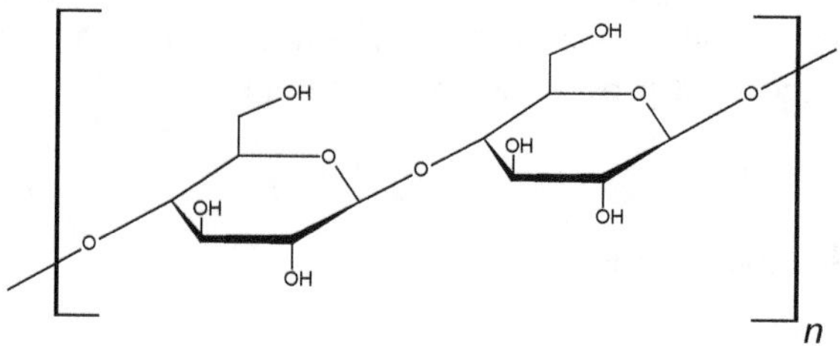

Fig. 1 The repeating unit of cellulose

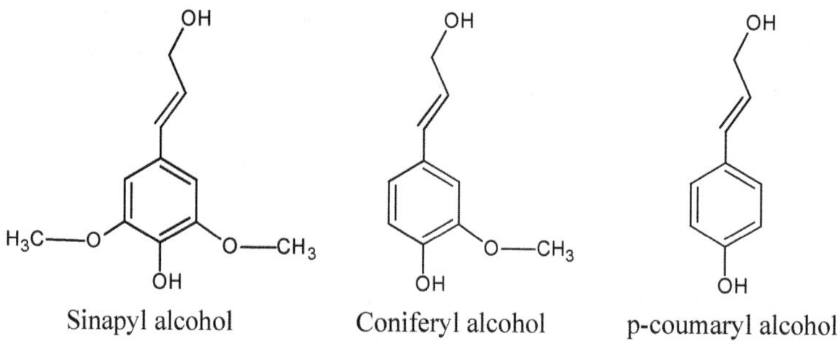

Fig. 2 Structures of the three major precursors of lignins

4 Surface Modification of Lignocellulosic Fiber

The use of natural fibers as reinforcement in polymer composites is attracting much interest due to its potential, however, the hydrophilic character of the fibers reduces the compatibility with the hydrophobic matrices resulting in composites with poor mechanical properties (Lu et al. 2014; Arbelaiz et al. 2005c). Most of the research related to lignocellulosic fibers and polyolefin matrices showed that the adhesion between the lignocellulosic fiber and polymers such as polypropylene (PP) or polyethylene (PE) was poor (Arbelaiz et al. 2005c; Merkel et al. 2014; Ranganathan et al. 2016). Other polymers such as poly(lactic acid) (PLA) presents moderate polarity induced by the presence of ester bonds (Orue et al. 2015; Jiang et al. 2012) and could form hydrogen bonds with the hydroxyl groups of lignocellulosic fibers. It was observed that the surface tension value of PLA is close to sisal fibers resulting in good fiber wettability with PLA matrix (Orue et al. 2015). Even though the wettability of lignocellulosic fiber with PLA polymer is good, the adhesion between them was not good enough to improve PLA strength after fiber addition (Orue et al. 2016b).

The adhesion between lignocellulosic fiber and polymer matrix can be improved (i) modifying the surface of fibers, (ii) modifying the polymer or (iii) modifying both of them at the same time. Surface modification is essential to reduce the hydrophilic character of the lignocellulosic fibers, improve the wettability between the fiber and matrix and to improve fiber/polymer matrix adhesion (Bledzki and Gassan 1999; Bledzki et al. 1996). In the literature, lignocellulosic fibers are subjected to several surface treatments such as chemical treatments (Arbelaiz et al. 2005c; Orue et al. 2015; Zou et al. 2012; Goriparthi et al. 2012; Chen et al. 2012; Sun et al. 2016; Datta and Kopczynska 2015; Paul et al. 2010b), physical treatments (Hou et al. 2014; Bozaci et al. 2013; Li et al. 2014), and biological treatments (Pickering et al. 2007; George et al. 2014). Chemical treatments provide the means of permanently modifying the nature of fiber cell walls by grafting polymer onto the fibers, crosslinking of the fiber cell walls, or by using coupling agents (Zhou et al. 2016; Xie et al. 2010). Physical treatments change the structural and surface properties of the lignocellulosic fiber by introducing surface crosslinking, modifying the surface energy and/or generating reactive free radicals and groups, and thereby influence the mechanical bonding to the matrix (Zhou et al. 2016). Biological treatments employ microorganisms, mainly white and soft-rot fungi, actinomycetes, and enzymes which degrade lignin through the action of lignin-degrading enzymes, such as peroxidases and laccases (Saritha et al. 2012). The main lignocellulosic surface treatments are summarized in Fig. 3.

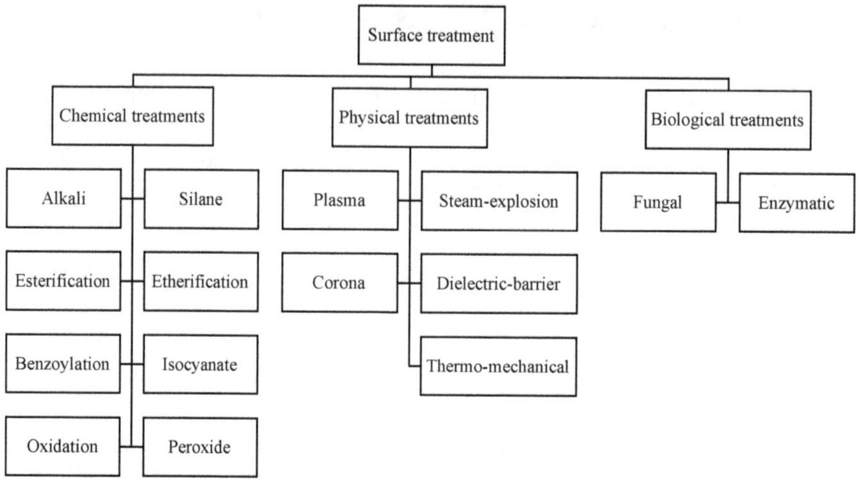

Fig. 3 Main lignocellulosic surface treatments

4.1 Chemical Treatments

4.1.1 Alkali Treatment

Alkali treatment is a simple and efficient method to improve lignocellulosic fiber/ polymer matrix adhesion (Orue et al. 2016b; Raj et al. 2011). Usually, in alkali treatments, lignocellulosic fibers were treated with NaOH or KOH solutions for a specified time to remove noncellulosic compounds, such as lignin, hemicelluloses and other organics (Orue et al. 2015, 2016b). The removal of noncellulosic compounds could create voids in the fiber structure (Orue et al. 2016b), and alkali treatment could depolymerize the native cellulose structure and expose cellulose crystallites (Li et al. 2007; Mohanty et al. 2001). Figure 4 presents the schematic view of the cellulose fiber structure, before and after an alkali treatment. The effect of alkali treatment on the properties of lignocellulosic fibers depends on the alkali treatment conditions, such as NaOH concentration and temperature (Gassan and Bledzki 1999). For example, severe alkali conditions can damage the fiber structure decreasing the properties of lignocellulosic fibers (Li et al. 2007; Wang et al. 2007). Therefore, all these conditions are important factors that must be optimized for each fiber type.

In previous works (Orue et al. 2015, 2016b) sisal fibers were pretreated with 2 wt% NaOH solution at 20 °C for overnight and after with 7.5 wt% NaOH solution under reflux at 100 °C for 90 min. The density values of alkali treated sisal fibers were higher than untreated fiber ones suggesting that the alkali treatment removed noncellulosic compounds increasing the cellulose fraction in sisal fiber. Although cellulose is the compound that gives the strength to the lignocellulosic fibers, Fig. 5 showed that the tensile strength and modulus values of alkali treated

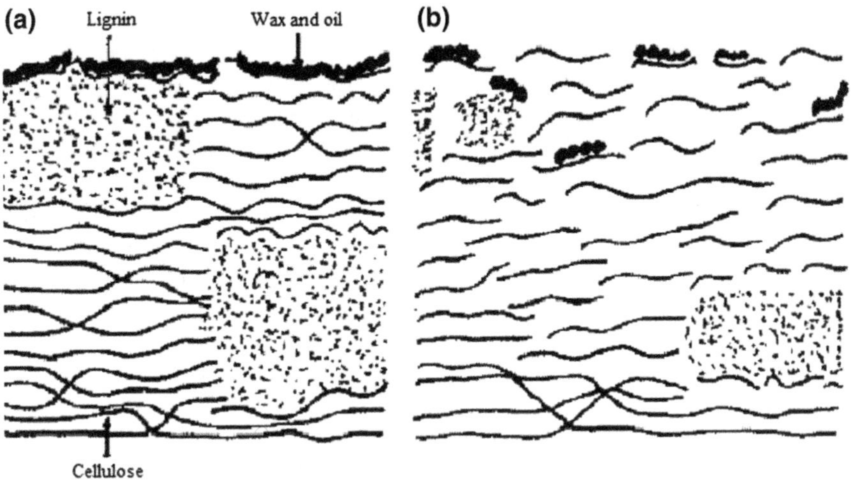

Fig. 4 Typical structure of (**a**) untreated and (**b**) alkali treated cellulose fiber. Reprinted from (Mwaikambo and Ansell 2002) with permission from John Wiley and Sons

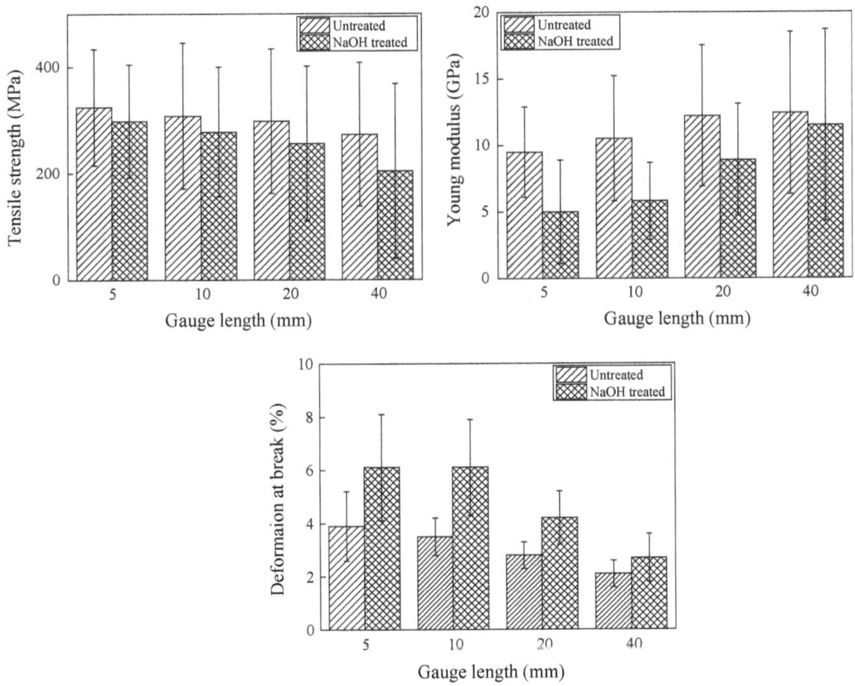

Fig. 5 Tensile strength, Young modulus, and deformation at break values of untreated and alkali treated sisal fiber at different gauge lengths

fibers were lower than untreated ones. A possible explanation could be that the removal of noncellulosic compounds could create voids in the sisal fiber structure resulting in lower tensile strength and modulus values (Orue et al. 2016b).

The removal of noncellulosic compounds was corroborated by Fourier Infrared Spectroscopy (FTIR) analysis in Fig. 6. After alkali treatment, the chemical composition of sisal fibers changed since the prominent bands of the raw fiber around 1740 and 1250 cm^{-1} were disappeared almost completely indicating that alkali treatment removed noncellulosic compounds. The band around 1740 cm^{-1} was ascribed to the acetyl and ester groups of hemicelluloses and aromatic components of lignin (Mondragon et al. 2014; De Rosa et al. 2010; Haiping et al. 2007) while the band at 1250 cm^{-1} was related to the C–O stretching vibration of hemicelluloses component (De Rosa et al. 2011) or aryl-alkyl ether compounds present in lignin (Mondragon et al. 2014). Moreover, thermogravimetric analysis (TGA) results indicated that the removal of noncellulosic compounds resulting in a better thermal stability of alkali treated fibers (Orue et al. 2015).

The optical microscopy images of untreated and alkali treated sisal fibers are shown in Fig. 7. It was observed that untreated sisal fibers showed a smooth surface, whereas, after alkali treatment, sisal fibers surface roughness was considerably increased and created fibers with smaller diameters probably due to the removal of noncellulosic compounds. Moreover, after treating sisal fibers with NaOH solution, slightly higher contact angles with a drop of water were obtained. Untreated sisal fibers showed a contact value of 68.2°, whereas alkali treated sisal fibers showed a value 75.4° suggesting that the alkali treated fibers were more hydrophobic than untreated ones thus improving the wettability with hydrophobic polymer matrices (Orue et al. 2015, 2016a).

In a previous work (Orue et al. 2016b), PLA/sisal fibers composites were compounded varying the loading of fibers from 20 to 40 wt%. Even though the tensile properties of alkali treated sisal fibers was lower than untreated ones, Table 2 showed that composites based on the alkali treated sisal fibers showed

Fig. 6 FTIR spectra of (**a**) untreated and (**b**) alkali treated sisal fibers

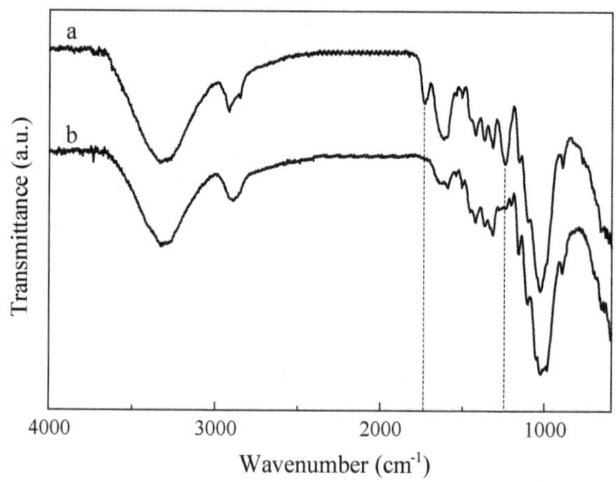

Fig. 7 Optical images of (**a**) untreated and (**b**) alkali treated sisal fibers. Reprinted from (Orue et al. 2015), Copyright (2015) with permission from Elsevier

Table 2 Tensile strength, Young modulus, and deformation at break values of PLA matrix and composites based on the untreated and alkali treated fiber

System	Fiber loading (%)	Strength (MPa)	Young modulus (GPa)	Elongation at break (%)
PLA	0	66.1 ± 1.7	3.9 ± 0.5	2.4 ± 0.3
Untreated	20	53.8 ± 0.9	5.5 ± 0.9	1.8 ± 0.4
	30	47.3 ± 1.7	5.9 ± 0.7	1.1 ± 0.3
	40	39.3 ± 3.4	6.4 ± 1.0	0.8 ± 0.1
NaOH	20	66.1 ± 3.2	6.5 ± 0.4	1.6 ± 0.3
	30	74.8 ± 3.5	7.2 ± 0.7	1.5± 0.1
	40	81.5 ± 1.4	8.1 ± 0.8	1.4 ± 0.1

higher tensile strength values than neat PLA matrix. This fact suggested that alkali treatment improved fiber/PLA matrix adhesion. One possible reason could be that in the injection process, the viscosity of melted polymer decreased considerably filling fiber voids and improving the mechanical interlocking adhesion between sisal fiber and PLA polymer matrix.

Scanning electron microscopy (SEM) micrographs of the fractured surface of composites based on the untreated sisal fibers showed pulled-out fibers and holes suggesting a poor interfacial adhesion between fibers and PLA. However, in SEM micrographs of composite based on alkali treated fibers hardly pulled-out fibers can be observed and it was difficult to distinguish fibers from the matrix suggesting that sisal fibers were coated with PLA polymer matrix and consequently fiber/matrix adhesion was improved.

In previous works (Arbelaiz et al. 2005a, b) flax fibers were soaked in a 20 wt% aqueous solution of sodium hydroxide for 1 h at room temperature. After that, flax fibers were washed for several times in distilled water and dried in an oven. It was observed that after alkalization, the surface roughness of the flax fibers was considerably increased improving the flax fiber/polypropylene interfacial shear strength values. Furthermore, surface energy measurements reported that after alkali treatment the polar component of surface energy of flax fibers decreased suggesting that alkali fibers became more hydrophobic. As a consequence, alkali treatment could improve the wettability with hydrophobic polymer matrices. In order to study the influence of fiber modification on the mechanical properties of composites, flax/PP composites were compounded with 30 wt% fiber content. It was observed that the addition of untreated flax fiber bundle decreased the tensile strength of neat PP, indicating lack of stress transfer from the matrix to the fiber. Even though, alkali treatment improved significantly the wettability and the interfacial shear strength determined by pull-out test, tensile, and flexural strength values were not changed after the addition of alkali treated flax fiber. It must be mentioned that the methods based on the single fiber tests to determine fiber/polymer adhesion could deviate significantly from the real composite performance.

Fernandes et al. (2013) treated sisal fibers in NaOH solution (5% w/v) for 2 h at room temperature and they observed by FTIR analysis and TGA the removal of noncellulosic compounds. In addition, X-ray diffraction (XRD) results showed that after removal lignin, hemicelluloses and other surface impurities the crystallinity index of alkali treated sisal fibers increased and consequently a small increase in density values was observed. They also observed a decrease of mechanical properties of alkali treated sisal fibers. They reinforced high-density polyethylene (HDPE)/cork blends with 10 wt% of untreated and alkali treated sisal fibers and they observed that composites based on the alkali treated fibers showed higher tensile strength and Young modulus values than untreated ones. They suggested that this fact could be related to the increase of fiber surface roughness after alkali treatment improving fiber/HDPE adhesion.

4.1.2 Silane Treatment

Silanes are considered versatile coupling agents that could improve the interface between lignocellulosic fiber and polymer matrix (Orue et al. 2016b; Le Moigne et al. 2014). Silane is a multifunctional molecule which can form a chemical link with the fiber surface through a siloxane bridge. A typical general structure of silane coupling agent is shown in Fig. 8.

During the treatment process of the fiber, silane forms silanols in the presence of water and by a condensation process, one end of silanol could react with the cellulose hydroxyl groups. Finally, when silane modified fibers are blended with the polymeric matrix, the other functional groups of silane react with the matrix functional group (Kabir et al. 2012; Zhang et al. 2005). Therefore, it is important to choose a silane agent which is suitable to react with functional groups of fiber and the polymeric matrix. On the contrary, the stress would not transfer efficiently from

$$R'\text{-}Si-(O-R)_3$$

Fig. 8 Typical general structure of silane coupling agent

$$R'{-}Si-(O-R)_3 + 3H_2O \rightarrow R'{-}Si-(OH)_3 + 3R-OH \qquad \text{Hydrolysis step}$$

$$R'{-}Si-(OH)_3 + Fiber-OH \rightarrow R'{-}Si-(OH)_2-O-Fiber + H_2O \qquad \text{Condensation step}$$

Fig. 9 Proposed reaction scheme for silane and lignocellulosic fiber

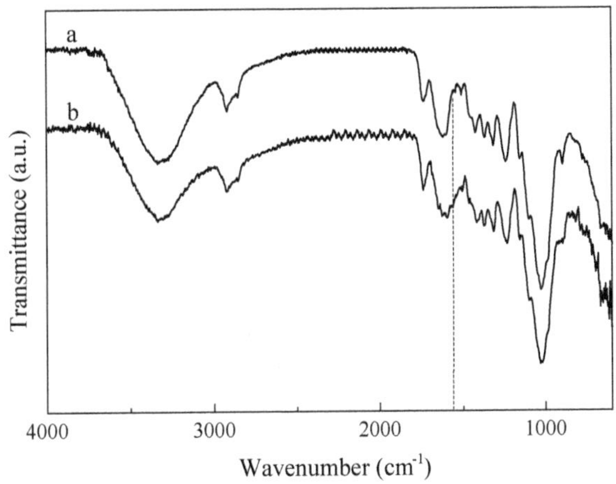

Fig. 10 FTIR spectra of (**a**) untreated and (**b**) silane treated sisal fibers

polymer matrix to lignocellulosic fibers and consequently, the mechanical properties of composites would not improve. The possible reaction scheme between silane agent and lignocellulosic fiber is shown in Fig. 9.

Silane chemical agent concentration, the medium used (water, organic solvent, or mixtures), the soaking time, the drying temperature, and time are all important factors to carry out silane treatment (Arbelaiz et al. 2016).

In previous works (Orue et al. 2015, 2016b) the effect of silane treatment on sisal fiber properties as well as on composite mechanical performance was studied. Sisal fibers were soaked in silane aqueous solution (2% w/v) under continuous stirring for 3 h. The pH of the solution was adjusted to 3–4 with glacial acetic acid and silane/fiber weight ratio used was 2:1 (w/w). 3-(2 aminoethylamino) propyltrimethoxysilane chemical agent was used as silane chemical agent and it was thought that the silane concentration employed could produce a multilayer polysiloxane network on the sisal fibers bundles (Abdelmouleh et al. 2002).

FTIR spectra of untreated and silane treated sisal fiber are shown in Fig. 10. It can be confirmed that silane treated fibers showed a new absorption band around 1560 cm^{-1} related to NH_2 bending vibration of organosilane agent. Moreover, it was observed in the thermogravimetric analysis that fibers treated with silane chemical agent showed a higher percentage of char which could be related to the presence of grafted silane (Rachini et al. 2012).

After treating sisal fibers with silane chemical agent, the contact angle with a drop of water increased from 68.2° to 88.2° suggesting that the silane treated fibers were more hydrophobic than untreated ones (Orue et al. 2015, 2016a). Figure 11 showed that after silane treatment, the tensile strength values of sisal fibers decreased slightly. This reduction could be related to the acid medium used for silane treatment that might decrease cellulose molecular weight resulting in lower mechanical properties (Orue et al. 2016b; Klemm et al. 1998).

As observed in Table 3, tensile strength values of composites based on silane treated sisal fibers increased slightly as fiber bundle content increased indicating that the adhesion between the polymer and fiber was improved (Orue et al. 2016b).

The improvement of adhesion between silane treated sisal fiber and PLA matrix was corroborated by pull-out test, being the interfacial shear strength value of silane treated fibers slightly higher than untreated fibers ones. This improvement was attributed to the chemical bond between PLA matrix and amino groups of silane agent linked to the fiber surface (Orue et al. 2015). In SEM, micrographs of composites based on the silane treated fibers, sisal fibers cannot be clearly distinguished suggesting that the fiber bundle adhesion was improved. However, a few amount of pulled-out fibers and holes were observed which meant that the adhesion was not very strong (Orue et al. 2016b).

Zou et al. (2012) studied the effect of silane treatment on the properties of short sisal fiber/PLA biocomposites. For this purpose, sisal fibers were soaked in 5% w/v

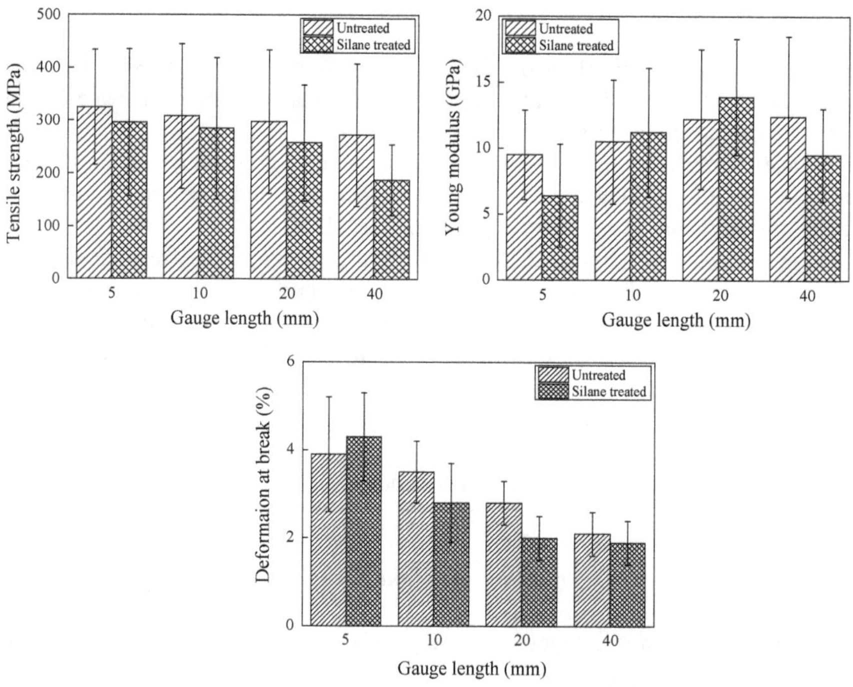

Fig. 11 Tensile strength, Young modulus, and deformation at break values of untreated and silane treated sisal fiber at different gauge lengths

Table 3 Tensile strength, Young modulus, and deformation at break values of PLA matrix and composites based on the untreated and silane treated fiber

System	Fiber loading (%)	Strength (MPa)	Young modulus (GPa)	Elongation at break (%)
PLA	0	66.1 ± 1.7	3.9 ± 0.5	2.4 ± 0.3
Untreated	20	53.8 ± 0.9	5.5 ± 0.9	1.8 ± 0.4
	30	47.3 ± 1.7	5.9 ± 0.7	1.1 ± 0.3
	40	39.3 ± 3.4	6.4 ± 1.0	0.8 ± 0.1
Silane	20	57.6 ± 2.9	4.4 ± 0.7	1.9 ± 0.3
	30	59.9 ± 1.5	6.4 ± 0.7	1.5 ± 0.3
	40	61.8 ± 3.1	7.5 ± 0.6	1.0 ± 0.2

γ-amine propyl triethoxysilane ethanol solution at room temperature for 2 h. The pH of the solution was adjusted to 4–5 with acetic acid and fibers were washed with ethanol and dried *under vacuum* at 70 °C before the preparation of the biocomposites. Composites with 10 wt% untreated and treated sisal fibers were compounded by compression molding technique and they observed that silane treatment improved the mechanical properties. They suggested that silane with ethoxy groups can be hydrolyzed in water and produced silanols which reacted with the hydroxyl groups of sisal fibers forming covalent bonds on the fiber surface. Furthermore, amine groups from the silane may further react with terminal groups of PLA matrix. The bridge formed at the interface led to the improvement of sisal fiber/PLA matrix adhesion.

In previous works, (Arbelaiz et al. 2005a, b) flax fibers were treated with vinyltrimethoxysilane (VTMO) chemical agent. It was observed that the polar component of surface energy was decreased considerably indicating that silane treated flax fibers could be better wetted out by nonpolar polymer matrices. However, when silane treated flax fiber/PP composites were compounded, the tensile and flexural strength were not improved suggesting that stress transfer from the matrix to the fiber occurred with a disappointing lack of efficiency. This fact could indicate that the vinyl group was not able to react with PP chains and consequently fiber/polymer matrix adhesion was not improved.

4.1.3 Esterification

Esterification with Acetyl Groups

Acetylation treatment of lignocellulosic fibers is an esterification reaction and it was reported that the treatment improved the lignocellulosic fiber/matrix adhesion (Li et al. 2007; Kabir et al. 2012; Kumar et al. 2011). Usually, the procedure included an alkali treatment initially, followed by acetylation treatment (Li et al. 2007; Kumar et al. 2011; Efanov and Ovchinnikov 2012). Among other reactive, acetic anhydride and acetic acid can be used (Paul et al. 2010b; De Rosa et al. 2011; Hill et al. 1998). The hydrolysis of acetic anhydride to acetic acid is a moderately highly exothermic fast reaction (Hirota et al. 2010) and the reaction with acetic anhydride involves the generation of acetic acid as by-product which must be removed from the lignocellulosic fibers (Hill et al. 1998). The hydroxyl groups of the lignocellulosic fibers react with the acetyl reducing the hydrophilic character of the fiber. The possible reaction between the hydroxyl groups of lignocellulosic fiber and carboxylic groups of acid is shown in Fig. 12.

The use of catalyst increases the rate of acetylation and the most common catalyst that can be used are pyridine, sulfuric acid, potassium, and sodium acetate. Nevertheless, strong acid catalyst conditions could damage the fiber structure (Kabir 2012). Therefore, selection of catalyst conditions is an important factor for the acetylation treatment.

De Rosa et al. (2011) studied the effect of acetic acid treatment on the mechanical properties of okra (*Abelmoschus esculentus*) fibers. They treated lignocellulosic fibers with 10% acetic acid and they observed that after the treatment the tensile strength and Young modulus values of okra fibers decreased considerably. The fracture surface of untreated and acetylated fiber was observed by SEM and they observed that the acetylation treatment tended to expose the lumen of lignocellulosic fibers and the acetylated fibers appear to be "cleaner" than raw materials. Untreated and acetylated fibers exhibited adsorption bands in the regions 1730–1745 and 1235–1240 cm^{-1} attributed to the C=O stretching of carbonyl in the ester and C–O stretching of acetyl groups, respectively. However, the low intensity of absorbance bands in the analyzed regions indicated that the fibers have a low degree of acetylation. Comparison of relative weight loss, defined as $(W_o - W_f)/W_o$ in which W_o is the weight of the sample at time 0 and W_f is the weight at the end of the test after drying at 70 °C for 2 h, confirmed that acetylation treatment reduced the hydrophilic character suggesting a better wettability with hydrophobic polymer resins could be achieved.

Zou et al. (2012) studied the effect of esterification treatment on the properties of short sisal fiber/PLA biocomposites. Chopped sisal fiber was soaked in acetic anhydride solution at room temperature for 2 h. The fibers were then filtered off and washed with distilled water until free from acid and dried *under vacuum* at 70 °C. After that, composites with 10 wt% treated sisal fibers and PLA polymer matrix were compounded by compression molding technique. They observed that the acetylation treatment improved the mechanical properties of composites suggesting that the anhydride groups covalently link with sisal fiber enhanced the interfacial adhesion.

$$Fiber - OH + R - COOH \rightarrow Fiber - O - CO - R + H_2O$$

Fig. 12 Proposed reaction for hydroxyl groups of lignocellulosic fiber and carboxylic groups of acid

Esterification with Fatty Acids

Fatty acids, such as stearic acid, or fatty acid salts are used to esterify the hydroxyl groups of lignocellulosic fibers (Saha et al. 2016). The hydroxyl groups of lignocellulosic fibers react with the carboxyl groups reducing the hydrophilic character of the fiber and improving the wettability between lignocellulosic fibers and hydrophobic polymer matrices. Therefore, the adhesion between lignocellulosic fibers and polymer matrix could be improved and better mechanical performance would achieve. The possible reaction between hydroxyl groups of lignocellulosic fibers and stearic acid is given in Fig. 13.

Paul et al. (2010b) studied the influence of esterification treatment on the mechanical properties of composites based on PP matrix and short banana fibers. Alkali pretreated banana fibers were soaked in 1% stearic acid in alcohol for 1 h and dried in an air oven at 60 °C for 1 h. They observed that the carboxyl groups of the stearic acid reacted with hydroxyl groups of the fiber through an esterification reaction, and hence, the treatment reduced the number of hydroxyl groups available for bonding with water molecules. As a consequence, they reported that the stearic acid treatment lowered the acidity and polarity values of banana fiber improving the wettability with hydrophobic polymer matrices. SEM micrographs of the stearic acid treated fiber surface revealed more fibrillation and roughness than untreated banana fibers. Therefore, the mechanical adhesion between banana fiber and PP could be improved. They characterized PP/banana fiber composites mechanical properties with 50% fiber loading and they observed that stearic acid treated fiber composites showed a slight enhancement of tensile and flexural properties compared to the composites based on the untreated fibers.

Esterification with Maleate Groups

Maleate coupling agents could provide an efficient link between the fiber surface and matrix. During grafting, maleic anhydride functional groups react with the hydroxyl groups of lignocellulosic fibers (Kabir et al. 2012). Maleic anhydride is not only used to modify fiber surface but also the polymer matrix, such as PP and PE, to achieve better interfacial bonding and mechanical properties in composites (Gassan and Bledzki 1997; Joseph et al. 2003; Lu and Chung 2000). For example, the PP and PE chains allow maleic anhydride to be cohesive and produce maleic anhydride grafted polypropylene (MAPP) and polyethylene (MAPE) copolymers, respectively. After that, the treatment of cellulose fibers with hot MAPP and MAPE copolymers provides covalent bonds across the interface (Bledzki and Gassan 1999; Li et al. 2007). The mechanism of reaction between maleate copolymer and

$$Fiber-OH+CH_3\left(CH_2\right)_{16}COOH\rightarrow Fiber-O-CO-\left(CH_2\right)_{16}CH_3+H_2O$$

Fig. 13 Proposed reaction for hydroxyl groups of lignocellulosic fiber and stearic acid

lignocellulosic fiber consisted on the activation of the copolymer by heating and the subsequent esterification of cellulose fiber (Bledzki et al. 1996). After the treatment, the surface energy of lignocellulosic fiber is increased to a level much closer to the surface energy of the matrix improving the wettability of fiber and esterification provides higher interfacial adhesion between fiber and polymer matrix (Li et al. 2007).

In previous works (Arbelaiz et al. 2005a, b) flax fibers were treated with maleic anhydride (MA) and with maleic anhydride polypropylene copolymer (MAPP). It was observed that untreated fiber showed a tensile strength and modulus of 802 MPa and 46.9 GPa, respectively. After MA treatment, tensile strength, and modulus decreased around 15 and 30%, respectively. However, MAPP treated flax fibers showed slightly higher tensile strength values whereas the Young modulus of MAPP treated fibers were similar to untreated flax fibers. It was suggested that the increments could be explained by the deposition of MAPP copolymer on the surface of the unit cells becoming the surface uniform and smooth. Contact angle measurements showed that after treating flax fibers, they became more hydrophobic being the polar component of surface energy similar to that for neat PP. Moreover, it was observed that after the treatment with MAPP, the interfacial shear strength (IFSS) value was improved around 15%. Flexural and tensile properties of flax/PP composites with 30 wt% fiber content showed that MAPP treatment increased flexural and tensile strength indicating an improvement of stress transfer from the matrix to flax fiber bundle. However, it was reported that MA treatment seemed not to act as stress transfer bridge between PP and flax fiber bundle. Even though the wettability was improved after MA treatment, composite tensile, and flexural results were not improved due to lack of adhesion efficiency.

Cisneros-Lopez et al. (2016) studied the effect of fiber surface treatment with maleic anhydride grafted polyethylene copolymer (MAPE) on the mechanical properties of PE/agave fiber composites. In this case, 1 wt% of MAPE was dissolved in 1,2,4-trichlorobenzene (TCB) at 90 °C. Then, alkali pretreated agave fibers were left for 30 min in the solution at the same temperature under high-intensity mixing. At least, the MAPE treated fibers were dried for 5 days in an oven at 80 °C for complete TCB removal. They observed that the MAPE treated fiber FTIR spectrum showed two strong bands at 2918 and 2852 cm^{-1} which are the characteristic bands of the vibration C–H groups of alkane associated with the polyethylene chains of MAPE that were grafted on the fiber surface (Benitez et al. 2013; Mohanty and Nayak 2006). SEM micrographs of MAPE treated fibers showed a smoother surface than untreated fibers with a more uniform texture confirming that MAPE was effectively grafted on the agave sisal surface. Moreover, they said that the significant decrease in fiber surface porosity observed as a result of MAPE treatment was also a sign that MAPE was able to penetrate this porosity and grafted with the alkali pre-treated fibers. Both phenomena will have a positive effect on fiber/matrix interactions in terms of adhesion, wettability, and dispersion (Jandas et al. 2013). After that, they compounded PE/agave fiber composites with a fiber content of 15 wt%. Composited based on the MAPE treated fibers showed significant improvements in tensile strength by up to 38% (from 13 to 18 MPa) compared with composites based on untreated fibers and 13% with respect to neat polymer matrix. Same behavior was observed in flexural and impact properties. They suggested that the improvements were related to better affinity, compatibility, and adhesion between agave fiber and PE polymer matrix.

4.1.4 Reaction with Free Radicals

Peroxide Treatment

Peroxide is a chemical compound with the specific functional group RO–OR containing the divalent ion bond O–O (Zhou et al. 2016). Lignocellulosic fiber surface can be modified with organic peroxides such as benzoyl peroxide, $(C_6H_5CO)_2O_2$, and dicumyl peroxide, $(C_6H_5C(CH_3)_2O)_2$ (Zhou et al. 2016; Li et al. 2007). Fibers were soaked with an organic peroxide in an acetone solution for about 30 min after alkali pretreatment (Paul et al. 2010b; Sreekala et al. 2000). Organic peroxides are highly reactive and decompose to create free radicals (RO•) which can be grafted onto cellulose and polymer matrix chain by reacting with the hydrogen groups of lignocellulosic fiber and polymer matrix (Zhou et al. 2016; Li et al. 2007). The free radical reaction initiated by peroxide compound between the polymer matrix and cellulose fibers is shown in Fig. 14.

Goriparthi et al. (2012) soaked jute fibers in 6% benzoyl peroxide acetone solution for about 30 min after alkali pretreatment and jute/PLA composites were compounded containing 50 wt% jute fibers. They observed that after the peroxide treatment, the tensile strength and moduli values of the composites were increased considerably. They suggested that the peroxide treatment produced many structural and chemical modifications on jute fiber surface improving the interfacial bonding mechanism between the fiber and the matrix. The same behavior was observed in flexural properties of peroxide treated composites. Izod impact strength of peroxide treated composites was slightly lower than the untreated composites. Debonding, pull-out, and fiber fractures are the mechanisms of energy absorption during impact and they mentioned that fiber pull-out requires higher energy than that of the fracture of fiber and debonding (Goriparthi et al. 2012). They suggested that high impact strength of untreated composites may be attributed to more fiber pull-outs due to weak interface between the fiber and the matrix. However, in the case of peroxide treated composites, fiber fracture occurred, being the reason for lower impact strength of peroxide treated composites compared to untreated composites.

Fig. 14 Proposed reaction scheme for the free radical reaction initiated by peroxide

$$RO-OR \rightarrow 2RO\bullet$$

$$RO\bullet + Polymer \rightarrow ROH + Polymer\bullet$$

$$RO\bullet + Cellulose \rightarrow ROH + Cellulose\bullet$$

$$Polymer\bullet + Cellulose\bullet \rightarrow Polymer - Cellulose$$

Acrylonitrile Grafting

Grafting acrylonitrile groups on fiber surface is an effective method of surface modification. Acrylonitrile initiates free radicals to react with the cellulose molecules by dehydrogenation and oxidation. Then, the activated free radical sites interact with the matrix. The resulting bond between the fiber and matrix enhances the interlocking efficiency at the interface (Kabir et al. 2012; Kalia et al. 2009).

Khan et al. (2015) grafted bleached jute fibers by acrylonitrile monomer (50 wt% of fiber) in the presence of $K_2S_2O_8$ as initiator (1 wt% of fiber), and $FeSO_4$ as a catalyst for 90 min at 70 °C in a water bath. Untreated and treated fibers were used for compounding jute/PLA composites by compression molding technique. In order to study the effect of acrylonitrile grafting, bleached jute composites were taken as reference material for comparison. They observed that the tensile, flexural, and impact strengths were increased by 14, 8, and 10%, respectively, in comparison of composites based on bleached fibers suggesting that the compatibility between the polymer matrix and jute fibers was improved after modification. SEM micrographs of the fractured surface of composites based on bleached fibers showed interfacial failure and fiber pull-out suggesting that the fiber/PLA adhesion was not very strong. In the case of acrylonitrile modified composites, the failure surface showed fibers completely covered with the matrix with no fiber pull-out. Therefore, they concluded that the increase of tensile and flexural strength of the composite was due to the improvement of the fiber/matrix interface.

Acrylate Grafting

Acrylate groups can be grafted on fiber surface by (i) reaction with free radicals (Li et al. 2007; Kalia et al. 2009) or (ii) an esterification reaction (Kabir et al. 2012). Depending on the polymer matrix used, the most interesting via should be used. In the first via, the acrylate groups can be grafted to cellulose molecule radicals. Cellulose can be treated with high-energy radiation to generate radicals together with chain scission (Li et al. 2007). In the second via, the acrylic acid can react with the cellulosic hydroxyl groups of the fiber and provide more access of reactive cellulose macro radicals to the polymerization medium. The carboxylic acids from coupling agents create ester linkages with the cellulose hydroxyl groups reducing the hydrophilic character of fiber and improving moisture resistance properties (Kabir et al. 2012). The reaction between the hydroxyl groups of lignocellulosic fibers and acrylic acid is given in Fig. 15. Afterward, the grafting of treated lignocellulosic fiber onto the matrix can be initiated by peroxides.

$$Fiber - OH + C_2H_3COOH \rightarrow Fiber - O - CO - C_2H_3 + H_2O$$

Fig. 15 Proposed reaction for hydroxyl groups of lignocellulosic fiber and acrylic acid

Zhang et al. (2013) immersed alkali pretreated wood fiber in a 0.3 M acrylic acid and 0.01 M benzoyl peroxide solution for 1 h at 85 °C. The volume ratio of water to benzene was 9:1 and benzoyl peroxide was used as an initiator. The acrylic acid treated fibers were soaked in a 6 M sodium chloride and 1 M sodium hydroxide solution or 15 min at room temperature. Then, wood fiber was removed and washed with an abundant amount of distilled water until the unreacted acrylic acid was eliminated. Finally, fibers were dried in an oven at 90 °C for 12 h. After that, the treated wood fiber was mixed with unsaturated polyester resin and composites were compounded. They observed that the flexural and tensile strength values increased as wood fiber content was increased suggesting that the chemical treatments could remove the impurities on the fiber surface and enhance the roughness improving the fiber/polymer adhesion. SEM micrographs of the impact-fractured surface of composites based on the untreated fiber evidenced a poor adhesion between the wood fiber and unsaturated polyesters. Nevertheless, they observed that after treating wood fibers with acrylic acid, SEM micrograph of the impact-fractured surfaces of composites suggested that fiber/polymer interfacial adhesion was improved.

Prasad et al. (2016) soaked alkali pretreated banana fibers in a 1% acrylic acid solution maintaining fiber/solution ratio of 1:15 (w/v) for 20 min. Then, they washed and dried the banana fibers in an air oven at 70 °C for 48 h to obtain acrylic acid treated fibers. By FTIR analysis they observed that the absorbance bands at about 2919 and 2854 cm^{-1} appeared to be slightly stronger for the acrylic acid treated fiber than untreated fiber, corresponding to the introduction of –CH– and $–CH_2–$groups by the acrylic acid treatment. Furthermore, they observed that the band near 1632 cm^{-1} for acrylic acid treated fibers was found to be broader, suggesting that some ester groups were grafted on banana fibers. Finally, they observed that the alkali treatment removed noncellulosic compounds. The absorbance band at about 1733 cm^{-1} related to the carbonyl of hemicelluloses, pectin and wax was not observed in the spectrum of acrylic acid treated fibers. In the same way, the change in the band near 1252 cm^{-1} was due to the C–O stretching vibration of acetyl groups of lignin which was related to the removal of lignin. The surface of untreated and treated banana fiber was studied by SEM. They observed that untreated banana fibers surface showed the presence of impurities, globular particles, wax, and fatty substances. However, the acrylic acid treated banana fiber surface appeared to be rough with the slight disintegration of the fiber along with fibrillation, which was attributed to the removal of noncellulosic compounds. After that, they compounded LDPE/banana fiber composite with 25 wt% fiber loading. After the acrylic acid treatment, the tensile strength of composites increased around 10% whereas the modulus of composites did not change compared to composites based on the untreated fibers. The increment of tensile strength was related to an improvement in the fiber wetting and bonding with LDPE matrix. In order to study the fractured surface of composites, SEM analysis was carried out. In SEM micrograph of untreated fiber composite, fibers seemed to be detached from the LDPE matrix and pulled-out fibers and voids were observed indicating poor interfacial bonding between fiber and matrix. Nevertheless, after acrylic acid treatment, no pull-out fibers and holes were observed, indicating better interlocking between banana fibers and the polymer matrix.

4.1.5 Etherification Reactions

Modification of cellulosic fibers by etherification reaction can change fiber properties and make it more useful and acceptable for some applications (Kalia et al. 2009; Mansour et al. 1994). Sodium hydroxide plays an important role in forming a charged intermediate species with the fiber, which allows the faster nucleophilic addition of epoxides, alkyl halides, and benzyl chloride (Kalia et al. 2009; Matsuda 1996).

Chen et al. (2012) added 10 g kenaf fiber and 100 ml of 40 wt% NaOH aqueous solution into a three neck round bottom flask with strong mechanically stirring and refluxing with water. The temperature was raised to 110 °C and 50 ml of benzyl chloride was added into the mixture and the reaction was allowed to proceed for different periods of time (90, 180 and 300 min). Then, benzylated kenaf fibers were washed with ethanol and water and dried in the vacuum oven at 60 °C. They observed that the chemical structure of kenaf fibers after the introduction of the benzyl group varied according to the reaction time. All benzylated kenaf fibers spectrum showed that the hydroxyl vibration absorption at about 3500 cm^{-1} was decreased after benzylation, while the absorptions at 1800–1950, 1600, 736, and 695 cm^{-1} increased, indicating the formation of mono-substituted benzene rings. Moreover, the band of benzylated kenaf at 1206–1207 cm^{-1} was assigned to the asymmetric and symmetric axial deformation of the C–O–C bonds of alkyl-aryl ether, which were not observed for the untreated kenaf fiber. In addition, they observed that the benzylation decreased considerably the hydrophilic character of the kenaf fibers because higher contact angle values with a drop of water were observed.

Sun et al. (2016) mixed 10 g of alkali pretreated bamboo fibers with 20 ml of benzyl chloride at 35 °C in a double glass reaction kettle equipped with a reflux condenser and a stirring bar. Then, the solution was heated to 120 °C and the reaction was carried out for 1 h. Finally, benzylated bamboo fibers were washed and dried in a vacuum oven at 73 °C at a pressure of 0.08 MPa for 12 h. Finally, bamboo fiber/PE composites panels containing 10, 20, and 30 wt% fiber loading were prepared using a hot press molding technique. After the benzylation treatment, they observed that the flexural and tensile strength values of composites increased considerably. The tensile and flexural strength of composites with a 30:70 fiber/PE mass ratio increased up to 24.2 and 26.7%, respectively, compared to the values obtained for the untreated bamboo fiber/PE composites. They suggested that the improvement of the mechanical properties of the benzyl treated fiber composites was mainly attributed to the enhancement of the interface compatibility.

4.1.6 Benzoylation Treatment

Benzoylation treatment, an important transformation in organic synthesis, is another treatment used to decrease the hydrophilic character of lignocellulosic fiber and improve fiber/matrix interfacial adhesion (Zhou et al. 2016; Kalaprasad et al. 2004). Prior to react lignocellulosic fiber with benzoyl groups, the lignocellulosic fiber should be initially pretreated with NaOH aqueous solution in order to activate and expose the hydroxyl groups on the fiber surface (Zhou et al. 2016; Kabir et al. 2012). Afterward, the lignocellulosic fiber was treated with benzoyl chloride.

Sampathkumar et al. (2014) activated the hydroxyl groups of areca natural fibers using 6 wt% NaOH solution. For this purpose, areca fibers were soaked in 6 wt% NaOH solution and agitated with benzoyl chloride for 15 min. Then, the treated areca fibers were soaked in ethanol solution for 1 h to remove benzoyl chloride and fibers were washed thoroughly using distilled water and dried in air. They observed that FTIR spectra of benzoylated areca fibers showed an absorption band in the range of 1400–1600 cm^{-1} due to the C=C stretching of aromatic rings in the treated fibers. The alkali pretreatment resulted in the absence of 1727 cm^{-1} band suggesting that the alkali pretreatment removed the noncellulosic compounds of areca fibers. Nevertheless, after benzoyl chloride treatment, they observed a carbonyl group absorption in the range of 1700 cm^{-1} suggesting the ester linkage in the benzoyl chloride treated areca fiber. The SEM micrograph of untreated areca fiber showed a network structure in which the fibrils were bound together by hemicelluloses and lignin. On the other hand, they observed that benzoyl chloride treated fibers showed a large number of pits and much more rough surface due to the removal of noncellulosic compounds. Finally, they reported that after treating with benzoyl chloride, the areca fibers reduced its hydrophilic character and became more compatible with the polymer matrix.

Paul et al. (2010b) soaked chopped banana fiber in 2 wt% NaOH solution for 90 min and agitated with benzoyl chloride for 30 min. Banana fibers were washed with water and dried in air oven at 70 °C. After that, they compounded PP/banana fiber composites with untreated and treated fiber with 50 wt% fiber loading and the mechanical properties of composites were studied. They observed that the tensile and flexural properties of benzoylated fiber composite were found to be higher than those untreated fiber composites. The tensile strength of benzoylated fiber composites improved around 13% whereas the flexural strength improved 6%. On the other hand, they observed that the Young modulus increased from 1521 to 1595 MPa while the flexural modulus improved from 1400 to 1520 MPa. The improvement in mechanical properties of composites based on the benzoylated treated fibers was attributed to the reduction in the hydrophilicity of fibers (Nair et al. 1996; Joseph et al. 2002). In another work, Paul et al. (2010a) studied the fractured surface of PP/banana composites by SEM analysis and they observed that tensile fractured surface of composites based on the benzoylated treated fibers showed a fiber breakage rather than fiber debonding due to better banana fiber/PP adhesion. In SEM micrograph of untreated banana/PP composites can be observed that the tensile rupture was accompanied by the debonding of the banana fibers leaving holes, which indicated a poor adhesion between PP matrix and untreated banana fibers.

4.1.7 Isocyanate Treatment

Isocyanate is reported to work as coupling agent in fiber reinforced composites (Paul et al. 1997; Joseph and Thomas 1996, Sreekala and Thomas 2003). The isocyanate functional group (–N=C=O) reacts with the hydroxyl groups of cellulose and lignins of fiber and a urethane linkage is formed (Li et al. 2007; Kabir et al. 2012). Isocyanate also reacts with the moisture present on the fiber surface and forms urea which can further react with the hydroxyl groups of the celluloses (George et al 2001). This secondary reaction results in higher moisture resistance properties of the fiber and provides better bonding with the matrix to enhance composite properties (Kalia et al. 2009; Kabir et al. 2012). The reaction between the fiber and isocyanate coupling agent is shown in Fig. 16, where R could be different chemical groups such as alkyl which react with the polymer matrix.

Datta and Kopczynska (2015) studied the effect of kenaf modification on morphology and mechanical properties of thermoplastic polyurethane materials. Among other treatments, modification of a blocked isocyanate was carried out. In order to obtain blocked isocyanate, the molten 4,4′-diphenylmethane diisocyanate (MDI) was mixed with methanol in a molar ratio of 1:1 for 30 min to obtain a 20 wt% solution of MDI and methanol in acetone. Then, fibers were soaked in this solution for one week and washed repeatedly with acetone and dried at 100 °C for 2 h. Spectra of blocked isocyanate treatment showed a band at 1595 cm^{-1} which was associated to the benzene rings of MDI structure and a band at 1541 cm^{-1} was related to the amide symmetric stretching vibration. SEM micrographs of untreated fibers showed a relatively smooth surface, and small particles of impurities attached to the surface were also observed. In order to study the effect of isocyanate treatment on kenaf/thermoplastic polyurethane composites, they compounded composites with 10 and 30 wt% fiber contents. They observed that isocyanate treated fibers improved the tensile strength values of composites with fiber loading of 10 wt%. However, with 30 wt% content of fibers, they observed that composite tensile strength was not improved. They reported that this fact was due to fiber agglomerations and poor adhesion between matrix and kenaf fibers.

$$R-NCO+Fiber-OH \rightarrow R-NH-CO-O-Fiber$$

Fig. 16 Proposed reaction between lignocellulosic fiber and isocyanate coupling agent

4.1.8 Oxidation Reactions

Permanganate Treatment

Permanganate treatment on natural fibers is conducted by potassium permanganate ($KMnO_4$) in acetone solution (Kabir et al. 2012). In permanganate treatment, highly reactive permanganate ions can react with cellulose hydroxyl groups. This treatment could enhance the chemical interlocking at the interface and also provide better adhesion with the matrix (Rahman et al. 2007). Furthermore, permanganate ions could also react with the hydroxyl groups located in lignin constituents reducing the hydrophilic character of the fiber (Kabir et al. 2012; Paul et al. 1997). Higher concentrations than 1% of $KMnO_4$ could degrade the fiber properties (Li et al. 2007; Paul et al. 1997).

Paul et al. (2010b) soaked alkali pretreated banana fiber in 0.5% $KMnO_4$ in acetone for 90 min. By SEM analysis, they found that permanganate could etch the fiber surface and made it physically rougher than untreated ones. In addition, they observed that the contact area between the fiber and the matrix was also increased after treatment as a result of the fiber roughness increment. In consequence, interfacial properties can be improved by mechanical interlocking. Empirical polarity parameters calculated in terms of Kamlet–Taft solvent polarity scale revealed that permanganate treatment lowered the acidity as well as the polarity of the banana fibers improving the wettability of permanganate treated fiber and polymer matrices. They said that the fiber/matrix interactions were dependent on the polarity parameters of the modified banana fiber surface and they suggested that lower polarity of the banana fibers led to better compatibility with PP matrix. Thus, they observed that the tensile and flexural properties were slightly increased for permanganate treated banana fiber/polypropylene (PP) composites.

Zou et al. (2012) studied the effect of permanganate treatment on the properties of short sisal fiber/PLA biocomposites. The chopped sisal fibers were soaked in 0.1% $KMnO_4$/acetone solution for 2 h. Fibers were then taken out and washed many times with distilled water. Finally, sisal fibers were dried *in vacuum* at 70 °C. Scanning electron microscopy (SEM) micrographs showed that the permanganate treatment roughened the sisal surface improving the interfacial bonding between permanganate treated sisal fiber and the polymer matrix. Composites with PLA matrix and 10 wt% of untreated and permanganate treated sisal fibers were prepared. They observed that permanganate treated sisal fiber reinforced composites showed an improvement in the tensile and impact properties compared to the composites based on untreated sisal fibers. Moreover, fractured surfaces of composites showed that the sisal fibers were tightly connected with PLA matrix suggesting that the adhesion between sisal fiber and PLA was improved after permanganate treatment.

Furfuryl Alcohol Modification

This modification is based on the selective oxidation by chlorine dioxide of guaiacyl and syringyl phenols of the lignin, generating *ortho-* and *para-* quinones able to react Diels-Alder reaction with furfuryl alcohol (Trindade et al. 2004). The latter were reacted with furfuryl alcohol, creating a coating around the fiber. This modification favored the fiber-matrix interaction at the interface but caused some fiber degradation that affected the mechanical properties (Trindade et al. 2005). The fiber modified with furfuryl alcohol exhibited degradation of hemicelluloses, but cellulose maintained most of its crystallinity (Trindade et al. 2004).

Saw et al. (2011) oxidized coir fiber with an aqueous chlorine dioxide solution, which was prepared by reactions between sodium chlorite and acetic acid in aqueous medium. After oxidation reaction, the yellow-red colored fibers were washed with distilled water until neutrality. Then, the oxidized coir fibers were impregnated with furfuryl alcohol and heated at 100 °C for 4 h in presence of N_2 flow. The excess of furfuryl alcohol was removed by soxhlet extraction using ethanol for 15 h. Finally, fibers were dried at 50 °C for 24 h. They observed that a broad adsorption band at 3700–3300 cm^{-1} region characteristic of the polymeric association of the hydroxyl groups and bonded –OH stretching vibration present in cellulose, hemicellulose, and lignins, decreased after the chemical treatment. They explained that the results reflected the decrease of phenolic/aliphatic hydroxyl groups in the fiber lignins after the oxidation reaction. Furthermore, they observed that after the modification of fibers, a decrease in intensity was observed for an aromatic band of lignin unit at 1606, 1512, and 1424 cm^{-1}. Scanning electron micrographs of untreated coir fibers showed the presence of a large number of regularly placed holes or pits and layer over layers of sheet substances like wax of fatty substances on the surface of unmodified fibers. After oxidation, the surfaces of the fibers became rougher due to the removal of the inter-cellulose binding materials and the amorphous waxy cuticle layer. Nevertheless, the SEM image of furfuryl alcohol grafted fibers showed lower surface roughness compared to oxidized treated fibers. Furthermore, to determine the surface roughness of the untreated and treated fiber, atomic force microscope was used. They observed that the untreated and oxidized treated coir fibers exhibited a root mean square roughness values of 71 and 135 nm, respectively, whereas the furfuryl alcohol grafted fiber exhibited a surface roughness of 85 nm proving that the surface of the fiber was covered with furfuryl alcohol. This results suggested that the mechanical adhesion between modified coir fibers and polymer matrix could be improved. They also observed that the oxidation and furfuryl alcohol modifications increased the water contact angle of coir fiber but, decrease the contact angle with other low-polar liquids like glycerol or ethylene glycol indicating the reduction of hydrophilic character of the modified fiber. This meant that the wettability between the modified coir fibers and hydrophobic polymer matrix could be improved.

In another work, Saw et al. (2013) studied the effect surface treatments on luffa fiber reinforced epoxy composites. Luffa fibers were firstly oxidized and then furfuryl alcohol was grafted on oxidized luffa fiber as detailed previously (Saw et al.

2011). After that, untreated and treated luffa fibers mats were separately impregnated with epoxy resin for fabricating the composites. In each composite, the weight ratio of matrix to fiber was maintained at 70:30. They observed that the chemical treatment of the fiber improved chemical bonding and helped it to withstand high tensile load by the composites made of modified luffa fibers. They observed that for furfuryl alcohol grafted fiber composites, the tensile strength and modulus values were increased by 100 and 123%, respectively, due to the wettability and the mechanical adhesion between modified luffa fiber and epoxy polymer matrix were improved. Finally, they found by SEM images that furfuryl alcohol grafted luffa fibers were well embedded in the epoxy matrix and many fewer fiber pull-out and holes were observed compared to that of composites based on untreated fibers suggesting that the furfurylation is an adequate surface treatment to improve the mechanical adhesion between lignocellulosic fiber and the epoxy polymer matrix.

4.2 *Physical Treatments*

Physical treatments do not change considerably the chemical composition of the lignocellulosic fibers but structural and surface properties of the fiber change considerably. Usually, physical treatments are used as a preparation stage for chemical treatments (Bataille et al. 1994). By physical treatments, fiber bundles can be separated into individual filaments or fiber surface can be modified (Mukhopadhyay and Fangueiro 2009). Methods like a steam explosion is used, when a separation of the bundles into individual filaments is required. However, plasma treatments, dielectric-barrier discharge technique or corona discharge treatments are employed when modification of the fiber surface is desired.

4.2.1 Steam Explosion Treatment

Steam explosion process separates the lignocellulosic fiber into its main components, namely cellulose, lignins, and hemicelluloses (Josefsson et al. 2002). This process involves heating lignocellulosic fibers at high temperature and pressures, followed by mechanical disruption of the pretreated materials by violent discharge into a collecting tank (Kestur et al. 2009). Steam explosion process can include catalysts at different concentrations and different conditions of temperatures, pressures, and times are used. The conditions used in steam explosion process depend on lignocellulosic fiber type (Kestur et al. 2009).

Hou et al. (2014) prepared lightweight composites based on PP and cotton stalk fibers treated with steam flash explosion (SFE) process. Treated cotton stalk fibers were carded with the flip cotton machine for two times and then blended with PP matrix to prepare composites using compression molding technique. They noticed that after treating SFE fibers with an alkali solution (SFE-AT), composite tensile strength and Young modulus values were 47.5 and 27.0% higher than the values obtained for composites reinforced with SFE treated fibers. The fractured surface of composites was studied by SEM and they observed that in composites reinforced with SFE treated fibers, cotton stalks fibers were pulled-out and fibers were hardly coated by PP matrix, which revealed a poor fiber-matrix adhesion. In contrast, in composites reinforced with SFE-AT fibers, cotton stalks fibers were coated by the PP matrix. Moreover, most of the fibers were fractured suggesting that the cotton stalk fibers were strained up to their breaking point. Therefore, they suggested that the interfacial adhesion was improved due to the removal of noncellulosic compounds and lower moisture regain of alkali treated fibers. SEM micrographs showed that SFE-AT cotton stalks fiber had a smaller diameter and more grooves on the surface compared with the non-alkali treated SFE fibers. Besides, SEM micrographs suggested that alkali treatment removed noncellulosic compounds from fiber surface (Wang et al. 2009) and steam explosion could dissolve some pectin and hemicelluloses (Ibrahim et al. 2010). On the other hand, PP composites reinforced SFE-AT fibers exhibited the smallest thickness swelling due to water absorption. They suggested that the water absorption of composites was mainly due to the diffusion of water molecules into the interface between cotton stalk fibers and PP. A better adhesion between PP and SFE-AT fibers was obtained and consequently, there was less space in the interfacial region where water molecules can diffuse.

4.2.2 Plasma and Corona Treatments

The plasma discharge can be generated by either corona treatment or cold plasma treatment. Both methods are considered plasma treatment when ionized gas has an equivalent number of positive and negative charged molecules that react with the surface of the present material (Young 1992). The difference between the corona treatments and cold plasma treatments is the frequency of the electric discharge. At high-frequency, cold plasma can be produced by microwave energy, while a lower frequency alternating current discharge at atmospheric pressure produces a corona plasma (Young 1992). Depending on the nature and composition of feed gases, a wide range of surface modifications can be achieved. One advantage of plasma treatment, among others, is the short time required for surface treatment.

Bozaci et al. (2013) studied the effect of plasma treatment on flax fiber surface properties and flax fiber/high-density polyethylene (HDPE) adhesion. Flax fibers were treated during 2 min by air and argon gas at different plasma powers; 100, 200, and 300 W, respectively. They observed that argon plasma treatment was slightly more harmful than air plasma treatment because worst tensile strength values were obtained. In fact, after air and argon plasma treatment of flax fibers at 300 W, tensile strength decreased at about 9 and 12%, respectively, respect to untreated fibers ones. The fiber surface observed by SEM showed a smooth surface for untreated flax fiber whereas plasma treated fiber surface exhibited tiny grains and roughness, increasing surface roughness when plasma power was increased.

They observed that argon plasma treated fibers had more grooves and grains than air plasma treated ones due to the higher etching tendency of the argon plasma. Consequently, air plasma treated flax fibers seemed to be smoother than that argon plasma treated flax fiber ones. Pull-out test results showed that after plasma treatment, flax fiber/HDPE interfacial shear strength (IFSS) values became higher than that of untreated flax fiber. They suggested that IFSS value improved due to the increase of roughness and contact area between the flax fiber and the HDPE polymer matrix.

Sever et al. (2011) studied the mechanical properties of jute/HDPE composites. Jute fibers were modified by oxygen plasma with different frequencies and discharge powers for 15 min. Jute/HDPE composites were prepared using two plies of jute fiber mats and three plies of HDPE sheets. Then, the composite laminate was heated beyond its melting temperature without applying any pressure. In the second step, the composite laminate was pressed at 195 °C at a pressure of 10 MPa for 15 min. For jute fiber/HDPE composites, they observed that after oxygen plasma treatment, the interlaminar shear strength (ILSS) value of composites was increased. In low frequency, ILSS value of oxygen plasma treated jute/HDPE composite increased up to 7.7 and 8.5 MPa, respectively, when 30 and 60 W of plasma power was applied. On the other hand, in radio frequency, the ILSS values of composites were enhanced when the plasma power rose up to 90 W increasing ILSS values of the composites by 65% for 30 W, 84% for 60 W, and 189% for 90 W in comparison with composites based on the untreated fiber ones. They suggested that the increment was related to the improved interfacial adhesion between the fiber surface and the HDPE matrix. The flexural and tensile strength values of jute fiber/HDPE composites improved after plasma treatment. In low frequency, when jute fibers were treated with oxygen plasma power at 60 W, the tensile, and flexural strength values of jute fiber/HDPE composites were increased around 31 and 45%, respectively. On the other hand, in radio frequency, they observed that the tensile and flexural strength values improved with increasing the plasma power. Even though jute fibers were treated with greater radio frequency plasma power, the tensile, and flexural strength of composites did not decrease suggesting that the interfacial adhesion between fiber surface and HDPE polymer matrix was improved. SEM micrographs of the fractured surface of untreated jute fiber/HDPE showed that jute fibers were pulled-out from the thermoplastic matrix and the fibers had a smooth and clean surface with no matrix adhered suggesting that interfacial adhesion between untreated jute fibers and HDPE matrix was very poor. In SEM micrographs of composites reinforced with oxygen plasma treated fibers, fiber surface was covered with large quantities of the matrix and a great number of fibers were appeared to be embedded in HDPE matrix, indicating a better fiber/matrix adhesion.

Sinha and Panigrahi (2009) studied the effect of plasma treatment on structure and wettability of jute fibers as well as on the flexural strength of jute/unsaturated polyester composites. Jute fibers were treated by argon at a plasma power of 20 W varying the exposing time from 5 to 15 min. They compounded unsaturated polymer resin composites with 15 wt% of untreated and plasma treated jute fibers.

They observed that the surface of untreated jute fiber was smooth while plasma treated fiber presented a rough surface. Moreover, they found that plasma treatment increased the contact angle of jute fiber with water, therefore decreasing its polarity. The decrease of the polar component was more pronounced when higher was the exposure time of plasma treatment. Thus, the wettability of jute fibers and polymer matrix may be increased. The increase of hydrophobicity was corroborated by FTIR analysis, since they observed that after plasma treatment the signal of characteristic bands of hydrophilic groups, such as carboxylic and hydroxyl groups were decreased. For these reasons, composites reinforced with plasma treated fibers during 10 min showed the highest flexural strength values. The flexural strength value improved about 14% compared to composites reinforced with untreated jute fibers. SEM micrograph of the fractured surface of composite reinforced with 15 min plasma treated fiber showed improved fiber/polymer adhesion. Nevertheless, the mechanical properties of composites were decreased respect to composites based on the 10 min treated fibers. This fact suggested that after 15 min plasma treatment jute fibers were degraded lowering the strength of the composite. In fact, SEM micrographs of 15 min plasma treated fiber showed the formation of pits on the fiber surface, corroborating that 15 min exposing time could damage the fiber.

4.2.3 Dielectric-Barrier Discharge Treatment

In dielectric-barrier discharge (DBD) treatment, glass, or ceramic barrier is placed between two electrodes connected to a very high-voltage alternating current (AC) generator. The advantage of this method is that the electrodes are not in close contact with the plasma (Mukhopadhyay and Fangueiro 2009). High-voltage alternating current produces random arc in a matter of milliseconds in the discharge gap which ionizes the gaseous molecules present within the gap and hence produce plasma. The conventional setting is a coaxial glass tube connected to the continuous alternating power source. The produced plasma depends on the gas pressure in the tube and hence the chance of collision between the molecules.

Li et al. (2014) studied the effect of helium plasma treatment on ramie fiber/polybutylene succinate (PBS) adhesion. Fibers were washed successively with acetone and distilled water followed by drying in an oven. Then, fibers were soaked in ethanol for 10 min with a mass gain of 5.0% before plasma treatment. The DBD was generated between two parallel copper electrodes. A power supply of 15.98 kHz frequency was applied to generate the plasma between the electrodes within a gap of 10 mm. In the process, helium was used as the treatment gas and treatment time was set up on 30 s. In addition, four voltages of 1.5, 3, 6, and 9 kV were chosen to treat ramie fibers. Surface morphologies of the untreated and treated ramie fibers were observed by SEM. Untreated ramie fibers had a relatively smooth yet naturally streaked surface while the treated fibers showed the different extent of plasma etching. Moreover, they observed that the fiber roughness increased with the increase of plasma treatment voltage. Many visible spots, small notched and large area of protuberances were observed especially for fibers treated with 9 kV. This fact suggested that treatment conditions could damage ramie fiber properties. The effect of different plasma treatment voltage on surface wettability of ramie fibers was investigated using dynamic contact angle analysis. They found that the advancing contact angle of untreated ramie fiber was 55.9° and plasma treated fiber with 1.5 kV voltage showed similar contact angle. Nevertheless, the contact angles of fibers treated with 3, 6, and 9 kV notably increased probably due to the plasma treatment induced grafting of ethyl groups onto the fiber surface. Additionally, they corroborated this suggestion by X-ray photoelectron spectroscopy (XPS) technique. Therefore, it was thought the fiber/matrix adhesion would be improved since the wettability of ramie fibers and the fiber surface roughness was increased. Thus, the IFSS values of ramie/PBS composites were determined using microbond pull-out test. Compared with untreated ramie fiber/PBS, the IFSS value for fiber treated with the lowest voltage, 1.5 kV, did not show significant differences with respect to untreated fiber/PBS system. Meanwhile, the IFSS value for fibers treated with 3, 6, and 9 kV raised about 29.6, 45.5, and 19.4%, respectively, compared to untreated fiber ones due to the combination of the better chemical compatibility and improved mechanical interlocking between the fiber and polymeric matrix.

Ragoubi et al. (2012) treated miscanthus fibers by corona discharge treatment (CDT). Fibers were treated in CDT device on a dielectric barrier technique. The treatment was assured by a low-frequency high-voltage generator (typically 15 kV, 50 Hz) and the samples were disposed between the electrodes and treated for several minutes. After the CDT, obtained XPS results evidenced that in treated samples there was an increase in oxygen content and a decrease in carbon content one. The increase of the O/C ratio was attributed to the surface oxidation generated by the corona treatment. On the other hand, SEM micrographs showed that after treating fiber for 15 min, some little cracks appeared on the surface of miscanthus fiber. For longer treatment time, the fibrils were pulled-off of the fibers resulting in the formation of some cavities that could be observed after 45 min. They observed that CDT produced many activated sites that could react with oxygen to give etching effect. In order to confirm this suggestion, they compounded composites based on miscanthus fiber using PP and PLA as matrices. For PP/treated miscanthus system, an increase in Young modulus, stress, and strength at yield values were observed compared with those values obtained for composites based on untreated fibers. They suggested that the wettability of miscanthus fibers was increased by CDT. Also, they suggested that the improvement of interfacial properties could be attributed mainly to a mechanical anchorage. The fractured surface of composites was observed by SEM and they reported that in PP composites reinforced with untreated miscanthus, the fiber surface appeared smooth and clean. At higher magnification, the pull-out fiber was observed indicating poor fiber/PP interfacial adhesion. On the other hand, fractured surfaced of composites based on the treated fibers showed that miscanthus fibers were covered with matrix and there were no signs of pull-out fiber which means that the interfacial adhesion between PP and miscanthus fiber was improved by corona treatment. Similar behavior was reported on the mechanical properties of PLA/miscanthus fibers after treating fibers by CDT.

4.3 Biological Treatments

Biological treatments are a promising technology due to its several advantages like low energy requirements and mild environmental conditions which make them eco-friendly and economically viable strategy (Kumar et al. 2009). Furthermore, biological modifications can selectively remove noncellulosic materials (George et al. 2014). Biological treatments with microorganisms, such as fungi, bacteria, and enzymes have been used for lignocellulosic fibers treatments (Buschle-Diller et al. 1999).

4.3.1 Fungal Treatment

Fungal treatments produce extracellular enzymes that degrade lignins, as well as extensive range of other noncellulosic materials (Li et al. 2009). It has been reported that fungal treatments produce an extensive system of hyphae which can make fine holes on the lignocellulosic fiber surface and may roughen the surface of the lignocellulosic fiber (Li et al. 2009; Daniel et al. 2004).

Pickering et al. (2007) modified hemp fibers using fungal treatments. Hemp fibers were sterilized in an autoclave for 15 min at 121 °C before fungal treatment. The fungi:hemp ratio was approximately 10:12 (mg/g) and incubation were carried out at 27 °C for 2 weeks. After fungal treatment, fibers were sterilized and washed for 10 min before being dried in an oven at 80 °C. The fungi used for hemp fiber treatment were *Phanerochaete sordida, Pycnoporus species, Schizophyllum commune, Absidia,* and *Ophiostomn floccosun*. The first three belong to white rot fungi group, whereas the fourth and the fifth belong to *zygomycetes* and *ascomycetes* groups, respectively. They observed that the untreated fiber surface appeared smooth and glossy, whereas after fungal treatment with white rot fungi, the surface of fibers led to less glossy and striations becoming more visible along the fiber. They observed by XRD analysis that after all fungal treatments, crystallinity values of fungal treated fibers increased slightly respect to untreated fibers supporting the removal of noncellulosic compounds. In addition, the removal of noncellulosic compound was corroborated by TGA analysis since after the fungal treatment hemp fiber were thermally more stable than untreated ones. However, they observed that after the fungal treatment, the tensile strength values of treated fibers were reduced respect to untreated fibers, especially, when hemp fibers were treated with *Schizophyllum commune* fungal. They observed that tensile strength reduced about 50% suggesting that the fiber strength reduction could be related to the cellulose degradation during fungal treatment. After the characterization of untreated and fungal treated fibers, they compounded composites with 40 wt% hemp fibers and PP matrix and they added 3 wt% MAPP coupling agent. Despite the tensile strength values of all fungal treated fibers were reduced respect to untreated fibers, composites reinforced with fungal treated fiber showed higher tensile strength values than composites reinforced with untreated hemp fibers. Thus, the highest tensile strength value was observed in composites based on fibers treated with *Schizophyllum commune*. Hemp fiber treated with *Schizophyllum commune* showed the lowest fiber tensile strength value, therefore, composites tensile strength data suggested that the interfacial bonding between PP and hemp fibers were greatly improved due to the removal of noncellulosic compounds and the increment of fiber surface roughness.

Li et al. (2009) carried out an analysis of hemp/PP composites using white rot fungal treatments. For fungal treatments, dried hemp fibers were sterilized using gamma radiation. Irradiated fibers were inoculated with white rot fungi for 2 weeks at 27 °C and water was added for all fungal treatments to give a moisture content of 60 wt%. The fungi used for hemp fiber treatment were *Phanerochaete sordid, Pycnoporus species,* and *Schizophyllum commune*, respectively. They observed that all fungal treatments reduced the amount of wax, pectin, hemicelluloses, and lignin in the hemp fiber. FTIR spectra of fungal treated fibers showed that the band around 1736 cm^{-1} attributed to the presence of the carboxylic ester in pectins showed a significant intensity reduction supporting the removal of pectins. In addition, the band around 1268 cm^{-1}, related to the COO stretching in lignins was disappeared in white rot fungi treated fibers. The surface morphology of untreated and treated fibers was characterized by SEM. Untreated hemp fiber showed a large amount of debris adhering to the surface of the fiber bundles, because they were coated with noncellulosic material. After fungal treatment, relatively cleaner surface was observed which supported the removal of noncellulosic compound. Nevertheless, they mentioned that the resolution of the available SEM was inadequate to detect fine holes caused by fungal hyphae attack on fungal treated fiber surfaces. The removal of noncellulosic compound was also corroborated by XRD analysis since the crystallinity values were improved after the white rot fungal treatment. The strength values of white rot fungi treated fibers were lower than untreated fibers. They observed that the tensile strength values of fungal treated hemp fiber decreased at least 23% and the highest decrease, around 38%, was observed for *Schizophyllum commune* treated fiber. They suggested that the decrease of tensile strength values was related to the creation of fine holes in the fiber surface. After that, they compounded PP/hemp composites with 40 wt% fiber content using 3 wt% MAPP coupling agent. Despite the fungal treatments reduced hemp fiber tensile strength value, the tensile strength values of composites reinforced with white rot fungi treated fibers were higher than composites based on the untreated ones. This fact suggested that the interfacial bonding adhesion between fiber and matrix seemed to be improved due to the removal of noncellulosic compounds and the increase of surface roughness.

4.3.2 Enzyme Treatment

Enzymes are biocatalysts that accelerate biomechanical reactions acting on a specific reactant called "substrate". All substrates have their own enzymes such as cellulase for cellulose, xylanase for hemicellulose, laccase for lignins, etc.

(Karaduman et al. 2012). Enzyme treatments are environmentally friendly and require mild operating conditions (George et al. 2016).

George et al. (2014) studied the surface morphology and thermal properties of lignocellulosic fibers treated with enzymes. Hemp and flax fibers were treated with different enzymes such as xylanase, xylanase + 10% cellulase, polygalacturonase, laccase, and pectinmethylesterase, respectively. The liquid to fiber ratio was maintained at 50:1 (v/w) to facilitate complete wetting of fiber. Enzymatic treatments were conducted for 90 min under constant agitation of 80 rpm at optimum pH and temperature conditions (see Table 4) and enzymes were deactivated by heating at 90 °C for 10 min.

They observed that after treating lignocellulosic fibers with xylanase + 10% cellulase enzymes, hemp, and flax fibers showed a reduction in the cellulosic content whereas lignocellulosic fibers treated with pectinases enzymes (polygalacturonase and pectinmethylesterase) showed a decrease in the hemicelluloses content. In addition, laccase treated flax fibers showed a significant reduction in the lignin content. It was observed from SEM micrographs that after enzyme treatment the elementary fibers became more visible compared to the surface of untreated fibers. On the other hand, they observed that a combination of xylanase and cellulase, and polygalacturonase treated samples exhibited an increase in surface roughness. The onset degradation temperature of enzyme treated hemp and flax fibers were improved due to the removal of pectin and hemicellulose materials. This improvement was achieved when lignocellulosic fibers were treated with xylanase and pectinases enzyme. However, no change in the onset degradation temperature was observed when lignocellulosic fibers were treated with laccase enzyme. On the other hand, they reported that the enzyme treatment resulted in higher surface hydrophilicity because of the removal of lignin and hemicellulose, thereby exposing the cellulosic backbone. In summary, enzyme treatment of lignocellulosic fibers can be used to selectively degrade components that limit the thermal stability and to modify the fiber surface characteristic.

George et al. (2016) studied the mechanical and moisture absorption of enzyme treated fibers/PP composites with 20 wt% of fiber content. They observed that after the enzyme treatment there was no change in mechanical properties. By TGA analysis, they observed that enzymatic treatment improved the thermal properties of the resulting composites. Furthermore, composites based on the enzyme treated fiber showed lower moisture absorption values than untreated fiber composites.

Table 4 Optimum pH and temperature conditions for each enzymatic reaction (George et al. 2014)

Enzyme	Xylanase	Xylanase +10% cellulase	Polygalacturonase	Laccase	Pectinmethylesterase
pH	7	6	4	7	5
Temperature (°C)	70	50	45	50	45

Table 5 Characteristic of each enzyme system with corresponding activity at optimum conditions (Hanana et al. 2015)

Enzyme	Xylanase	Laccase	Pectinase
pH	7	5	4.8
Temperature (°C)	55	50	50

Xylanase + cellulase treatments provided the greatest reduction in moisture both for hemp and flax fiber composites. After 4 weeks of immersion in water at room temperature, they observed that the xylanase + cellulase treatment decreased the average moisture absorption of composites based on the hemp fiber around 250% whereas the average moisture absorption of flax fibers reduced around 200%. Finally, in order to study the fiber/matrix adhesion, the fractured surface of the composite was observed by SEM. Composites based on the untreated fibers showed pulled-out fibers due to the poor interfacial adhesion. However, composites based on the enzyme treated samples revealed a better distribution of fibers indicating a better fiber dispersion and fiber/matrix adhesion.

Hanana et al. (2015) treated alfa fibers with several enzymes. The enzyme treatments were carried out after alfa fiber were treated with 0.5 M NaOH solution in autoclave for 20 min at 120 °C with a pressure of 1 bar. The optimal pH and temperature conditions of enzyme treatment are listed in Table 5. For all experiments, the liquid to fiber ratio was maintained at 150:1 (v/w) to facilitate complete wetting of fiber and different enzyme reaction times were used, 2, 4, and 8 h, respectively.

In enzyme treated fibers, the cellulose micro-fibrils were visible indicating that the lignin and other noncellulosic materials were greatly removed from the fiber surface becoming fiber surface rougher. They reported that pectinase enzyme treatment was the most effective for separating alfa fiber into smaller and single fiber due to the removal of materials from the fiber surface and the degradation of polygalacturonan. In consequence, they observed that pectinase enzyme treatment improved the mechanical properties of composite based on the alfa fibers. The tensile strength values of composites based on the pectinase treated fiber improved around 30% compared to alkali treated composites. However, the tensile strength values of composites based on the laccase and xylanase treated fibers did not observe important changes.

5 Conclusion

The use of natural fibers as reinforcement in polymer composites is attracting a lot of interest due to its mechanical properties, processing advantages, and environmental benefits. However, the hydrophilic character of the fibers lowers the compatibility with the hydrophobic matrix which results in poor mechanical properties of the composites. Therefore, it is necessary to modify the fiber surface morphology in order to reduce the hydrophilic character of the fibers and improve the fiber/

matrix adhesion. The fiber treatments could modify tensile properties of treated fibers, being often lower than untreated fibers. Nevertheless, composites based on the treated fibers could show higher tensile strength values than composites based on untreated fibers due to the adhesion improvement between the lignocellulosic fiber and polymer matrix after the surface treatments. The adhesion improvement for each fiber/polymer system depends on the fiber treatment type chosen as well as the chemical composition and morphological features of the composite components. In composites based on the lignocellulosic fibers, usually, the chemical treatments are used for fiber surface modification, such as alkali, silane, and acetylation among others. Inadequate treatment conditions could damage fiber structure, decreasing considerably fiber mechanical properties and consequently, the optimization of surface treatment conditions is a key factor to prepare composites with superior mechanical performance. Chemical treatments are the most used to modify fiber surface, however, physical and especially biological treatments, could be promising alternatives to improve lignocellulosic fiber/matrix adhesion because they are more environmentally friendly treatments than chemical ones. In most cases, physical and biological treatments are combined with chemical treatments in order to make more effective treatments.

References

Abdelmouleh M, Boufi S, Salah A, Belgacem MN, Gandini A (2002) Interaction of silane coupling agents with cellulose. Langmuir 18:3203–3208

Allaby M (1998) A dictionary of plant sciences, 2nd edn. Oxford University Press, Oxford

Arbelaiz A, Cantero G, Fernández B, Mondragon I, Gañán P, Kenny JM (2005a) Flax fiber surface modifications: effects on fiber physico mechanical and flax/ polypropylene interface properties. Polym Compos 26:324–332

Arbelaiz A, Fernandez B, Cantero G, Llano-Ponte R, Valea A, Mondragon I (2005b) Mechanical properties of flax fiber/polypropylene composites. Influence of fiber/matrix modification and glass fiber hybridization. Composites, Part A 36(2):637–644

Arbelaiz A, Fernandez B, Ramos JA, Retegi A, Llano-Ponte R, Mondragon I (2005c) Mechanical properties of short flax fiber bundle/polypropylene composites: influence of matrix/fiber modification, fiber content, water uptake and recycling. Compos Sci Technol 65(10): 1582–1592

Arbelaiz A, Trifol J, Peña-Rodriguez C, Labidi J, Eceiza A (2016) Modification of Poly(lactic acid) matrix by chemically modified flax fiber bundles and Poly(ethylene glycol) plasticizer. In: Visakh PM, Sigrid L (eds) Polyethylene-based biocomposites and bionanocomposites. ISBN: 978-1-119-03845-0

Bataille P, Dufourd M, Sapieha S (1994) Copolymerization of styrene on to cellulose activated by corona. Polym Int 34(4):387–391

Benitez AN, Monzón MD, Angulo I, Ortega Z, Hernandez PM, Marrero MD (2013) Treatment of banana fiber for use in the reinforcement of polymeric matrices. Measurement 46(3): 1065–1073

Bledzki AK, Gassan J (1999) Composites reinforced with cellulose based fibres. Prog Polym Sci 24:221–274

Bledzki AK, Reihmane S, Gassan J (1996) Properties and modification methods for vegetable fibers for natural fiber composites. J Appl Polym Sci 59:1329–1336

Bozaci E, Sever K, Sarikanat M, Seki Y, Demir A, Ozdogan E (2013) Effects of the atmospheric plasma treatments on surface and mechanical properties of flax fiber and adhesion between fiber—matrix for composite materials. Compos B 45(1):565–572
Buschle-Diller G, Fanter C, Loth F (1999) Structural changes in hemp fibers as a result of enzymatic hydrolysis with mixed enzyme systems. Text Res J 69(4):244–251
Chen C, Cho M, Kim B, Nam J, Lee Y (2012) Thermo plasticization and characterization of kenaf fiber by benzylation. J Ind Eng Chem 18(3):1107–1111
Cisneros-Lopez EO, Perez-Fonseca AA, Fuentes-Talavera FJ, Anzaldo J (2016) Rotomolded polyethylene-agave fiber composites: effect of fiber surface treatment on the mechanical properties. Polym Eng Sci 56(8):856–865
Daintith J (2000) A dictionary of chemistry, 4th edn. Oxford University Press, Oxford
Daniel G, Volc J, Niku-Paavola ML (2004) Cryo-FE-SEM & TEM immune-techniques reveal new details for understanding white-rot decay of lignocellulose. C R Biol 327(9–10):861–871
Datta J, Kopczynska P (2015) Effect of kenaf fibre modification on morphology and mechanical properties of thermoplastic polyurethane materials. Ind Crops Prod 74:566–576
De Rosa IM, Kenny JM, Puglia D, Santulli C, Sarasini F (2010) Morphological, thermal and mechanical characterization of okra (*Abelmoschus esculentus*) fibres as potential reinforcement in polymer composites. Compos Sci Technol 70:116–122
De Rosa IM, Kenny JM, Maniruzzaman M, Moniruzzaman Md, Monti M, Puglia D, Santulli C, Sarasini F (2011) Effect of chemical treatments on the mechanical and thermal behaviour of okra (*Abelmoschus esculentus*) fibres. Compos Sci Technol 71:246–254
Efanov MV, Ovchinnikov PV (2012) Acetylation of alkali treated peat. Solid Fuel Chem 46(2): 138–141
Egües I (2013) Lignocellulosic materials as new renewable resources: extraction, purification and potential use of hemicelluloses. Ph.D. thesis, Faculty of Engineering, Gipuzkoa, Spain
Fernandes EM, Mano JF, Reis RL (2013) Hybrid cork-polymer composites containing sisal fibre: morphology, effect of the fibre treatment on the mechanical properties and tensile failure prediction. Compos Struct 105:153–162
Focher B (1992) Physical characteristics of flax fibre. In: Sharma HSS, Van Sumere CF (eds) The biology and processing of flax. M. Publications, Belfast, UK, pp 11–32
Gassan J, Bledzki AK (1997) The influence of fiber-surface treatment on the mechanical properties of jute-polypropylene composites. Compos A 28(12):1001–1005
Gassan J, Bledzki AK (1999) Alkali treatment of jute fibers: relationship between structure and mechanical properties. J Appl Polym Sci 71(4):623–629
George J, Sreekala MS, Thomas S, George J, Sreekala MS, Thomas S (2001) A review on interface modification and characterization of natural fiber reinforced plastic composites. Polym Eng Sci 41(9):1471–1485
George M, Mussone PG, Bressler DC (2014) Surface and thermal characterization of natural fibres treated with enzymes. Ind Crops Prod 53:365–373
George M, Mussone PG, Alemaskin K, Chae M, Wolodko J, Bressler DC (2016) Enzymatically treated natural fibres as reinforcing agents for biocomposite material: mechanical, thermal, and moisture absorption characterization. J Mater Sci 51(5):2677–2686
Goriparthi BK, Suman KNS, Mohan N (2012) Effect of fiber surface treatments on mechanical and abrasive wear performance of polylactide/jute composites. Compos A 43(10):1800–1808
Haiping Y, Rong Y, Hanping C, Dong HL, Chuguang Z (2007) Characteristics of hemicellulose, cellulose and lignin pyrolysis. Fuel 86:1781–1788
Hanana S, Elloumi A, Placet V, Tounsi H (2015) An efficient enzymatic-based process for the extraction of high mechanical properties alfa fibres. Ind Crops Prod 70:190–200
Hill CAS, Abdul Khalil HPSA, Hale MD (1998) A study of the potential of acetylation to improve the properties of plant fibres. Ind Crops Prod 8:53–63
Hirota WH, Rodrigues RB, Sayer C, Giudici R (2010) Hydrolysis of acetic anhydride: non-adiabatic calorimetric determination of kinetics and heat exchange. Chem Eng Sci 65:3849–3858

Hou X, Sun F, Yan D, Xu H, Dong Z, Li Q (2014) Preparation of lightweight polypropylene composites reinforced by cotton stalk fibers from combined steam fl ash-explosion and alkaline treatment. J Cleaner Prod 83:454–462

Ibrahim MM, Agblevor FA, El-Zawawy WK (2010) Isolation and characterization of cellulose and lignin from steam-exploded lignocellulosic biomass. BioResources 5(1):397–418

Jandas PJ, Mohanty S, Nayak SK (2013) Surface treated banana fiber reinforced poly (lactic acid) nanocomposites for disposable applications. J Cleaner Prod 52:392–401

Jiang A, Xi J, Wu H (2012) Effect of surface treatment on the morphology of sisal fibers in sisal/polylactic acid composites. J Reinf Plast Compos 31:621–630

Josefsson T, Lennholm H, Gellerstedt G (2002) Steam explosion of aspen wood. Characterisation of reaction products. Holzforschung 56(3):289–297

Joseph K, Thomas S (1996) Effect of chemical treatment on the tensile properties of short sisal fibre-reinforced polyethylene composites. Polymer 37(23):5139–5149

Joseph PV, Joseph K, Thomas S (2002) Short sisal fiber reinforced polypropylene composites: the role of interface modification on ultimate properties. Compos Interfaces 9:171–205

Joseph PV, Joseph K, Thomas S, Pillai CKS, Prasad VS, Groeninckx G, Sarkissova M (2003) The thermal and crystallization studies of short sisal fiber reinforced polypropylene composites. Compos A 34(3):253–266

Kabir MM (2012) Effects of chemical treatments on hem fibre reinforced polyester composites. Ph. D. thesis, Faculty of Engineering and Surveying, University of Southern Queensland, Toowoomba, Australia

Kabir MM, Wang H, Lau KT, Cardona F (2012) Chemical treatments on plant-based natural fiber reinforced polymer composites: an overview. Compos B 43(7):2883–2892

Kalaprasad G, Francis B, Thomas S, Kumar CR, Pavithran C, Groeninckx G, Thomas S (2004) Effect of fibre length and chemical modifications on the tensile properties of intimately mixed short sisal/glass hybrid fibre reinforced low density polyethylene composites. Polym Int 53 (11):1624–1638

Kalia S, Kaith BS, Kaur I (2009) Pretreatments of natural fibers and their application as reinforcing material in polymer composites—a review. Polym Eng Sci 49(7):1253–1272

Karaduman Y, Gokcan D, Onal L (2012) Effect of enzymatic pretreatment on the mechanical properties of jute fiber-reinforced polyester composites. J Compos Mater 47(10):1293–1302

Kestur SG, Ramos LP, Wypych F (2009) Comparative study of Brazilian natural fibers and their composites with other. In Thomas S, Pothan LA (eds) Natural fibre reinforced polymer composites: from Macro to Nanoscale. Old City Publishing, Inc. Philadelphia, USA

Khan GMA, Shaikh H, Alam MS, Gafur MA (2015) Effect of chemical treatments on the physical properties of non-woven Jute/PLA. BioResources 10(4):7386–7404

Kim JK, Pal K (2010) Surface modifications in WPC with pre-treatment methods. In: Kim JK, Pal K (eds) Recent advances in the processing of wood-plastic composites. Springer, Berlin, pp 23–57

Klemm D, Philipp B, Heinze T, Heinze U, Wagenknecht W (1998) Comprehensive cellulose chemistry: fundamentals and analytical methods, vol 1. Wiley-VCH, Weinheim

Kumar P, Barrett DM, Delwiche MJ, Stroeve P (2009) Methods for pretreatment of lignocellulosic biomass for efficient hydrolysis and biofuel production. Ind Eng Chem Res 48(8):3713–3729

Kumar R, Obrai S, Sharma A (2011) Chemical modifications of natural fiber for composite material. Chem Sin 2(4):219–228

Le Moigne N, Longerey M, Taulemesse J, Bénézet J, Bergeret A (2014) Study of the interface in natural fibres reinforced poly (lactic acid) biocomposites modified by optimized organosilane treatments. Ind Crops Prod 52:481–494

Li X, Tabil LG, Panigrahi S (2007) Chemical treatments of natural fiber for use in natural fiber-reinforced composites: a review. J Polym Environ 15:25–33

Li Y, Pickering KL, Farrell RL (2009) Analysis of green hemp fibre reinforced composites using bag retting and white rot fungal treatments. Ind Crops Prod 29(2–3):420–426

Li Y, Zhang J, Cheng P, Shi J, Yao L, Qiu Y (2014) Helium plasma treatment voltage effect on adhesion of ramie fibers to polybutylene succinate. Ind Crops Prod 61:16–22

Lu B, Chung TC (2000) Synthesis of maleic anhydride grafted polyethylene and polypropylene, with controlled molecular structures. J Polym Sci, Part A-1: Polym Chem 38:1337–1343
Lu T, Liu S, Jiang M, Xu X, Wang Y, Wang Z, Gou J, Hui D, Zhou Z (2014) Effects of modifications of bamboo cellulose fibers on the improved mechanical properties of cellulose reinforced poly (lactic acid) composites. Compos B 62:191–197
Mansour OY, Nagaty A, El-Zawawy WK (1994) Variables affecting the methylation reactions of cellulose. J Appl Polym Sci 54(5):519–524
Matsuda H (1996) Chemical modification of solid wood. In: Hon DNS (ed) Chemical modification of lignocellulosic materials. Marcel Dekker Inc, New York, pp 159–184
Merkel K, Rydarowski H, Kazimierczak J, Bloda A (2014) Processing and characterization of reinforced polyethylene composites made with lignocellulosic fibres isolated from waster plant biomass such as hemp. Compos B 67:138–144
Mohanty S, Nayak SK (2006) Mechanical and rheological characterization of treated jute-HDPE composites with different morphology. J Reinf Plas Compos 25(13):1419–1439
Mohanty AK, Misra M, Drzal LT (2001) Surface modifications of natural fibers and performance of the resulting biocomposites: an overview. Compos Interfaces 8(5):313–343
Mondragon G, Fernandes S, Retegi A, Peña C, Algar I, Eceiza A, Arbelaiz A (2014) A common strategy to extracting cellulose nanoentities from different plants. Ind Crops Prod 55:140–148
Mukhopadhyay S, Fangueiro R (2009) Physical modification of natural fibers and thermoplastic films for composites—a review. J Thermoplas Compos Mater 22(2):135–162
Mwaikambo LY, Ansell MP (2002) Chemical modification of hemp, sisal, jute, and kapok fibers by alkalization. J Appl Polym Sci 84:2222–2234
Nair KCM, Diwan SM, Thomas S (1996) Tensile properties of short sisal fiber reinforced polystyrene composites. J Appl Polym Sci 60:1483–1497
Orue A, Jauregi A, Peña-Rodriguez C, Labidi J, Eceiza A, Arbelaiz A (2015) The effect of surface modifications on sisal fiber properties and sisal/poly (lactic acid) interface adhesion. Compos B 73:132–138
Orue A, Eceiza A, Peña-Rodriguez C, Arbelaiz A (2016a) Water uptake behavior and Young modulus prediction of composites based on treated sisal fibers and poly (lactic acid). Materials 9(5):400–414. https://doi.org/10.3390/ma9050400
Orue A, Jauregi A, Unsuain U, Labidi J, Eceiza A, Arbelaiz A (2016b) The effect of alkaline and silane treatments on mechanical properties and breakage of sisal fibers and poly (lactic acid)/sisal fiber composites. Compos A 84:186–195
Paul A, Joseph K, Thomas S (1997) Effect of surface treatments on the electrical properties of low-density polyethylene composites reinforced with short sisal fibers. Compos Sci Technol 57(1):67–79
Paul SA, Sinturel C, Joseph K, Mathew GDG, Pothen LA, Thomas S (2010a) Dynamic mechanical analysis of novel composites from commingled polypropylene fiber and banana fiber. Polym Eng Sci 50(2):384–395
Paul SA, Joseph K, Mathew GDG, Pothen LA, Thomas S (2010b) Influence of polarity parameters on the mechanical properties of composites from polypropylene fiber and short banana fiber. Compos A 41(10):1380–1387
Pickering KL, Li Y, Farrell RL, Lay M (2007) Interfacial modification of hemp fiber reinforced composites using fungal and alkali treatment interfacial modification of hemp fiber reinforced composites using fungal and alkali treatment. J Biobased Mater Bioenergy 1:109–117
Prasad N, Kumar V, Shishir A (2016) Banana fiber reinforced low—density polyethylene composites: effect of chemical treatment and compatibilizer addition. Iran Polym J 25(3): 229–241
Praveen KM, Thomas S, Grohens Y, Mozetic M, Junkar I, Primc G, Gorjanc M (2016) Applied surface science investigations of plasma induced effects on the surface properties of lignocellulosic natural coir fibres. Appl Surf Sci 368:146–156
Rachini A, Le Troedec M, Peyratout C, Smith A (2012) Chemical modification of hemp fibers by silane coupling agents. J Appl Polym Sci 123(1):601–610

Ragoubi M, George B, Molina S, Bienaimé D, Merlin A, Hiver J, Dahoun A (2012) Effect of corona discharge treatment on mechanical and thermal properties of composites based on miscanthus fibres and polylactic acid or polypropylene matrix. Compos A 43(4):675–685

Rahman MM, Mallik AK, Khan MA (2007) Influences of various surface pretreatments on the mechanical and degradable properties of photografted oil palm fibers. J Appl Polym Sci 105 (5):3077–3086

Raj G, Balnois E, Baley C, Grohens Y (2011) Role of polysaccharides on mechanical and adhesion properties of flax fibres in flax/PLA biocomposite. Int J Polym Sci 2011:1–11

Ranganathan N, Oksman K, Nayak SK, Sain M (2016) Structure property relation of hybrid biocomposites based on jute, viscose and polypropylene: The effect of the fibre content and the length on the fracture toughness and the fatigue properties. Compos A 83:169–175

Saba N, Tahir P, Jawaid M (2014) A review on potentiality of nano filler/natural fiber filled polymer hybrid composites. Polymers 6:2247–2273

Saha P, Chowdhury S, Roy D, Adhikari B, Kuk J (2016) A brief review on the chemical modifications of lignocellulosic fibers for durable engineering composites. Polym Bull 73(2): 587–620

Sampathkumar D, Punyamurthy R, Bennehalli B, Ranganagowda RP, Venkateshappa SC (2014) Natural areca fiber: surface modification and spectral studies. J Adv Chem 10(10):3263–3273

Saritha M, Arora A, Lata (2012) Biological pretreatment of lignocellulosic substrates for enhanced delignification and enzymatic digestibility. Indian J Microbiol 52(2):122–130

Saw SK, Sarkhel G, Choudhury A (2011) Applied surface science surface modification of coir fibre involving oxidation of lignins followed by reaction with furfuryl alcohol: characterization and stability. Appl Surf Sci 257(8):3763–3769

Saw SK, Purwar R, Nandy S, Ghose J, Sarkhel G (2013) Fabrication, characterization, and evaluation of luffa cylindrica fiber reinforced epoxy composites. BioResources 8(4):4805–4826

Sever K, Erden S, Gülec HA, Seki Y, Sarikanat M (2011) Oxygen plasma treatments of jute fibers in improving the mechanical properties of jute/HDPE composites. Mater Chem Phys 129(1–2): 275–280

Sinha E, Panigrahi S (2009) Effect of plasma treatment on structure, wettability of jute fiber and flexural strength of its composite. J Compos Mater 43(17):1791–1802

Sreekala MS, Thomas S (2003) Effect of fibre surface modification on water-sorption characteristics of oil palm fibres. Compos Sci Technol 63(6):861–869

Sreekala MS, Kumaran MG, Joseph S, Jacob M, Thomas S (2000) Oil palm fibre reinforced phenol formaldehyde composites: influence of fibre surface modifications on the mechanical performance. Appl Compos Mater 7(5):295–329

Sun E, Sun F, Dong Y (2016) Interface morphology and thermoplasticization behavior of bamboo fibers benzylated with benzyl chloride. Surf Interface Anal 48:64–72

Trindade WG, Hoareau W, Razera IAT, Ruggiero R, Frollini E, Castellan A (2004) Phenolic thermoset matrix reinforced with sugar cane bagasse fibers: attempt to develop a new fiber surface chemical modification involving formation of quinones followed by reaction with furfuryl alcohol. Macromol Mater Eng 289:728–736

Trindade WG, Hoareau W, Megiatto JD, Razera IAT, Castellan A, Frollini E (2005) Thermoset phenolic matrices reinforced with unmodified and surface-grafted furfuryl alcohol sugar cane bagasse and curaua fibers: properties of fibers and composites. Biomacromol 6(5):2485–2496

Wang B, Panigrahi S, Tabil L, Crerar W (2007) Pre-treatment of flax fibers for use in rotationally molded biocomposites. J Reinf Plast Compos 26(5):447–463

Wang K, Jiang JX, Xu F, Sun RC (2009) Influence of steaming pressure on steam explosion pretreatment of *Lespedeza* stalks (*Lespedeza crytobotrya*): Part 1. Characteristics of degraded cellulose. Polym Degrad Stab 94(9):1379–1388

Xie Y, Hill CAS, Xiao Z, Militz H, Mai C (2010) Silane coupling agents used for natural fiber/ polymer composites: a review. Compos A 41(7):806–819

Young R (1992) Activation and characterization of fiber surfaces for composites. In: Rowell R, Shulz TP, Narayan R (eds) Emerging technologies for materials and chemicals from biomass. American Chemical Society, Washington D.C., pp 115–135

Yu Y, Huang X, Yu W (2014) A novel process to improve yield and mechanical performance of bamboo fiber reinforced composite via mechanical treatments. Compos B 56:48–53

Zhang SM, Liu J, Zhou W, Cheng L, Duo XD (2005) Interfacial fabrication and property of hydroxyapatite/polylactide resorbable bone fixation composites. Curr Appl Phys 5:516–518

Zhang H, Cui Y, Zhang Z (2013) Chemical treatment of wood fiber and its reinforced unsaturated polyester composites. J Vinyl Addit Technol 19(1):18–24

Zhou Y, Fan M, Chen L (2016) Interface and bonding mechanisms of plant fibre composites: an overview. Compos B 101:31–45

Zou H, Wang L, Gan H, Yi C (2012) Effect of fiber surface treatments on the properties of short sisal fiber/poly(lactic acid) biocomposites. Polym Compos 33:1659–1666

Mechanical and Thermal Properties of Less Common Natural Fibres and Their Composites

Fabrizio Sarasini

Abstract Interest in the utilization of renewable resources is increasing because of depleting natural resources, environmental awareness and economic considerations. Natural fibre composites (biocomposites) have already attracted significant attention from the academic and industry (mainly automotive) fields. Natural fibres like kenaf, hemp, sisal, flax and jute have been the subject of extensive investigations for composite applications and therefore the focus of this chapter is on less common lignocellulosic fibres which are inexpensive and abundantly available at present. These fibres will be presented in terms of extraction methods, chemical, morphological, thermal and mechanical properties with a view to assessing their suitability as biofibres in composite applications. The introduction of these less common natural fibres in thermoplastic and thermosetting matrices is reviewed and compared with the traditional and much more common natural fibres, while the very last part of the chapter is dedicated to the development of cellulose-based nanocomposites, which is perceived as one of the most promising research fields related to lignocellulosic-based products.

Keywords Less common natural fibres · Agricultural residues
Biobased composites · Nanocellulose extraction · Chemical composition
Mechanical properties · Thermal stability

Contents

F. Sarasini (✉)
Department of Chemical Engineering Materials Environment,
Sapienza-Università di Roma, Via Eudossiana 18, 00184 Rome, Italy
e-mail: fabrizio.sarasini@uniroma1.it

S. Kalia (ed.), *Lignocellulosic Composite Materials*, Springer Series on Polymer and Composite Materials, https://doi.org/10.1007/978-3-319-68696-7_4

1 Introduction

Over the last decades, a resurgent interest in biocomposites has been recorded. Biocomposites are generally defined as composites derived from a polymer matrix and fibres coming from renewable resources, in general, natural fibres of plant origin. The interest in natural fibres was at first mainly academic but soon after natural fibre composites gained an increased attention in several industrial fields, being the automotive and the construction sectors the key drivers. This resurgent interest has been motivated by serious environmental concerns related not only to the unsustainable production processes of synthetic fibres and polymers, but also to the limited recyclability of conventional composites and their end of life disposal options. Natural fibres are usually classified as vegetal and animal ones, depending on their respective origin (Bledzki and Gassan 1999; John and Thomas 2008). Among the animal fibres, the most famous are silk and wool, which are widely diffused in the textile industry. As regards the use of natural fibres as reinforcement in polymer composites, the vegetal ones show the most promising results. Two are the main sources of vegetal fibres suitable for being used in composite materials: agricultural waste products or byproducts (such as bagasse from sugarcane or cereal straw) and plants that are intentionally cultivated for the extraction of fibres (such as textile fibres). Vegetal fibres can be extracted from many different parts of the plant, but mainly from the leaves (such as in the case of date palm, banana, agave, sisal) or from the bast (such as in the case of Malvaceae or Linaceae). Some attempts have also been made for leguminosae, grasses and Urticaceae due to their availability and spontaneity. Bast fibres, in particular hemp, flax, and kenaf, are generally considered to be the most promising because of their properties and availability. It is not surprising that they have been the subject of extensive investigations that led to several specific reviews (Summerscales et al. 2010a, b, 2013; Shahzad 2011; Akil et al. 2011; Yan et al. 2014; George et al. 2016). More recently, due to the growing industrial demand for vegetable fibres, the level of production is experiencing difficulties in meeting the demand. Moreover, there is an increasing quantity of byproducts coming from commodity crops, which are expensive and renewable resources suitable for fibre production. According to Huda and co-workers (Huda et al. 2007), about 2000 million tonnes of byproducts are produced worldwide every year from the major crops such as corn, wheat, rice, soybeans and sugarcane. Most byproducts are composed of considerable amounts of cellulose that can be extracted in the form of fibres and used as potential fillers in composites. This huge availability of fibres from byproducts of currently limited use can definitely enhance the sustainability of the fibre industry while adding value and increasing the income from the agricultural crops. Moreover, problems associated with the disposal of such byproducts after harvest could also be reduced if they can find suitable industrial applications. In addition to fillers obtained by byproducts, the integration of new natural fibres with acceptable properties can contribute to meet the demand of natural and renewable materials. In this framework, the present chapter deals with the review of properties and attributes of less common natural fibres envisaged as potential reinforcement in composites.

2 Morphological, Chemical and Physical Properties of Less Common Natural Fibres

The present paragraph addresses the physico-chemical properties of several natural fibres which are suitable to be used as reinforcement in polymer matrix composites. For better clarity, the chemical composition and the density of several less common natural fibres are reported and compared with those of commonly used natural fibres in Table 1.

2.1 Okra

Okra (*Abelmoschus esculentus L.*) is a member of the Malvaceae family which is currently grown throughout tropical Asia, Africa, the Caribbean and the southern United States. It represents the only vegetable crop in the Malvaceae family whose products are used in the food sector. Okra plant can be as high as 2 m and has leaves 10–20 cm long and broad, with lobes ranging from 5 to 7. The edible part of the plant is represented by the immature young seed pods. Aside from seed pods, this plant is also specifically cultivated for the edible oil that can be extracted from the seeds, in addition to the ripe seeds which can be roasted and ground, and used as a substitute for coffee. Therefore, this plant is not specifically grown for the fibres that can be obtained from the stem but this could represent an additional environmental advantage if one considers that the stems are usually treated as an agricultural waste product (Arifuzzaman Khan et al. 2009; Hakeem et al. 2014).

Table 1 Physico-chemical properties of common and less common natural fibres

Category	Fibre type	Species	Cellulose (%)	Hemicellulose (%)	Lignin (%)	Density (g/cm^3)	Source
Leaf	Sisal	*Agave sisalana*	56.5–78	5.6–16.5	8–14	1.45	Bledzki et al. (1996), Malkapuram et al. (2009), Fuqua et al. (2012)
	New Zealand Flax	*Phormium Tenax*	45.1–72.0	30.1	11.2	–	Carr et al. (2005), De Rosa et al. (2010c), Fuqua et al. (2012)
	Abaca	*Musa textilis*	56–63	20.8	7–12.4	1.5	Sun et al. (1998), Mohanty et al. (2005), Fuqua et al. (2012)
	Henequen	*Agave fourcroydes*	77.6	4–8	13.1	1.2	Malkapuram et al. (2009), Fuqua et al. (2012)
	Pineapple	*Ananas comosus*	70–82	15–19	5–12	1.44	Saha et al. (1990), Malkapuram et al. (2009), Fuqua et al. (2012)
	Banana	*Musa acuminata*	63–64	10–19	5	1.35	Mohanty et al. (2005), Fuqua et al. (2012)
	Curaua	*Ananas erectifolius*	70.2–73.6	9.9–18.3	7.5–9.3	1.38	Hoareau et al. (2004), Corrêa et al. (2010), Fuqua et al. (2012)
	Date palm	*Phoenix dactylifera*	33.9	26.1	27.7	0.92	Kaddami et al. (2006), Fuqua et al. (2012)
	Sansevieria	*Sansevieria cylindrica*	79.7	10.13	3.8	0.91	Sreenivasan et al. (2011)
	Buriti	*Mauritia flexuosa L.f.*	65–71*		21–27	0.63–1.12	Fangueiro and Rana (2016), Ornaghi Júnior et al. (2016)
Bast	Flax	*Linum usitatissimum*	71–81	18.6–20.6	2.2–3	1.51–1.54	Bledzki et al. (1996), Fuqua et al. (2012)
	Hemp	*Cannabis sativa*	70.2–74.4	17.9–22.4	3.7–5.7	1.48	Bledzki et al. (1996), Fuqua et al. (2012)

(continued)

Table 1 (continued)

Category	Fibre type	Species	Cellulose (%)	Hemicellulose (%)	Lignin (%)	Density (g/cm^3)	Source
	Jute	*Corchorus capsularis*	61–73.2	13.6–20.4	12–16	1.3–1.45	Bledzki et al. (1996), Fuqua et al. (2012)
	Ramie	*Boehmeria nivea*	68.6–76.2	13.1–16.7	0.6–1	1.45	Bledzki et al. (1996), Fuqua et al. (2012)
	Artichoke	*Cynara cardunculus L.*	75.3 ± 1.2	–	4.3 ± 0.5	1.58	Fiore et al. (2011)
	–	*Arundo donax L.*	43.2	20.5	17.2	1.17	Fiore et al. (2014b)
	Okra	*Abelmoschus esculentus*	60–70	15–20	5–10	0.88	Shamsul Alam and Arifuzzaman Khan (2007), Arifuzzaman Khan et al. (2009), Srinivasababu et al. (2015)
	Veldt grape	*Cissus quadrangularis*	82.73	7.96	11.27	1.21–1.22	Indran and Raj (2015), Mayandi et al. (2015a), (2016)
	Isora	*Helicteres Isora Linn.*	74.8	–	23.0	1.35–1.39	Joshy et al. (2007a), Mathew et al. (2007)
	Thespesia	*Thespesia Lampas*	60.63	26.64	12.70	–	Reddy et al. (2014)
	Marshmallow	*Althaea officinalis L.*	44.6	13.5	2.7	1.18	Sarikanat et al. (2014)
	Chakshir	*Ferula communis*	53.3	8.5	1.4	1.24	Seki et al. (2013)
	–	*Lygeum spartum L.*	–	–	–	1.50	Belouadah et al. (2015)

(continued)

Table 1 (continued)

Category	Fibre type	Species	Cellulose (%)	Hemicellulose (%)	Lignin (%)	Density (g/cm^3)	Source
	Hop	*Humulus lupulus L.*	84	–	6	–	Reddy and Yang (2009)
	–	*Prosopis juliflora*	61.65	16.14	17.11	0.58	Saravanakumar et al. (2013)
Fruit/seed	Coir	*Cocos nucifera*	32–47	0.3–20	31–45	1.15–1.45	Bledzki et al. (1996), Hill and Abdul Khalil (2000), Fuqua et al. (2012)
	Cotton	*Gossypium sp.*	82.7–92	5.7–6	0	1.5–1.6	Malkapuram et al. (2009), Fuqua et al. (2012)
	Borassus	*Borassus flabellifer L.*	45.67–53.40	29.60–32.76	17–21.53	0.78–0.80	Saravanan and Pallavi (2008), Reddy et al. (2009a) Obi Reddy et al. (2013)
Grasses	Alfa (Esparto)	*Stippa tenacissima*	45.4	38.5	14.9	0.89	Hattalli et al. (2002), Fuqua et al. (2012)
	Straw-Wheat		28.8–48.8	35.4–39.1	17.1–18.6	1.49	Kaushik et al. (2010), Fuqua et al. (2012)
	Bamboo	*Bambuseae ge.*	48.2–60.8	25.1	2.1–32.2	0.6–1.1	Jain et al. (1992), Fuqua et al. (2012)
	Napier grass or Elephant grass	*Pennisetum purpureum*	45.66	33.67	20.60	0.88–1.08	Rao et al. (2007), Reddy et al. (2012), Ridzuan et al. (2016a)
Root	Veldt grape	*Cissus quadrangularis*	77.17	11.02	10.45	1.51	Indran et al. (2014)

* Holocellulose (cellulose + hemicelluloses)

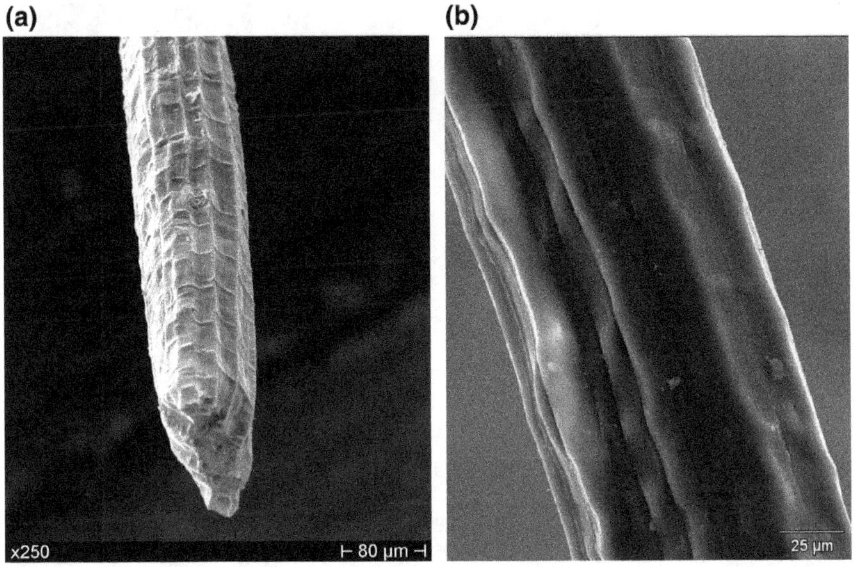

Fig. 1 SEM micrographs of okra fibre **a** cross section and **b** longitudinal surface

The fibres can be obtained from the stems of okra plants by water retting for about 15–20 days (Arifuzzaman Khan et al. 2009; De Rosa et al. 2010b).

Okra fibres exhibit the typical polygonal shape of most natural fibres as can be seen in Fig. 1 with a diameter in the range 40–180 μm and cell wall thickness and lumen diameter in the range 1–10 and 0.1–20 μm, respectively (De Rosa et al. 2010b). As regards the chemical composition, the amount of cellulosic and non-cellulosic constituents (Table 1) is similar to that found in other common bast and leaf fibres.

2.2 *Borassus*

These fibres are extracted from dried ripened fruits after being dipped in water for 2 weeks and subsequent removal of the black skin. Under this skin, usually, two different types of fibres can be found, long coarse and short fine fibres. The former is present in fruit nut edge to edge whilst the latter are well-adhered to the shell of the fruit (Reddy et al. 2009a; Obi Reddy et al. 2013). Figure 2 shows the Borassus tree, the fruit bunch and the extracted coarse and fine fibres whilst Fig. 3 reports the SEM micrographs of single borassus fibre in terms of longitudinal surface and cross section.

Based on 100 single fibres, Reddy et al. (2009a) showed that the average length, diameter, and aspect ratio of the coarse fibres are 20, 0.028 and 610 cm, respectively whereas for fine fibres these parameters are 12, 0.013 and 556, respectively.

Fig. 2 Photographs of **a** Borassus tree **b** fruit bunch **c** coarse fibres and **d** fine fibres [Reprinted with permission by Elsevier (Obi Reddy et al. 2013)]

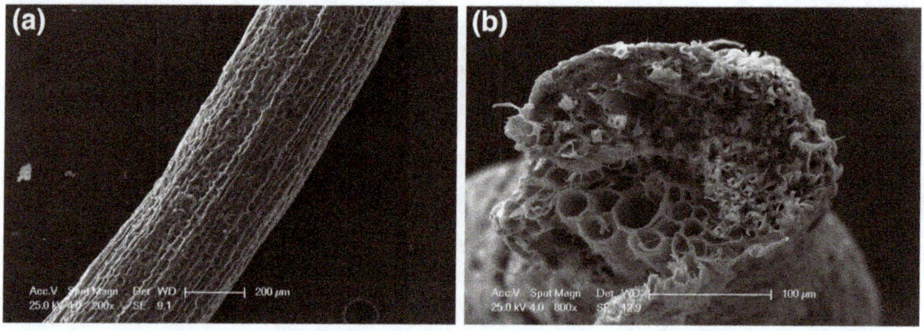

Fig. 3 SEM micrographs of borassus fibre **a** longitudinal surface and **b** cross section

The chemical composition, reported in Table 1, was found to be similar to that of other fruit fibres, such as coir, in terms of lower amount of cellulose and lower degree of crystallinity compared to bast or leaf fibres. The degree of crystallinity was estimated to be equal to 27.46 and 32.0 for coarse and fine fibres, respectively (Reddy et al. 2009a). It is also to be mentioned the notably low density of these fibres which is in the range of 0.778–0.800 g/cm^3 (Saravanan and Pallavi 2008).

2.3 *Arundo donax*

Giant reed (*Arundo donax L.*) belongs to the subfamily *Arundinoideae* of the Gramineae family (Andreu-Rodriguez et al. 2013). This plant is thought to be native to Asia but it is now common in southern Europe, North Africa, the Middle East, Australia, North and South America (Boose and Holt 1999). This plant can grow to 8–9 m with hollow stems of 2–3 cm and a growth rate of 0.3–0.7 m per

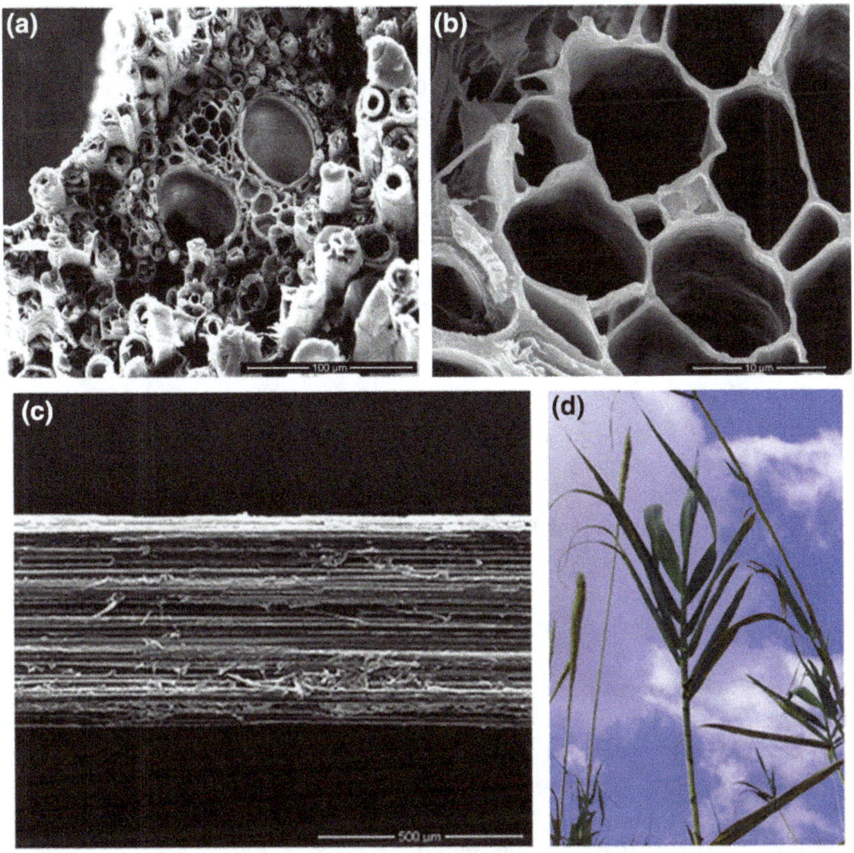

Fig. 4 SEM micrographs of **a–b** cross section and **c** longitudinal surface of arundo fibres along with the typical **d** Arundo donax plant. [Reprinted with permission from Elsevier (Fiore et al. 2014b; Scalici et al. 2016)]

week in ideal conditions (Andreu-Rodriguez et al. 2013). Arundo donax has been cultivated for thousands of years for horticultural and medicinal purposes and, most notably, for production of reeds for musical instruments (Mariani et al. 2010), paper (Ververis et al. 2004) and energy feedstock (Angelini et al. 2005, 2009). Many of these uses have been currently discarded and Arundo donax is considered as an invasive and unwanted plant able to alter wildlife, affect flood control and create serious fire hazards (Thornby et al. 2007). Physical removal is the primary means of controlling this weed even if it is highly ineffective because of prolific asexual reproduction from an extensive rhizome system (Boose and Holt 1999). Another way to cope with the widespread diffusion and related disposal issues is to obtain fibres to be used in polymer matrix composites. Typical Arundo donax plant and morphology of fibres extracted from the stem are reported in Fig. 4. In this case, the stems are at first separated from the foliage and then cut into small parts and dried. After this phase, the fibres are extracted from the stem by mechanical separation. In particular, the outer part of the culm is manually decorticated with the help of blade so as to obtain thin strips from which fibres can be easily separated with the aid of a scalpel (Fiore et al. 2014b). These fibres (Table 1) show a chemical composition with relatively high lignin content whilst hemicelluloses amount is close to that of okra and flax fibres. In contrast, the cellulose content is relatively low compared to other bast fibres.

Interestingly, the same authors (Scalici et al. 2016)) investigated also the possibility of obtaining fibres from the leaves. In this case, leaves were separated from the culms and fibres were successfully extracted by mechanical separation through a combing process with a metal teeth brush.

2.4 Isora

Isora is a bast fibre obtained from the bark of *Helicteres isora* plant by retting process (particularly effective when carried out in running water) whilst root stems and fruits of the plant are commonly used for medicinal applications. Recent research results suggested that this plant is a rich source of bioactive compounds such as polyphenols, tannins and alkaloids that exhibit therapeutic effects. Recently, the fibres of the plant have attracted considerable interest and it is now known that the best type of fibre is obtained when the plants are 1–1.5 years old while plants older than 2 years yield coarse and brittle fibres. Stalks can be harvested annually for fibre extraction from regenerated shoots (Mathew et al. 2007). Isora is a cellulose-rich fibre (Table 1) thus suggesting a potential ability to act as reinforcement in polymer matrix composites.

2.5 Napier Grass Fibres

Napier grass (also called elephant grass) belongs to the Poaceae family of *Pennisetum purpureum* species and needs a very little supplement of nutrients for growth. Native to subtropical Africa, now it is introduced into most tropical and subtropical countries. It is a robust perennial with a vigorous root system. Culms are usually 180–360 cm high, branched upwards with Leaf-sheaths glabrous or with tubercle-based hairs. Leaf-blades are 20–40 mm wide, margins thickened and shiny. Fibres can be extracted from the stem internodes using the common water retting process. Figure 5 shows photographs of the plant and of the extracted fibres.

It can be harvested 3–4 months after planting and can continue at an interval of 6–8 weeks for up to 5 years with a dry biomass yield per hectare per annum of 40 tons (Reddy et al. 2012). These fibres are characterized by a relatively low cellulose content which does not suggest remarkable mechanical properties even if a low

Fig. 5 **a** *Pennisetum purpureum* grass and **b** the extracted *P. purpureum* fibres [Reprinted with permission from Elsevier (Ridzuan et al. 2016a)]

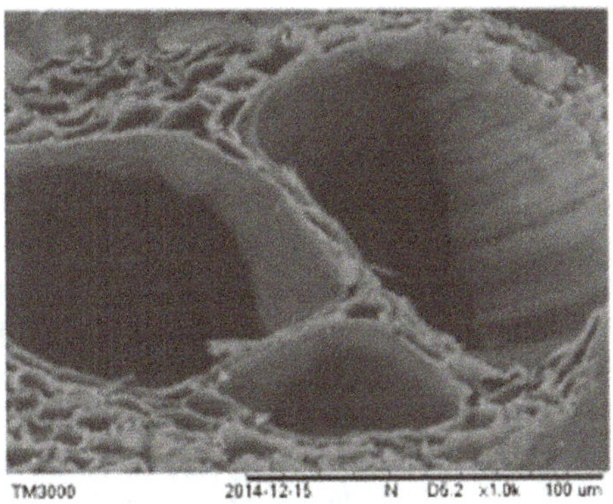

Fig. 6 SEM micrograph of the cross section of *Pennisetum purpureum* fibre [Reprinted with permission by Elsevier (Ridzuan et al. 2016a)]

density has been reported (Rao et al. 2007). The morphology is the typical multicellular structure highly porous in nature, as can be seen in Fig. 6.

2.6 *Cissus Quadrangularis*

Cissus quadrangularis is a perennial plant belonging to the *Vitaceae* (grape) family that is probably native to India or Srilanka, but is also found in Africa, Arabia and South Asia (Indran et al. 2014). This plant has medicinal value (Panthong et al. 2007; Bhujade et al. 2012; Sapsrithong et al. 2012) and is rich in fibre content in its roots as well as in the stem. This perennial climber can reach a height of around 2.5 m and has quadrangular sectioned stems with internodes of 1.5–1.8 cm width and 10–15 cm length (Indran et al. 2014) (Fig. 7).

Fig. 7 *Cissus quadrangularis* plant with its extracted **a** root and **b** stem fibres [Reprinted with permission from Elsevier (Indran et al. 2014; Indran and Raj 2015)]

Stem and root fibres can be extracted through water soaking to allow for microbial degradation. After being washed and dried in natural sunlight, combing was used with a metal brush to obtain long and uniform fibres. From a chemical point of view, stem and root fibres were found to exhibit similar percentages of main constituents, i.e. cellulose, hemicelluloses and lignin. In particular, both fibres showed a high content of cellulose (higher than 70%) which ensures good mechanical properties and a relatively low hemicellulose content (about 10%) which is the main hydrophilic constituent in natural fibres. It is also worth noting the low wax content (less than 0.2%) which could be responsible for a better bonding with polymer matrices. A low microfibril angle (about 5°) was found for both fibres with a degree of cellulose crystallinity around 50% which is in line with values of other bast and leaf fibres (Sreenivasan et al. 2011). These fibres exhibited a surface morphology characterized by the presence of shallow pores and impurities which increase the surface roughness of the fibres, as can be clearly noted in Fig. 8.

3 Thermal and Mechanical Properties of Less Common Natural Fibres

This paragraph addresses the thermal and mechanical properties of several natural fibres which are suitable to be used as reinforcement in polymer matrix composites. For better clarity, the mechanical properties of several less common natural fibres are reported and compared with those of commonly used natural fibres in Table 2.

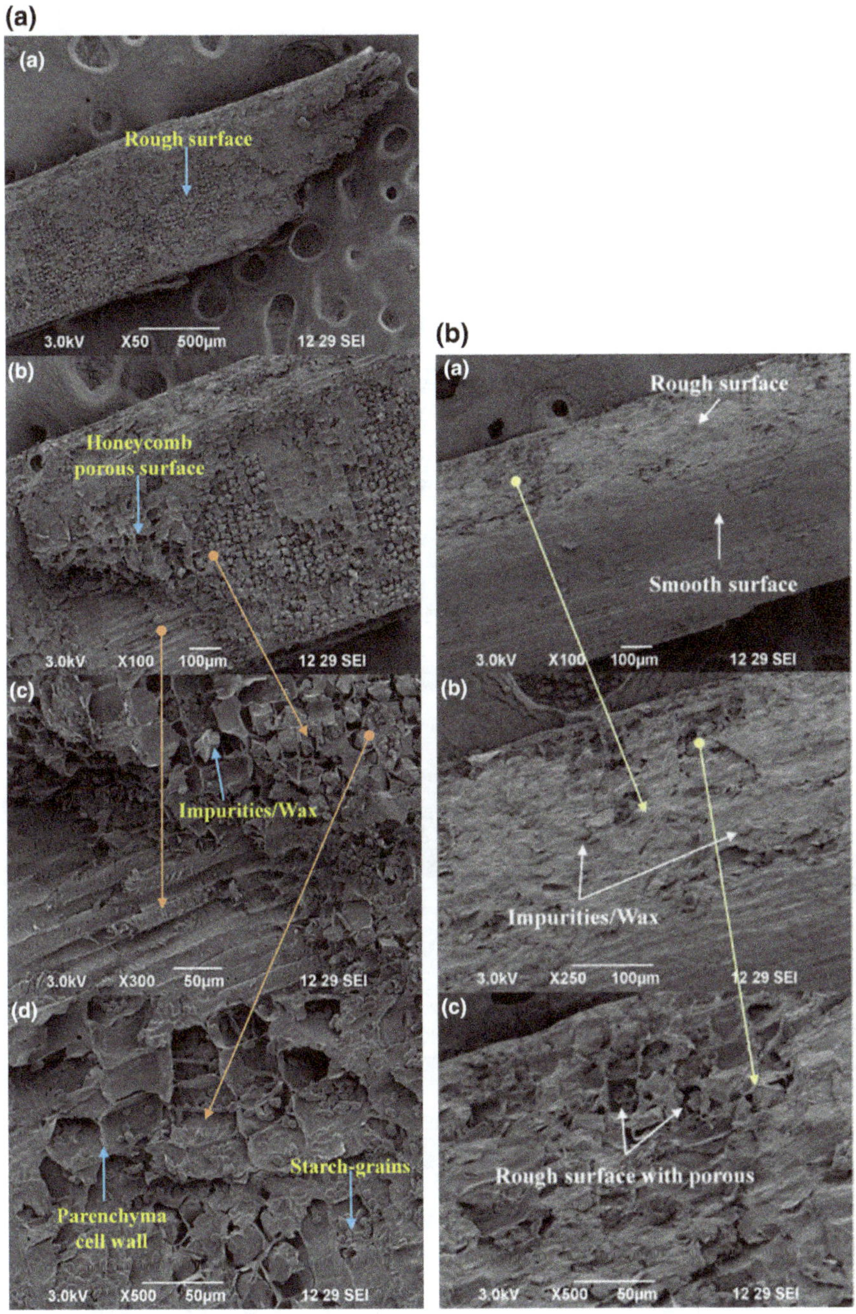

Fig. 8 SEM micrographs of *Cissus quadrangularis* fibres extracted from **a** stem and **b** root [Reprinted with permission from Elsevier (Indran et al. 2014; Indran and Raj 2015)]

Table 2 Mechanical properties of common and less common natural fibres

Category	Fibre type	Diameter (μm)	Tensile strength (MPa)	Young's modulus (GPa)	Elongation at break (%)	Source
Leaf	Sisal	50–200	80–640	1.46–15.8	3–7	Jayaraman (2003), Mishra et al. (2004), Fuqua et al. (2012)
	Abaca	28	756	31.1	2.9	Shibata et al. (2002)
	Henequen	180	500	13.2	4.8	Herrera-Franco and Valadez-Gonzalez (2005)
	Pineapple	20–80	413–1627	34.5–82.5	0.8–1.6	Mishra et al. (2004)
	Banana	50–280	529–914	7.7–32	1.8–3.7	Fuqua et al. (2012)
	Curaua	9–10	913	30	3.9	Gomes et al. (2007)
	Date palm	100–1000	170–275	5–12	5–10	Al–Khanbashi et al. (2005)
	Phormium tenax	100–140	465–770	23–27	2–5	De Rosa et al. (2010a)
	Sansevieria	80–90	630–670	5–7.5	10–12	Munawar et al. (2006), Sreenivasan et al. (2011)
	Buriti	150–170	735 ± 168	27 ± 5	2.7 ± 0.4	Ornaghi Júnior et al. (2016)
Bast	Flax	30–110	450–1500	27.6–38	1.5–3.2	Barkoula et al. (2009), Malkapuram et al. (2009), Arias et al. (2013)
	Hemp	53.7	690–873	9.93	1.6–4.7	Graupner et al. (2009), Fuqua et al. (2012)
	Jute	25–200	393–773	2.5–26.5	1–2	Malkapuram et al. (2009), Fuqua et al. (2012)
	Kenaf	43.3–140	223–624	11–14.5	2.7–5.7	Malkapuram et al. (2009), Fuqua et al. (2012)
	Ramie	34	400–938	24.5–128	1.2–3.8	Angelini et al. (2000), Goda et al. (2006b)
	Okra	40–180	234–380	5–13	1.8–2.1	De Rosa et al. (2010b), (2011)
	Arundo donax	450–700	248	9.4	3.24	Fiore et al. (2014b)
	Veldt grape (1)	770–870	2300–5479	56–234	3.75–11.14	Indran and Raj (2015)
	Veldt grape (2)	–	61.42 ± 17.3	1.1 ± 0.3	5.6 ± 1.37	Mayandi et al. (2015a)
	Isora	10	561–600	19–22	5–6	Joshy et al. (2007a), Mathew et al. (2007)
	Thespesia	64	573 ± 15	61.2 ± 12	0.79 ± 0.11	Reddy et al. (2014)

(continued)

Table 2 (continued)

Category	Fibre type	Diameter (μm)	Tensile strength (MPa)	Young's modulus (GPa)	Elongation at break (%)	Source
	Marshmallow	156–194	415.2 ± 11.5	65.4 ± 7.2	3.9	Sarikanat et al. (2014)
	Chakshir	90–300	475.6 ± 15.7	52.7 ± 3.7	4.2 ± 0.2	Seki et al. (2013)
	Lygeum spartum	180–433	64.63–280.03	4.47–13.27	1.49–3.74	Belouadah et al. (2015)
	Hop	–	4.1 ± 1.9 (g/den)	161 ± 57 (g/den)	3.3 ± 1.2	Reddy and Yang (2009)
	Proscopis juliflora	–	558 ± 13.4	–	1.77 ± 0.04	Saravanakumar et al. (2013)
	Artichoke	100–300	100–300	5–15	2.5–3	Fiore et al. (2011)
Fruit/ seed	Cotton	15.6–21	287–800	1.1–12.6	6–9.7	Malkapuram et al. (2009), Fuqua et al. (2012)
	Coir	40–450	106–175	1.27–6	15–59.9	Satyanarayana et al. (1986), Malkapuram et al. (2009)
	Borassus	130–280	50.9–65.2	1.22–4.91	41.2–47.2	Reddy et al. (2009a)
Grasses	Straw-Wheat	84–94	59–140	3.7–4.8	–	Sain and Panthapulakkal (2006, Fuqua et al. (2012)
	Alfa (Esparto)	–	350	22	5.8	Fuqua et al. (2012)
	Bamboo	88–125	140–441	11–36	1.3–8	Okubo et al. (2004), Fuqua et al. (2012)
	Napier grass	210–270	73–75	5.68–6.8	1.4–2.8	Reddy et al. (2012), Ridzuan et al. (2016a)
Root	Veldt grape	610–725	1857–5330	68–203	3.57–8.37	(Indran et al. 2014)
Manmade	E-glass	9.8–17	2000–3500	70	0.5–4.8	Thomason (1999), Fuqua et al. (2012)
	Aramid	12	3000–3150	63–130	3.3–3.7	Fuqua et al. (2012)
	Carbon	5–17	4000	230–240	1.4–1.8	Fuqua et al. (2012)

3.1 Okra

Comparing the chemical composition of okra fibres to that of other bast or leaf fibres (Table 1), similarities are evident that suggest also comparable thermal behaviour and stability. It is well-known that a key limiting factor in the use of natural fibres as reinforcement in biocomposites is their low thermal stability. The thermal stability of okra fibres was investigated by thermogravimetric analysis by De Rosa et al. (2010b) in nitrogen atmosphere (250 ml/min) from 30 to 600 °C at a heating rate of 10 °C/min. In Fig. 9 are reported the typical TG and DTG curves of okra fibres.

The thermogravimetric signal showed three weight loss steps, while the decomposition occurred in two main stages. An initial weight loss (~8%) was observed between 30 and 110 °C which was related to the vaporization of the water, whilst the onset degradation for the okra fibres was found to occur after 220 °C. Above this temperature, the thermal stability gradually decreased. In particular, the first stage T_I (220–310 °C) was due to the thermal depolymerization of hemicellulose, pectin and the cleavage of glycosidic linkages of cellulose (weight loss of 16.1%) whilst the second one (T_{II}) occurred in the range (310–390 °C) and corresponded to the degradation of α-cellulose in the fibre (weight loss of 60.6%). In general, the decomposition of lignin, owing to its complex structure, tends to occur within the whole temperature range. These results are also confirmed by DTG curve where the maximum decomposition rates for the weight losses are shown. Furthermore, it has been observed a residual weight percentage of 7.6% which can be ascribed to carbonaceous residues and possible undegraded fillers. From these results, it can be concluded that the thermal stability of okra fibres is comparable to that of many vegetable fibres (Wielage et al. 1999; Ouajai and Shanks 2005; Yao

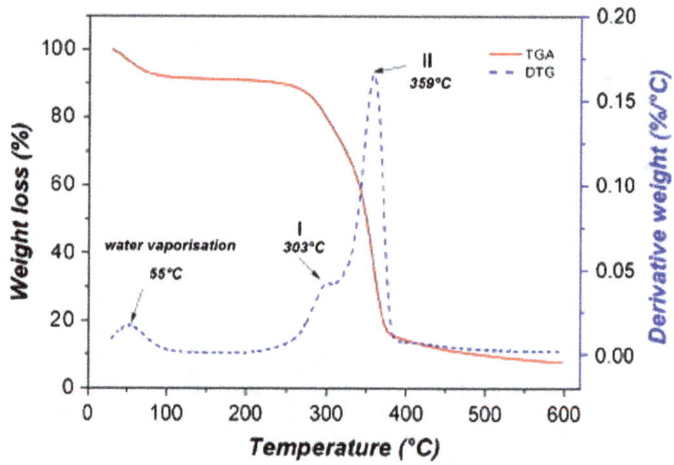

Fig. 9 TG and DTG curves of okra fibres [Reprinted with permission by Elsevier (De Rosa et al. 2010b)]

et al. 2008). As regards the mechanical properties, okra fibres exhibited a brittle behaviour and also a large scatter in data from single fibre tensile tests which was successfully modelled with a two-parameter Weibull distribution (Weibull modulus equal to 1.91) (De Rosa et al. 2010b). From Table 2 it can be concluded that comparable mechanical behaviour with soft bast fibres, such as kenaf, and leaf fibres, such as date palm, henequen and sisal, is evident. From these results, the suitability of okra fibres as reinforcement in polymer matrix composites can be easily inferred (Hakeem et al. 2014). In an attempt to increase the interfacial interaction with polymer matrices, De Rosa et al. (2011) investigated the effects of several chemical treatments on the thermal and mechanical properties of okra fibres. As a general comment, all chemical treatments caused some decrease in mechanical properties, particularly of Young's modulus. In particular, treatments involving acetylation resulted in a lower level of fibre degradation, whilst bleaching, scouring and particularly permanganate treatment appeared all, at some extent (and in growing order), damaging to the fibres. As regards the thermal stability, acetylation and sodium sulphate treatments decreased the thermal stability of the fibre, while in the case of bleached fibres there was no significant difference in the thermal behaviour of the treated fibres (both for acetylation and sodium hydroxide treatment).

3.2 *Borassus*

Thermal stability of fine and coarse borassus fibres was investigated in (Reddy et al. 2009a) using a nitrogen atmosphere at a heating rate of 10 °C/min from 50 to 600 °C. The authors found no significant differences between fine and coarse fibres suggesting their use as reinforcement in thermoplastic matrix materials whose processing temperature is below 270 °C. A slightly higher thermal stability was reported for borassus fibres previously treated in a 5% aqueous sodium hydroxide solution at room temperature, maintaining a liquor ratio of 30: 1 for 30 min. A marginal difference in the thermal stability of untreated and alkali-treated borassus fibres for different treatment durations (1, 4, 8 and 12 h) in a 5% aqueous sodium hydroxide solution at room temperature was reported by (Obi Reddy et al. 2013). Alkali treatment was found not to affect in a significant way the thermal stability of the fibres even if the mechanical properties were remarkably enhanced after this treatment. First of all the authors highlighted the better mechanical properties of fine fibres compared to coarse fibres and secondly the alkali treatment up to a duration of 8 h increased tensile strength, Young's modulus and elongation at break with maximum improvements of 41, 69 and 40%, respectively. A treatment of 12 h was found to be detrimental for the mechanical properties due to degradation of cellulose occurring at longer durations. It is to be noted that despite the low mechanical properties of borassus fibres, the elongation at break is significantly higher compared to bast and leaf fibres, thus suggesting their potential use in hybrid systems aiming at increasing the ductility of composites while preserving considerable strength and stiffness.

3.3 Arundo Donax

The thermal stability of Arundo donax fibres extracted from the stem of the plant was investigated in (Fiore et al. 2014b) from 30 to 750 °C at a heating rate of 10 °C/min in an air atmosphere. The results of the thermogravimetric analysis are shown in Fig. 10.

In addition to the first peak (40–115 °C), due to vaporization of absorbed water, the DTG curve of arundo fibres showed four additional peaks. Thermal degradation of arundo fibres began at 275 °C (onset degradation temperature) with the first degradation peak occurring at about 295 °C due to the thermal depolymerization of hemicelluloses and pectin, and the glycosidic linkages of cellulose (12% weight loss). The second peak (320 °C) is ascribed to the degradation of α-cellulose (70% weight loss) with similar degradation occurring at 308.2, 298.2 and 309.2 °C for hemp, jute and kenaf fibres, respectively (Yao et al. 2008). The two small peals at 435 °C and 518 °C are instead due to oxidative degradation of the charred residue. Similar thermal stability was found also for arundo fibres extracted from the leaves of the plant, even if these fibres were reported to be stable until around 210 °C and no additional peaks were detected after the degradation peak of the α-cellulose at 340 °C, as this time the analysis was performed in inert atmosphere (Fig. 11) (Scalici et al. 2016).

Plasma treatment did neither affect the thermal behaviour nor the mechanical properties of arundo leaf fibres (Scalici et al. 2016) which exhibited lower mechanical properties compared to fibres extracted from the stem. Also, in this

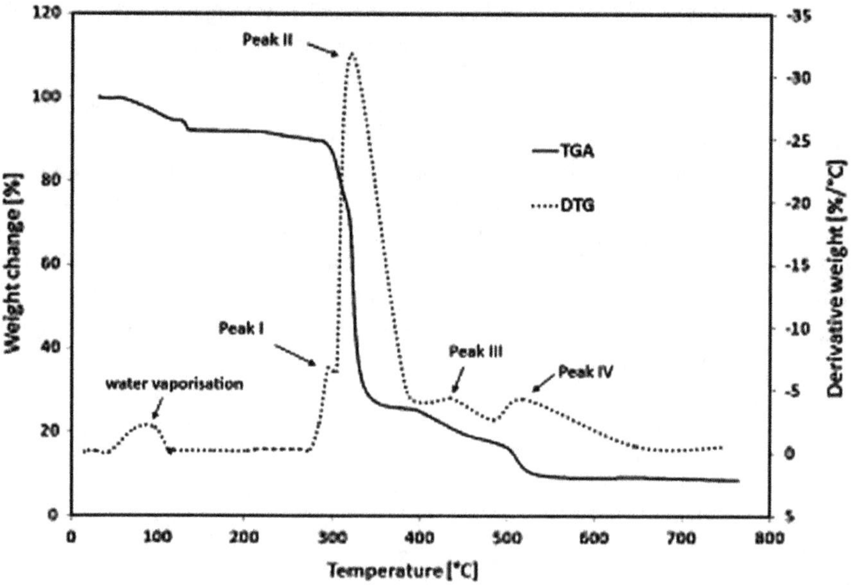

Fig. 10 TG and DTG curves of arundo fibres extracted from stem [Reprinted with permission from Elsevier (Fiore et al. 2014b)]

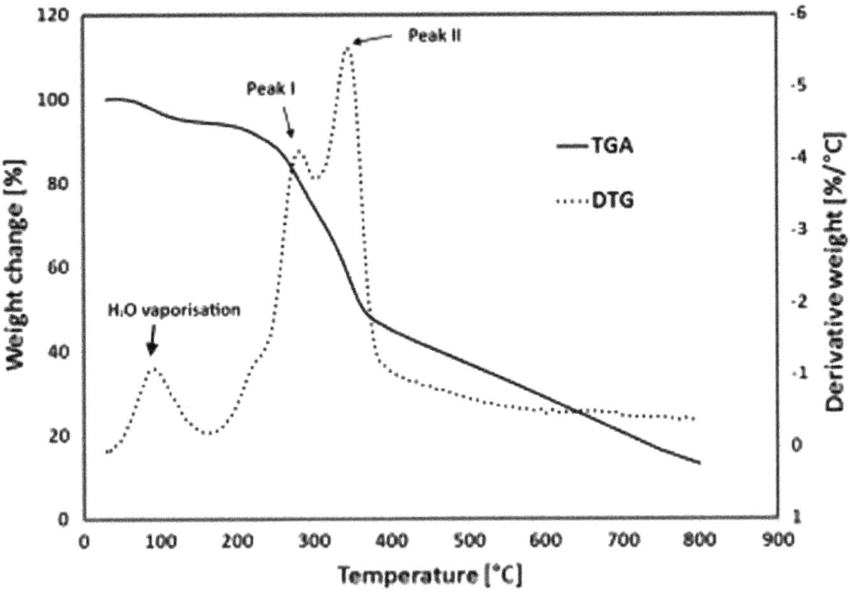

Fig. 11 TG and DTG curves of arundo leaf fibres [Reprinted with permission from Elsevier (Scalici et al. 2016)]

case, the mechanical properties were interpreted based on a two-parameter Weibull distribution (Weibull modulus equal to 1.42 and 3.12 for leaf and stem fibres, respectively). It is also evident that the mechanical properties of arundo fibres are comparable to those of other natural fibres currently investigated as potential reinforcement in polymer matrix composites. Moreover, the microfibril angle of arundo fibres (i.e. 7.37°) is also similar to one of the other natural fibres such as jute (8.1°), flax (5°) and hemp (6.2°), thus confirming their suitability as reinforcement in polymer matrices.

3.4 Isora

Isora fibres exhibited an enhanced thermal stability when treated with alkali and this was achieved for two different treatment durations and NaOH concentrations, namely 1% aqueous alkali at 30 °C for 4 h (Joshy et al. 2007a) and 5% NaOH for 4 h (Mathew et al. 2007). Major weight loss of untreated fibres took place at about 360 °C but this temperature was raised to about 380 °C by mercerization. Thermal stability was also enhanced by a silane treatment even if this effect was less evident compared to the alkali treatment (Mathew et al. 2007). These surface treatments were beneficial to the thermal stability but affected in a negative way the mechanical properties of single isora fibres. In particular, tensile strength decreased from 561 MPa for untreated fibres to 500 and 474 MPa after alkali and silane treatments, respectively (Mathew et al. 2007). Also, elongation at break was reported to be slightly influenced by the surface treatments whilst no information was provided for Young's modulus. The decrease in mechanical properties was ascribed by the authors to the partial removal of lignin after chemical treatments. Despite the negative influence of surface treatments, Isora fibres still show a satisfactory combination of strength and ductility (~4–5%) to suggest their use in composite materials.

3.5 Napier Grass Fibres

Mechanical and thermal properties of Napier grass fibres have been the subject of several investigations. Haameem et al. (2014) recently determined that the maximum tensile strength (106.24 MPa) of *Pennisetum purpureum* single fibres was achieved with 10% alkali treatment (for 24 h) with 2.19 GPa modulus of elasticity. However, this is in contrast with the results of Reddy et al. (2012) who determined that the maximum tensile strength (136 MPa) and Young's modulus (10.5 GPa) were achieved with 5% alkali treatment (for 30 min at room temperature). Ridzuan et al. (2016a) performed an extensive investigation of the effects of different alkaline treatments. Alkali solutions of 5, 7, 10, 12 and 15% were applied to the fibres for 24 h at room temperature. The effects of different alkali concentration on tensile strength and Young's modulus are reported in Fig. 12. With increasing concentration, the authors noted a reduction in fibre diameter and in moisture

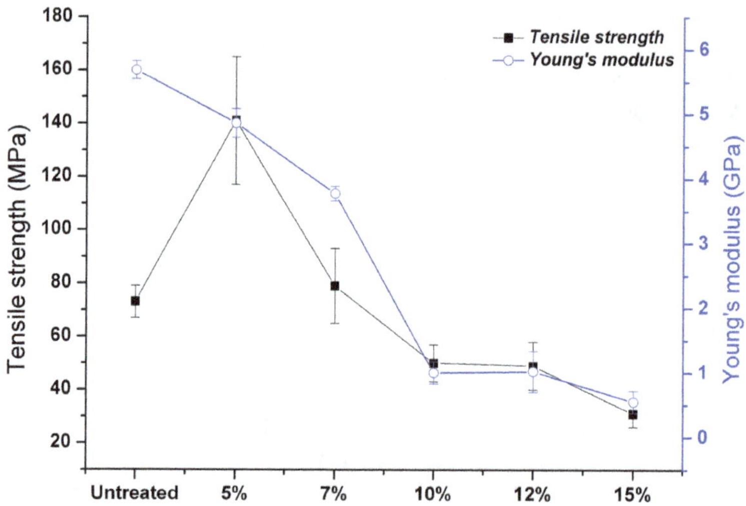

Fig. 12 Tensile strength and Young's modulus of Napier grass fibres as a function of alkali concentration (Ridzuan et al. 2016a)

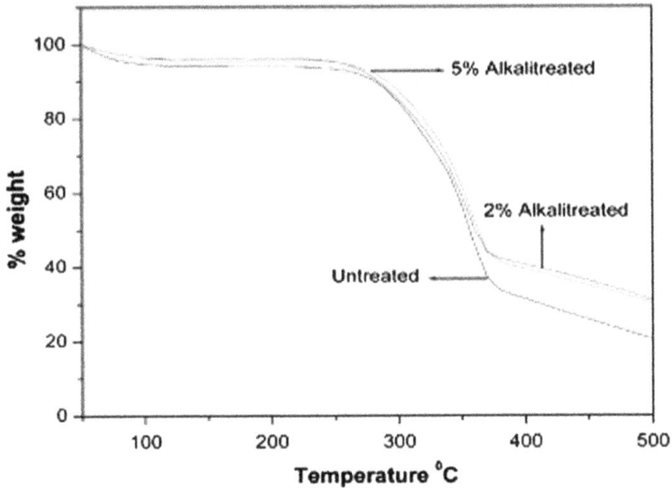

Fig. 13 Thermograms of untreated and alkali-treated Napier grass fibres [Reprinted with permission from Elsevier (Reddy et al. 2009b)]

content which were ascribed to the partial removal of hemicelluloses and the compression of the lumen structure of fibres.

From Fig. 12 it is evident that the 5% alkali-treated fibres exhibit the highest tensile strength, followed by 7% treated, untreated, 10% treated, 12% treated and 15% treated fibres, respectively. The authors reported also a decrease in Young's modulus and an increase in elongation at break with increasing alkali concentration, effects confirmed also for other types of natural fibres (Goda et al. 2006a). Thermal analysis of untreated and alkali-treated Napier grass fibres highlighted an increase of thermal stability with increasing alkali concentration, as can be seen in Fig. 13.

Also, the crystallinity of Napier grass fibres was found to increase after alkali treatment, thus suggesting the use of 5%-alkali-treated fibres in polymer matrix composites.

3.6 *Cissus Quadrangularis*

Fibres extracted from the root of the plant were investigated by Indran et al. (2014). As previously mentioned, these fibres were found to have a high cellulose content (~77%), a high degree of crystallinity (56.6%) and quite low microfibril angle (5.89°), all properties that suggest high mechanical properties. In fact, the authors reported quite high mechanical properties for these fibres, which were obtained through single fibre tensile tests at five different gauge lengths (from 10 to 50 mm). Tensile strength was in the range 2500–3100 MPa while Young's modulus in the range 90–130 GPa which seem to outperform the other natural fibres, being comparable to the absolute values of glass fibres. These results are quite surprising, even considering that the authors found an increasing tendency of both strength and Young's modulus with increasing gauge length. Tensile strength and Young's modulus are indeed highly dependent on gauge length but usually, at short gauge lengths the strength values are higher and decrease with increasing fibre length (Defoirdt et al. 2010; De Rosa et al. 2010a; Bezazi et al. 2014; Sarasini et al. 2015). The dependence of failure strength on clamping length is a well-known characteristic of all fibres, which can be explained by the distribution of defects within the fibres. The longer the stressed length, the higher the number of defects present in the stressed fibre portion, contributing to the weakening of its structure. As regards the thermal stability, an initial weight loss was reported at around 89 °C, which was attributed to the removal of moisture content from the fibre. The fibre was thermally stable up to around 230 °C. After 230 °C and up to 330 °C a weight reduction, corresponding to the thermal decomposition of the hemicellulose and glycosidic linkages of cellulose, occurred. Two additional small peaks were observed at 481 and 696 °C due to degradation and pyrolysis of lignin, respectively. The same authors reported also the characterization (Indran and Raj 2015) of fibres extracted from the stem of the same plant. Also, in this case, the fibres showed extremely high mechanical properties, with tensile strength in the range 2300–4200 MPa and Young's modulus in the range 56–130 GPa.

The thermal stability was found to be higher compared to that of fibres extracted from the root even if the degradation pattern was similar, as confirmed by the thermogram reported in Fig. 14. These results are in contrast with what reported by Mayandi et al. (2015a) who, while comparing the properties of *Cissus quadrangularis* and *Phoenix reclinata* natural fibres, reported a high value of cellulose

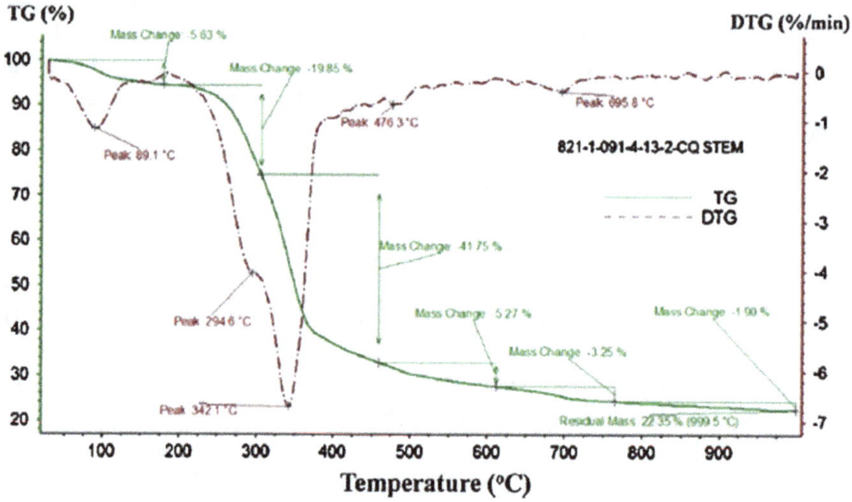

Fig. 14 TG and DTG curves of the *Cissus quadrangularis* stem fibres [Reprinted with permission from Elsevier (Indran and Raj 2015)]

($\sim$80%) but significantly lower mechanical properties (at 50 mm gauge length), namely 61.42 MPa and 1.1 GPa for tensile strength and Young's modulus, respectively. The strain to failure was indeed comparable with the one reported by Indran and Raj (Indran and Raj 2015), i.e. $\sim$5%.

4 Mechanical Properties of Less Common Natural Fibre Reinforced Composites

Mechanical properties of natural fibre composites have been the subject of extensive investigation and therefore a large amount of literature is available on this topic. To provide a sound and comprehensive comparison of data for composites reinforced with less common natural fibres and those with common natural fibres, details of mechanical properties obtained including tensile strength and stiffness, as well as impact strength are reported in Table 3. A graphical overview of the range of strength, stiffness, specific stiffness and specific strength compared with those for glass fibre reinforced plastics is shown in Fig. 15.

A first consideration that can be drawn from Fig. 15 is that specific strength and stiffness are comparable to those of glass fibre reinforced composites even if the absolute strength is currently well below that of synthetic counterparts. However, the tensile properties of various composites reinforced with less common natural fibres considered in the present study are in good agreement with those of composites made from other natural fibres as presented by earlier investigators (Table 3). In addition, one of the value-added methods for the use of these lignocellulosic biomass is the production of cellulose nanocrystals (CNC) or whiskers to develop green nanocomposites (Hubbe et al. 2008; Abdul Khalil et al. 2012). This suggests that the use of these less common natural fibres as the source of CNC may be a viable alternative to using them in the form of technical fibres (Kalia et al. 2011). CNC can be extracted from a variety of cellulosic sources, such as plants, bacteria and algae. Various physical (high-pressure homogenization), chemical (acid hydrolysis) (Prasad Reddy and Rhim 2014; Rhim et al. 2015) and biological (enzymatic hydrolysis) (Henriksson et al. 2007) methods have been reported for the isolation of nanocellulose, but among them, acid hydrolysis is the most widely used method to prepare cellulose nanowhiskers. Fortunati et al. (2013a) demonstrated that the same hydrolysis parameters used for extraction of cellulosic fraction starting from microcrystalline material (Fortunati et al. 2012) can be successfully adapted for hydrolysis starting from a technical fibre, such as okra. In this particular case, a two-step procedure was implemented, in which the holocellulose produced by the action of a first chemical treatment was then exposed to the action of sulphuric acid for further hydrolysis. The cellulose nanocrystals obtained were employed as reinforcement in a poly (vinyl alcohol) (PVA) matrix by solvent casting. Cellulose nanocrystals were able to increase the crystallinity degree and the Young's modulus of PVA matrix. Reddy et al. (2016) recently reported the

Table 3 Comparison of mechanical properties of less common natural fibre composites with other common natural fibre reinforced composites

Fibre	Matrix	Fibre content (%) (*w* for weight and *v* for volume)	Tensile strength (MPa)	Young's modulus (GPa)	Charpy (c) or Izod (i) impact strength (kJ/m^2 or J/m)	Notes	References
Sisal (unidirectional)	Epoxy	50 (v)	~270	~9	–	Compression moulding/untreated	Rong et al. (2001)
Sisal (unidirectional)	Epoxy	50 (v)	~340	~5.8	–	Compression moulding/alkali-treated	Rong et al. (2001)
Sisal (unidirectional)	Epoxy	50 (v)	~260	~4	–	Compression moulding/acetylated	Rong et al. (2001)
Flax (unidirectional)	Epoxy	42 (v)	280	35	–	Resin transfer moulding/enzyme extracted	Oksman (2001)
Flax (unidirectional)	Epoxy	32 (v)	132	15	–	Resin transfer moulding/traditionally retted	Oksman (2001)
Flax (unidirectional)	Epoxy	40 (v)	133	28	–	Autoclave/untreated	Van de Weyenberg et al. (2003)
Phormium tenax (unidirectional)	Epoxy	55 (w)	223	16.8	–	Compression moulding/untreated	Le and Pickering (2015)
Sisal (unidirectional)	Epoxy	46 (v)	211	19.7	–	Resin transfer moulding/untreated	Oksman et al. (2002)
Flax (unidirectional)	Epoxy	32 (v)	132	15	–	Resin transfer moulding/untreated	Oksman et al. (2002)
Flax (unidirectional)	Epoxy	42 (v)	378	39.9	–	Autoclave (prepregs)	Baets et al. (2014)
Alfa (aligned)	Polyester	32 (v)	118	10.2	–	Wet lay up/alkali treated then bleached	Brahim and Cheikh (2007)
Flax (random)	Polyester	30 (v)	61	6.3	13.2 (c)	Vacuum infusion/untreated	Rodriguez et al. (2005)
Flax (random)	Acrylic	30 (v)	52	6.3	15.0 (c)	Vacuum infusion/untreated	Rodriguez et al. (2005)
Sisal (random)	Polyester	30 (v)	24	5.3	12.2 (c)	Vacuum infusion/untreated	Rodriguez et al. (2005)

(continued)

Table 3 (continued)

Fibre	Matrix	Fibre content (%) (*w* for weight and *v* for volume)	Tensile strength (MPa)	Young's modulus (GPa)	Charpy (c) or Izod (i) impact strength (kJ/m^2 or J/m)	Notes	References
Sisal (random)	Acrylic	30 (v)	31	3.4	12.7 (c)	Vacuum infusion/untreated	Rodriguez et al. (2005)
PALF (random)	Polyester	30 (w)	52.9	2.29	24.2 (c)	Compression moulding/untreated	Uma Devi et al. (1997)
PALF (random)	Polyester	30 (w)	55.4	1.46	–	Compression moulding/alkali treated	Uma Devi et al. (1997)
Flax (aligned)	PP	51 (v)	287.5	28.7	–	Compression moulding/untreated	Madsen and Lilholt (2003)
Flax (unidirectional)	PP	43 (v)	251	26.9	–	Compression moulding/untreated	Madsen and Lilholt (2003)
Hemp (unidirectional)	PP	42 (v)	215	21.1	–	Compression moulding/untreated	Madsen et al. (2009)
E-glass (unidirectional)	PP	35 (v)	700	26.5	–	Compression moulding	Shah (2013)
Flax (2D-random)	Epoxy	22 (v)	60	9.2	13.09 (c)	Autoclave moulding/alkali treated (1%)	George et al. (1999)
Flax (2D-random)	Epoxy	22 (v)	62.54	9.83	8.41 (c)	Autoclave moulding/3-aminopropyltriethoxysilane	George et al. (1999)
Flax (2D-random)	Epoxy	22 (v)	53.40	7.86	10.47 (c)	Autoclave moulding/untreated	George et al. (1999)
Flax (3D-random)	PP	30 (w)	27	1.7	–	Injection moulding/untreated	Arbelaiz et al. (2005)
Flax (3D-random)	PP	30 (w)	36.9	2.13	–	Injection moulding/5 wt% MAPP	Arbelaiz et al. (2005)
Hemp (3D-random)	PP	30 (w)	30	1.5	–	Injection moulding/alkyl ketene dimmer	Mutje et al. (2006)

(continued)

Table 3 (continued)

Fibre	Matrix	Fibre content (%) (*w* for weight and *v* for volume)	Tensile strength (MPa)	Young's modulus (GPa)	Charpy (c) or Izod (i) impact strength (kJ/m^2 or J/m)	Notes	References
E-glass (3D-random)	PP	30 (w)	49	2.2	–	Injection moulding	Mutje et al. (2006)
Jute (3D-random)	PP	60 (w)	74	11	195 (i)	Injection moulding/3 wt% MAPP	Rana et al. (1998)
Ramie (3D-random)	PLA	30 (v)	52.5	–	10 (i)	Two-roll mill/untreated	Yu et al. (2010)
Ramie (3D-random)	PLA	30 (v)	59.3	–	17 (i)	Two-roll mill/3-Aminopropyltriethoxy silane	Yu et al. (2010)
Ramie (3D-random)	PLA	30 (v)	66.8	–	22.5 (i)	Two-roll mill/alkali treated (5% w/v)	Yu et al. (2010)
Flax (2D-random)	PHB	30 (v)	40	4.2	–	Compression moulding/untreated	Bodros et al. (2007)
Okra (2D-random)	*Starch*	*25 (w)*	*2.21*	*0.24*	–	*Compression moulding/untreated*	Guleira et al. (2016)
Okra (3D-random)	*PLA*	*30 (w)*	*58.8*	*4.6*	–	*Injection moulding/untreated*	Fortunati et al. (2013b)
Okra (3D-random)	*PLA*	*30 (w)*	*55.3*	*3.86*	–	*Injection moulding/alkali treated*	Fortunati et al. (2013b)
Borassus fine fibres (3D-random)	*PP*	*15 (w)*	*28*	*2.5*	*27.5 (i)*	*Injection moulding/alkali treated (4%) + 5 wt% MAPP*	Sudhakara et al. (2013a)
Borassus fine fibres (3D-random)	*PP*	*5 (w)*	*23*	*1.4*	*25 (i)*	*Injection moulding/alkali treated (4%) + 5 wt% MAPP*	Sudhakara et al. (2013b)
Borassus fine fibres (3D-random)	*HDPE*	*20 (w)*	*20.1*	*0.53*	*80 (i)*	*Injection moulding/untreated*	Maheswari et al. (2013)
Borassus fine fibres (3D-random)	*HDPE*	*20 (w)*	*23.2*	*0.64*	*95 (i)*	*Injection moulding/5 wt% MA-g-HDPE*	Maheswari et al. (2013)

(continued)

Table 3 (continued)

Fibre	Matrix	Fibre content (%) (*w* for weight and *v* for volume)	Tensile strength (MPa)	Young's modulus (GPa)	Charpy (c) or Izod (i) impact strength (kJ/m^2 or J/m)	Notes	References
Borassus fibres (2D-random)	*Polyester*	*54 (w)*	*39.30*	*1.43*	*56.3 (c)*	*Compression moulding/untreated*	Velmurugan and Manikandan (2007)
Arundo donax (3D-random)	*Epoxy*	*15 (w)*	*37.5*	*3.04*	–	*Casting*	Fiore et al. (2014c)
Arundo donax (2D-random)	*Polyester*	*40 (v)*	*16.35*	*2.37*	–	*Compression moulding/untreated*	Chikouche et al. (2015)
Arundo donax (2D-random)	*Polyester*	*40 (v)*	*25.60*	*2.91*	–	*Compression moulding/alkali treated (6%)*	Chikouche et al. (2015)
Arundo donax (2D-random)	*Bio-based epoxy*	*5 (w)*	*68*[a]	*2.7*[a]	–	*Compression moulding/untreated*	Scalici et al. (2016)
Arundo donax (2D-random)	*Bio-based epoxy*	*5 (w)*	*72*[a]	*3.0*[a]	–	*Compression moulding/plasma treated*	Scalici et al. (2016)
Arundo donax (2D-random)	*PLA*	*10 (w)*	*37.3*	*3.85*	–	*Extrusion + compression moulding*	Fiore et al. (2014a)
Isora (unidirectional)	*Epoxy*	*51 (v)*	*200*	*6.5*	*240 (i)*	*Compression moulding/untreated*	Joshy et al. (2007a)
Isora (unidirectional)	*Epoxy*	*51 (v)*	*215*	*7.8*	–	*Compression moulding/alkali treated (1%)*	Joshy et al. (2007a)
Isora (2D-random)	*Polyester*	*24 (v)*	*36*	*1.6*	*120 (i)*	*Compression moulding/untreated*	Joshy et al. (2008)
Isora (2D-random)	*Polyester*	*24 (v)*	*47*	*2.8*	*110 (i)*	*Compression moulding/acetylated*	Joshy et al. (2008)
Isora (2D-random)	*Polyester*	*24 (v)*	*50*	*3.8*	*140 (i)*	*Compression moulding/alkali-treated + 5% isooctylphenoxypolyethoxyethanol*	Joshy et al. (2008)
Isora (unidirectional)	*Polyester*	*45 (v)*	*190*	*7.6*	*360 (i)*	*Compression moulding/untreated*	Joshy et al. (2007b)

(continued)

Table 3 (continued)

Fibre	Matrix	Fibre content (%) (*w* for weight and *v* for volume)	Tensile strength (MPa)	Young's modulus (GPa)	Charpy (c) or Izod (i) impact strength (kJ/m^2 or J/m)	Notes	References
Isora (unidirectional)	*Polyester*	*45 (v)*	230	8.4	–	*Compression moulding/alkali treated (1%)*	Joshy et al. (2007b)
Isora (3D-random)	*Natural rubber*	*15 (w)*	*14*	*2.71 MPa*	–	*Two-roll mixing/untreated*	Mathew and Joseph (2007)
Napier grass (unidirectional)	*Polyester*	*31 (v)*	*118*	*2.17*	–	*Compression moulding chemically extracted (NaOH)*	Rao et al. (2007)
Napier grass (2D-random)	*Polyester*	*25 (v)*	*18.17*	*2*	–	*Compression moulding/untreated*	Haameem et al. (2016)
Napier grass (2D-random)	*Epoxy*	*30 (v)*	*21*	*1.9*	–	*Vacuum infusion/untreated*	Ridzuan et al. (2016b)
Cissus quadrangularis (2D-random)	*Polyester*	*40 (w)*	*27.34*	*1.19*	*39.91 (i)*	*Compression moulding/untreated*	Mayandi et al. (2016)
Cissus quadrangularis (2D-random)	*Polyester*	*40 (w)*	*33.41*	*1.84*	*43.1 (i)*	*Compression moulding/alkali treated (5%)*	Mayandi et al. (2016)
Cissus quadrangularis (2D-random)	*Polyester*	*40 (w)*	*35.62*	*2.20*	*44.75 (i)*	*Compression moulding/$Ca(OH)_2$ treated*	Mayandi et al. (2016)
Cissus quadrangularis (2D-random)	*Polyester*	*40 (w)*	*31.38*	*1.75*	*41.28 (i)*	*Compression moulding/NaOH + silane treated (trichloro vinyl silane)*	Mayandi et al. (2016)
Cissus quadrangularis (2D-random)	*Polyester*	*40 (w)*	*27.34*	*1.19*	*39.91 (i)*	*Compression moulding/untreated*	Mayandi et al. (2015b)

[a] Flexural properties

PP = polypropylene

PLA = poly-lactic acid

PHB = poly (3-hydroxylbutyrate

HDPE = High-density polyethylene

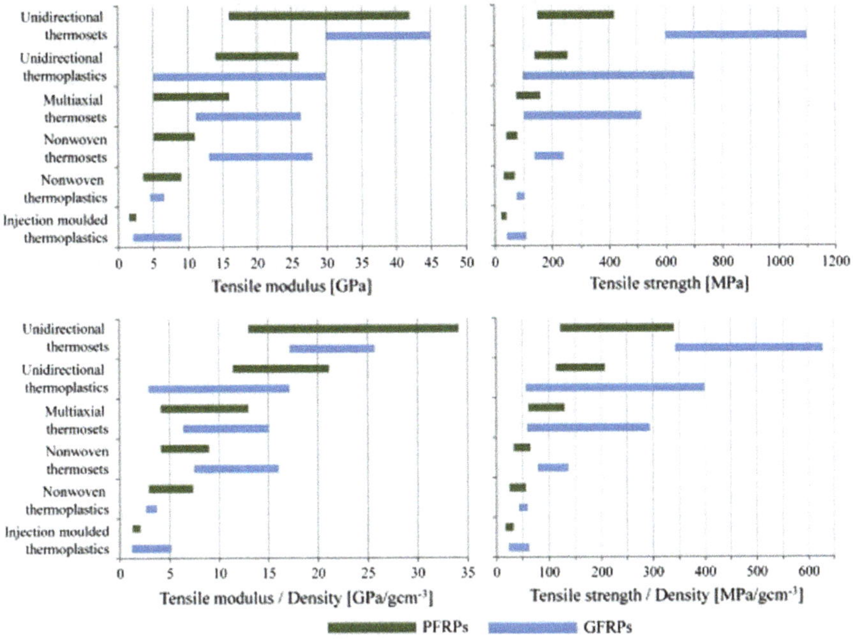

Fig. 15 Comparison of the absolute and specific tensile properties of plant fibre reinforced plastics (PFRPs) with E-glass reinforced plastics (GFRPs) [Reprinted with permission from Elsevier (Shah 2014)]

extraction of cellulose microfibers from Borassus fibres by acidified chlorination with alkalization and subsequent acid hydrolysis. The values of crystallinity calculated for raw fibres and cellulose microfibres were found to be 40.6 and 77.3%, respectively. Thermal stability of the cellulose microfibres was considerably improved with the onset decomposition at 250 °C, 40° higher than the one of the raw fibres. Chirayil et al. (2014a) successfully isolated cellulose nanofibrils from isora fibres by steam explosion method. Fibres were first treated with 2 wt% NaOH in an autoclave and kept under 25 psi pressure and at a temperature of 110 °C for 1 h. Then the pressure was released immediately and the fibres were washed in water. Then the fibres were bleached (NaOH and acetic acid), washed and dried before being subjected to acid hydrolysis using 10% oxalic acid under a pressure of 25 psi in an autoclave for 15 min. The pressure was released immediately and the process was repeated eight times. The fibres were finally taken out, washed, suspended in water and homogenized under continuous stirring. Atomic force microscopy indicated a width of 20 ± 10 nm with an aspect ratio of 15 for isora nanofibrils. These nanofibrils exhibited also a high crystalline nature (90%) and good thermal stability. Isora nanofibrils were used as a reinforcement in a polyester matrix (Chirayil et al. 2014b) and the resulting nanocomposites exhibited a percentage increase in tensile strength at 0.5 wt% of nanofibrils of 57% compared to the neat matrix. The presence of isora nanofibrils at a concentration of 0.5 wt% was found also to increases the T_g of the polyester matrix by 10 °C. In this case, a strong polymer chain confinement effect associated with network formation among nanofibrils and hydrogen bonding interaction between nanofibrils and polyester chains elucidated the superior mechanical and barrier properties of the nanocomposites. Lu et al. (2014) used *Pennisetum sinese* Roxb as a cellulose source for the extraction of cellulose nanocrystals. *Pennisetum sinese* is a kind of hybrid of *Pennisetum purpureum* and *Pennisetum americanum*, a perennial grass that forms in robust bamboo-like clumps. Cellulose was extracted from *Pennisetum sinese* with alkaline hydroxide and then bleached with sodium hypochlorite. From this cellulose, CNCs were obtained through hydrolysis with 50% sulfuric acid at 65 °C under ultrasonic treatment with continuous stirring. The extracted CNC showed a rod-like shape with a diameter in the range 20–30 nm and a length of 200–300 nm, forming an interconnected web-like structure with a crystallinity index of about 77.3%, thus confirming its application in polymer nanocomposites.

5 Conclusions and Future Perspectives

Several unusual natural fibres have been investigated and their thermal and mechanical behaviour compared with the one typical of common natural fibres such as sisal, hemp, flax and jute. The results suggest that there is scope in continuing the integration of new natural fibres in polymer composites with a view to contributing to increase the production of natural reinforcements and expand their use into new applications. Despite the promising and in some cases comparable results with standard natural fibres, a more detailed investigation of the mechanical properties of single fibres as well as the effects of surface treatments on the behaviour of single plant fibres is strongly advised and necessary. Another issue that needs to be addressed is the fabrication of hybrid composites by combining all these lignocellulosics to achieve further performance improvements. There is also a strong need to address the durability of the resulting composites along with the biodegradability when aiming to 100% bio-based composites application, especially when dealing for instance with semi-structural parts of exterior panels for future automotive applications. More aspects have to be considered such as reproducibility of these composites' properties and their long life cycle, all topics that are lacking in the current available literature. Despite these concerns, the use of these fibres in composite applications can definitely establish an entirely new value chain for the crops.

References

Abdul Khalil HPS, Bhat AH, Ireana Yusra AF (2012) Green composites from sustainable cellulose nanofibrils: A review. Carbohydr Polym 87:963–979. doi:10.1016/j.carbpol.2011.08.078

Akil HM, Omar MF, Mazuki AAM et al (2011) Kenaf fiber reinforced composites: a review. Mater Des 32:4107–4121. doi:10.1016/j.matdes.2011.04.008

Al-Khanbashi A, Al-Kaabi K, Hammami A (2005) Date palm fibers as polymeric matrix reinforcement: fiber characterization. Polym Compos 26:486–497

Andreu-Rodriguez J, Medina E, Ferrandez-Garcia MT et al (2013) Agricultural and industrial valorization of *Arundo donax* L. Commun Soil Sci Plant Anal 44:598–609. doi:10.1080/00103624.2013.745363

Angelini L, Lazzeri A, Levita G et al (2000) Ramie (*Boehmeria nivea* L.) and Spanish Broom (*Spartium junceum* L.) fibres for composite materials: agronomical aspects, morphology and mechanical properties. Ind Crops Prod 11:145–161

Angelini LG, Ceccarini L, Bonari E (2005) Biomass yield and energy balance of giant reed (*Arundo donax* L.) cropped in central Italy as related to different management practices. Eur J Agron 22:375–389. doi:10.1016/j.eja.2004.05.004

Angelini LG, Ceccarini L, Nassi o Di Nasso N, Bonari E (2009) Comparison of *Arundo donax* L. and *Miscanthus x giganteus* in a long-term field experiment in Central Italy: Analysis of productive characteristics and energy balance. Biomass Bioener 33:635–643. doi:10.1016/j.biombioe.2008.10.005

Arbelaiz A, Fernández B, Cantero G et al (2005) Mechanical properties of flax fibre/polypropylene composites. Influence of fibre/matrix modification and glass fibre hybridization. Compos Part A Appl Sci Manuf 36:1637–1644. doi:10.1016/j.compositesa.2005.03.021

Arias A, Heuzey M-C, Huneault MA (2013) Thermomechanical and crystallization behavior of polylactide-based flax fiber biocomposites. Cellulose 20:439–452

Arifuzzaman Khan GM, Shaheruzzaman M, Rahman MH et al (2009) Surface modification of okra bast fiber and its physico-chemical characteristics. Fibers Polym 10:65–70

Baets J, Plastria D, Ivens J, Verpoest I (2014) Determination of the optimal flax fibre preparation for use in unidirectional flax-epoxy composites. J Reinf Plast Compos 33:493–502. doi:10.1177/0731684413518620

Barkoula N, Garkhail S, Peijs T (2009) Effect of compounding and injection molding on the mechanical properties of flax fiber polypropylene composites. J Reinf Plast Compos 29:1366–1385

Belouadah Z, Ati A, Rokbi M (2015) Characterization of new natural cellulosic fiber from *Lygeum spartum* L. Carbohydr Polym 134:429–437. doi:10.1016/j.carbpol.2015.08.024

Bezazi A, Belaadi A, Bourchak M et al (2014) Novel extraction techniques, chemical and mechanical characterisation of *Agave americana* L. natural fibres. Compos Part B Eng 66: 194–203. doi:10.1016/j.compositesb.2014.05.014

Bhujade AM, Talmale S, Kumar N et al (2012) Evaluation of *Cissus quadrangularis* extracts as an inhibitor of COX, 5-LOX, and proinflammatory mediators. J Ethnopharmacol 141:989–996. doi:10.1016/j.jep.2012.03.044

Bledzki A, Gassan J (1999) Composites reinforced with cellulose based fibres. Prog Polym Sci 24:221–274

Bledzki A, Reihmane S, Gassan J (1996) Properties and modification methods for vegetable fibers for natural fiber composites. J Appl Polym Sci 59:1329–1336

Bodros E, Pillin I, Montrelay N, Baley C (2007) Could biopolymers reinforced by randomly scattered flax fibre be used in structural applications? Compos Sci Technol 67:462–470. doi:10.1016/j.compscitech.2006.08.024

Boose AB, Holt JS (1999) Environmental effects on asexual reproduction in *Arundo donax*. Weed Res 39:117–127. doi:10.1046/j.1365-3180.1999.00129.x

Ben Brahim S, Ben Cheikh R (2007) Influence of fibre orientation and volume fraction on the tensile properties of unidirectional Alfa-polyester composite. Compos Sci Technol 67: 140–147. doi:10.1016/j.compscitech.2005.10.006

Carr D, Cruthers N, Laing R, Niven B (2005) Fibers from three cultivars of New Zealand flax (*phormium tenax*). Text Res J 75:93–98

Chikouche MDL, Merrouche A, Azizi A et al (2015) Influence of alkali treatment on the mechanical properties of new cane fibre/polyester composites. J Reinf Plast Compos 34: 1329–1339. doi:10.1177/0731684415591093

Chirayil CJ, Joy J, Mathew L et al (2014a) Isolation and characterization of cellulose nanofibrils from *Helicteres isora* plant. Ind Crops Prod 59:27–34. doi:10.1016/j.indcrop.2014.04.020

Chirayil CJ, Joy J, Mathew L et al (2014b) Nanofibril reinforced unsaturated polyester nanocomposites: morphology, mechanical and barrier properties, viscoelastic behavior and polymer chain confinement. Ind Crops Prod 56:246–254. doi:10.1016/j.indcrop.2014.03.005

Corrêa AC, Morais Teixeira E, Pessan LA, Mattoso LHC (2010) Cellulose nanofibers from curaua fibers. Cellulose 17:1183–1192. doi:10.1007/s10570-010-9453-3

De Rosa IM, Kenny JM, Maniruzzaman M et al (2011) Effect of chemical treatments on the mechanical and thermal behaviour of okra (*Abelmoschus esculentus*) fibres. Compos Sci Technol 71:246–254. doi:10.1016/j.compscitech.2010.11.023

De Rosa IM, Kenny JM, Puglia D et al (2010a) Tensile behavior of New Zealand flax (*Phormium tenax*) fibers. J Reinf Plast Compos 29:3450–3454. doi:10.1177/0731684410372264

De Rosa IM, Kenny JM, Puglia D et al (2010b) Morphological, thermal and mechanical characterization of okra (*Abelmoschus esculentus*) fibres as potential reinforcement in polymer composites. Compos Sci Technol 70:116–122. doi:10.1016/j.compscitech.2009.09.013

De Rosa IM, Santulli C, Sarasini F (2010c) Mechanical and thermal characterization of epoxy composites reinforced with random and quasi-unidirectional untreated *Phormium tenax* leaf fibers. Mater Des 31:2397–2405. doi:10.1016/j.matdes.2009.11.059

Defoirdt N, Biswas S, De Vriese L et al (2010) Assessment of the tensile properties of coir, bamboo and jute fibre. Compos Part A Appl Sci Manuf 41:588–595. doi:10.1016/j.compositesa.2010.01.005

Fangueiro R, Rana S (eds) (2016) Natural fibres: advances in science and technology towards industrial applications. Springer, Netherlands, Dordrecht

Fiore V, Botta L, Scaffaro R et al (2014a) PLA based biocomposites reinforced with *Arundo donax* fillers. Compos Sci Technol 105:110–117. doi:10.1016/j.compscitech.2014.10.005

Fiore V, Scalici T, Valenza A (2014b) Characterization of a new natural fiber from *Arundo donax* L. as potential reinforcement of polymer composites. Carbohydr Polym 106:77–83. doi:10.1016/j.carbpol.2014.02.016

Fiore V, Scalici T, Vitale G, Valenza A (2014c) Static and dynamic mechanical properties of *Arundo Donax* fillers-epoxy composites. Mater Des 57:456–464. doi:10.1016/j.matdes.2014.01.025

Fiore V, Valenza A, Di Bella G (2011) Artichoke (*Cynara cardunculus* L.) fibres as potential reinforcement of composite structures. Compos Sci Technol 71:1138–1144. doi:10.1016/j.compscitech.2011.04.003

Fortunati E, Armentano I, Zhou Q et al (2012) Multifunctional bionanocomposite films of poly (lactic acid), cellulose nanocrystals and silver nanoparticles. Carbohydr Polym 87:1596–1605. doi:10.1016/j.carbpol.2011.09.066

Fortunati E, Puglia D, Monti M et al (2013a) Cellulose nanocrystals extracted from okra fibers in PVA nanocomposites. J Appl Polym Sci 128:3220–3230. doi:10.1002/app.38524

Fortunati E, Puglia D, Monti M et al (2013b) Okra (*Abelmoschus esculentus*) fibre based PLA composites: mechanical behaviour and Biodegradation. J Polym Environ. doi:10.1007/s10924-013-0571-5

Fuqua MA, Huo S, Ulven CA (2012) Natural fiber reinforced composites. Polym Rev 52:259–320. doi:10.1080/15583724.2012.705409

George J, Verpoest I, Ivens J (1999) Mechanical properties of flax fibre reinforced epoxy composites. Die Angew Makromol Chemie 272:41–45. doi:10.1002/(SICI)1522-9505(19991201)272:1<41:AID-APMC41>3.0.CO;2-X

George M, Chae M, Bressler DC (2016) Composite materials with bast fibres: structural, technical, and environmental properties. Prog Mater Sci 83:1–23. doi:10.1016/j.pmatsci.2016.04.002

Goda K, Sreekala M, Gomes A et al (2006a) Improvement of plant based natural fibers for toughening green composites-effect of load application during mercerization of ramie fibers. Compos Part A Appl Sci Manuf 37:2213–2220

Goda K, Sreekala MS, Gomes A et al (2006b) Improvement of plant based natural fibers for toughening green composites—effect of load application during mercerization of ramie fibers. Compos Part A Appl Sci Manuf 37:2213–2220. doi:10.1016/j.compositesa.2005.12.014

Gomes A, Matsuo T, Goda K, Ohgi J (2007) Development and effect of alkali treatment on tensile properties of curaua fiber green composites. Compos Part A Appl Sci Manuf 38:1811–1820. doi:10.1016/j.compositesa.2007.04.010

Graupner N, Herrmann AS, Mussig J (2009) Natural and man-made cellulose fibre-reinforced poly (lactic acid) (PLA) composites: an overview about mechanical characteristics and application areas. Compos Part A Appl Sci Manuf 40:810–821

Guleira A, Singha AS, Rana R (2016) Mechanical, thermal, morphological, and biodegradable studies of Okra cellulosic fiber reinforced starch-based biocomposites. Adv Polym Technol n/a–n/a. doi:10.1002/adv.21646

Haameem JAM, Abdul Majid MS, Afendi M et al (2016) Mechanical properties of Napier grass fibre/polyester composites. Compos Struct 136:1–10. doi:10.1016/j.compstruct.2015.09.051

Haameem MJA, Abdul Majid MS, Haslan M et al (2014) Effects of alkaline treatments on the tensile strength of Napier grass fibres. Appl Mech Mater 695:340–343. doi:10.4028/www.scientific.net/AMM.695.340

Hakeem KR, Jawaid M, Rashid U (eds) (2014) Biomass and bioenergy. Springer International Publishing, Cham

Hattalli S, Benaboura A, Castellan A (2002) Adding value to Alfa grass (*Stippa tenacissima* L.) soda lignin as phenolic resins. Polym Degrad Stab 75:259–264

Henriksson M, Henriksson G, Berglund LA, Lindström T (2007) An environmentally friendly method for enzyme-assisted preparation of microfibrillated cellulose (MFC) nanofibers. Eur Polym J 43:3434–3441. doi:10.1016/j.eurpolymj.2007.05.038

Herrera-Franco PJ, Valadez-Gonzalez A (2005) A study of the mechanical properties of short natural fiber reinforced composites. Compos Part B Eng 36:597–608

Hill C, Abdul Khalil H (2000) Effect of fiber treatments on mechanical properties of coir or oil palm fiber reinforced polyester composites. J Appl Polym Sci 78:1685–1697

Hoareau W, Trindade W, Siegmund B et al (2004) Sugar cane bagasse and curaua lignins oxidized by chlorine dioxide and reacted with furfuryl alcohol: characterization and stability. Polym Degrad Stab 86:567–576

Hubbe M, Rojas O, Lucia L, Sain M (2008) Cellulosic nanocomposites: a review. bioresources 3:929–980

Huda S, Reddy N, Karst D et al (2007) Nontraditional biofibers for a new textile industry. J Biobased Mater Bioenergy 1:177–190. doi:10.1166/jbmb.2007.022

Indran S, Raj RE (2015) Characterization of new natural cellulosic fiber from *Cissus quadrangularis* stem. Carbohydr Polym 117:392–399. doi:10.1016/j.carbpol.2014.09.072

Indran S, Raj RE, Sreenivasan VS (2014) Characterization of new natural cellulosic fiber from *Cissus quadrangularis* root. Carbohydr Polym 110:423–429. doi:10.1016/j.carbpol.2014.04.051

Jain S, Kumar R, Jindal U (1992) Mechanical behaviour of bamboo and bamboo composite. J Mater Sci 27:4598–4604

Jayaraman K (2003) Manufacturing sisal/polypropylene composites with minimum fibre degradation. Compos Sci Technol 63:367–374

John M, Thomas S (2008) Biofibres and biocomposites. Carbohydr Polym 71:343–364. doi:10.1016/j.carbpol.2007.05.040

Joshy MK, Lovely M, Rani J (2007a) Effect of alkali treatment on properties of unidirectional isora fibre reinforced epoxy composites. 36:259–266. doi:10.1179/174328907X229682

Joshy MK, Mathew L, Joseph R (2008) Influence of Fiber surface modification on the mechanical performance of isora-polyester composites. Int J Polym Mater 58:2–20. doi:10.1080/00914030802461824

Joshy MK, Mathew L, Joseph R (2007b) Studies on interfacial adhesion in unidirectional isora fibre reinforced polyester composites. Compos Interfaces 14:631–646. doi:10.1163/156855407782106429

Kaddami H, Dufresne A, Khelifi B et al (2006) Short palm tree fibers-thermoset matrices composites. Compos Part A Appl Sci Manuf 37:1413–1422

Kalia S, Dufresne A, Cherian BM et al (2011) Cellulose-based Bio- and Nanocomposites: a review. Int J Polym Sci 2011:1–35. doi:10.1155/2011/837875

Kaushik A, Singh M, Verma G (2010) Green nanocomposites based on thermoplastic starch and steam exploded cellulose nanofibrils from wheat straw. Carbohydr Polym 82:337–345

Le TM, Pickering KL (2015) The potential of *harakeke* fibre as reinforcement in polymer matrix composites including modelling of long *harakeke* fibre composite strength. Compos Part A Appl Sci Manuf 76:44–53. doi:10.1016/j.compositesa.2015.05.005

Lu Q, Tang L, Wang S et al (2014) An investigation on the characteristics of cellulose nanocrystals from *Pennisetum sinese*. Biomass Bioenerg 70:267–272. doi:10.1016/j.biombioe.2014.09.012

Madsen B, Lilholt H (2003) Physical and mechanical properties of unidirectional plant fibre composites—an evaluation of the influence of porosity. Compos Sci Technol 63:1265–1272. doi:10.1016/S0266-3538(03)00097-6

Madsen B, Thygesen A, Lilholt H (2009) Plant fibre composites—porosity and stiffness. Compos Sci Technol 69:1057–1069. doi:10.1016/j.compscitech.2009.01.016

Maheswari CU, Reddy KO, Muzenda E et al (2013) A comparative study of modified and unmodified high-density polyethylene/borassus fiber composites. Int J Polym Anal Charact 18:439–450. doi:10.1080/1023666X.2013.814027

Malkapuram R, Kumar V, Negi YS (2009) Recent development in natural fiber reinforced polypropylene composites. J Reinf Plast Compos 28:1169–1189

Mariani C, Cabrini R, Danin A et al (2010) Origin, diffusion and reproduction of the giant reed (*Arundo donax* L.): a promising weedy energy crop. Ann Appl Biol 157:191–202. doi:10.1111/j.1744-7348.2010.00419.x

Mathew L, Joseph KU, Joseph R (2007) Isora Fibre. J Nat Fibers 3:13–27. doi:10.1300/J395v03n04_02

Mathew L, Joseph R (2007) Mechanical properties of short-isora-fiber-reinforced natural rubber composites: effects of fiber length, orientation, and loading; alkali treatment; and bonding agent. J Appl Polym Sci 103:1640–1650. doi:10.1002/app.25065

Mayandi K, Rajini N, Pitchipoo P et al (2015a) A comparative study on characterisations of *Cissus quadrangularis* and *Phoenix reclinata* natural fibres. J Reinf Plast Compos 34:269–280. doi:10.1177/0731684415570045

Mayandi K, Rajini N, Pitchipoo P, et al (2016) Properties of untreated and chemically treated *Cissus Quadrangularis* natural fibers and their composites with polyester as the matrix. Polym Compos n/a–n/a. doi:10.1002/pc.24011

Mayandi K, Rajini N, Pitchipoo P et al (2015b) Mechanical performance of *Cissus quadrangularis*/polyester composite. Mater Today Commun 4:222–232. doi:10.1016/j.mtcomm.2015.08.001

Mishra S, Mohanty AK, Drzal LT et al (2004) A review on pineapple leaf fibers, sisal fibers and their biocomposites. Macromol Mater Eng 289:955–974. doi:10.1002/mame.200400132

Mohanty A, Misra M, Drzal L (eds) (2005) Natural fibers, biopolymers, and biocomposites. Taylor & Francis, Boca Raton (FL)

Munawar SS, Umemura K, Kawai S (2006) Characterization of the morphological, physical, and mechanical properties of seven nonwood plant fiber bundles. J Wood Sci 53:108–113. doi:10.1007/s10086-006-0836-x

Mutje P, Gironès J, Lòpez A et al (2006) Hemp Strands: PP composites by injection molding: effect of low cost physico-chemical treatments. J Reinf Plast Compos 25:313–327. doi:10.1177/0731684406059784

Obi Reddy K, Uma Maheswari C, Shukla M et al (2013) Tensile and structural characterization of alkali treated *Borassus* fruit fine fibers. Compos Part B Eng 44:433–438. doi:10.1016/j.compositesb.2012.04.075

Oksman K (2001) High quality flax fibre composites manufactured by the resin transfer moulding process. J Reinf Plast Compos 20:621–627. doi:10.1177/073168401772678634

Oksman K, Wallström L, Berglund LA, Filho RDT (2002) Morphology and mechanical properties of unidirectional sisal-epoxy composites. J Appl Polym Sci 84:2358–2365. doi:10.1002/app.10475

Okubo K, Fujii T, Yamamoto Y (2004) Development of bamboo-based polymer composites and their mechanical properties. Compos Part A Appl Sci Manuf 35:377–383

Ornaghi Júnior HL, De Oliveira Moraes AG, Poletto M et al (2016) Chemical composition, tensile properties and structural characterization of buriti fiber. Cellul Chem Technol 5:15–22

Ouajai S, Shanks RA (2005) Composition, structure and thermal degradation of hemp cellulose after chemical treatments. Polym Degrad Stab 89:327–335. doi:10.1016/j.polymdegradstab.2005.01.016

Panthong A, Supraditaporn W, Kanjanapothi D et al (2007) Analgesic, anti-inflammatory and venotonic effects of *Cissus quadrangularis* Linn. J Ethnopharmacol 110:264–270. doi:10.1016/j.jep.2006.09.018

Prasad Reddy J, Rhim J-W (2014) Isolation and characterization of cellulose nanocrystals from garlic skin

Rana AK, Mandal A, Mitra BC et al (1998) Short jute fiber-reinforced polypropylene composites: effect of compatibilizer. J Appl Polym Sci 69:329–338. doi:10.1002/(SICI)1097-4628(19980711)69:2<329:AID-APP14>3.0.CO;2-R

Rao KMM, Prasad AVR, Babu MNVR et al (2007) Tensile properties of elephant grass fiber reinforced polyester composites. J Mater Sci 42:3266–3272. doi:10.1007/s10853-006-0657-8

Reddy KO, Ashok B, Reddy KRN et al (2014) Extraction and characterization of novel Lignocellulosic fibers From *Thespesia Lampas* Plant. Int J Polym Anal Charact 19:48–61. doi:10.1080/1023666X.2014.854520

Reddy KO, Guduri BR, Rajulu AV (2009a) Structural characterization and tensile properties of Borassus fruit fibers. J Appl Polym Sci 114:603–611. doi:10.1002/app.30584

Reddy KO, Maheswari CU, Dhlamini MS, Kommula VP (2016) Exploration on the characteristics of cellulose microfibers from Palmyra palm fruits. Int J Polym Anal Charact 21:286–295. doi:10.1080/1023666X.2016.1147799

Reddy KO, Maheswari CU, Reddy DJP, Rajulu AV (2009b) Thermal properties of Napier grass fibers. Mater Lett 63:2390–2392. doi:10.1016/j.matlet.2009.08.035

Reddy KO, Maheswari CU, Shukla M, Rajulu AV (2012) Chemical composition and structural characterization of Napier grass fibers. Mater Lett 67:35–38. doi:10.1016/j.matlet.2011.09.027

Reddy N, Yang Y (2009) Properties of natural cellulose fibers from hop stems. Carbohydr Polym 77:898–902. doi:10.1016/j.carbpol.2009.03.013

Rhim JW, Reddy JP, Luo X (2015) Isolation of cellulose nanocrystals from onion skin and their utilization for the preparation of agar-based bio-nanocomposites films. Cellulose 22:407–420. doi:10.1007/s10570-014-0517-7

Ridzuan MJM, Abdul Majid MS, Afendi M et al (2016a) Characterisation of natural cellulosic fibre from *Pennisetum purpureum* stem as potential reinforcement of polymer composites. Mater Des 89:839–847. doi:10.1016/j.matdes.2015.10.052

Ridzuan MJM, Abdul Majid MS, Afendi M et al (2016b) Moisture absorption and mechanical degradation of hybrid *Pennisetum purpureum*/glass–epoxy composites. Compos Struct 141:110–116. doi:10.1016/j.compstruct.2016.01.030

Rodriguez E, Petrucci R, Puglia D et al (2005) Characterization of composites based on natural and glass fibers obtained by vacuum infusion. J Compos Mater 39:265–282. doi:10.1177/0021998305046450

Rong MZ, Zhang MQ, Liu Y et al (2001) The effect of fiber treatment on the mechanical properties of unidirectional sisal-reinforced epoxy composites. Compos Sci Technol 61:1437–1447. doi:10.1016/S0266-3538(01)00046-X

Saha S, Das B, Ray P et al (1990) SEM studies of the surface fracture morphology of pineapple leaf fibres. Text Res J 60:726–731

Sain M, Panthapulakkal S (2006) Bioprocess preparation of wheat straw fibers and their characterization. Ind Crops Prod 23:1–8

Sapsrithong T, Kaewprem W, Tongumpai S et al (2012) Cissus quadrangularis ethanol extract upregulates superoxide dismutase, glutathione peroxidase and endothelial nitric oxide synthase expression in hydrogen peroxide-injured human ECV304 cells. J Ethnopharmacol 143:664–672. doi:10.1016/j.jep.2012.07.027

Sarasini F, Tirillò J, Puglia D et al (2015) Effect of different lignocellulosic fibres on poly (ε-caprolactone)-based composites for potential applications in orthotics. RSC Adv 5:23798–23809. doi:10.1039/C5RA00832H

Saravanakumar SS, Kumaravel A, Nagarajan T et al (2013) Characterization of a novel natural cellulosic fiber from *Prosopis juliflora* bark. Carbohydr Polym 92:1928–1933. doi:10.1016/j.carbpol.2012.11.064

Saravanan D, Pallavi N (2008) Investigations into structural aspects of Borassus flabellifer L (*palmyrah palm*) fruit fibres. J Text Inst. doi:10.1080/00405000701570914

Sarikanat M, Seki Y, Sever K, Durmuşkahya C (2014) Determination of properties of Althaea officinalis L. (*Marshmallow*) fibres as a potential plant fibre in polymeric composite materials. Compos Part B Eng 57:180–186. doi:10.1016/j.compositesb.2013.09.041

Satyanarayana KG, Sukumaran K, Kulkarni A et al (1986) Fabrication and properties of natural fibre reinforced polyester composites. Composites 17:329–333

Scalici T, Fiore V, Valenza A (2016) Effect of plasma treatment on the properties of *Arundo Donax* L. leaf fibres and its bio-based epoxy composites: a preliminary study. Compos Part B Eng 94:167–175. doi:10.1016/j.compositesb.2016.03.053

Seki Y, Sarikanat M, Sever K, Durmuşkahya C (2013) Extraction and properties of Ferula communis (*chakshir*) fibers as novel reinforcement for composites materials. Compos Part B Eng 44:517–523. doi:10.1016/j.compositesb.2012.03.013

Shah DU (2014) Natural fibre composites: comprehensive Ashby-type materials selection charts. Mater Des 62:21–31. doi:10.1016/j.matdes.2014.05.002

Shah DU (2013) Developing plant fibre composites for structural applications by optimising composite parameters: a critical review. J Mater Sci 48:6083–6107

Shahzad A (2011) Hemp fiber and its composites—a review. J Compos Mater 46:973–986. doi:10.1177/0021998311413623

Shamsul Alam M, Arifuzzaman Khan GM (2007) Chemical analysis of okra bast fiber (*Abelmoschus Esculentus*) and its physico-chemical properties. J Text Apparel Technol Manag 5:1–9

Shibata M, Takachiyo K-I, Ozawa K et al (2002) Biodegradable polyester composites reinforced with short *abaca* fiber. J Appl Polym Sci 85:129–138. doi:10.1002/app.10665

Sreenivasan V, Somasundaram S, Ravindran D et al (2011) Microstructural, physico-chemical and mechanical characterisation of *Sansevieria cylindrica* fibres-An exploratory investigation. Mater Des 32:453–461

Srinivasababu N, Thomas MJJ S, A ST and LP, et al (2015) An overview of okra fibre reinforced polymer composites. IOP Conf Ser Mater Sci Eng 83: 012003. doi:10.1088/1757-899X/83/1/012003

Sudhakara P, Jagadeesh D, Wang Y et al (2013a) Fabrication of *borassus* fruit lignocellulose fiber/PP composites and comparison with jute, sisal and coir fibers. Carbohydr Polym 98:1002–1010. doi:10.1016/j.carbpol.2013.06.080

Sudhakara P, Kamala Devi AP, Venkata Prasad C et al (2013b) Thermal, mechanical, and morphological properties of maleated polypropylene compatibilized *Borassus* fruit fiber/polypropylene composites. J Appl Polym Sci 128:976–982. doi:10.1002/app.38135

Summerscales J, Dissanayake N, Virk A, Hall W (2010a) A review of bast fibres and their composites. Part 2—composites. Compos Part A Appl Sci Manuf 41:1336–1344. doi:10.1016/j.compositesa.2010.05.020

Summerscales J, Dissanayake N, Virk A, Hall W (2010b) A review of bast fibres and their composites. Part 1—fibres as reinforcements. Compos Part A Appl Sci Manuf 41:1329–1335

Summerscales J, Virk A, Hall W (2013) A review of bast fibres and their composites: Part 3—modelling. Compos Part A Appl Sci Manuf 44:132–139. doi:10.1016/j.compositesa.2012.08.018

Sun R, Fang J, Goodwin A et al (1998) Fractionation and characterization of polysaccharides from *abaca* fibre. Carbohydr Polym 37:351–359

Thomason J (1999) The influence of fibre properties on the performance of glass fibre reinforced polyamide 6,6. Compos Sci Technol 59:2315–2328

Thornby D, Spencer D, Hanan J, Sher A (2007) L-DONAX, a growth model of the invasive weed species, *Arundo donax* L. Aquat Bot 87:275–284. doi:10.1016/j.aquabot.2007.06.012

Uma Devi L, Bhagawan SS, Thomas S (1997) Mechanical properties of pineapple leaf fiber-reinforced polyester composites. J Appl Polym Sci 64:1739–1748

Van de Weyenberg I, Ivens J, De Coster A et al (2003) Influence of processing and chemical treatment of flax fibres on their composites. Compos Sci Technol 63:1241–1246. doi:10.1016/S0266-3538(03)00093-9

Velmurugan R, Manikandan V (2007) Mechanical properties of palmyra/glass fiber hybrid composites. Compos Part A Appl Sci Manuf 38:2216–2226. doi:10.1016/j.compositesa.2007.06.006

Ververis C, Georghiou K, Christodoulakis N et al (2004) Fiber dimensions, lignin and cellulose content of various plant materials and their suitability for paper production. Ind Crops Prod 19:245–254. doi:10.1016/j.indcrop.2003.10.006

Wielage B, Lampke T, Marx G et al (1999) Thermogravimetric and differential scanning calorimetric analysis of natural fibres and polypropylene. Thermochim Acta 337:169–177. doi:10.1016/S0040-6031(99)00161-6

Yan L, Chouw N, Jayaraman K (2014) Flax fibre and its composites—a review. Compos Part B Eng 56:296–317. doi:10.1016/j.compositesb.2013.08.014

Yao F, Wu Q, Lei Y et al (2008) Thermal decomposition kinetics of natural fibers: activation energy with dynamic thermogravimetric analysis. Polym Degrad Stab 93:90–98. doi:10.1016/j.polymdegradstab.2007.10.012

Yu T, Ren J, Li S et al (2010) Effect of fiber surface-treatments on the properties of poly(lactic acid)/ramie composites. Compos Part A Appl Sci Manuf 41:499–505. doi:10.1016/j.compositesa.2009.12.006

Lignocellulosic Fibres Reinforced Thermoset Composites: Preparation, Characterization, Mechanical and Rheological Properties

Hind Abdellaoui, Rachid Bouhfid and Abou El Kacem Qaiss

Abstract Due to the biodegradability, renewability and important specific properties, the lignocellulosic fibres become a celebrity in the composite materials fields. This interest is essentially forms elements of the strategic outline of sustainable development and environmental respect. In this context, this chapter is obviously focused on thermoset composites reinforced by lignocellulosic fibres. We present in this chapter the different thermoset matrices and a deep depiction of lignocellulosic fibres from their morphological structures to their mechanical properties. Then, we present a review of the largest part of an eloquent work of literature that have paid attention to the exploration of the mechanical and rheological properties of composite lignocellulosic fibres and realm of application of this new category of materials.

Keywords Lignocellulosic fibers · Composite · Thermoset · Mechanical
Rheological · Biodigradability

Contents

H. Abdellaoui · R. Bouhfid (✉) · A. E. K. Qaiss
Moroccan Foundation for Advanced Science, Innovation and Research (MASCIR), Laboratory of Polymer Processing, Institute of Nanomaterials and Nanotechnology, Rabat, Morocco
e-mail: r.bouhfid@mascir.com

S. Kalia (ed.), *Lignocellulosic Composite Materials*, Springer Series on Polymer and Composite Materials, https://doi.org/10.1007/978-3-319-68696-7_5

1 Environmental Context

"We do not inherit the earth, we borrow it from our children", as Antoine de Saint-Exupéry defends its vision to protect the planet, conserve resources and save it from all usurpation which destroys it more and more. Currently, with the depletion of fossil resources and many natural phenomena that reflect the anger of the earth, an appeal to preserve nature tenderly by exploiting its renewable resources to be useful to the human needs and comfort, proves irremediably necessary. The preservation of the earth brings to the introduction of sustainable development at all levels: social, economic and especially environmental.

The sustainable development, as defined by the World Commission on Environment and Development of the United Nations in 1983, is the "Social and economic advance to assure human beings a healthy and productive life, but one that did not compromise the ability of future generations to meet their own needs". According to this definition, environmental and social effects are at the same level of importance as the economic benefits. Hence, reducing the exploitation of harmful substances, the implementation of appropriated health and security procedures and the necessity of design and manufacture of ecofriendly materials, is unquestionably the right remedy (Vilaplana et al. 2010).

The materials design must consider the option of renewable resources as raw materials for their production. For the composite materials produced by using natural components, the central issue lies in the design of products that are structurally and functionally stable on their application (Bodros et al. 2007; Alves et al. 2010).

The Composites manufactured by natural constituents must meet the following requirements (Vilaplana et al. 2010; Lucas et al. 2008):

(a) Use of renewable and/or biodegradable,
(b) Elimination of environmental and toxicological effects (emissions of degradation components, additives or fillers…) that may possibly take place at any stage of the product life cycle,
(c) Implementation of an option for waste managing (recycling, composting and incineration) to guarantee the return of the energetic cost of the product.

Several research studies have been available since the years 2000 on composite materials with natural constituents, including various matrices of biological original and natural lignocellulosic reinforcements, their preparation, modification, properties and their applications. In this chapter, the focus is granted to thermosetting composite reinforced by lignocellulosic fibres.

2 Introduction

In recent years, tremendous interest was expressed for composites and the integration of these materials in many industrial fields. The composite utilization rate varies for certain sectors from 60 to 100% as in the areas of aerospace, military and civilian. Comparing with traditional materials, the composites possess significant advantages that ensure competition. They make possible the design lightweight structures and complex shapes. The composites are commonly elaborated with an organic matrix and glass or carbon fibres (Abdellaoui and Echaabi 2014).

Currently, in the face of more stringent economic and environmental constraints, using the lignocellulosic fibres to reinforce the organic matrix composites, in particular reinforcement of the thermoset matrix appears as an alternative choice. Lignocellulosic fibres are increasingly attractive due to their good specific properties, biodegradability, renewable, abundant, and low cost. These advantages are relevant for structural applications in various fields, including the automotive industry, which requires lightweight materials with high performance, an environmental respect, and the material cost reduction (Alves et al. 2010).

Several lignocellulosic fibres like flax, jute, kenaf, coir, sisal and doum are studied by researchers to replace glass fibres in thermoset composites (Malha et al. 2013; Abdellaoui et al. 2015; Silva et al. 2008; Defoirdt et al. 2010; Le Duigou et al. 2010). However, the hydrophilic nature of lignocellulosic fibres thanks to the existence of hydroxyl groups constitutes a restriction to the excellent interfacial adhesion with the hydrophobic polymer matrix. Good interfacial adhesion has a positive influence on the rheological and mechanical properties of composites, being strongly correlated to the interface properties. Various investigations have already done on surface fiber modification to improve the adhesion with the synthetic polymer. It has been confirmed that appropriate treatment applied (alkali treatment, silane…). On the fibres may enhance the compatibility of natural fibres and the polymer matrix. Different methods of surface fiber modification were mentioned in the literature as the alkaline treatment, silane treatment, acetylation, the compatibilizers, etc.

A series of studies realized on composite lignocellulosic fibres tell that the mechanical properties of composites are not only influenced by improving the interfacial fiber-matrix adhesion. Certainly, the interfacial adhesion is of crucial interest, but there are other factors that impact on the progress of the mechanical performance of the composite such as the content and orientation fiber, reinforcement architecture, hybridization (natural fiber/natural fiber or traditional synthetic fiber/natural fiber), etc. (Abdellaoui et al. 2015; Defoirdt et al. 2010; Essabir et al. 2014; Gassan and Bledzki 1999; Umer et al. 2011; Ben Brahim and Ben Cheikh 2007). Enhancement of mechanical and rheological properties of composite carefully imposes a mastery of these implementation parameters.

This chapter is in its entirety, an exploratory study of thermoset composites reinforced by lignocellulosic fibres, which their development now exceeds the embryonic stage. A summary of the different thermosetting matrices correlated with a descriptive representation of lignocellulosic fibres (structure, advantages and disadvantages, and properties…) leads to the composite manufacturing field, in which we treat different manufacturing processes, and characterization tests of lignocellulosic fibres composite. Afterwards, a literary analysis of the mechanical properties of lignocellulosic fibres reinforcing the composite, led to many studies dealing with enhancement the mechanical behavior of composites, respecting the variability of certain manufacturing parameters. As achievement, some applications of lignocellulosic fibres composite are cited in this chapter, especially applications in niche areas such as the automobile industry and the building sectors.

3 Thermosetting Matrices

Thermosetting matrices have the function of binding the reinforcing fibres, distribute loads, bring the thermal resistance of the structure and provide the desired shape to the ultimate product. The general composite characteristics are associated with the nature of the matrix, particularly for long-term keeping. The Table 1 summarized several properties of the commonly used thermosetting matrices (Regional centre of animation in advanced materials 2004).

The thermosetting structure has the shape of a three-dimensional network that laying (double bond polymerization) to cure in the form of permanently during heating. The transformation is irreversible. Some matrices are commonly used:

Unsaturated polyester matrix is the frequently used resin in composite applications, particularly in the marine industry. Many applications like yachts, dinghies and workboats are manufactured by composite with glass fiber and polyester matrix. The unsaturated polyester matrix contains an unsaturated monomer (Fig. 1),

Table 1 Comparative study of certain properties of polyester, epoxy and vinyl ester matrices (Regional centre of animation in advanced materials 2004; Megiatto et al. 2009)

Properties	Polyester	Epoxy	Vinyl ester
Density (Mg^3)	1.2–1.5	1.1–1.4	1.2–1.4
Young's modulus (GPa)	2–4.5	3–6	3.1–3.8
Tensile strength (MPa)	40–90	35–100	69–83
Compressive strength (MPa)	90–250	100–200	–
Tensile elongation to break (%)	2	1–6	4–7
Water absorption 24 h at 20 °C	0.1–0.3	0.1–0.4	–

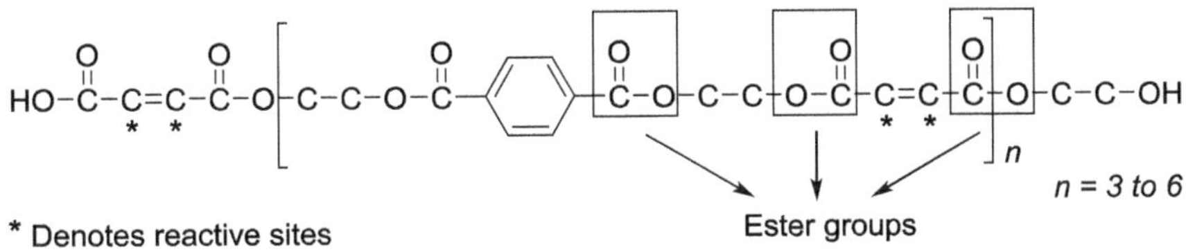

Fig. 1 Chemical structure of polyester

usually styrene. This monomer is formed by a cross-linking at room temperature by adding an accelerator and an organic peroxide catalyst. This matrix is distinguished by a good chemical resistance and less onerous compared to other TD matrices, but it is flammable and hygienic conditions are less suitable. There are different varieties of unsaturated polyester (Regional centre of animation in advanced materials 2004; Ray et al. 2002), which are:

- Orthophthalic: the most common;
- Isophtalic: giving better resistance to moisture;
- Chlorinated: providing a self-extinguishing;
- Bisphenol: possessing excellent chemical and thermal characteristics.

Epoxy matrices are a big family of high performance resins. It's consequence from the polyaddition of epichlorohydrin with a polyhydric alcohol (Fig. 2). Two epoxy matrix classes are distinguished based on hardener (hot or cold) and the temperature resistance (120–130 or 180–200 °C).

Epoxy is distinguished by good mechanical, thermal and chemical strength, which led to its exploitation in the field of aircraft. However, it is a bit expensive (Sarkar et al. 2015).

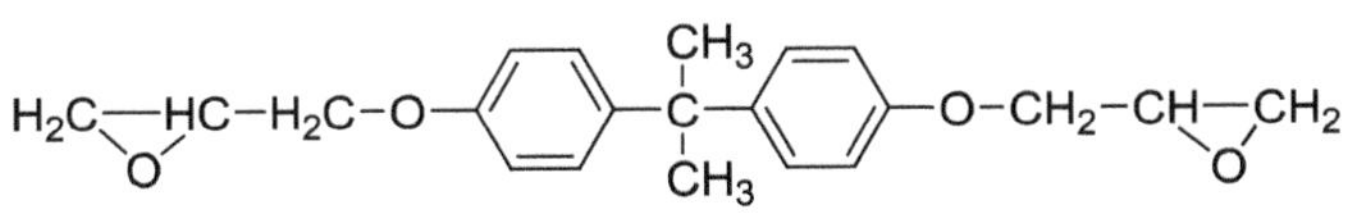

Fig. 2 Chemical structure of epoxy

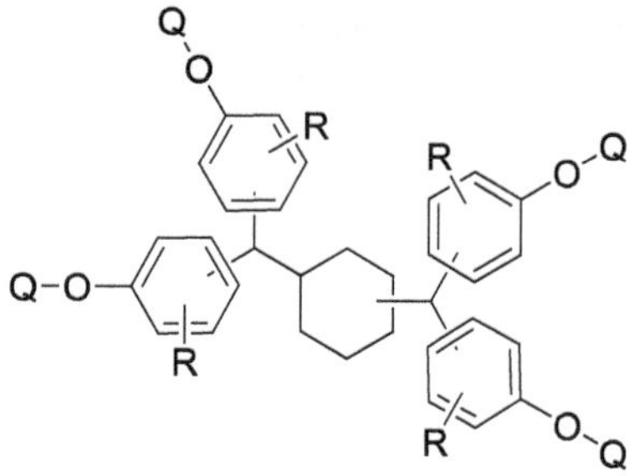

Fig. 3 Chemical structure of polyphenol

Phenolic matrices are resultant from the polycondensation of phenol and formaldehyde (Fig. 3). It has good fire resistance and low chemical emissions during the cure phase which is relatively fast. However, it remains fragile, susceptible to humidity and difficult to color (Regional centre of animation in advanced materials 2004).

Vinylester matrices are considered as variant of polyester produced based on acrylic acid (Fig. 4). They combine the ease of manufacturing and rapid curing characteristics of polyester matrix and the admirable chemical resistance and thermal and mechanical properties of epoxy matrix. They have good fatigue resistance and excellent corrosion behavior, but they are combustible (Stocchi et al. 2007).

There are other kinds of thermosetting matrices, but they are seldom used in industrial applications like (Regional centre of animation in advanced materials 2004):

Polyurethanes and polyureas matrices In the composite application, are mainly used elastomeric formulas whose low viscosity makes a good fill of the mold. The components are distributed to the state liquid prepolymers:

- Polyols + polyisocyanates = Polyurethane;
- Polyethers + Polyamines = Polyureas.

They are distinguished by good chemical and aging resistance, but their mechanical strength is reduced.

Polyamides matrices are mainly employed in high performance composites HP, when looking for a good stability at high temperatures (>250 °C). They still remain very expensive and difficult to implement.

Bis-maleimides matrices less used. They are highly resistant to both impact and temperature, but are difficult to implement.

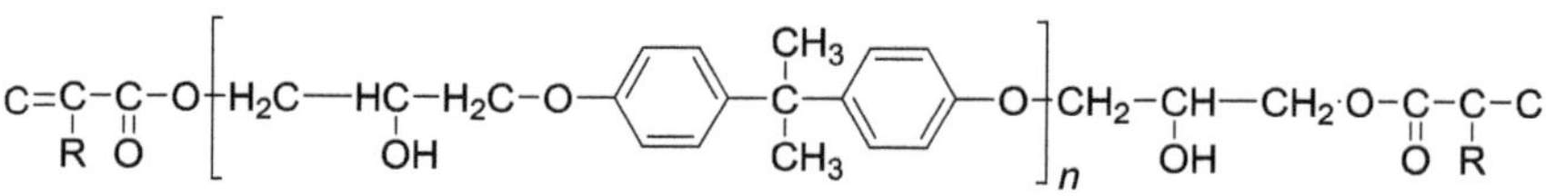

Fig. 4 Chemical structure of vinyl ester

4 Cellulosic Fibres

4.1 Origin of Natural Fibres

Natural fibres are employed for nearly 3000 years in composite materials, for example, in very old Egypt (time of the Pharaohs), where straw and clay were mixed for making walls. During the last decade, the composite reinforced natural fiber had a great interest from the academia and various industries (Saheb and Jog 1999). There are an extensive variety of natural fibers that may be used for reinforcement or as fillers for thermosetting matrices (Gassan and Bledzki 1999; Jeong et al. 2011; Pizzi et al. 2009).

Natural fibres are subdivided depending on their origin and provenance (plant, animal or mineral) as summarized in Fig. 5:

- Vegetable fibres, which are lignocellulosic fibres contain three varieties depending on their origin (John and Thomas 2008):
  - Fibres extracted from seminal seed bristles, such as cotton kapok,
  - Bast fibres from plant stems such as jute, flax, hemp, ramie…
  - Hard fibres derived from the leaves like sisal, from the trunks as manila, from fruit envelope as coconut.

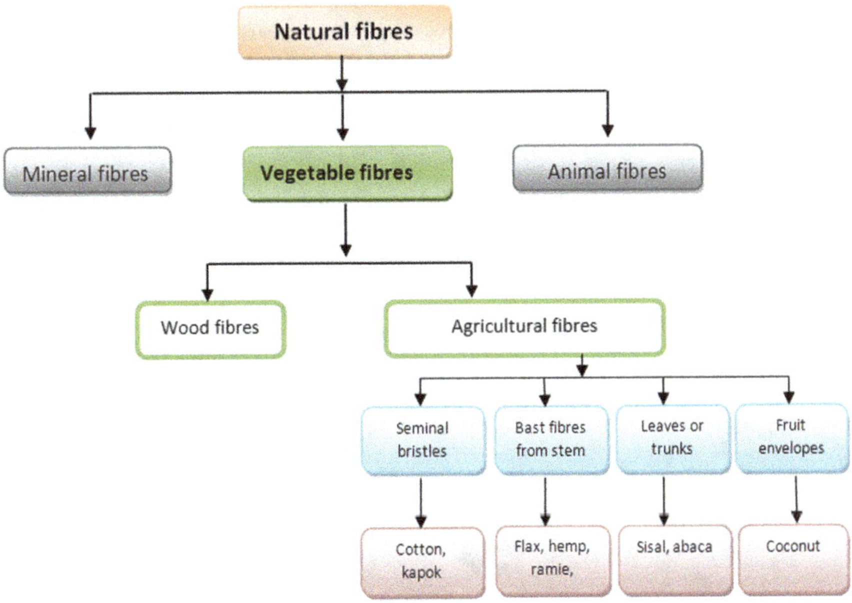

Fig. 5 Different origin of natural fiber (Abdelmouleh et al. 2007)

- Animal fibres are extracted from animal bristles as fleece, and secretion as silk.
- Mineral fibres like basalt.

All Lignocellulosic fibres are comprised of cellulose, while animal fibres are comprised of proteins.

4.2 *Structure of Lignocellulosic Fibres*

The structure of natural fibres, especially bast fibres (from stems), can be analyzed at various levels (Charlet et al. 2007) as exposed in Fig. 6.

- At the macroscopic scale, from the outside inwards, the stem consists of bark, phloem, xylem and a central void. Phloem is comprised of almost 30 fiber bundles.
- At the Mesoscopic scale, the cross section of a bundle contains between 10 and 40 fibres which are linked together mostly by pectins.
- At the Microscopic scale, a fiber is made of concentric layers which differ in their thickness, chemical composition and their structure. Each one of layers is composed of microfibrils.
- A nanoscale, the microfibrils are formed of cellulose chains embedded in an amorphous matrix of lignin, hemicellulose, and pectin.

Into plants, fiber contributes to the essential functions for its development: support and conduction. The fiber morphology can be governed by the seasons and

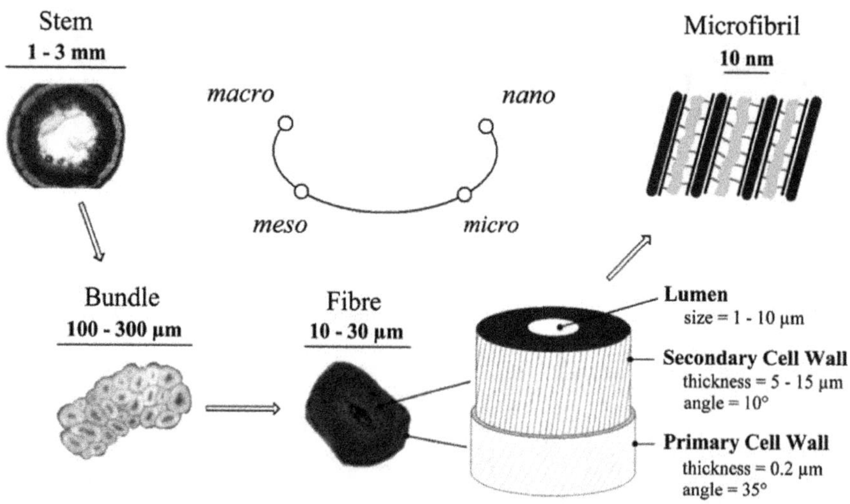

Fig. 6 Multi-scale composite structure of lignocellulosic fiber (Charlet et al. 2007)

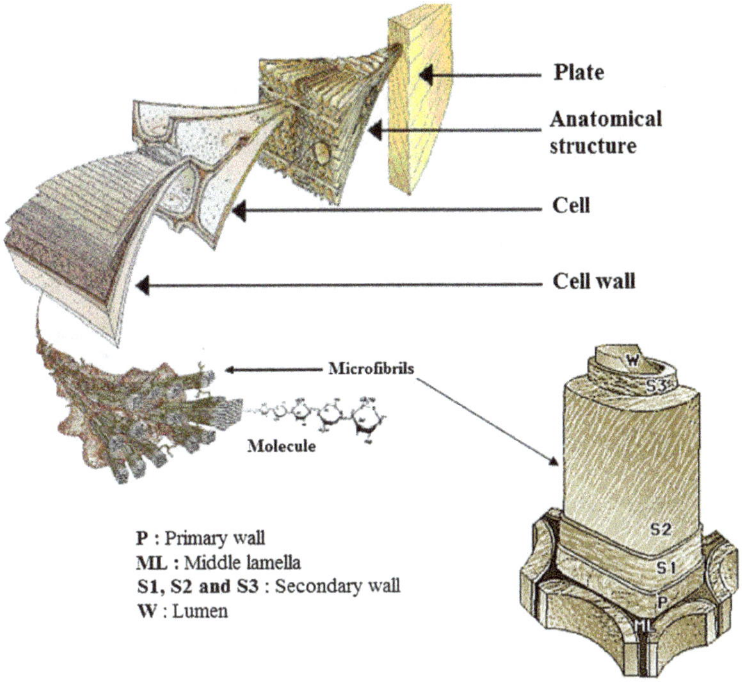

Fig. 7 Structure of lignocellulosic fibres (multi-scale observations)

the constraints affecting the plants (Charlet et al. 2007; Harish et al. 2009). The fiber contains the following parts (Fig. 7):

- The middle lamella M: thickness of 0.5 to 2 μm in the periphery, which is common to neighboring cells. It is mainly composed of lignin (70%) associated with other organic components such as hemicellulose, pectin, and cellulose.
- The primary wall P: very thin (0.03–0.1 μ). It is made of a fibrillar structure coated with a combination of lignin, pectin, and hemicellulose (Fig. 8).
- The secondary wall, the main component of the fiber which comprises three distinct parts:
  - The outer wall S1: Thin 0.1–0.2 um, wherein the fibrils are wound in parallel helices and of opposite direction.
  - The central wall S2: thicker (0.5–8 um), the fibrils are arranged in parallel helices. The fiber stiffness is detached to the thickness of this wall.
  - The inner wall or tertiary S3: thin (0.04–0.1 μ), is adjacent to lumen W. between the fibres, there are spaces that are partially composed of lignin and hemicellulose.

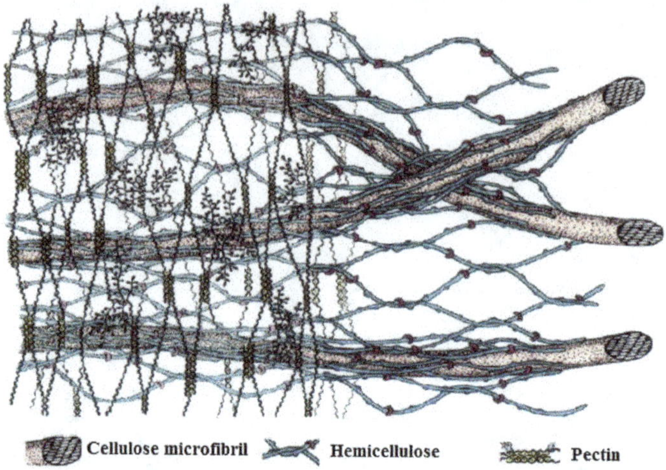

Fig. 8 The primary wall: Three-dimensional structure

4.3 Chemical Composition of Cellulosic Fibres

Lignocellulosic fibres consist mainly of lignin, cellulose, and hemicellulose. Other components such as proteins, pectins, starch and inorganic salts are also existing but in low proportion. The holocellulose proportion (cellulose and hemicellulose) and lignin varies significantly following the origin of the natural fiber (wood, jute, sisal, kenaf, etc.) (Saheb and Jog 1999). The chemical composition of the fibres depends considerably on climatic conditions, growth condition, age and nature of the fiber. Table 2 gathers the chemical composition of different natural fibres (Saheb and Jog 1999; Xie et al. 2010).

Table 2 The composition of several lignocellulosic fibres (Saheb and Jog 1999)

Fibres	Cellulose (%)	Lignin (%)	Hemicellulose (%)	Pectin (%)
Stem fibres				
Flax	71	2.2	18.6–20.6	2.3
Kenaf	31–57	15–19	21.5–23	–
Jute	45–71.5	12–26	13.6–21	0.2
Hemp	57–77	3.7–13	14–22.4	0.9
Ramie	68.6–91	0.6–0.7	5–16.7	1.9
Leaves fibres				
Abaca	56–63	7–9	15–17	–
Sisal	47–78	7–11	10–24	10
Henequen	77.6	13.1	4–8	–

4.3.1 Cellulose

The cellulose is considered as a natural polymer consisting of molecules formed by long chains. These chains are mainly composed of *D*-anhydroglucopyranoses ($C_6H_{11}O_5$) linked by bonds β-(1,4)-glycosidic in position C_1 and C_4 (Fig. 9). The degree of polymerization is estimated about 10,000, which varies depending on lignocellulosic fiber. Each repeated unit is composed of three hydroxyl groups which the crystallinity is conditioned by the aptitude of these groups to make hydrogen bonds (Gilberto et al. 2010).

The chains of cellulose are combined to form basic structures called microfibrils as elucidated by a study carried by transmission electron microscope (Saheb and Jog 1999). The microfibrils diameter (Fig. 10) is average of 2–20 nm which each microfibril is considered as crystals chains linked amorphous fields (John and Thomas 2008). The cellulose properties are strongly correlated to the high density of hydrogen bonds that develop between chains. Comparing to other amorphous components of the fiber, the cellulose has a crystalline structure at a high proportion. Crystalline cellulose is treated as a polymer having the highest elasticity modulus is about 136 GPA compared to that of fibreglass 75 GPA (Gilberto et al. 2010).

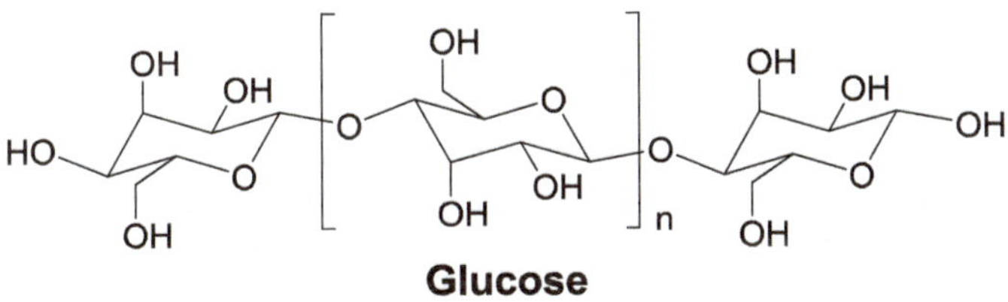

Fig. 9 Cellulose molecule (Gilberto et al. 2010)

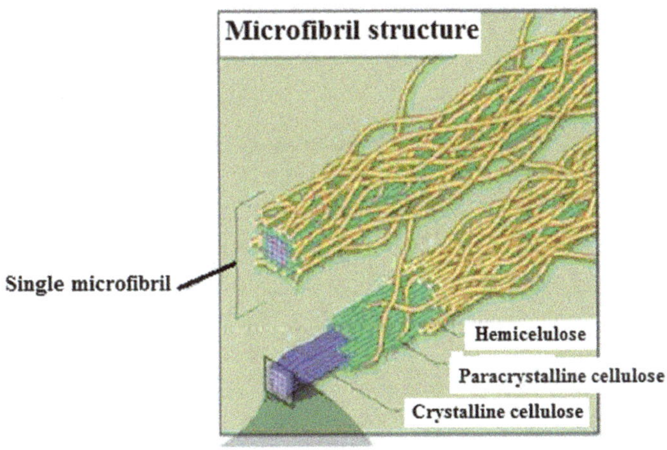

Fig. 10 Structure of the cellulose microfibril

Table 3 Chemical composition of lignocellulosic fibres (Saheb and Jog 1999)

	Cellulose content (%)	Microfibrillar angle (degree)	Diameter (μm)	Length (mm)
Flax	64–71	10	5–76	4–77
Ramie	83	7.5	16–126	40–250
Hemp	78	6.2	10–51	5–55
Jute	61–71	8	25–200	–
Sisal	67–78	20	7–47	0.8–8
coir	43	45	12–24	0.3–1

Cellulose is located in the wall of the fiber and is the main abundant organic substance in nature, with an estimated global production of about 1.3×10^{10} tons per year. To scale a tree, this corresponds to the development of 10 g of cellulose per day. The cellulose is highly hydrophilic, and without water solubility character. It is resistant to strong alkalis, but easily hydrolyzed by acids (John and Thomas 2008). Also, for a given percentage of cellulose, more the microfibrillar angle is small and more the fiber stiffness is important (Saheb and Jog 1999). The Table 3 shows, for different fibres, the cellulose content, microfibril angle, and fiber size.

4.3.2 Hemicellulose

Hemicellulose comprises a group of polysaccharides containing a combination of cycles at 5 and 6 carbons such as represented in the example of Fig. 11. The hemicelluloses are distinguished by their solubility in dilute alkaline solutions. Typically, a solution of 10 g of sodium hydroxide is required to extract hemicelluloses remaining in the medium neutralization (Saheb and Jog 1999; John and Thomas 2008).

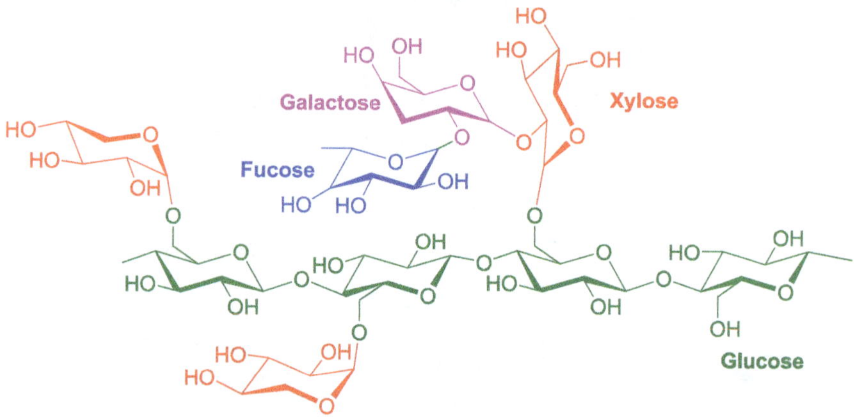

Fig. 11 Structure of xyloglucans

Hemicellulose has a chemical structure that varies following the plant, the cell type, location in the wall and the maturity of the tissue. Among the commonly known hemicellulose are the mannans, xylans, galactans, the most studied are xyloglucans (John and Thomas 2008).

Hemicellulose cellulose differs in some aspects (Saheb and Jog 1999), which are:

- Hemicellulose consisted of neutral sugars: xylose, arabinose, galactose, glucose, mannose and uronic acids, while cellulose contains only 1,4- β-D-glucopyranose units.
- In nature, it has a degree of polymerization between 50 and 300, while that of the primitive cellulose is 10–100 times higher.

Structurally, if the fiber is treated as a composite material, where the cellulose is the fiber reinforcement and the lignin is the matrix, the hemicellulose acts as compatibilizer at the interface between the cellulose and the lignin elements (Saheb and Jog 1999).

4.3.3 Lignin

Lignin is an aromatic polymer extremely heterogeneous. It's composed of a three-dimensional molecule consisting of phenyl propane units (Fig. 12). Primary units (monolignols) constituting lignin are coniferyl alcohol, sinapyl alcohol and *p*-coumaryl alcohol (Fig. 13). During the lignifications process, vegetable phenoloxidases such as laccases are involved and allow the polymerization of the various elementary units. Once synthesized, the lignin is related to the various polysaccharides to make a constitutive matrix of the wall of plant (John and Thomas 2008).

Also, it is essential to reminder that in the plant, lignin is correlated to the polysaccharide by covalent bonds which improve adhesiveness between the matrix and the cellulosic fibres. These aromatic polymers appear in secondary tissues of higher plants. They constitute a three dimensional network which gives the wall its impermeability and its resistance against microbial attack. The lignin is completely amorphous and hydrophobic. It is not hydrolyzed by acids, but soluble in a hot sodium hydroxide solution (Bodros et al. 2007).

4.3.4 Pectins

Pectins are acidic polysaccharides, which the main chain is composed of uronic acid monomers bonded to 1–4 as presented in Fig. 14. Regularly, rhamnose molecules are inserted between these monomers by links 1–2 and 1–4. Some of these rhamnose units bear side chains composed of neutral monosaccharide from which galactose and arabinose are the most abundant. The macromolecule of pectin

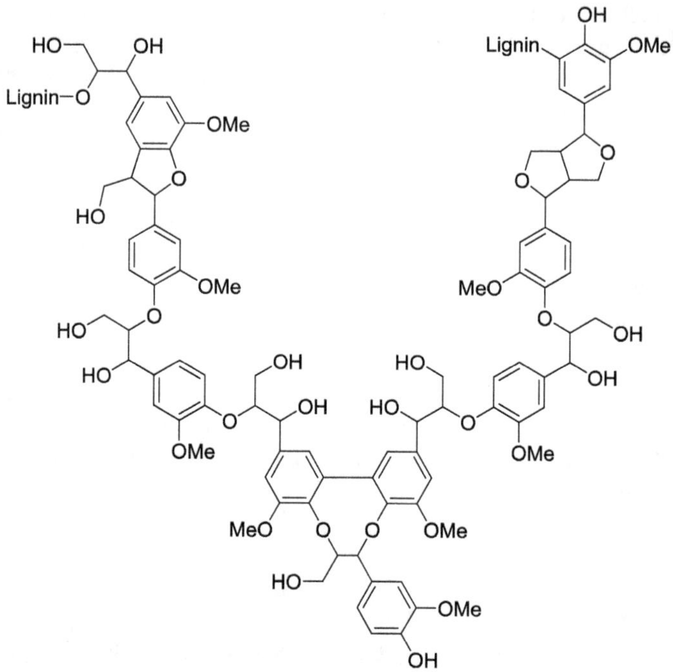

Fig. 12 Schematic illustration of the lignin (Saheb and Jog 1999)

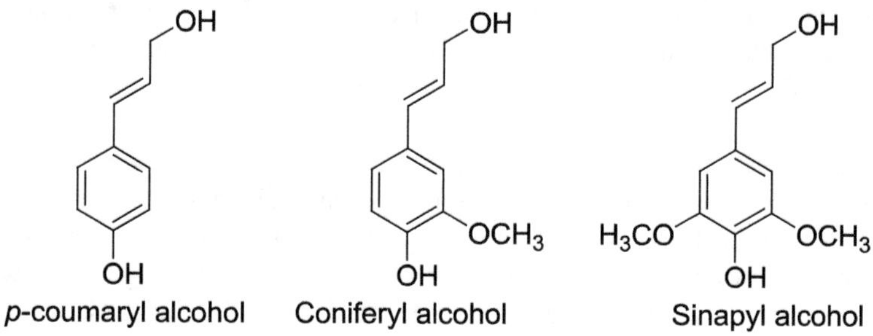

Fig. 13 Structure precursors of lignin (Saheb and Jog 1999)

appears as a zigzag. This arrangement gives it special properties. It provides flexibility to plants (John and Thomas 2008).

4.3.5 Waxes

The waxes are constituents of natural fibres, which could be extracted with organic solutions. These materials consist of diverse types of water-insoluble alcohols and

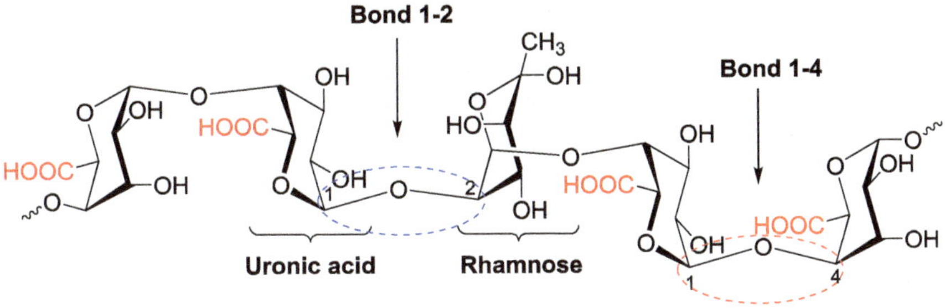

Fig. 14 Structure of a polygalacturonic acid chain (Saheb and Jog 1999)

various acids such as palmitic acid, stearic acid and oilseed acid (John and Thomas 2008).

4.4 The Lignocellulosic Fiber Preparation

Fiber preparation is performed depending to the following four steps: retting, scutching, combing, and dubbing/stretching (Basu et al. 2009).

The retting: is a natural reaction carried out thanks to the weather. After grubbing, the stems of the plant are placed on the soil in the form of windrows to undergo the action of natural microorganisms. Degradation of pectin cements is favored by alternating warmth and humidity. That provides a first separation of the fibres and wood.

The scutching: consists in extracting the fiber from the stems by crushing and threshing straws. The mechanical actions exerted on the fibres are sorting, threshing, grinding between two fluted rollers, stretching from differential rolling.

The combing: is to keep the fiber wicks in a succession of increasingly fine combs.

The dubbing/stretching: conglomerate of fibres permits them to take a ribbon of several meters useable by spinning. Ribbons are officially approved by the mixture of fibres from different sources.

4.5 Advantages and Disadvantages

Lignocellulosic fibres are an attractive substitute to synthetic fibres, particularly glass fibres, owing to their environmentally promising characteristics. The fibres possess various advantages as reinforcement of composite materials. However, some disadvantages may be a brake at the industrial development (Vilaplana et al. 2010; Bodros et al. 2007; John and Thomas 2008).

The main advantages of lignocellulosic fibres are:

- Lignocellulosic fibres are determined by low density, which ensures high specific mechanical properties when it is compared with synthetic fibres.
- Economical production, i.e. fabrication imposing little energy.
- They present safer handling and non abrasive with few requirements for apparatus.
- They are renewable and biodegradable thanks to natural organisms.
- The lignocellulosic fibres have a carbon dioxide neutral character, particularly while they are combusted or composted.
- Hollow cell structure provides good acoustic insulation properties.
- Worldwide availability remains a subsidiary factor.

Notwithstanding the advantages mentioned above, the lignocellulosic fibres possess some disadvantages, which are:

- Lignocellulosic fibres present a polar and hydrophilic aspect owing to the existence of hydroxyl and other oxygen groups containing in the fiber. The matrices are mostly non polar and hydrophobic. This results in poor dispersion of fibres in the matrix.
- High moisture absorption, which the lignocellulosic fibres absorb variably between 5 and 10%. This brings to reduced mechanical properties and decreases dimensional stability of the composites.
- Lignocellulosic fibres acquire little microbial resistance and susceptibility to decay. These properties are serious problems during the transport, storage, processing and implementation of composites.
- The variation in properties into the same fiber can create problems in producing composites with uniform properties.
- The lignocellulosic fibres are susceptible to degradation at high temperature of manufacturing, which limits the choice of matrix. The thermal degradation of fibres is a two stages process, during the temperature range 220–280 °C and another in the range 280–300 °C. The low temperature degradation of manufacturing process is related with degradation of hemicellulose, whereas, the high temperature process is owing to lignin.

It is obvious that the advantages compensate the disadvantages. Most of the deficiencies have corrective measures in the form of treatments or modifications in the fiber.

4.6 Weave Architecture of Reinforcement

The development of textile structures such as weaving, braiding and knitting has allowed the realization of composites having remarkable mechanical properties (John and Thomas 2008).

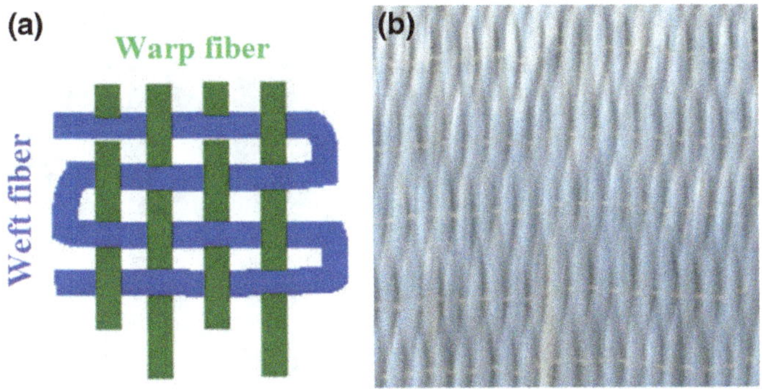

Fig. 15 **a** Weave structure and **b** Unidirectional weave

The weave structure elaborated by the entanglement of the warp fibres (0°) and weft fibres (90°) in a regular model as illustrated in Fig. 15a.

A description of various architectures commonly found in weaving structures:

Unidirectional weave: in a UD fabric, the fibres are assembled parallel compared relatively to each other with a very light weft (Fig. 15b).

Plain weave: each warp fiber moved over and below each weft fiber alternately. The fabric is symmetrical with good planicidad and relative rigidity. However, the fabric is not very deformable at the implementation. Many successive intersections generate significant crimp, which resulted in a diminution in mechanical properties.

Serge weave: each warp fiber runs over two or more weft fibers weft fiber, and each fiber weft runs over two or more warp fiber. Serge structure is distinguished by a smooth surface and slightly higher mechanical properties. Also the structure is very flexible relative to the plain weave (Fig. 16b).

Satin weave: each warp fiber passes over a number (n − 1) of weft fibres alternately. Satin structure is flexible with a high degree of drape. Thus, satin has good mechanical properties, but it remains an asymmetric structure (Fig. 16c).

Basket weave: it is mainly the similar as the plain weave, exception that two or more warp fibres pass alternately in excess of two or more weft fibres. This structure is very flexible, but it is poorly stable (Fig. 16d).

It is as well possible to make hybrid structures reinforcements by weaving fibres of different natures or, by superimposing fabrics or layers of different fiber reinforcements.

4.7 Types of Lignocellulosic Fibres

Being an interesting renewable resource, the lignocellulosic fibres extracted from vegetable plants will have an important role in the conversion to a sustainable economy and more environmentally friendly. There are different kinds of natural

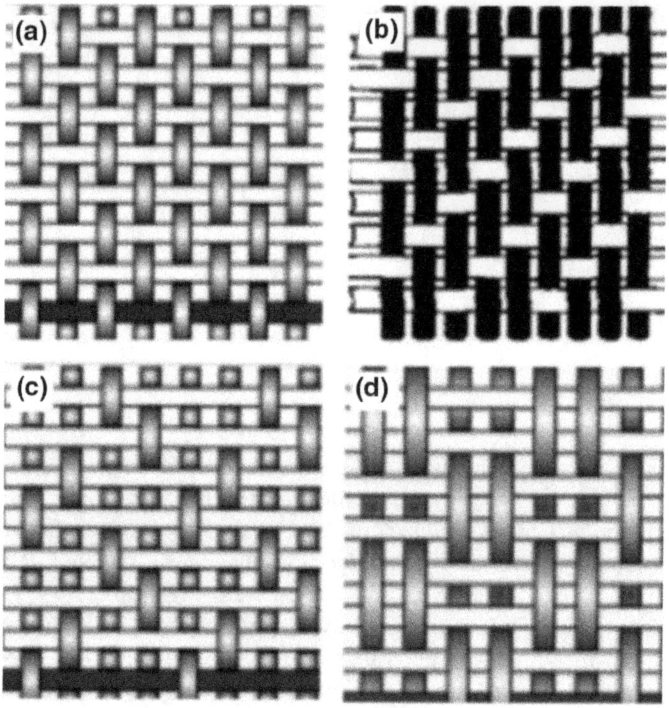

Fig. 16 **a** Plain weave, **b** Satin weave, **c** Serge weave and **d** Basket weave

fibres like doum, sisal kenaf, jute, flax, etc. these fibres are different in botanical origin, climate, instead of growth (Mwaikambo and Ansell 2002). A description of certain types of lignocellulosic fibres:

Flax fibres: The Flax (Linum Usitatissimum) is an yearly plant of the Linaceae family (Fig. 17a), cultivated in Europe and Asiatic countries as China. Flax fibres, length 40–50 mm, are contained in the stem of a length of 60–120 cm and a diameter of 1–3 mm. flax fibres are 2–3 times more resistant than cotton fibres. They have a low elasticity, but are highly absorbent and good conductors of heat. Currently, these fibres are applied in papermaking (Le Duigou et al. 2010; Charlet et al. 2010).

Hemp fibres: The hemp (Cannabis sativa) is an yearly plant of the cannabaceae family (Fig. 17b), with a height of 1–3 m, cultivated in countries with temperate climate such as Eastern Europe, in France and in Italy. Bast fiber is flattened and slightly woodier than flax. It can absorb until 30% of moisture and holds almost 95% of UV rays. It burns without leaving any waste. Hemp fiber is applied in textiles, geotextiles and erosion control fabric (Amaducci et al. 2008).

Sisal fibres: The sisal (Agave sisalana) or Heineken is a plant of the Agavaceae family (Fig. 17c). It is a tropical plant, mainly grown in Brazil, Tanzania and Kenya. sisal fibres are mined from the leaves of the plant, in which each leaves

Fig. 17 **a** Flax plant, **b** Hemp plant, **c** Sisal plant, **d** Jute plant, **e** Kenaf plant, **f** Cotton plant, **g** Abaca plant, **h** Ramie plant, **i** Coir plant, and **j** Doum plant

contains 1000–1200 packets with an average fiber length of 3 mm. the fiber is very resistant. The sisal fibres are applied for manufacturing ropes, fabrics and carpets (Silva et al. 2008).

Jute fibres: The Jute (Corchorus capsularis) of the Malvaceae family (Fig. 17d) is a plant grown in tropical regions, mainly in Bangladesh and India. Jute fibres are extracted from a stem from a height of 4–6 m and a diameter of about 3 cm; the fibres have a length of 2 mm and a width of 20 μm. The ultimate fiber is very short and lignified. Jute fibres are used in papermaking and rope (WinklerPrins 2006).

Kenaf fibres: The Kenaf (Hibiscus cannabinus L.) of the Malvaceae family (Fig. 17e) is an herbaceous plant whose stem reaches 3 m high and almost 2 cm in diameter. It is grown in tropical regions and in South America. The kenaf fiber of 1.5 mm in length is between that of hardwood fibres and the resinous fibres. The kenaf fibres are intended only for the papermaking field (Mwaikambo and Ansell 2002).

Cotton fibres: The Cotton (Gossypium) is a shrub of the Malvaceae family (Fig. 17f), from 1 to 1.5 m high. It is grown in warm areas of America, Africa, Asia and Europe (Spain). The cotton fibres come from cotton seeds. this fiber is employed for its good mechanical and physical characteristics, including luxury papers, bank notes and textiles (Saheb and Jog 1999).

Abaca fibres: The abaca (Musa textilis) or manila abaca is a family of Musaceae that grows to Philippines (Fig. 17g). Abaca fiber is mined from the oblong leaves and stems. It has a length of 6 mm and a diameter of 22 μ. The abaca fibres are principally used for rope-making, and also in the composition of filter papers and teabags (Regional centre of animation in advanced materials 2006).

Ramie fibres: The ramie (Boehmeria nivea L.) of the nettles family is a shrubby plant (Fig. 17h), native to the Far East. The main producers of ramie fiber are China and Japan. The ramie fiber has an average length between 40 and 250 mm and an average diameter of 45 μ. These fibres are similar to flax fibres, but are still longer and more rigid. Ramie fibres are produced for artisanal uses (Regional centre of animation in advanced materials 2006).

Coir fibres: The coconut (Cocos nucifera) is a species of palm of the Cocoeae family (Fig. 17i). It is acclimated in most tropical countries. The coir fibres come from the fibrous layer that surrounds the coconut. The coir fibres are highly resistant and insulating (Defoirdt et al. 2010).

Doum fibres: The doum (chamaerops L) is a small palm of the Coryphoideae family (Fig. 17j). It is originally from the regions bordering of the western Mediterranean Sea. The doum fibres are extracted from the leaves of the plant. Doum mainly is applied for decorating gardens (Malha et al. 2013).

4.8 Lignocellulosic Fibres Properties

Lignocellulosic fibres properties Structure, defects, cell dimensions, and the chemical composition of lignocellulosic fibres are of the key variables that influence the natural fiber. Usually, the tensile measurements (strength and the Young's modulus) of lignocellulosic fibres increase with increasing cellulose (Saheb and Jog 1999; Singh et al. 2008). The major properties of lignocellulosic fibres are gathers in Table 4.

Table 4 The physical properties of several natural fibres (Regional centre of animation in advanced materials 2006)

	Density (g/cm^3)	Tensile stress (MPa)	Young's modulus (GPa)	Specific Young's modulus (GPa)	Deformation (%)
Flax	1.45	500–900	50–70	48	1.5–4
Hemp	1.48	350–800	30–60	41	1.6–4
Kenaf	1.3	400–700	25–50	38	1.7–2.1
Jute	1.3	300–700	20–50	38	1.2–3
Bamboo	1.4	500–740	30–60	36	2
Sisal	1.5	300–500	10–30	20	2–5
Coir	1.2	150–180	4–6	5	20–40
Glass fiber	2.5	1200–1800	72	29	2.5
Carbon fiber	1.4	4000	235	168	2

Comparing with properties of synthetic fibres, it is obvious that the glass fibres modulus is greater than that of lignocellulosic fibres. In contrast, if specific properties are considered, it will be noted that the specific modulus of lignocellulosic fiber is comparably high that the synthetic fiber modulus.

4.9 Lignocellulosic Fiber Modification

The fiber-matrix interface quality is very significant for using the natural fibres as fibrous reinforcement for the thermosetting polymers. However, the lignocellulosic fibres present an incompatibility with hydrophobic polymers. Consequently, it has proved necessary to treat the fiber surface in order to enhance this interface. The core idea is to amend the nature of the fiber by adding functional groups that have more affinity with the polymer matrix (Mwaikambo and Ansell 2002).

The lignocellulosic fibres may be modified by chemical or physical methods. Physical methods, such as stretching, calendering, heat treatment, laser, gamma rays, UV, plasma, used to change the surface and structural properties of the fiber. Chemical modifications include treatment with sodium hydroxide, the silane (silicon alkoxide functionalized), acetic acid, or based molecules of benzoyl, isocyanate, triazine or imidazolidinone, etc. (Mwaikambo and Ansell 2002; Saheb and Jog 1999).

Chemical coupling method stays one of the chemical methods, which allows improving the interfacial adhesion. The chemical composition of the coupling agents allows their reaction with the surface of the fibres by forming a bridge of chemical bond that associates the fibres with the matrix. In general, the coupling agents are molecules that having two functions, the first function reactive with the -OH groups of the cellulose and the second with the functional groups of the matrix (Xie et al. 2010).

To enhance the compatibility between two incompatible materials, it is possible to incorporate another material having intermediate properties. The most frequently used methods are the alkali treatment firstly often followed by silanization or acetylation or adding a compatibilizers (Malha et al. 2013; Corrales et al. 2007; Xie et al. 2010). Table 5 shows the values of various mechanical properties in function of the treatment of the fiber.

4.9.1 Alkaline Treatment

Treatment with NaOH or also called mercerizing is one of the used treatments to treat natural fiber. According to ASTM 1965, that is a process which consists to subject the vegetable fiber in an interaction with a concentrated aqueous solution of a strong base, so as to cause substantial swelling resulting in variations in the fine structure, the size, morphology and the mechanical properties (Mwaikambo and Ansell 2002).

Table 5 Mechanical properties of untreated and treated sisal fiber (Sood et al. 2015)

Treatment	Tensile strength (g/tex)	Tensile modulus (10^3 g/tex)	Elongation at break (%)
Untreated fiber	30.7	1.18	2.5
Benzol treatment	38.8	0.99	3.7
Acetic acid treatment	9.3	0.39	2.6
Alkaline treatment	31.7	0.53	7.5
Acetate treatment	33.2	0.35	8.3
Heat treatment	42	1.22	3.5
Alkaline followed by thermal treatment	27.6	0.7	4.7
Thermal followed by alkaline treatment	25.7	0.71	4.4

The major modification by the alkali treatment is the decrease of hydrogen bonds between the cellulose and the two others constituents: hemicellulose and lignin. Also, this treatment banishes wax, pectin and a part of the lignin and hemicellulose, which are amorphous structures, and therefore they lead to a low tensile strength.

Besides, after the soda treatment, many –OH groups are present on the fiber surface (Fig. 18) and the fibres may be simply treated with other chemical constituents such as silanes, acetic acid and others (Mwaikambo and Ansell 2002).

NaOH concentration should be considered when treating fibres, as it affects the degree of swelling, and thus the degree of cellulose conversion. In one study, when the bamboo fibres are treated in alkaline solution at a concentration less than 5%, the density of bamboo fiber increases owing to the elimination of most of hemicellulose and lignin. This allows approximation of the cellulose microfibrils,

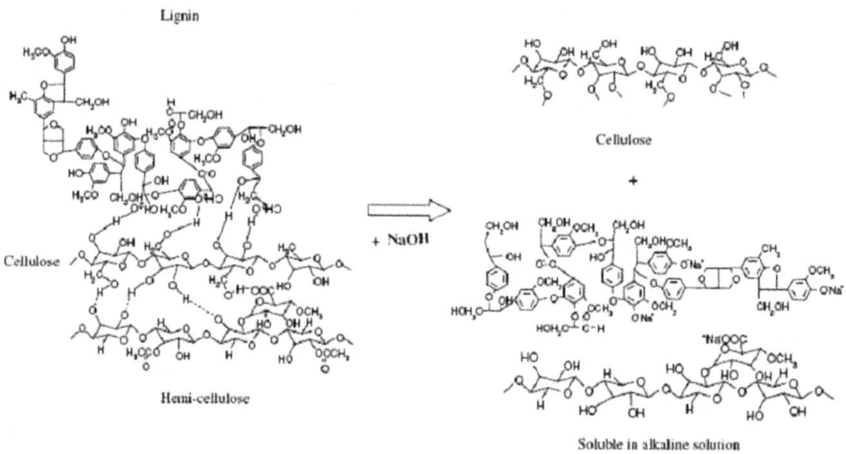

Fig. 18 Reaction of alkali treatment (Mwaikambo and Ansell 2002)

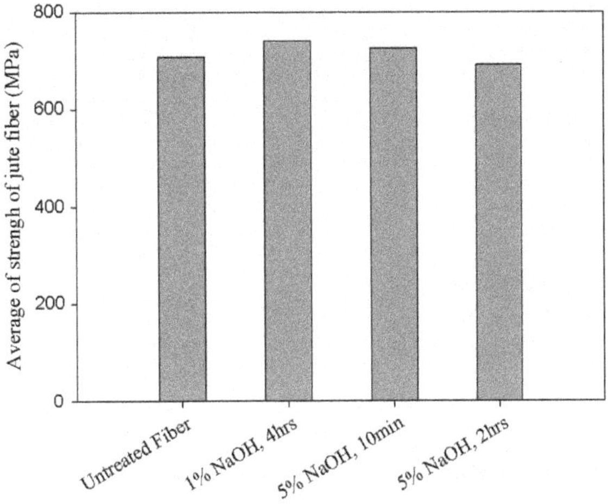

Fig. 19 Tensile test on treated jute fibres with different solutions of sodium hydroxide (Loan 2006)

leading to a more compact and increasing the density. At a NaOH concentration of above 5%, the density decreases as a result of the disappearance of other substances sensitive to alkalis, this brings to the creation of microvoids (Bessadok et al. 2007; Kim and Netravali 2010).

The alkali treatment influences the thermal and physical properties of the fibres natural. Loan studied the modification of the jute fiber surface. They proved that the treatment with soda (NaOH 1%, 4 h) improves the fiber strength (Fig. 19). Due to the alkaline treatment, lignin and hemicellulose are removed; the interfibrillary region is less dense and less rigid, allowing the fibrils to reorganize in the sense of traction. When the fibres are stretched, into such arrangements between the fibrils would result in a better load sharing (Loan 2006).

Altogether, the alkali treatment may affect the fiber in two ways: (1) it enhances the surface roughness, and (2) it increases the quantity of cellulose which covers the fiber surface, expanding the number of possible reaction sites (OH groups). (Kim and Netravali 2010; Loan 2006) Therefore, the effect of alkali treatment is durable on the mechanical behavior of fibres, particularly on their strength and rigidity. After the alkali treatment, the fibres are often treated with silane, acetic acid, etc., to further improve the adhesion fiber—matrix.

4.9.2 Treatment with Silanes

Silanes or silica functionalized alkoxide, represented by R'-Si(OR)$_3$, are hydrophilic compounds with different groups associated with the silicon such that one end interacts with the matrix and the other with the hydrophilic fiber, acting as a bridge between them. The silane absorption is greatly dependent on a certain quantity of factors, such as hydrolysis time, functionality, the temperature and pH. The silicon alkoxides are able of forming bonds with the hydroxyl groups.

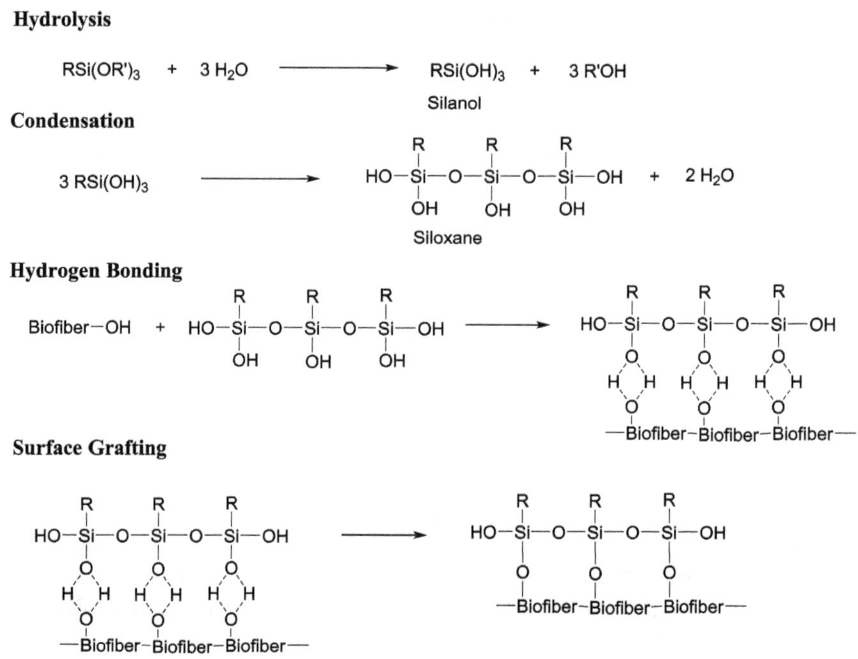

Fig. 20 Chemical reactions of treatment with a functionalized silica alkoxide (Seki 2009)

Silanols may form polysiloxanes by reaction with the hydroxyl groups of the fiber (Seki 2009). The chemical reaction is schematized in Fig. 20.

Silane solutions in a blend of water and ethanol at a concentration of 0.033% were used by Valadez et al. (Seki 2009) to treat the fibres. They found that the effect of treatment by the silane resulted in a higher tensile resistance than the alkali treatment.

According to Nguyen Huy Tuong et al. (Seki 2009), when the mass concentration of silane is increased by 0.5–1%, the shear strength is increased at the interface. However, no variation was noted when the silane concentration is raised from 1 to 2%. With the silane concentration increasing, the alkoxides groups react with the bamboo fiber and the surface is more active. But when the silane concentration is 2%, the silane concentration contained in the fiber surface cannot rise because this one is saturated. The silane not improves the surface activation. An amount of 1% silane is sufficient to cover the entire surface of the bamboo fibres.

4.9.3 Acetylation

The acetylation allows modifying the natural fibres surface which makes it more hydrophilic. It has been employed to decrease swelling of the wood in water. The concept of the method is to react the hydroxyl groups (–OH) of the fiber with the acetyls groups (CH_3–CO–), and thus make the surface more hydrophobic fiber. The hydroxyl groups that react are hemicellulose and lignin and amorphous part of the cellulose. The hydroxyl groups of crystalline regions in the fibres are tightly compacted with strong intermolecular bonds, and are inaccessible to chemical reagents (Bessadok et al. 2007). The acetylation of the OH group is shown below:

– Acetylation with an acid catalyst:

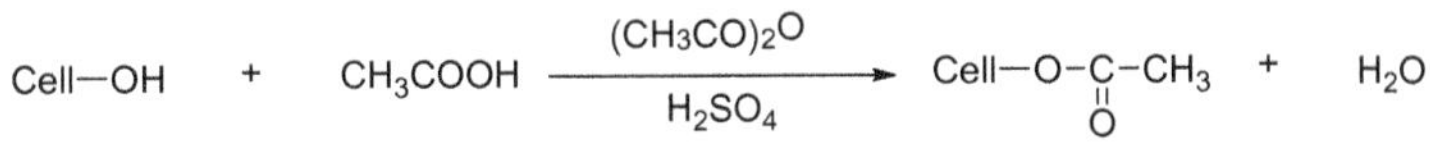

– Acetylation without acid catalyst:

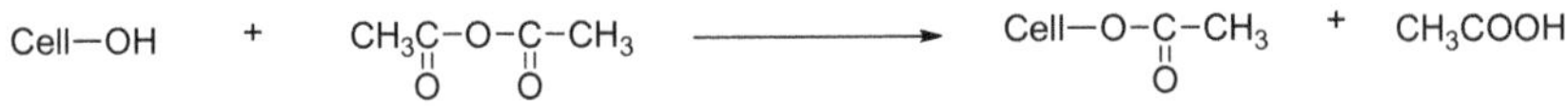

Because acetic acid does not respond sufficiently to the reaction with the cellulose, it is estimable to employ acetic anhydride. However, acetic anhydride is not a good swelling agent for the cellulose, and then to accelerate the reaction, the cellulose is first steeped in acetic acid and then treated with acetic anhydride for 1–3 h. In composite reinforced by natural fibres, acetylation of the hydroxyls groups swells the wall of plant cells significantly reducing hygroscopic nature of cellulose fibres. This brings to dimensional stability of the composites, because no water absorption can lead to swelling or shrinkage of the composite material (Bessadok et al. 2007; Saheb and Jog 1999; Seki 2009).

Rana et al. (Rana et al. 1997) worked on the thermal properties and morphology of acetylated jute fiber surface. The thermogravimetric analysis curves of acetylated jute fibres show that the decomposition of the hemicellulose has disappeared, and that the losing weight allied with the main degradation step (373 °C) increased. The main decomposition temperature increased from 363 to 373 °C, indicating that the thermal stability of acetylated jute fibres is higher than that of the untreated fibres. The same conclusions were made by Albano et al. (Albano et al. 1999) during the study acetylation sisal fibres

4.9.4 Compatibilizers

Among the organic coupling agents, anhydrides such as maleic anhydride, acetic anhydride, succinic anhydride and phthalic anhydride are the most employed coupling agents for the reinforced natural fibres composite. Acetic anhydride, succinic anhydride and phthalic anhydride have two functional groups (e.g. –COO– carboxylate), which may be related to lignocellulosic fibres by esterification and hydrogen bonding (Fig. 21) (Malha et al. 2013; Bessadok et al. 2007).

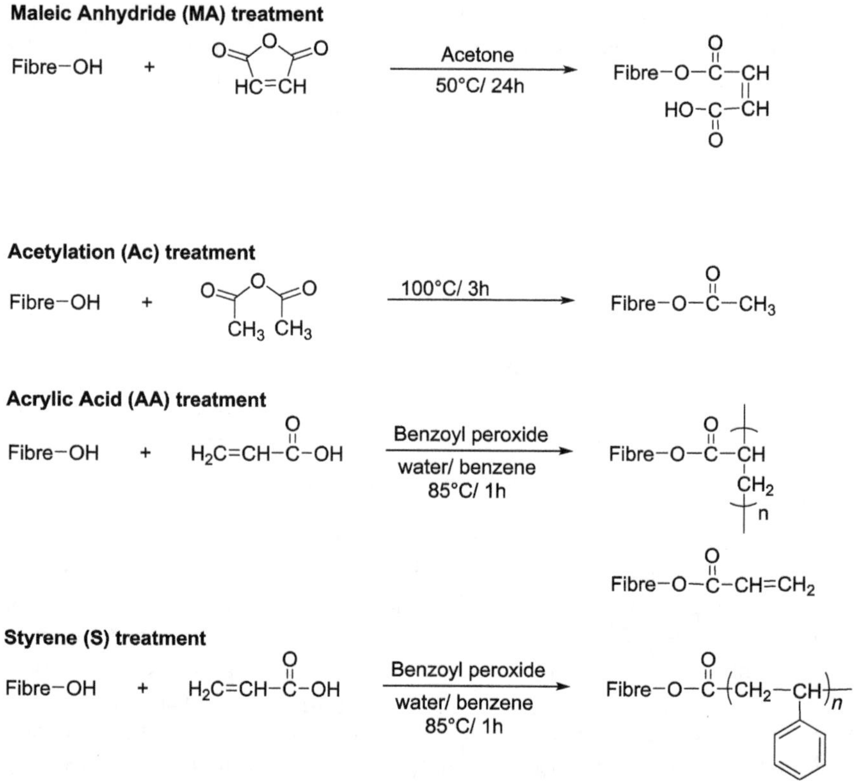

Fig. 21 Scheme of the chemical treatments: maleic anhydride, styrene, acrylic acid and acetic anhydride (Bessadok et al. 2007)

Maleic anhydride is an unsaturated carbonyl compound containing a carbon-carbon double bond (–C = C–) and two carboxylate groups (–COO–). This structure considerably increases the reactivity of the carbon-carbon bond with the hydrophobic matrix, ensuing in a reaction or strong adhesion at the interface (Malha et al. 2013).

Mohammad (Beg 2007) studied polypropylene/kraft fibres composites (10–60% by weight of fibres) containing polypropylene modified with maleic anhydride (MAPP) as a coupling agent (1–4% by weight). The optimal result is obtained with 50% by weight of fibres and 3% by weight of MAPP. The stress is increased with fiber content to 50% by weight of fibres, but a decrease is observed for 60% by weight of fibres, which could be due to the increased viscosity, ensuing in poor distribution of the fibres. The increase in stress and the elastic modulus of the samples containing MAPP seems to be due to better interfacial bonding.

5 Thermoset Composites Reinforced Lignocellulosic Fiber

The record of plastic reinforced fibres began in 1908 with the cellulose fiber in phenolic compounds, passing subsequently through the melamine fiber to the glass fiber. The massive and exorbitant use of oil resources has evoked the development efforts of biodegradable plastics. This is based on plants of natural origin and agricultural products that can compete in markets dominated by synthetic product (Vilaplana et al. 2010).

The manufacture of materials at 100% of natural origin as substitute for synthetic products is not an economically profitable solution. A more practicable and optimal solution would be to combine synthetic biodegradable products and natural products. Biopolymers or synthetic polymers reinforced natural fibres are a gainful option to glass fiber composites. Scientists are studying the different possibilities of combining the lignocellulosic fibres with polymer matrices, including thermoset matrices (John and Thomas 2008).

5.1 Classification of Composites

5.1.1 Green Composites

Green composites are fully biodegradable composite, combining lignocellulosic fibres with biodegradable resins. Green composites are attractive because they are environment friendly, wholly degradable and durable. At the ending of their life, they be simply able to be removed or composted undoubtedly without detracting the environment (Vilaplana et al. 2010; Bodros et al. 2007).

The biodegradable polymers are classified according to four families. The first family's agro-polymer (e.g. polysaccharides) obtained from biomass fractionation. The second and the third family are the polyesters obtained respectively by fermentation from biomass or genetically adapted plants (e.g. polyhydroxyalkanoate) and synthesis from monomers (e.g. polylactic acid). The fourth family are totally polyesters synthesized by petrochemical process (e.g. polyesteramine). Many of these polymers are commercially available (Saheb and Jog 1999).

5.1.2 Thermoset Composites

For thermoset composites, the lignocellulosic fibres are combined with phenolic matrices, epoxy, and polyester resins to shape composites. Thermosetting polymers are three-dimensional structures containing reactive groups, which help the development of the interface. These thermosetting matrices, owing to their structure, acquire during crosslinking a form which cannot subsequently be changed by rising the temperature. The most developed applications for automotive parts (body parts: hood, tailgate), recreation (boat hulls) (Megiatto et al. 2009).

Most resins employed in composite applications are thermosetting, which the epoxy resins are the most accustomed. Thermosets may moreover be charged with organic fillers like inorganic carbonates or natural fillers like clay to purpose of increasing its mechanical properties. In a study of jute fibres reinforced polyester composites, it is found that the mechanical properties were influenced by the chemical bond between the jute fiber and the polyester matrix; and the same results obtained with the use of epoxy and vinylester matrix (Gassan and Bledzki 1999; Stocchi et al. 2007; Sangthong et al. 2009).

5.1.3 Hybrid Composites

The integration of several types of fibres into one matrix allows the fabrication of hybrid composites. Hybrid composites have the advantage of using several types of fibres that are complementary in function of mechanical, chemical or physical properties. Accordingly, a equilibrium of cost and performance could be achieved by dint of the implementation of these composites (Khan et al. 2009).

The properties of the hybrid composites depend essentially on the fiber, the length of the individual fibres, the orientation and arrangement of the fibres. The resistance of the hybrid composite is also function of the strain at break of each fiber. Hybrid composite can be realized by synthetic fibres and natural fibres in a matrix or by mixture of two kinds of natural fibres in the matrix. Hybridization with the glass fibres provides a way for improving the mechanical properties of the composite natural fiber (Jawaid et al. 2010).

5.2 *Processing of Lignocellulosic Fiber Composites*

The majority of frequent processes for manufacturing composites with lignocellulosic fibres are contact molding, compression molding, injection and infusion. These processes involve the implementation of the unhardened composite in a mold so that the material can be shaped into the final piece. Some other methods like centrifugal casting, filament winding, pultrusion, are used for thermoset composites, but the employ of these processes for lignocellulosic fibres composites still impracticable (Abdellaoui et al. 2015).

In thermosetting matrices, the fibres are used as mats or woven structures (unidirectional, fabric, satin…). Also, the type of employed process influences directly the properties of the material by the porosity rate, mastery of reinforcement orientation and the quality of the impregnation.

5.2.1 Hand Lay up

This is a manual process, performed at an ambient temperature and without pressure. The reinforcements are deposited on the mold and impregnated with liquid thermosetting resin, catalyzed and accelerated, to be formed gradually by airing system (bubble withdraw and brushes). After the cure of the resin as revealed in the Fig. 22, the part is separated from the mold and trimmed (Regional centre of animation in advanced materials 2004).

This process is applied for small series from 1 to 1000 pieces/year. It is branded by obtaining the piece with one smooth gelcoated surface, and average to high mechanical properties. However, it gives a low production rate per mold and the homogeneity of the part strongly accorded to the operator which manufactures the part. It is used in the fields of transport auto body, public works (formwork), swimming pools and water sports (Regional centre of animation in advanced materials 2004).

5.2.2 Injection Molding—RTM

Injection molding of liquid resin RTM (Resin Transfer Molding) occurs between two rigid mold parts (male and female). The reinforcement (mats, perform, optionally tissues) is disposed in the mold. Once it securely closed, resin, accelerated and catalyzed, is injected under low pressure (1.5–4 bars) through the reinforcement to complete the fill of the cavity as schematized in Fig. 23. After curing the resin, the mold is opened to remove the shaped piece (Abdellaoui and Echaabi 2014; Le Riche et al. 2003; Deleglise et al. 2006).

This process is distinguished by the fabrication of relatively complex forms with excellent precision, elevated production rate and good working conditions. However, the reinforcement rate is limited (30–40%) and need to conduct post molding finishes. it is applied primarily in auto body parts for passenger vehicles and utility vehicles, cowlings, tennis racket, bicycle fork, and various industrial parts (Deleglise et al. 2006).

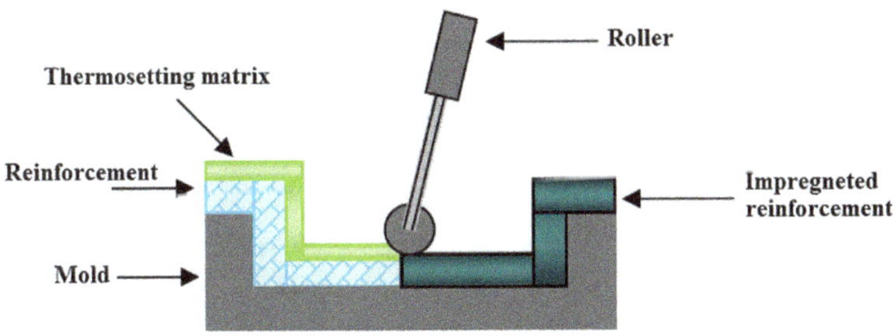

Fig. 22 Han lay-up process

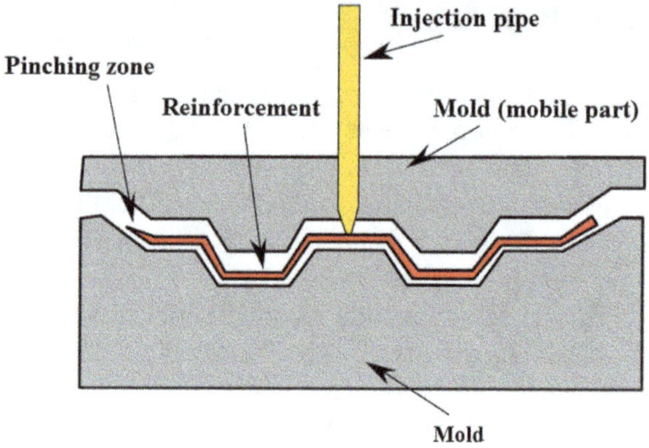

Fig. 23 Resin transfer molding process

This process is considered environmentally friendly thanks to the use of the closed mold, which prevents the release of noxious odors thermosetting resins (styrene).

5.2.3 Infusion

The principle of the infusion remains comparable to the RTM except that the upper mold is replaced by a flexible tarpaulin. The resin is injected and distributed on reinforcements until complete impregnation of them. The Fig. 24 reports the infusion process. Injection occurs at room temperature, even if the resin can be heated during injection to optimize the conditions of polymerization and viscosity. The vacuum is drawn from the sheet in order to accelerate the diffusion of the resin on the reinforcement and reduce levels of porosity (Park and Kang 2003; Han et al. 2003).

Infusion allows to acquire a significant reinforcement rate which varies following to the depression applied to the part and to ensure reproducibility of the parts manufacturing conditions (Park and Kang 2003).

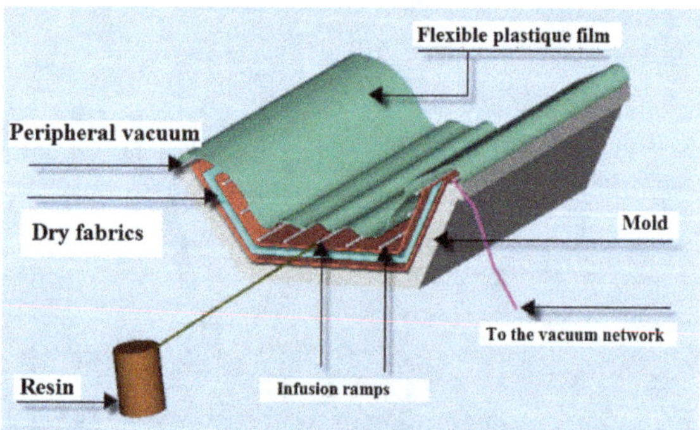

Fig. 24 Infusion process (Regional centre of animation in advanced materials 2004)

5.2.4 Compression Molding

Molding using a press to compression between the rigid parts of the mold, initially without external heat supply. Open mold, the reinforcement is laid on the lower part of the mold and then the resin provided with a highly reactive catalyst system is poured loose on the reinforcement (Abdellaoui et al. 2015; Bickerton and Abdullah 2003) as shown in the Fig. 25.

The mold is closed under pressure (2–4 bars) causes distribution of the resin inside the mold cavity and impregnating the reinforcement. Curing of the resin is gradually accelerated by the mold temperature rise thanks to the exothermicity of the reaction, allowing a fast release (Kelly and Bickerton 2009).

This process is distinguished by a elevated production rate, and normal working conditions. However, it is limited to the dimensions and some complex shapes. It is mainly applied for cowlings and bins (Regional centre of animation in advanced materials 2004).

5.3 Characterization of Lignocellulosic Fiber Composites

For composite as for all engineering materials, the mechanical properties are fundamental because they allow the dimensioning and designing structure that must continue to last.

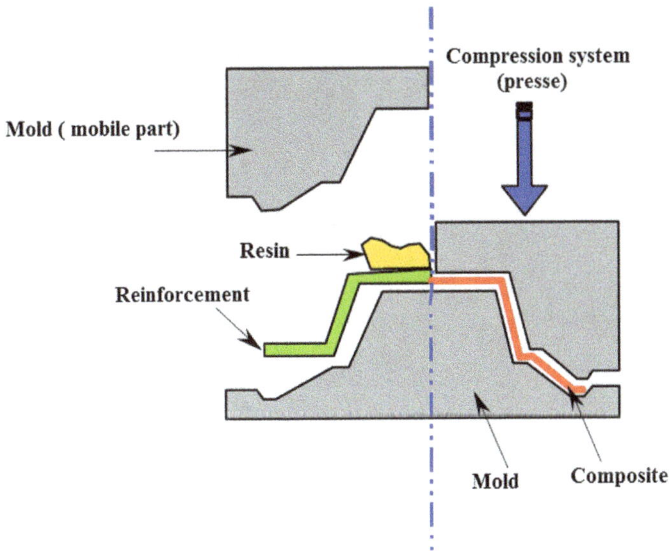

Fig. 25 Compression molding (Regional centre of animation in advanced materials 2004)

To characterize the mechanical properties of the composite with different lignocellulosic fibres, it is required to carry out certain tests such as tensile, bending, and torsion tests. Before proceeding to mechanical testing, it is as well important to study the density of the manufactured structures

5.3.1 Physical Control

The mainstream of thermoset composite structures are laminates structures at various numbers of layers. Experimentally, it is useful to define the intrinsic characteristics of manufactured parts, which may be summarized in experimental density, reinforcement and resin content. The void content in the structure is also considered by calculating the experimental and theoretical densities (Abdellaoui et al. 2015).

(a) Mass content

- Reinforcement mass content:

$$M_f = \frac{\text{Reinforcement mass contain in composite}}{\text{Total mass of composite}} \tag{1}$$

- Resin mass content:

$$M_m = \frac{\text{Resin mass contain in composite}}{\text{Total mass of composite}} \tag{2}$$

With:

$$M_f(\%) + M_m(\%) = 100\% \tag{3}$$

(b) Experimental density:

$$\rho_{\exp} = \frac{\text{Weight of composite}}{\text{Volume of composite}} \tag{4}$$

(c) Theoretical density:

$$\rho_{\text{theor}} = \rho_f V_f + \rho_m V_m \tag{5}$$

Where V_m and V_f are the matrix and fiber volume fraction, respectively.

Table 6 Average characteristics of developed composite

Composite	M_f (%)	M_m (%)	ρ_{Exp} (g/cm^3)	$\rho_{Théo}$ (g/cm^3)	V_f (%)	V_m (%)	ϑ_{void} (%)
	49	51	1.03	1.23	41	59	16.2

$$V_f = \frac{M_f/\rho_f}{M_f/\rho_f + M_m/\rho_m} \quad \text{and} \quad V_m = \frac{M_m/\rho_m}{M_m/\rho_m + M_f/\rho_f}$$

where ρ_f and ρ_m are the fiber and matrix density, respectively.

The void content is calculated by:

$$\rho_{\text{Exp}} = (1 - \vartheta_{\text{void}})\rho_{\text{Theo}} + \vartheta_{\text{void}} \cdot \rho_{\text{void}} \tag{6}$$

With

$$\vartheta_{\text{void}} \cdot \rho_{\text{void}} = 0$$

Then

$$\vartheta_{\text{void}} = 1 - \frac{\rho_{\text{Exp}}}{\rho_{\text{Theo}}} \tag{7}$$

In a novel study (Abdellaoui et al. 2015), they fabricated a laminates based on epoxy resin reinforced with jute fibres by a compression process. For the various laminates, they calculated the theoretical and experimental densities which the following Table 6 summarizes the obtained results.

The void content reaches moderately almost of 16.2% in the manufactured composites. The presence of the void in the composite structure generates a stress concentration, leading to an early phenomenon of delamination. Also, this included void can affect the mechanical characteristics of elaborated composite (Abdellaoui et al. 2015).

5.3.2 Flexural Test

The bending test is used to test the composite flexural strength, which results in a deformation in the form of a curvature.

The principle of this test is to measure a displacement (arrow y) from a force (F) in the centre of two supports. Relationships from the materials resistance used to make the connection between the two variables (y and F), the geometry of the part (L, h and b) and the material characteristics (σ, ε and E). The bending curves ($\sigma = f(\varepsilon)$) are used to define the flexural modulus, the stress at break and strain at break (Regional centre of animation in advanced materials 2004).

The rectangular specimen, supported as a beam, is subjected to bending in the centre of the range (3 three-point bending) at a constant rate until failure of the sample as mentioned in the Fig. 26. The different characteristics can be assessed by means of the following relation the following relationships (Jawaid et al. 2011):

$$\sigma_f = \frac{3.F.D}{2.b.h^2} \quad (8)$$

$$\varepsilon_f = \frac{6.h.Y}{D^2} \quad (9)$$

$$E = \frac{D^3.F}{4.b.h^3.y} \quad (10)$$

5.3.3 Tensile Test

The tensile test used to measure the level of material strength at break. It consists of placing a sample between the chuck jaws of a tensile testing machine (Abdellaoui et al. 2015), which pulls the sample until it breaks as demonstrated in the Fig. 27.

The machine software registers the elongation and the applied force, which are then converted into deformation and stress.

$$\sigma = \frac{F}{S} \quad (11)$$

and

$$\varepsilon = \frac{\Delta L}{L} \quad (12)$$

The tensile test gives some several values as:

The Young's modulus, the tensile strength, and elongation at break measures the aptitude of a material to elongate under load before rupture.

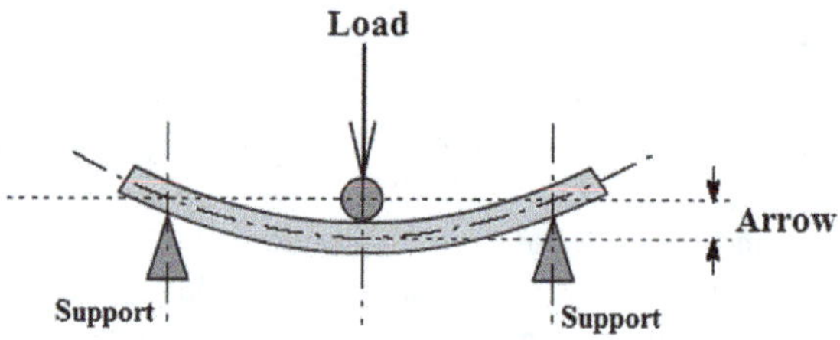

Fig. 26 Flexural test (Regional centre of animation in advanced materials 2004)

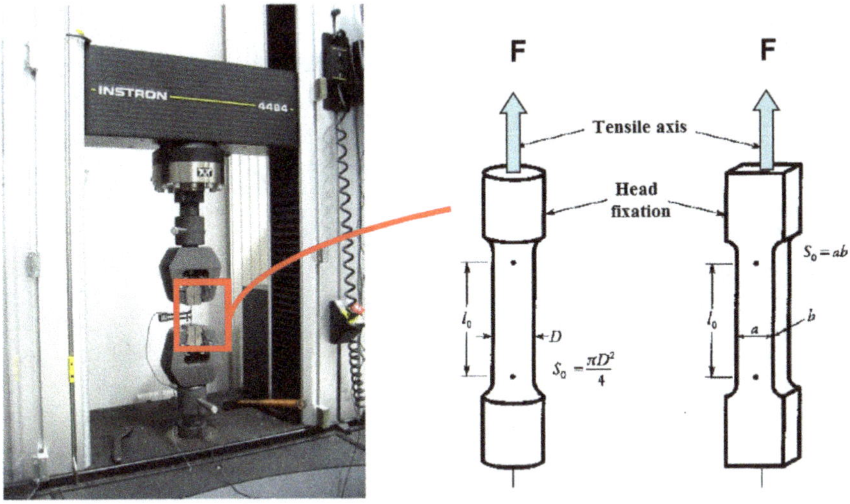

Fig. 27 Tensile test (Abdellaoui et al. 2015)

5.3.4 Rheological Test

The rheology allows characterizing the viscoelasticity of viscoelastic material (polymers and composites) through a dynamic torsional test. Contrary to the simple torsional test, the dynamic torsional test is performed on a Rheometer ARES using rectangular torsion mode with prismatic samples (Abdellaoui et al. 2015) as revealed in Fig. 28.

The test consists firstly in select strain amplitude through a sweep deformation and fixing the frequency. Thereafter, a frequency sweep is performed by fixing the strain at the selected amplitude value. The torsion complex modulus G^*, obtained by oscillatory tests performed in frequency sweep mode (0.1–40 Hz) (Abdellaoui et al. 2015), is given by:

$$G^* = \sqrt{G'^2 + G''^2} \tag{13}$$

With G' and G'' are respectively the storage modulus indicating the elastic part and the loss modulus indicating the viscous part of the viscoelastic material. This module is comparable to the simple torsion in the case of low frequency (0.1 Hz).

Thus, the loss factor can be calculated using:

$$\tan \delta = \frac{G''}{G'} \tag{14}$$

It measures the ratio of energy dissipated by amortization and the elastic energy stored during the dynamic deformation.

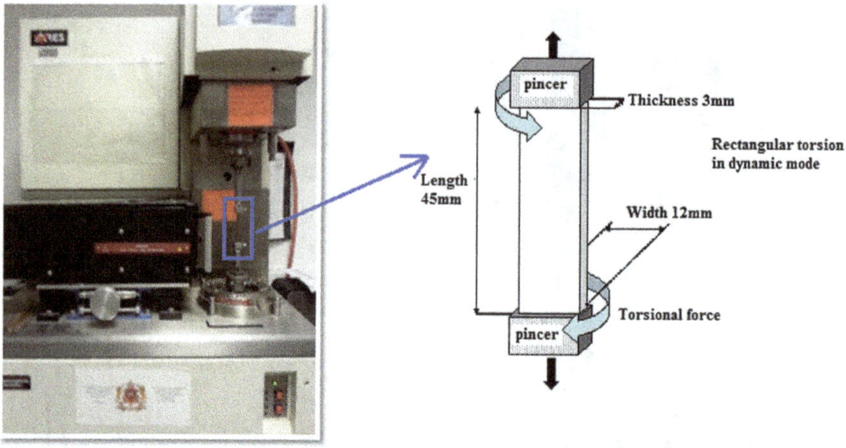

Fig. 28 Dynamic torsional test (Abdellaoui et al. 2015)

The viscoelastic properties of the material depend mainly on the time, frequency and temperature, which temperature effect is the reverse of that frequency.

6 Mechanical and Rheological Properties of Lignocellulosic Fibres Composites

The mechanical strength of the thermoset matrices increases appropriately with the lignocellulosic fiber content, simply because the mechanical properties of the fibres are greatly superior to the properties of the matrix (Abdellaoui et al. 2015). But there is a limit to this phenomenon since the fibres must be absolutely covered with polymer to be effective (to have a good adhesion). There have been numerous studies on the mechanical properties of the lignocellulosic fiber reinforced the thermoset composites (Saheb and Jog 1999). A brief review following the mainly important works carried out on the thermoset composites reinforced lignocellulosic fibres.

6.1 Flexural Properties

Flexural properties are essentially depending on the fiber content, the void fraction and the fiber-matrix interfacial adhesion.

Mishra et al. (Mishra and Biswas 2013) have treated composites fabricated with bidirectional jute fiber mat and epoxy resin by the method of hand lay-up, which the aim is to study the fiber loading effect on the mechanical properties. The results of three-point bending test show a diminution in the flexural properties of specimen with 12 wt% fiber loading. This diminution is mainly caused by a fragile interfacial bonding and existence of voids in the composite structure. The flexural strength and modulus increase with the augment in the fiber loading after 12 wt% fiber loading. The maximal flexural properties are obtained at 48 wt% fiber loading, which the flexural modulus is increased by 37% in comparison to virgin epoxy. The jute fiber inclusions improve the load bearing capacity and ability to withstand bending of the composite. The results are shown in the Fig. 29.

In other recent innovative study, Rout et al. (2001) have fabricated composites on the basis of polyester reinforced by unidirectional non woven coir mat, which the fiber content in composite was changed from 0 to 33%. It was noted that the optimal mechanical properties of coir/polyester composite were earned at 25 wt% fiber loading. The flexural strength of composite at 25 wt% of fiber loading was superior to that of the virgin polyester matrix. Depending on the authors, that at lower fiber loading a fragile interface was produced because of the poor wetting of the fibres in the matrix. However, at elevated fiber loading (17 and 25 wt%), adequate fiber loading in the composite led to better wetting. The effect various loading fiber on the flexural strength is specified in Table 7.

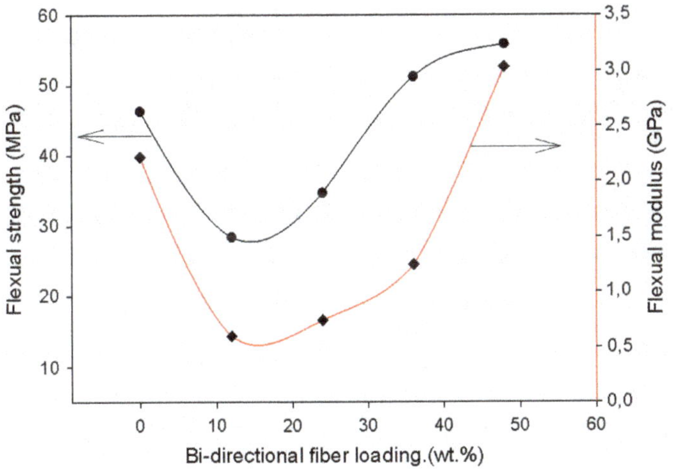

Fig. 29 Effect of fiber loading on the flexural measurements (Mishra and Biswas 2013)

Table 7 Flexural strength at different fiber loading (Rout et al. 2001)

Fiber (wt%)	Flexural strength (MPa)
0	26.60
10	15.22
17	21.30
25	26
33	24.55

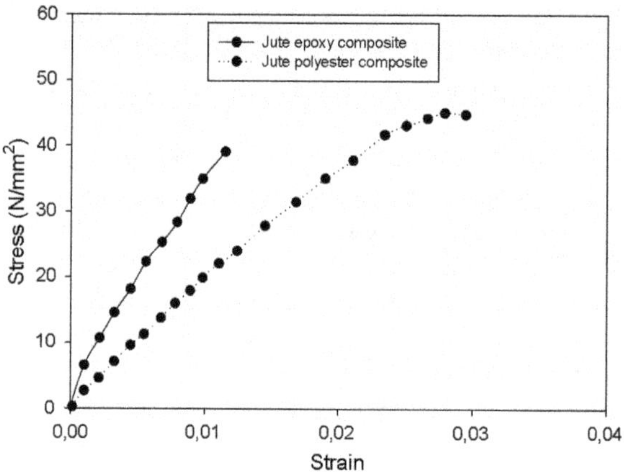

Fig. 30 Flexural stress-strain curves for jute/epoxy and jute/polyester composites (Gopinath et al. 2014)

In a comparative study, Gopinath et al. (2014) have prepared composites with epoxy and polyester reinforced by jute fibres of length 5–6 mm. The composites are synthesized with 18 wt% of fiber and 82 wt% of resin. The three points bending results carried out on two types of composites jute/epoxy and jute/polyester show those two composites were deemed to be brittle in nature as it underwent a sudden failure as shown in the Fig. 30. As revealed in results, the jute reinforced epoxy composite indicated good flexural strength compared to the jute reinforced polyester composites. It appears that the jute reinforced epoxy composite has a higher strain to failure and ultimate strength than the other reinforcement impregnated by cross-linking resins.

6.2 *Tensile Properties*

The tensile properties are especially depending on the volume fraction of fiber, number of layers, fibre direction, grammage, and yarn size.

Mir et al. (2009) have carried out laminates structures elaborated with jute fibres and epoxy by the infusion process. The fabrics are stacked in the following sequence [0°] S, [90°] S and [+ 45°/−45°] S, which the behavior of fabrics is strongly influenced by weft and warp weaving. The tensile results reveal that the manufactured composites have an elastic behavior before the appearance of a clean break as indicated in the Fig. 31. This rigid behavior occurs in both the chain and the weft direction. The only difference remains the value at which the rupture is done. This value is 5400 N in the chain direction and 7800 N in the weft direction as demonstrated in Fig. 32.

In the equal work of Mishra and Biswas (2013) mentioned above, they studied the tensile properties of epoxy reinforced bidirectional jute fabrics. The results point out that the increasing of fiber content in the matrix brings to a raise of the tensile strength and modulus. There is explained by a proper conduction and distribution of

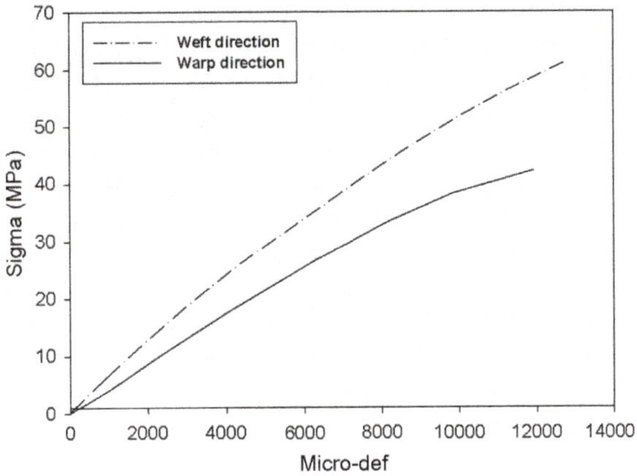

Fig. 31 Tensile curve (Mir et al. 2009)

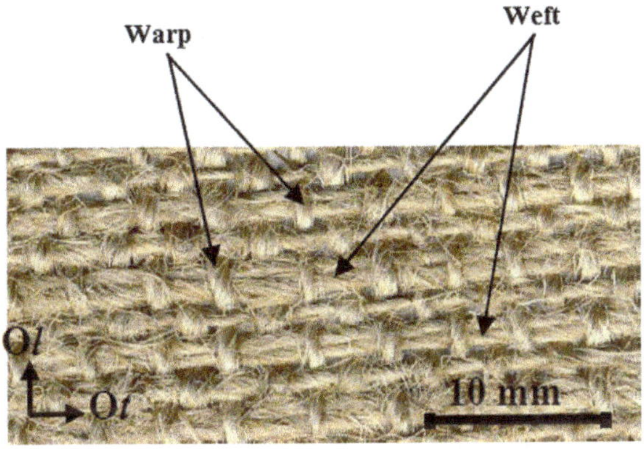

Fig. 32 View of the jute fabric structure (Mir et al. 2009)

the applied stress by the epoxy resin resulting in higher strength. According to authors, the bidirectional jute fiber composite can bear higher load before failure compared to virgin epoxy. The tensile strength changes from 43 to 110 MPa and the tensile modulus from 0.15 to 4.45 GPa with the fiber changes from 0 to 48 wt% as demonstrated in Fig. 33.

In another study, Roe and Ansell (John and Thomas 2008) are also studied the tensile properties of composites based jute/polyester by the press molding process. The fabricated samples allow a maximum loading of fiber volume fraction at 0.6. At this value, the polyester resin forms an intimate link with the jute fibres, above which the quantity of resin was insufficient to impregnate totally the fibres. At this optimal fiber volume fraction, the composite tensile strength was 250 MPa and the Young's modulus was 35 GPa.

For Sanadi et al. (Saheb and Jog 1999), they have treated the mechanical properties of polyester reinforced sunhemp fibres. The tensile measurements

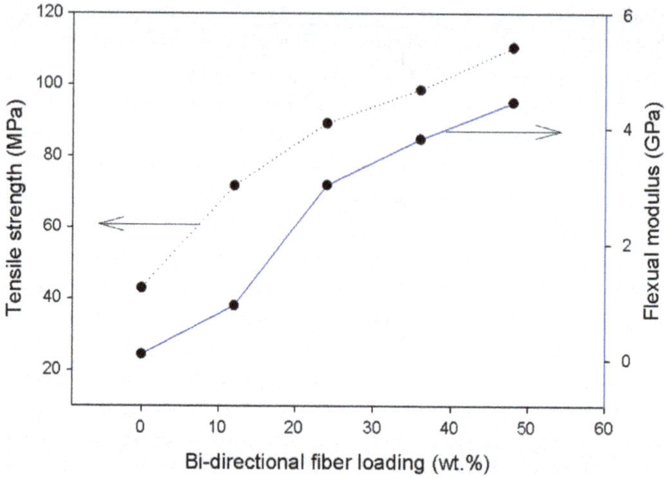

Fig. 33 Effect of fiber loading on the tensile measurements (Mishra and Biswas 2013)

(tensile strength and young's modulus) augment linearly with raising the fiber content following the mixing law. The enhancement of the tensile properties was noted only when the fiber content exceeds to 19%.

As was noted in literature, the best part of research is conducted on composites reinforced by jute fiber, because they are one of lignocellulosic fibres widely employed as reinforcement of composites.

6.3 Rheological Properties

The rheological properties of lignocellulosic fiber composites are determined by a dynamic torsional test. Like the bending and tensile properties, the torsional properties are also depending on the fiber loading and the interfacial adhesion fiber-matrix. Unfortunately, the torsional properties of composites based on thermoset matrix and lignocellulosic fiber are rarely considered by researchers in literature.

In a new study, Abdellaoui et al. (2015) have fabricated laminates structures on basis of epoxy and jute fibres by a compression process. Laminates are performed at different number of layers (1, 3, 5 and 7) layers. The progression of the torsion modulus G* and the loss factor tan δ according to frequency (frequency range between 0.1 and 40 Hz) and the number of layers is publicized in Figs. 34 and 35. The results explain that the complex modulus G^* increases significantly with increasing frequency (from 0.1 to 40 Hz). This increasing permits to deduct that the manufactured composite behave as a solid at elevated frequencies. So through the deformation phase, the fibres can spin satisfactorily as to be transporting part of

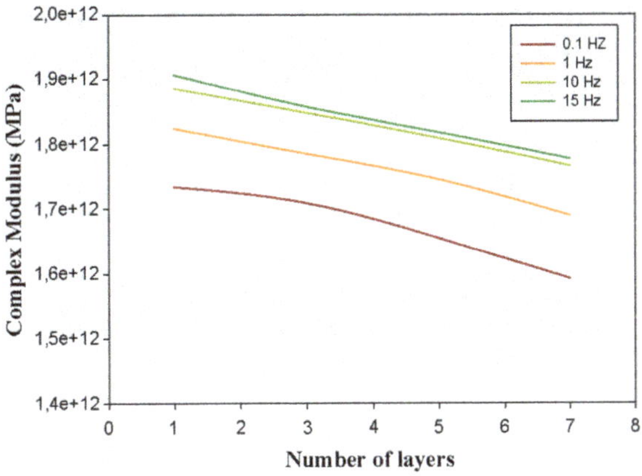

Fig. 34 Complex modulus G* for epoxy/jute composite (Abdellaoui et al. 2015)

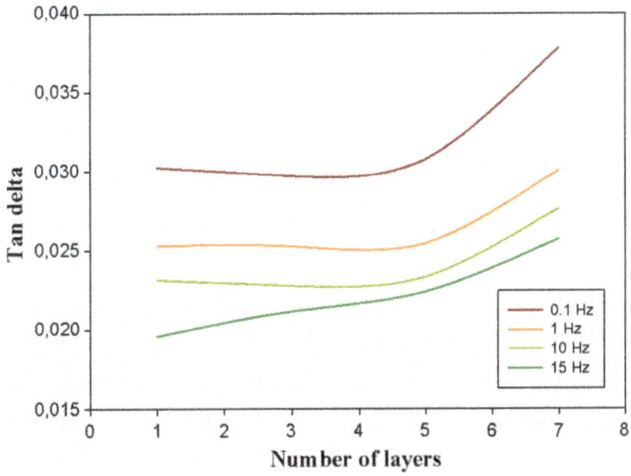

Fig. 35 Loss factor tan δ of manufactured laminates (Abdellaoui et al. 2015)

the load. Also, the complex modulus diminished slightly with increasing number of layers from 1 to 7. According to authors, at a higher number of layers, the adhesion between fibres and matrix diminished, which increase fibre–fibre interaction leading to decrease in the shear stress modulus at the interface fibre–matrix.

The tan δ is used to indicate the rate of the elastic portion and that of the viscous portion in the produced composites. According to Fig. 34, it is obvious that the viscous portion doesn't exceed 4% in the laminates. This progression validates that the composite behave as an elastic solid material. This elasticity chiefly owes to the elastic behaviour of jute fibres (Abdellaoui et al. 2015).

7 Improvement of Mechanical Properties of Lignocellulosic Fibres Composites

In composite structures, it's important to enhance the mechanical properties of equally matrix and fiber. The tensile strength is more susceptible to the matrix properties, whereas the fiber properties affect more the modulus.

The mechanical properties can be improved thanks to strong interfacial adhesion fiber-matrix, fiber orientation, and wettability of fibres, hybridization of fibres and the used manufacturing process. Also, the fiber length can affect greatly the strong of composites when this one exceed to the critical length (Saheb and Jog 1999).

7.1 *Effect of Fiber Treatment*

The modification of the hydrophilic surface of the lignocellulosic fiber ensures good compatibility with the hydrophobic matrix. Thereafter, the excellent interfacial adhesion improves the mechanical properties of the produced composite structures.

In one study,

$$\text{Cell–OH} + CH_3COOH \xrightarrow[H_2SO_4]{(CH_3CO)_2O} \text{Cell–O–}\underset{\underset{O}{\|}}{C}\text{–}CH_3 + H_2O$$

Seki (2009) has fabricated composites by using unsaturated polyester and epoxy reinforced by jute fibres. The adhesion among jute fiber and thermoset matrix was improved by alkaline and oligomeric siloxane treatments of jute fibres. The tensile results divulge that the alkalized jute/epoxy composite (AJE) shows better tensile strength that the jute/epoxy composite (JE). The alkali treatment can lead to 8% raise in tensile strength, owing to the deletion of pectin and waxes. The Young's modulus of Alkalized jute/Epoxy composite augmented by 17% due to alkali treatment as compared to that of Jute/Epoxy composite as elucidated in the Fig. 36. Young's modulus of Jute/Epoxy composite was 48% elevated as compared with that of jute/ polyester (Jute/Polyster) composite.

Concerning the siloxane treatment, the results confirm that the Young's modulus of Alkalized siloxane treated jute/epoxy (ASJE) composite has increased to 9.5% as compared with that of Alkalized jute/Epoxy composite, while the alkalized siloxane treated jute/polyester (ASJP) composite has improved by 8.4% compared to that of Alkalized Jute/Epoxy composite. The siloxane treatment allows enhancing the interfacial adhesion fiber-matrix, and then, improving the mechanical properties of composites.

In other case, Ray el al. (2001) have prepared composites with jute fiber and vinylester matrix. The jute fibres were treated in alkaline solution with 5% NaOH designed for four different times (2, 4, 6 and 8 h). The results tell that the flexural

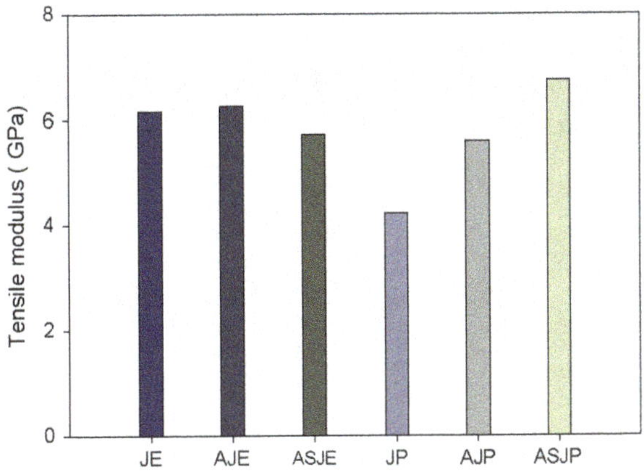

Fig. 36 Young's modulus of treated jute composites (Seki 2009)

properties of elaborated composites were improved with the use of treated jute fibres. The maximum upgrading in the flexural properties was noted at 4 h of alkali treatment. According to the authors, this enhance in fiber strength properties is owing to the reduction of the linear density of the fibres on alkali treatment and also to the closer packing of the cellulose chains and molecular reorientation on prolonged soaking in the alkali solutions (4–8 h). The results are summarized in Table 8.

For the coir/polyester composites, Rout et al. (2001) have improved the interfacial adhesion of composites by a chemically modification of coir fiber through many methods as alkali treatment, cyanoethylation, dewaxing, bleaching, and vinyl polymer grafting. The mechanical properties of a variety of elaborated composites are resumed in Table 9. As exposed by various studies of literature, all methods lignocellulosic fiber surface modification can get better the mechanical properties of the prepared composites.

7.2 *Effect of Hybridization*

The hybridization implies the integration of several different types of fibres in the same matrix in order to enhance the mechanical properties of biocomposites. The lignocellulosic fiber hybrid composite permits to incorporate natural fiber/natural fiber or natural fiber/glass fiber in the thermoset matrix.

Many researchers are focused on the study of hybrid composite reinforced by lignocellulosic fiber and glass fiber as Mishra et al. (2003). They proceeded to manufacture hybrid composites with properties similarly comparable to the synthetic composites based on glass fiber. Mishra et al. (Singh and Gupta 2005) have fabricated hybrid composites on basis of polyester matrix reinforced by Pineapple leaf fibres (PALF)/glass fibres and sisal fibres/glass fibres in the aim to study the effect of the glass fibres loading on the enhancement of the mechanical

Table 8 Flexural properties of Untreated/Alkali treated jute fibres and vinylester composites (Rout et al. 2001)

Jute volume fraction (%)	Fiber stage	Modulus (GPa) Mean value	Flexural strength (MPa) Mean value
0	–	2.91	120.70
8	Untreated	4.22	106.3
	Treated 2 h	3.44	96.27
	Treated 4 h	4.20	120.20
	Treated 6 h	3.96	101.8
	Treated 8 h	3.13	93.97
23	Untreated	7.35	145.70
	Treated 2 h	8.06	157.70
	Treated 4 h	9.38	172.70
	Treated 6 h	8.54	155.40
	Treated 8 h	7.13	145.80
30	Untreated	10.03	180.60
	Treated 2 h	10.99	189.40
	Treated 4 h	12.85	218.15
	Treated 6 h	12.49	195.90
	Treated 8 h	11.17	197.50
35	Untreated	11.89	199.10
	Treated 2 h	12.70	205.20
	Treated 4 h	14.69	238.90
	Treated 6 h	14.89	232.00
	Treated 8 h	12.32	204.20

properties of biocomposites. The results divulge that the tensile strength of PALF-polyester composite is significantly improved by incorporating the glass fiber, which the adding of glass fiber by about 8.6 wt% increases the ultimate tensile strength of hybrid composite by average 66%. However, further increase in weight fraction of glass fiber to 12.9 wt% decreases the tensile strength of PALF/glass fiber hybrid polyester composite by about 10% as illustrated in Fig. 37.

Table 9 Mechanical properties of chemically modified coir (17 wt%)/polyester composites (Mishra et al. 2003)

Sample	Tensile strength (MPa)	Flexural strength (MPa)
Untreated	21.30	51.49
Dewaxed	22.94	55.75
Alkali treatment		
2%	26.8	59.01
5%	24.70	60.45
10	17.34	49.3
Cyanoethylation		
50 °C	22.58	53.86
60 °C	26.30	54.13
70 °C	24.27	53.09
Bleaching		
65 °C	20.32	61.65
75 °C	19.27	57.80
85 °C	17.91	55.32
PMMA grafting		
5%	25.01	63.68
10%	21.58	57.26
20%	14.81	37.81

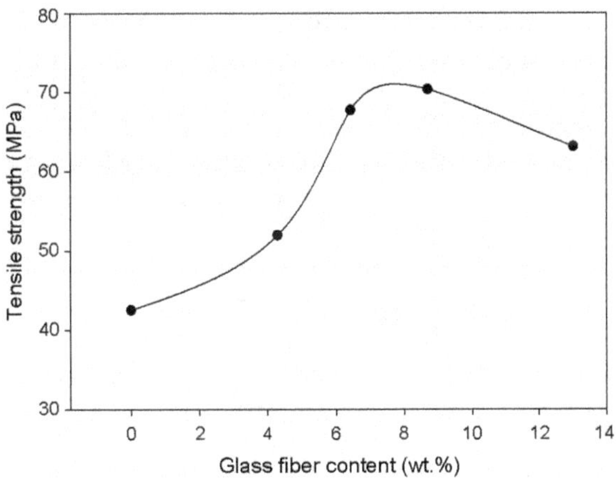

Fig. 37 Curve of Tensile strength versus glass fiber content in case of hybrid PAFL/glass fiber reinforced polyester composite (Jawaid et al. 2011)

Concerning, the hybrid sisal fiber/glass fiber reinforced the polyester composite (Singh and Gupta 2005), the integration of glass fibrin composite increased the tensile strength of the sisal fiber/glass fiber reinforced polyester hybrid composite as compared to composites with sisal-polyester, which saturated at 5.7 wt% of glass fiber loading as represented in the Fig. 38. The enhancement of tensile strength of the composite at low wt% of glass fiber loading will be caused by the fact that at

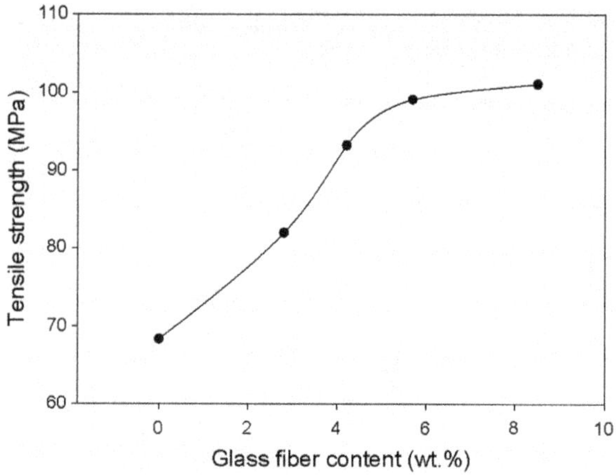

Fig. 38 Curve of Tensile strength versus glass fiber content in case of hybrid sisal/glass fiber reinforced polyester composite (Jawaid et al. 2011)

these particular compositions, the sisal fiber can efficiently transmit the load from the glass fiber.

The same case for the work of Bakar et al. (John and Thomas 2008) they study the tensile properties of hybrid oil palm fiber/glass fiber reinforced epoxy matrix. The hybridization of the oil palm fiber and glass fiber increased the tensile measurements (strength, Young's modulus and elongation at break) of the composites.

In other hand, a several studies were performed on the hybrid composite by combining two kinds of lignocellulosic fibres. Jawaid et al. (2011) have prepared hybrid composite based on tri layer empty fruit bunches (EFB)/jute woven (JW) reinforced epoxy hybrid composite. The composites were made-up by hand layup process. Composites realized with Pure EFB and pure jute is also prepared in order to prove the hybridization effect on getting better the mechanical properties of the composite. The Fig. 39 represents the hybrid composite stacking. The results illustrate that the pure jute woven fabric composite reveals elevated tensile strength and modulus compared to the pure EFB composites. This is because the woven jute structure is more strong and rigid than the mat structure of EFB. The hybridization of oil palm empty fruit (EFB) with the jute woven permits an increase of tensile strength and modulus of the pure EFB composite. This is owing to the position of jute woven fabric at the two surfaces (Fig. 39). The tensile measurements (strength and modulus) of the hybrid composite JW/EFB/JW is softly higher that hybrid composite EFB/JW/EFB as showed in the Fig. 40.

7.3 *Effect of Fiber Stacking*

The fiber stacking sequence effect is increasingly studied because of its influence on the mechanical properties of composite structures.

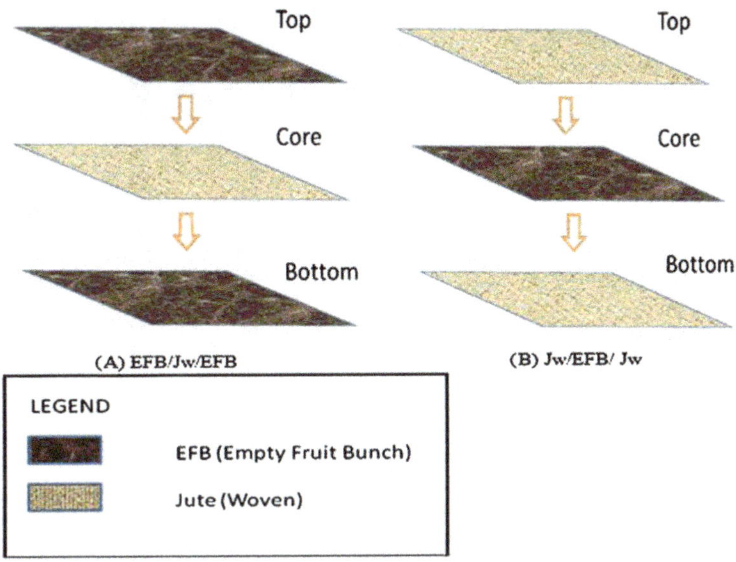

Fig. 39 Different layer stacking of hybrid composite (Jawaid et al. 2011)

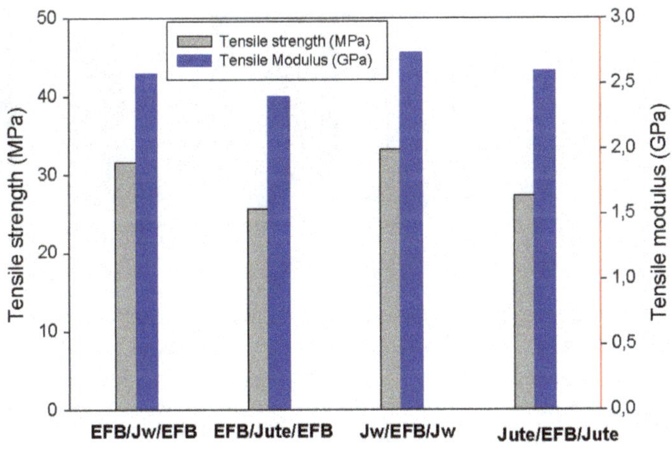

Fig. 40 Curve of tensile strength and modulus of jute woven and chopped jute fibre hybrid composites (Jawaid et al. 2011)

For example, in the work of Abdellaoui et al. (2015), they realized laminated composite at different number of layers (1, 3, 5, 7) and in two stacking sequences [0°] and [45°]. The composites are cut in two orientations 0° and 45°. The obtained various cases are as follows: [0°/0°/0°], [45°/45°/45°], [0°/45/°0/45°] and [45°/0°/45°/0°]. The results demonstrate that more the layers are oriented at 0° more the tensile strength and Young's modulus, are high. This is explained by the load

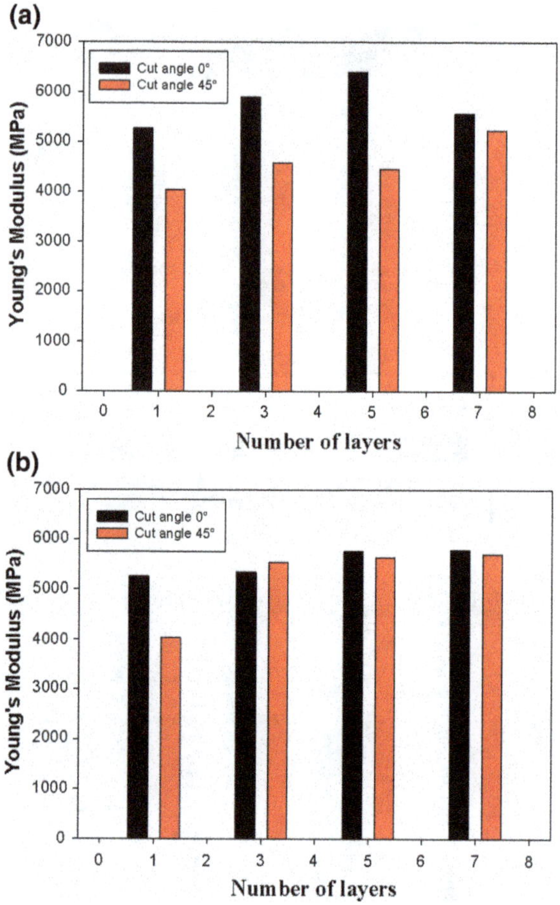

Fig. 41 Young's modulus of laminated at different stacking sequences (Abdellaoui et al. 2015)

that is maintained by the length of the unidirectional fibres in the longitudinal sense of fibres, i.e. (0°), while out of this direction (45°), the axial load forms an angle with yarns in both sides. Also, an augment in Young's modulus was noted at 5 layers, which may be due relatively to the perfect adhesion between jute fibres and matrix compared to other cases as remarked in the Fig. 41.

In other work, the investigators (Jawaid et al. 2011) were studied epoxy hybrid composites reinforced by empty fruit bunch (EFB) mat and jute woven (JW). The obtained samples were carried out as: [EFB mat/EFB mat/ EFB mat], [JW/JW/JW], [EFB mat/JW/EFB mat], and (JW/EFB mat/JW]. The tensile results divulge that the composite realized with only jute woven indicated an elevated tensile strength and Young's modulus compared to the pure EFB composite. This is because the woven structure is stiffer than the mat structure. Also, the hybrid composite [EFB mat/JW/EFB] mat show an increase of tensile measurement compared to [EFB mat/EFB mat/ EFB mat]; and the hybrid structure [JW/EFB mat/JW] is slightly

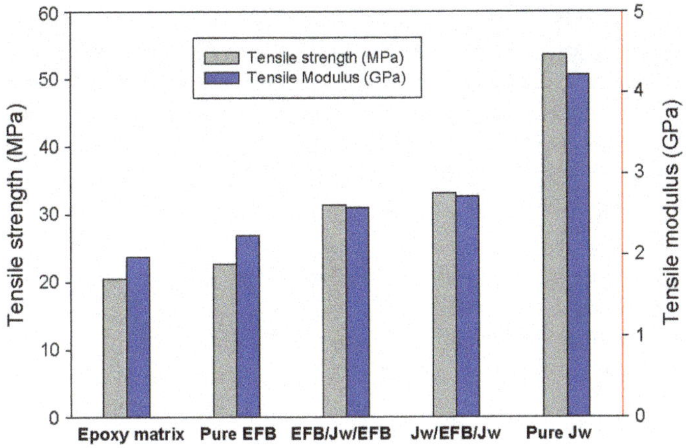

Fig. 42 Tensile measurements of hybrid EFB mat/woven jute fiber composites (Jawaid et al. 2011)

strong than the [EFB mat/JW/ EFB mat] as represented in Fig. 42. Researchers highlighted that stacking model of the different elements in hybrid laminates structures occupy a significant function in influencing the mechanical properties of the hybrid composites (Jawaid et al. 2011).

8 Application of Lignocellulosic Fiber Composites

The lignocellulosic fibres attracted more interest as alternate industrial materials, especially with wood fiber, which their use has spread to virtually all areas (automotive, construction, furniture). Thermoset composites reinforced by wood fibres are applied in many applications in desks, docks, windows frames, and panel components (John and Thomas 2008). Pipes, panels and pultruded profiles fabricated with polyester resin are also very popular.

Nowadays, a revival of the use of lignocellulosic fibres as reinforcement in technical applications is in progress, mainly in the automobile manufacturing and the building sector (John and Thomas 2008).

8.1 Automobile Industry

The car market is in perpetual growth in terms of quantity, quality and variety of products, where energy efficiency is seen as a key factor for future growth. A 25% reduction in vehicle weight is equivalent to a saving of about 250 million barrels of oil and the reduction of CO_2 emission of about 220 billons pounds per year.

Then, the use of natural fibres is an alternative of great significance due to their low weight and lower costs, significant specific characteristics (Alves et al. 2010).

The automobile industry is one of the biggest users of lignocellulosic fibres composites in interior applications like door panels and boot trim (Alves et al. 2010). The Figure 43 shows the various auto parts from lignocellulosic fiber composite. The first applications were introduced in the 1930s and 1940s when Henry Ford began with the use of hemp fibres to reinforce soya resin to produce body panels (Alves et al. 2010; John and Thomas 2008).

Virtually, all major automakers in Germany (Daimler Chrysler, Mercedes, Volkswagen, Audi, BMW, Ford and Opel) now use the lignocellulosic fibres composites in various applications. Cotton fibres impregnated with polyester resin were used in car body 'Trabant' since the 70 s by German automakers. An auxiliary of the Mercedes-Benz company has launched this concept with 'Beleem project'

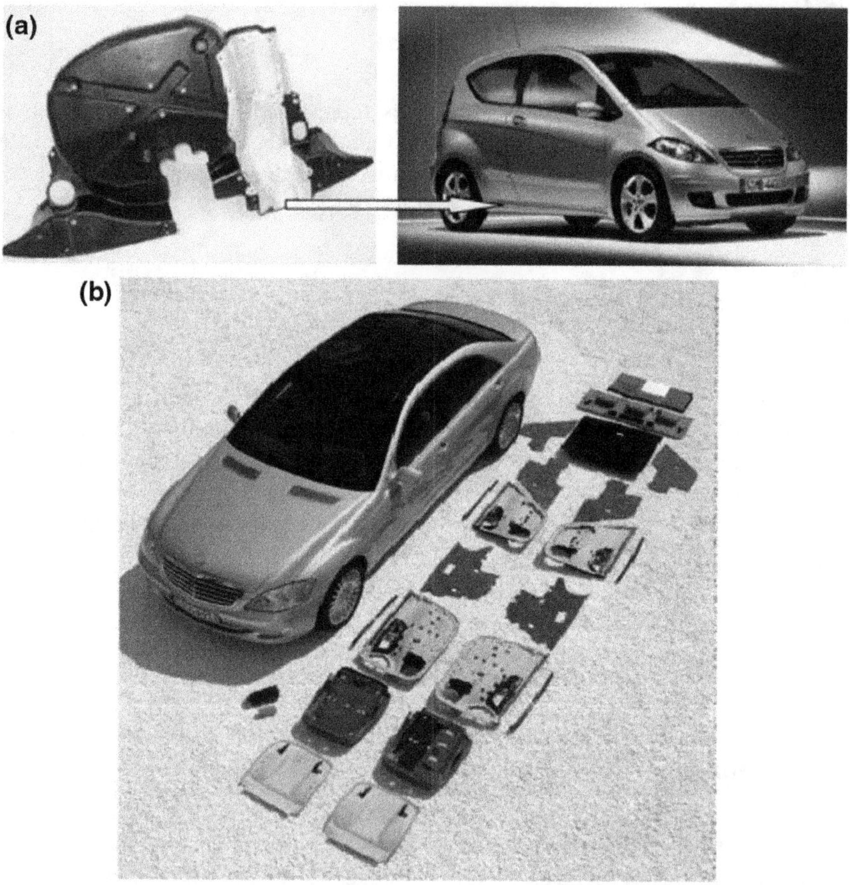

Fig. 43 **a** Parts of Mercedes A class based on composites with banana fibres. **b** Automotive components based on different lignocellulosic fiber composites (John and Thomas 2008)

based in Sao Paulo, Brazil. In this case, the coconut fibres were employed in utilitarian vehicles. Mercedes also used jute door panels in the E-Class vehicles in 1996. Dailmer-Benz explored the idea of substituting glass fibres by natural fibres in automotive components since 1991. Actually, Dailmer-Chrysler has increased its research and development in based composites reinforced polyester flax fibres for external applications (Alves et al. 2010; John and Thomas 2008).

8.2 Building Sector

The lignocellulosic fiber composites have enormous potential as alternative material in the building market and furnishings. Various lignocellulosic fibres based composite products like laminates/panels, door frames, roofing and shutters, have been produced.

Laminates/panels
The laminates panels have fabricated by using polyester, phenolic or polyurethane matrix reinforced by sisal, jute or coir fibres with non woven or woven structures (Fig. 44). The laminates are elaborated by compression molding process (Singh and Gupta 2005).

Door shutters
The door shutters have prepared by bonding the jute/sisal laminates face sheets with stiff foam cores, which the core made from polyester-polyurethane hybrid resin (Fig. 45a). Also, the jute composite doors can be realized by using jute stick blocks as a core and jute/polyester laminates as a face sheet (Fig. 45b) (Singh and Gupta 2005).

Jute pultruded door frames
Composite profiles have prepared from jute hessian fabrics and phenolic resin by the pultrusion process. The cured profiles contain 50–60% jute fibres (Fig. 46) (Singh and Gupta 2005).

Fig. 44 Laminates panels (Jawaid et al. 2011)

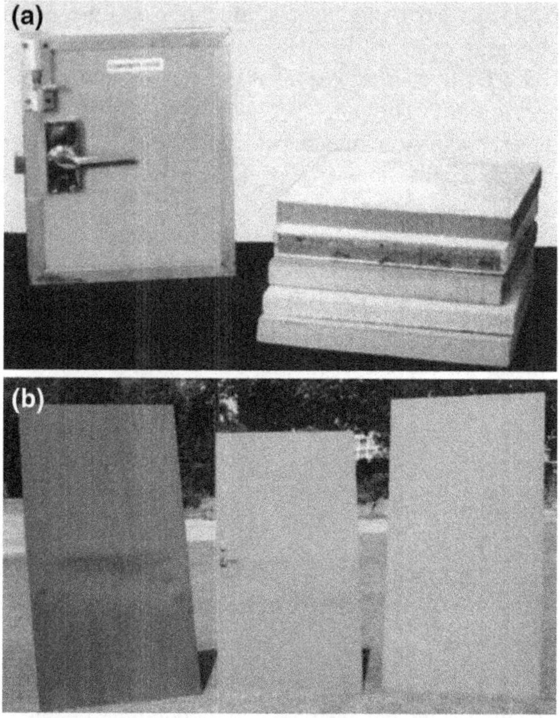

Fig. 45 Jute/Sisal doors: **a** Prototype of door and panel, **b** Composite door shutters (Jawaid et al. 2011)

Fig. 46 Jute frames (Jawaid et al. 2011)

Fig. 47 Sisal/polyester roofing sheet (Jawaid et al. 2011)

Roofing sheets
The roofing sheets have prepared from chopped sisal strand mats fibrous filler, antiaging agent and unsaturated polyester resin (Fig. 47). In many situations, the roofing represents more than 25–30% of the total construction cost of a low cost house (Singh and Gupta 2005).

9 Conclusion

This chapter covers the thermoset composites reinforced by lignocellulosic fibres which experienced their rise in recent years thanks to a diversity of environmental considerations. A review of the most extensive works performed on mechanical and rheological properties of thermoset composites reinforced by lignocellulosic fibres, has been elucidated while previously explaining the most extensively used manufacturing processes and characterizing tests for these composites in expanding phase. Thus, this chapter holds its entirety thanks to a complete elucidation of the morphological structure, chemical composition, advantages and disadvantages and mechanical properties of lignocellulosic fibres, and also a reminder of the diverse thermosetting matrices used in composites.

References

Abdellaoui H, Echaabi J (2014) Rheological models for modeling the viscoelastic behavior in liquid composite molding processes (LCM) review. J Reinf Plast Compos 33:714–732

Abdellaoui H, Bensalah H et al (2015a) Fabrication, characterization and modelling of laminated composites based on woven jute fibres reinforced epoxy resin. J Mater 68:104–113

Abdellaoui H, Bouhfid R et al (2015b) Experimental and modeling study of viscoelastic behaviour of woven dried jute under compressive stress. J Reinf Plast Compos 34:405–420

Abdelmouleh M, Boufi S et al (2007) Short natural-fibre reinforced polyethylene and natural rubber composites: effect of silane coupling agents and fibres loading. Compos Sci and Technol 67:1627–1639

Albano C, Gonzalez J et al (1999) Thermal stability of blends of polyolefins and sisal fiber. J Polymer Degrad Stab 66:179–190

Alves C, Ferrão PMC, Silva AG et al (2010) Ecodesign of automotive components making use of natural jute fiber composites. J Clean Prod 18:313–327

Amaducci S, Zatta A et al (2008) Influence of agronomic factors on yield and quality of hemp (Cannabis sativa L.) fibre and implication for an innovative production system. F Crop Res 107:161–169

Basu G, De SS, Samanta AK (2009) Effect of bio-friendly conditioning agents on jute fibre spinning. Ind Crops Prod 29:281–288

Beg MDH (2007) The improvement of interfacial bonding, weathering and recycling of wood fibre reinforced polypropylene composites. Doctoral thesis, University of Waikato, Hamilton, New Zealand

Ben Brahim S, Ben Cheikh R (2007) Influence of fibre orientation and volume fraction on the tensile properties of unidirectional Alfa-polyester composite. Compos Sci Technol 67(1):140–147

Bessadok A, Marais S et al (2007) Effect of chemical treatments of Alfa (Stipa tenacissima) fibres on water-sorption properties. Compos Sci Technol 67:685–697

Bickerton S, Abdullah MZ (2003) Modeling and evaluation of the filling stage of injection/compression moulding. Compos Sci Technol 63:1359–1375

Bodros E, Pillin I et al (2007) Could biopolymers reinforced by randomly scattered flax fibre be used in structural applications? Compos Sci Technol 67:462–470

Charlet K, Baley C et al (2007) Characteristics of Hermès flax fibres as a function of their location in the stem and properties of the derived unidirectional composites. Compos Part A Appl Sci. Manuf 38:1912–1921

Charlet K, Jernot JP et al (2010) Scattering of morphological and mechanical properties of flax fibres. Ind Crops Prod 32:220–224

Corrales F, Vilaseca F et al (2007) Chemical modification of jute fibers for the production of green-composites. J Hazard Mater 144(3):730–735

Defoirdt D, Biswas S et al (2010) Assessment of the tensile properties of coir, bamboo and jute fibre. Compos Part A Appl Sci Manuf 41:588–595

Deleglise M, Binetruy C et al (2006) Simulation of LCM processes involving induced or forced deformations. Compos Part A Appl Sci Manuf 37:874–880

Essabir H, Bensalah MO et al (2014) Fabrication and characterization of apricot shells particles reinforced high density polyethylene based bio-composites: mechanical and thermal properties. J Biobased Mater Bioenergy 8:344–351

Gassan J, Bledzki AK (1999) Possibilities for improving the mechanical properties of jute/epoxy composites by alkali treatment of fibres. J. Compos Sci Technol 59:1303–1309

Gilberto S, Bras J, Dufresne A (2010) Cellulosic Bionanocomposites: a review of preparation, properties and applications. Polym J 2:728–765

Gopinath A, Kumar S, Elayaperumal A (2014) Experimental investigations on mechanical properties of jute fiber reinforced composites with polyester and epoxy resin matrices. Proced engin 97:2052–2063

Han N, Suh S et al (2003) Resin film infusion of stitched stiffened composite panels. Compos Part A Appl Sci Manuf 34:227–236
Harish S, Michael DP et al (2009) Mechanical property evaluation of natural fiber coir composite. Mater Charact 60:44–49
Jawaid M, Abdul Khalil HPS, Abu Bakar A (2010) Mechanical performance of oil palm empty fruit bunches/jute fibres reinforced epoxy hybrid composites. Mater Sci Eng, A 527:7944–7949
Jawaid M, Abdul Khalil HPS, Abu Bakar A (2011) Woven hybrid composites: Tensile and flexural properties of oil palm-woven jute fibres based epoxy composites. Mater Sci Eng A 528:5190–5195
Jeong E, Lim JW et al (2011) Effects of physicochemical treatments of illite on the thermo-mechanical properties and thermal stability of illite/epoxy composites. J Ind Eng Chem 17:77–82
John M, Thomas S (2008) Biofibres and biocomposites. Carbohydr Polym 71:343–364
Kelly P, Bickerton S (2009) A comprehensive filling and tooling force analysis for rigid mould LCM processes. Compos Part A Appl Sci Manuf 40:1685–1697
Khan MA, Ganster J et al (2009) Hybrid composites of jute and man-made cellulose fibers with polypropylene by injection moulding. Compos Part A Appl Sci Manuf 40:846–851
Kim JT, Netravali AN (2010) Mercerization of sisal fibers: effect of tension on mechanical properties of sisal fiber and fiber-reinforced composites. Compos Part A Appl Sci Manuf 41:1245–1252
Le Duigou A, Davies P et al (2010) Interfacial bonding of Flax fibre/Poly(l-lactide) bio-composites. Compos Sci Technol 70:231–239
Le Riche R, Saouab A, Bréard J (2003) Coupled compression RTM and composite layup optimization. Compos Sci Technol 63:2277–2287
Loan DTT (2006) Iinvestigation on jute fibres and their composites based on polypropylene and epoxy matrices. Doctoral thesis, Dresden university, German
Lucas N, Bienaime C, Belloy C et al (2008) Polymer biodegradation: mechanisms and estimation techniques. Chemosphere 73:429–442
Malha M, Nekhlaoui S et al (2013) Mechanical and thermal properties of compatibilized polypropylene reinforced by woven doum. J Appl Polym Sci 130:4347–4356
Megiatto JD, Silva CG et al (2009) Thermoset matrix reinforced with sisal fibers: Effect of the cure cycle on the properties of the biobased composite. Polym Test 28:793–800
Mir A, Zitoune R et al (2009) Caractérisation mécanique et thermomécanique d'un stratifié Jute/ époxy Mechanical and thermomechanical characterization of jute/epoxy laminate, proceedings of the JNC 16, Toulouse, 2009
Mishra V, Biswas S (2013) Physical mechanical properties of bi-directional jute fiber epoxy composites. J Proced Engin 51:561–566
Mishra S, Mohanty AK et al (2003) Studies on mechanical performance of biofibre/glass reinforced polyester hybrid composites. Compos Sci Technol 63:1377–1385
Mwaikambo LW, Ansell MP (2002) Chemical modification of hemp, sisal, jute, and kapok fibers by alkalization. J Appl Polym Sci 84:2222–2234
Park J, Kang MK (2003) A numerical simulation of the resin film infusion process. Compos Struct 60:431–437
Pizzi A, Kueny K et al (2009) High resin content natural matrix–natural fibre biocomposites. Ind Crops Prod 30:235–240
Rana AK, Basak RK, Mitra C et al (1997) Studies of Acetylation of jute using simplified procedure and its characterization. J Appl Polym Sci 64(8):1517–1523
Ray D, Sarkar BK et al (2001) The mechanical properties of vinylester resin matrix composites reinforced with alkali-treated jute fibres. Compos Part A 32:119–127
Ray D, Sarkar BK et al (2002) Dynamic mechanical and thermal analysis of vinylester resin matrix composites reinforced with untreated and treated jute fibres. Compos Sci Technol 62:911–917
Regional centre of animation in advanced materials (2004) Glossary of composite materials
Regional centre of animation in advanced materials (2006) Glossaire des matériaux composites renforcés de fibres d'origine renouvelable

Rout J, Misra M, Tripathy SS et al (2001) The influence of fibre treatment on the performance of coir-polyester composites. Compos Sci technol 61:1303–1310

Saheb DN, Jog JP (1999) Natural fiber polymer composites: a review. Adv Polym Technol 18:351–363

Sangthong S, Pongprayoon T, Yanumet N (2009) Mechanical property improvement of unsaturated polyester composite reinforced with admicellar-treated sisal fibers. Compos Part A Appl Sci Manuf 40:687–694

Sarkar P, Modak N et al (2015) Reciprocationg friction behavior of Al Particulate glass fiber reinforced epoxy composite. Mater Today 2:2708–2717

Seki S (2009) Innovative multifunctional siloxane treatment of jute fiber surface and its effect on the mechanical properties of jute/thermoset composites. Mater Sci Eng A 508:247–252

Silva FDA, Chawla N et al (2008) Tensile behavior of high performance natural (sisal) fibers. Compos Sci Technol 68:3438–3443

Singh B, Gupta M (2005) Natural fiber composites for building applications. Natural fibres, biopolymers, and biocomposites. Taylor & Francis, United strates of America, pp 261–290

Singh S, Mohanty AK et al (2008) Renewable resource based biocomposites from natural fiber and polyhydroxybutyrate-co-valerate (PHBV) bioplastic. Compos Part A Appl Sci Manuf 39:875–886

Sood M, Dharmpal D, Gupta VK (2015) Effect of fiber chemical treatment on mechanical properties of sisal fiber/recycled HDPE composite. Mater Today 2:3149–3155

Stocchi A, Lauke B et al (2007) A novel fiber treatment applied to woven jute fabric/vinylester laminates. Compos Part A Appl Sci Manuf 38:1337–1343

Umer R, Bickerton S et al (2011) The effect of yarn length and diameter on permeability and compaction response of flax fibre mats. Compos Part A Appl Sci Manuf 42(7):723–732

Vilaplana F, Strömberg E et al (2010) Environmental and resource aspects of sustainable biocomposites. Polym Degrad Stab 95:2147–2161

WinklerPrins AMG (2006) Jute cultivation in the Lower Amazon, 1940–1990: an ethnographic account from Santarém, Pará, Brazil. J Hist. Geogr 32:818–838

Xie Y, Hill CAS et al (2010) Silane coupling agents used for natural fiber/polymer composites: a review. Compos Part A Appl Sci Manuf 41:806–819

Pineapple Leaf Fiber: From Waste to High-Performance Green Reinforcement for Plastics and Rubbers

Nanthaya Kengkhetkit, Thapanee Wongpreedee and Taweechai Amornsakchai

Abstract Various kinds of plant natural fibers have been studied. Many of these are purposely grown for the fiber while some are derived from agricultural waste. A few types of plant natural fibers have been produced on a commercial scale. With various current problems facing us all today, the needs of plant natural fibers are even greater. There remains some fiber containing agricultural wastes which are underutilized. One of these is pineapple leaf waste. Pineapple leaf fiber (PALF) is known to possess high mechanical properties and can be obtained from pineapple leaf waste using different extraction methods. However, most of these methods are not suitable for large-scale production of the fiber for industrial uses. In addition, PALF produced with these methods is normally large in size, and this limits its applications. Recently, a novel method for the extraction of PALF has been presented. The method allows short and fine PALF to be produced. This PALF has diameter as small as 3 μm and with a cut length of 6 mm, the aspect ratio (length to diameter ratio) could be up to 2000. These characteristics make this PALF very suitable for the effective reinforcement of both plastics and rubbers. It will be demonstrated how to best utilize this PALF. This PALF could be surface treated or used in conjunction with compatibilizer or adhesion promoter as in other cellulose fibers. Recent progress will be presented and potential applications will be reviewed.

N. Kengkhetkit
Faculty of Sciences and Agricultural Technology, Rajamangala University of Technology Lanna, Phitsanulok Campus, Muang, Phitsanulok 65000, Thailand
e-mail: K.nanthaya@gmail.com

T. Wongpreedee
Program in Materials Science and Engineering, Faculty of Science, Mahidol University, Salaya, Nakhon Pathom 73170, Thailand
e-mail: Thapanee.w@gmail.com

T. Amornsakchai (✉)
Center of Sustainable Energy and Green Materials, Department of Chemistry and Center of Excellence for Innovation in Chemistry, Faculty of Science, Mahidol University, Salaya, Nakhon Pathom 73170, Thailand
e-mail: Taweechai.amo@mahidol.ac.th

S. Kalia (ed.), *Lignocellulosic Composite Materials*, Springer Series on Polymer and Composite Materials, https://doi.org/10.1007/978-3-319-68696-7_6

Keywords Pineapple field waste · Pineapple leaf waste · Pineapple leaf fiber Green composite · Agricultural waste · Natural fiber

Abbreviations

AMI	Amino silane
CB	Carbon black
CV	Conventional vulcanization
HDT	Heat distortion temperature
LD	Longitudinal direction
MAPP	Maleic anhydride-grafted polypropylene
NBR	Acrylonitrile butadiene rubber
NFM	Nonfibrous material
NR	Natural rubber
PA6	Polyamide 6 or Nylon 6
PA6/66	Polyamide 6/66 or Nylon 6/66
PALF	Pineapple leaf fiber
PALMF	Pineapple leaf microfiber
PP	Polypropylene
San	Santoprene
SAXS	Small angle X-ray scattering
SEM	Scanning electron microscope
Si	Silica
TD	Transverse direction
TPALF	NaOH treated pineapple leaf fiber
UPALF	Untreated pineapple leaf fiber
WAXS	Wide angle X-ray scattering
WGL	Whole ground leaves

Contents

1 Introduction

Pineapple is grown in Africa, tropical America, and Asian countries with total plantation area of approximately 1 million hectares. The biggest pineapple fruit producing countries in the world are Thailand and the Philippines. Thailand's pineapple cultivation area is about 100,000 ha. Figure 1 displays a pineapple cultivation area in Rayong province, east of Thailand. A large amount of pineapple leaf waste is left in the field waiting for farmers to deal with after the fruits have been harvested. Figure 2 displays different ways which farmers deal with their pineapple leaf waste in the field. Although different methods exist for the extraction of pineapple leaf fiber (PALF), the utilization of pineapple leaf waste is still very limited. Mostly, PALF is used for textile materials and paper for handcrafts.

Pineapple leaf fiber (PALF) is known to possess high mechanical properties, and its properties rank among the top of plant natural fibers as shown in Table 1. Properties of some synthetic fibers are included in the bottom part of the table for comparison. It is clear that PALF's properties are comparable to that of glass and aramid fibers, especially when compared in terms of specific properties (property divided by density). Glass fiber is normally used to reinforce plastic materials for improved mechanical properties while aramid fiber is gaining more acceptances for rubber reinforcement. Thus, considering that PALF is obtained from abundant waste and it has very attractive mechanical properties, PALF is a good candidate for both plastic and rubber reinforcements.

So far, PALF obtained with conventional extraction methods has been studied as reinforcement for various types of polymer matrices including plastics (Chollakup et al. 2011; George et al. 1995; Kaewpirom and Worrarat 2014; Luo and Netravali 1999; Threepopnatkul et al. 2009), rubbers (Lopattananon et al. 2006, 2011; Lopez-Manchado and Arroyo 2002), and epoxy resin (Devi et al. 1997; Payae and Lopattananon 2008). A few excellent reviews on this subject can be found

Fig. 1 Pineapple cultivation area in Rayong province, east of Thailand

Fig. 2 Different methods of dealing with pineapple leaf waste. Top: waste leaves are pushed to the side of the field and left to dry; bottom: waste leaves are chopped into smaller pieces (with a machine) and left to dry in the field before tillage for the next cycle

Table 1 The properties of natural fibers and synthetic fibers (adapted from Satyanarayana et al. 1982; Kalia et al. 2009)

Fibers	Young's modulus (GPa)	Tensile strength (MPa)	Elongation at break (%)	Diameter (μm)	Density (g/cm^3)
Ramie	61.4–128	400–938	1.2–3.8	–	1.5
PALF	34.5–82.5	413–1627	0.8–1.6	20–80	1.44
Flax	27.6	45–1100	2.7–3.2	–	1.5–3
Jute	13–26.5	393–773	7–8	20–200	1.3–1.45
Sisal	9.4–22	468–640	3–7	50–200	1.45
Banana	7.7–20.0	529–754	1.0–3.5	–	1.35
Cotton	5.5–12.6	287–800	7–8	–	1.5–1.6
Coir	4–6	131–175	15–40	100–450	1.15
Hemp	–	690	1.6	–	–
E-glass	70	2000–3500	2.5	–	2.5
S-glass	86	4570	2.8	–	2.5
Aramid	63–67	3000–3150	3.3–3.7	–	1.4
Carbon	230–240	4000	1.4–1.8	–	1.7

(Leao et al. 2015; Mishra et al. 2004; Sapuan et al. 2011). However, these PALF reinforced composites did not display any distinct properties compared to other fibers. Thus, opportunities for the utilization of PALF in polymer matrix composites remain high if a proper fiber extraction and composite preparation methods are used. Here, we will show that with a new type of fiber extraction method, short and fine PALF suitable for both plastic and rubber reinforcements could be obtained. It will also be shown how to best utilize the fiber.

2 A Novel Fiber Extraction Method

Traditional methods for PALF extraction include manual or hand scraping, machine decortication, and retting. PALF obtained from these methods is in the form of long fiber bundles. The long fiber is not suitable for reinforcement of polymer composites unless continuous reinforcement is required. The most suitable and convenient form to use is short fiber. Recently, a novel extraction method for short PALF was reported (Kengkhetkit and Amornsakchai 2012).

2.1 Description of the Process

Fresh pineapple leaves are chopped into pieces of 6 mm in length and then milled with stone grinder or other suitable machines. The soft tissue will be ground into paste while fiber bundles will be broken into smaller bundles or even single fiber with the same length. The material in this stage is composed of fibrous and nonfibrous materials and is called whole ground leaf or WGL. After drying, the WGL is sieved with a relatively coarse screen to separate the fibrous and nonfibrous materials. The former is PALF and the latter NFM. PALF tangles to form a bulky mass and stay on the screen while NFM stays as separate small particles and fall through the screen. Figure 3 displays steps in the process for the extraction of short PALF. Unlike conventional methods for which long and perfect leaves are preferred, the new method can receive any type of leaves. The only requirement is that the leaves must not be dry. Otherwise, PALF will lose its strength (Kengkhetkit and Amornsakchai 2012).

2.2 Characteristic of PALF

The morphology of PALF obtained by mechanical grinding method is shown in Fig. 4. It is seen that the PALF is a mixture of small microfibers and bundles of microfibers as shown in Fig. 4a. High magnification image of fiber bundle end in Fig. 4b clearly reveals elementary microfibers of size approximately 3 μm (Kengkhetkit and Amornsakchai 2012).

PALF obtained from the grinding method described above is finer than that obtained from other conventional methods. Kengkhetkit and Amornsakchai (2012) compared the diameter of PALF obtained from two conventional methods,

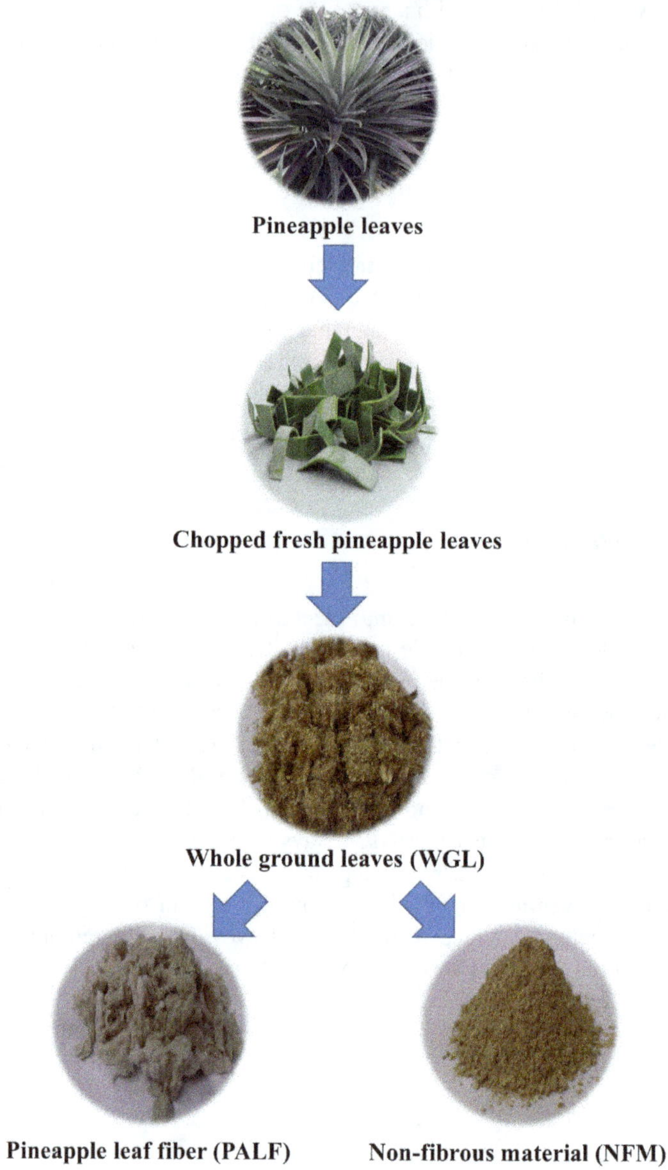

Fig. 3 Steps in the process for the extraction of short PALF

i.e., hand scraping (S-PALF) and water retting (R-PALF), and that obtained from mechanical methods, i.e., ball-milling (BM-PALF) and milling (M-PALF). The results are shown in Fig. 5. It is seen that mechanical methods give PALF with high fraction of finer diameter. Although ball-milling gives the highest fraction of finer PALF, it is much slower than normal milling machine.

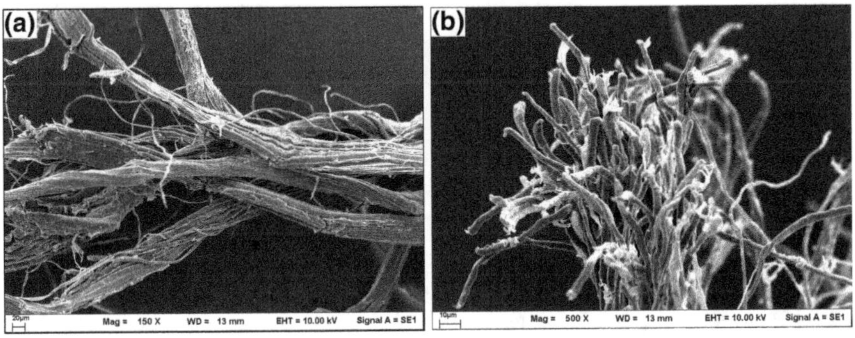

Fig. 4 SEM images of PALF obtained by mechanical grinding method at low and high magnifications

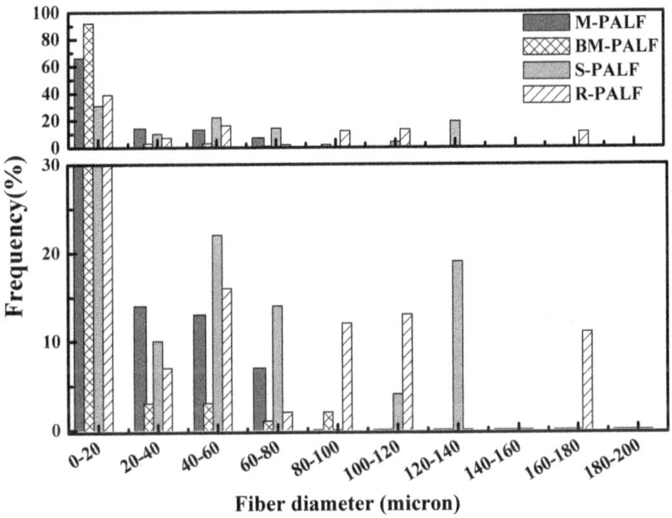

Fig. 5 The distribution of diameter of PALF prepared by different extraction methods. (adapted from Kengkhetkit and Amornsakchai 2012) For codenames see text

Chemical compositions including lignin, hemicellulose, and cellulose content of whole ground leaves (WGL), short PALF, and nonfibrous material (NFM) are shown in Table 2. The fibrous material or PALF displays high cellulose content.

The above PALF can be further treated with sodium hydroxide and mechanical forces from high-speed mixer to produce micro-size PALF or pineapple leaf microfiber (PALMF) as shown in Fig. 6. This PALMF has much narrower size distribution than normal PALF discussed above. It also has very high aspect ratio which could be up to 2000 (cut length of 6 mm and diameter of 3 μm). Since both PALF and PALMF have high mechanical properties and also very high aspect ratio, they both have high potential as reinforcement for polymer matrix composites.

Table 2 Chemical compositions of whole ground leaves (WGL), short PALF, and nonfibrous material (NFM) (adapted from Kengkhetkit and Amornsakchai 2014)

	Lignin (%)	Hemicellulose (%)	Cellulose (%)
Whole ground leaves (WGL)	4.57 ± 0.13	34.38 ± 0.66	46.65 ± 0.83
Fibrous material (PALF)	1.98 ± 0.08	19.80 ± 0.21	70.98 ± 0.93
Nonfibrous material (NFM)	7.70 ± 0.16	36.53 ± 0.23	43.69 ± 0.78

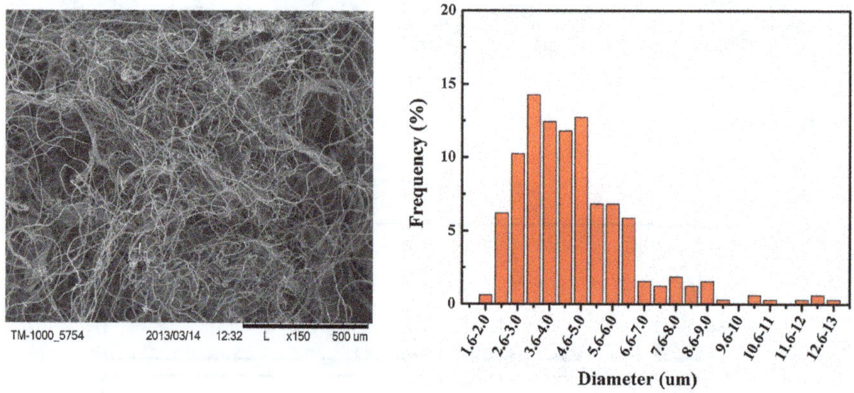

Fig. 6 SEM images of PALMF (left) and its size distribution (right)

3 PALF as an Effective Reinforcement for Polymer Matrix Composites

It is known that reinforcing efficiency of any filler will depend on its size, shape, and aspect ratio (the ratio of the largest dimension to the smallest dimension). Particulate fillers with aspect ratio close to unity have the lowest reinforcing efficiency while those with platelet or fibrous shapes would have much greater reinforcing efficiencies. Thus, when using PALF (with characteristics described above), one could expect that it will be an effective reinforcement for polymer matrix composites. The degree of reinforcement is highest when the fiber is uniaxially aligned and lowest when it is randomly oriented. Thus, it is clear that for maximum reinforcement, one should aim for uniaxial alignment of PALF.

3.1 *Orienting the Fiber for Maximum Reinforcement*

PALF can be incorporated into polymer matrix using conventional methods that are appropriate for the matrix. For example, melt mixing with plastic materials can be carried out in any type of mixer, be it twin screw extruder, internal mixer, or even heated two-roll mill to get a homogeneous mix. A similar machine can be used for

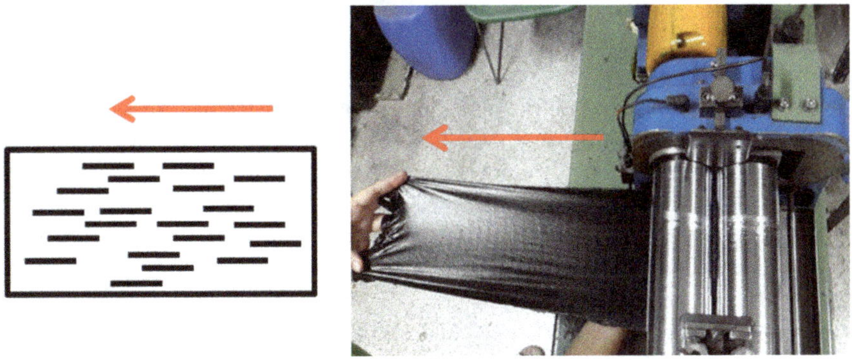

Fig. 7 The step of PALF alignment on a two-roll mill during the preparation of uniaxially aligned PALF reinforced rubber composite

rubber matrices. To get proper fiber orientation in the matrix, two-roll mill proof to be a very convenient tool.

The most important step in getting the fiber align in the machine direction is the taking out step for both plastic and rubber matrices. A very narrow nip between the two rollers is set. The rollers should be run at low speed. The mix is then passed through the nip. This will force the fiber to align along the machine direction. Gently pull the mix with slight tension to keep the fiber align as shown in Fig. 7. The compound will be in the form of a thin and long sheet (or prepreg) ready for the next compression molding step.

It is important that the fiber alignment is kept the same during subsequent compression molding. This is achieved by stacking the thin sheet of materials such that its dimensions are very close to that of the mold cavity, and the thickness is slightly greater than the required thickness. If needed, other types of fiber alignment such as angle-ply or cross-ply can be prepared.

When successful fiber alignment is obtained, the composite will display very high anisotropic properties, i.e., its properties in two directions will be very different. It will have a very high modulus in the machine direction and low modulus in the transverse direction.

3.2 *Rubber Matrix Composites*

Normally, rubbers are reinforced with particulate fillers such as carbon black, silica, or calcium carbonate. In certain applications where excessive deformation is undesirable, rubbers are reinforced with fabric. Short fibers have also been studied as reinforcement for rubbers as there are several advantages. With proper fiber orientation, fiber reinforced rubber has very unique properties of very high stretching stress at low strains. Since properties of the rubber matrix depend on many factors such as rubber type, filler type and content, and cross-link type and density, it is conceivable that fiber reinforced rubber with very wide range of properties can be prepared.

PALF reinforced acrylonitrile butadiene rubber (NBR) was the first rubber composite reported by our group (Wisittanawat et al. 2014a). NBR was chosen to demonstrate the effectiveness of PALF in the reinforcement because it is a polar rubber. Formulations and mixing sequences for the preparation of PALF reinforced rubbers are shown in Tables 3 and 4, respectively. NBR-PALF composites have very different stress–strain behavior from that of NBR. Stress–strain behavior of NBR-PALF in the longitudinal direction (LD) is very different from that in the transverse direction (TD), indicating a high degree of PALF alignment, as shown in Fig. 8. Stress in the LD rises sharply when the composite is stretched and then fails at the relatively low strain. Modulus and tensile strength increase with increasing PALF content while elongation at break decreases. Stress–strain behavior in the TD does not change much with the addition of PALF.

Failure of NBR-PALF composites shown in Fig. 8 dominates by fiber pull-out (Wisittanawat et al. 2014a). This has led Wisittanawat et al. to incorporate an adhesion promoter in order to improve the adhesion between the fiber and the rubber matrix. As expected, modulus, tensile strength, and tear strength of the composites increased. Failure of the composites changed to fiber breakage. However, the composites still failed at very low strain, which could limit their applications.

In order to solve the low elongation at break of the NBR-PALF composites, hybrid filler system was utilized. It was hypothesized that PALF would provide

Table 3 Formulations of rubber compounds

Ingredient	Amount (phr)
Rubber	100
ZnO	5
Steric acid	2
PALF	10, 20, 30
CBS	1
Sulfur	2

Table 4 Steps of mixing of NBR and NR composites (adapted from Wisittanawat et al. 2014b; Wongpreedee and Amornsakchai 2015)

Ingredient	Mixing time (min)	Ingredient	Mixing time (min)
NBR	0–1	NR	0–2
Sulfur	1–2	PALF	2–8
PALF	2–11	ZnO and stearic acid	8–11
ZnO and stearic acid	11–17	CBS	11–13
CBS	17–20	Sulfur	13–15
Total	20	Total	15

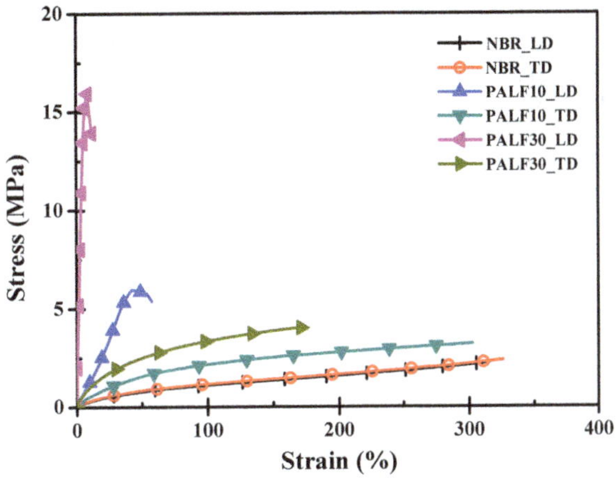

Fig. 8 Stress–strain curve of NBR and NBR-PALF composites containing 10 and 30 phr of PALF in longitudinal direction (LD) and transverse direction (TD) (adapted from Wisittanawat et al. 2014a)

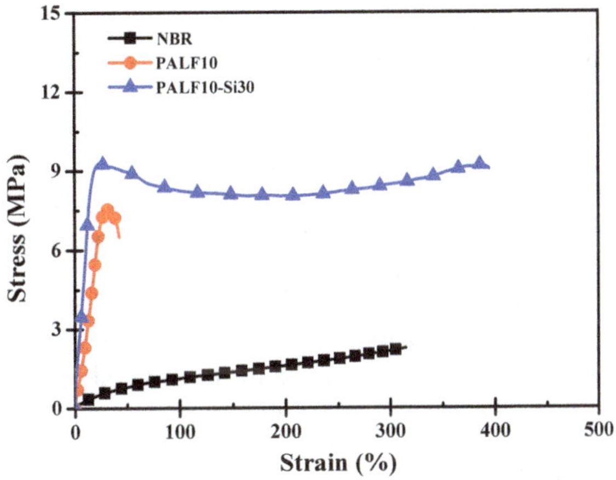

Fig. 9 Stress–strain curves of NBR and NBR-PALF-Si composites (adapted from Wisittanawat et al. 2014b)

reinforcement at low elongation while the other filler would provide reinforcement at high elongation (Wisittanawat et al. 2014b). Silica was selected as the hybrid filler as it is well known for effective reinforcement for rubbers. Figure 9 displays selected results extracted from the work reported by Wisittanawat et al. It is clearly seen that addition of silica improves the elongation at break of the composites significantly. The modulus at low strain is slightly improved due to the increased in stiffness of the rubber matrix (with silica). Thus, it can be concluded that by using hybrid filler system of PALF-silica, rubber composites with very distinct properties can be obtained.

Later work on the use of carbon black as hybrid filler confirms the above concept (Prukkaewkanjana et al. 2015). This is shown in Fig. 10. It is now seen that NBR-PALF10-CB30 displays stress–strain curve with three distinct zones. A very sharp rise in stress is seen in the low strain zone. The end of the first zone marks the

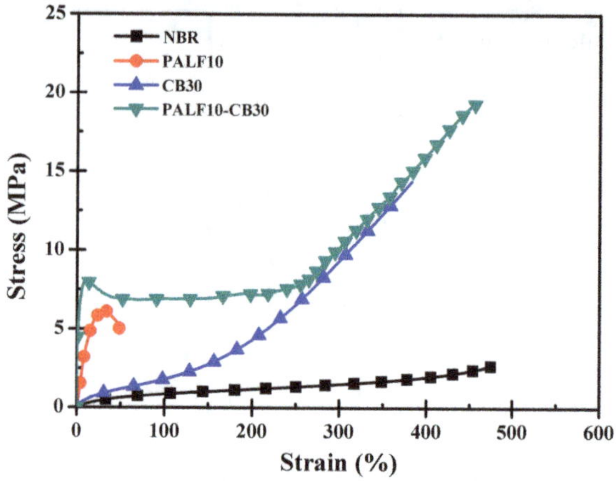

Fig. 10 Stress–strain curves of NBR and NBR-PALF-CB composites (adapted from Prukkaewkanjana et al. 2015)

position at which pulling out of the reinforcing fiber (PALF) occurs. In the second zone, the stress remains roughly constant with increasing strain. In this region, the fibers are still carrying some loads via frictional stress transfer. In the final zone, the stress upturn is seen. The result of NBR-CB30 indicates that the final zone is the characteristic of the CB filled NBR. It is now seen that with proper combination of fiber and particulate filler, reinforced rubber with very high stretching stress at low elongation and high tensile strength and high elongation at break could be prepared. The shape of the stress–strain curve of the composite can be adjusted by changing the amount of particulate filler.

Recently, Pittayavinai et al. have shown, by using NR-PALF-CB system, that the stress–strain behavior of rubber matrix composites can be adjusted by changing either CB content or cross-linking density (Pittayavinai et al. 2016). For NR system, the adhesion between PALF and the matrix is not as good as NBR. As a result, the maximum stress reached in the first zone is lower than that observed in NBR system. The stress–strain curve still displays three stress zones as shown in Fig. 11. With increasing cross-link density, the length of the second stress zone becomes shorter, and the stress upturn occurs at a lower elongation at break.

Wongpreedee and Amornsakchai studied NR-PALF composites using two types of PALF, i.e., untreated or as-prepared PALF (UPALF) and sodium hydroxide treated PALF (TPALF) (Wongpreedee and Amornsakchai 2015). Both types of PALF gave composites with similar stress–strain curves in the low strain region as shown in Fig. 12. At high strains, however, composite containing TPALF exhibited stress upturn at lower strains than that with UPALF. Synchrotron wide-angle X-ray scattering technique was used to follow the crystallization of NR during stretching of the composites. The presence of PALF caused the onset of strain-induced crystallization to occur at much lower strain than normal NR. The early stress upturn in the system containing treated PALF was attributed to better stress transfer to the fiber.

From the results shown above, it is clear that PALF is an effective reinforcement for rubber matrix composites. Since fiber pull-out still be a major failure

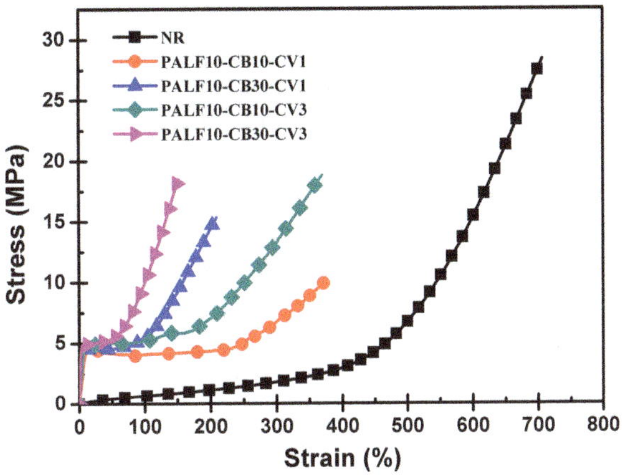

Fig. 11 Stress–strain curves of NR and NR-PALF-CB composites (adapted from Pittayavinai et al. 2016)

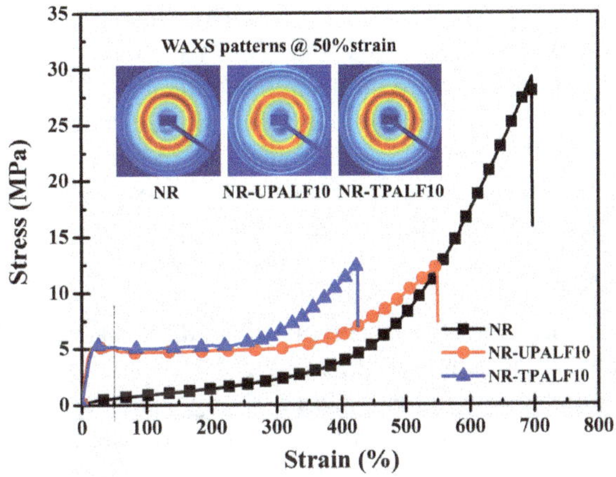

Fig. 12 Stress–strain curves of NR and NR-PALF composites. Insets are WAXS patterns of indicated composites stretched to 50% strain (adapted from Wongpreedee and Amornsakchai 2015)

mechanism, there are still rooms for the improvement in low strain properties. This could be achieved by improving the adhesion between the fiber and the rubber matrix.

Rubber composites with the above characteristics can be useful in many different applications where flexibility is still needed but large or excessive deformation is not acceptable. Examples of products with such requirements are power transmission belts, conveyor belts, pressure hoses, roller covering rubbers, and tires.

3.3 *Plastic Matrix Composites*

The use of cellulosic fillers such as sawdust or wood flour in plastic matrix composites has received wide acceptance, especially in wood replacement applications.

Since this type of cellulosic filler has particulate shape, large improvement in mechanical properties cannot be expected. The main objective is to reduce the use of expensive oil-derived plastics.

Cellulosic or natural fiber, due to its shape, could provide much greater reinforcement than sawdust. For this reason, the use of natural fiber for the reinforcement of plastic matrix is getting greater attention. There are many reviews on this (Karnani et al. 1997; Mishra et al. 2004; Sapuan et al. 2011). Natural fiber reinforced plastic can be used in various areas of applications such as aerospace, automotive parts, construction, marine, electronic component, and sports equipment.

PALF possesses high mechanical properties. Thus, it can be used to reinforce plastic matrix composites like other natural fibers. It has been shown that all parts of pineapple leaf including WGL, PALF, and NFM prepared by the novel fiber extraction method present in Sect. 2 could be used (Kengkhetkit and Amornsakchai 2014). However, different degree of reinforcement would be obtained as shown in Fig. 13 for PP composites. It is clear that PALF provides the highest degree of reinforcement, followed with WGL and NFM. Therefore, only the use of PALF will be covered in more detail in this section.

PALF has been tested as reinforcement for nylon, which requires much higher processing temperature than PP. Due to practical problem with nylon on two-roll mill, the mix was prepared with a laboratory twin screw extruder and then injection molded. PALF could be aligned along the flow direction. The results for PA6 and PA6/66 systems filled with PALF show the improvement in both tensile modulus and tensile strength as shown in Fig. 14.

When PALF is incorporated in thermoplastic PP and Santoprene (San) and aligned according to the method described in Sect. 3.1, it displays very effective reinforcement in both matrices. Longitudinal properties are improved significantly while transverse properties are relatively unaffected as shown in Fig. 15. Thus, it can be assumed that a high degree of fiber alignment is also obtained in these systems.

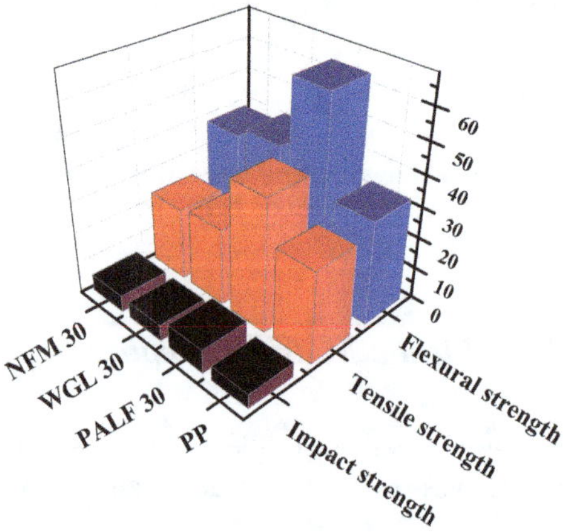

Fig. 13 Mechanical properties of PP-PALF, PP-WGL and PP-NFM composites containing 30% of pineapple leaf materials (adapted from Kengkhetkit and Amornsakchai 2014)

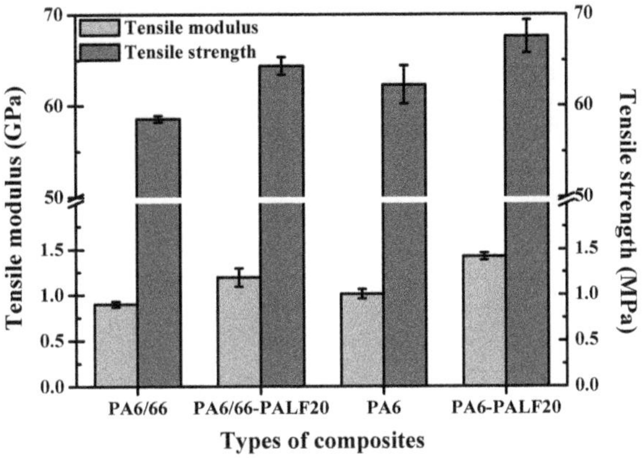

Fig. 14 Tensile properties of PA6/66, PA6/66-PALF composite, PA6, and PA6-PALF composite (adapted from Nopparut and Amornsakchai 2016; Panyasart et al. 2014)

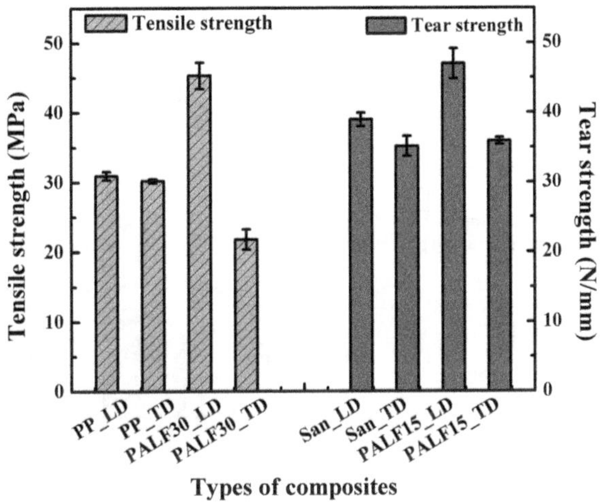

Fig. 15 Strength performance of PP-PALF composites and San-PALF composites both in longitudinal and transverse directions (adapted from Kalapakdee and Amornsakchai 2014; Kengkhetkit 2014)

The presence of PALF in thermoplastic composite not only improves ambient properties but also improves high-temperature properties. This is seen from the increase in heat distortion temperature (HDT) of both PP composites and nylon composites as shown in Fig. 16. This can be considered very distinct improvement and is not normally reported for natural fiber reinforced plastics. The improvement in HDT would certainly widen the applications of PALF reinforced plastics in high-temperature environments.

Since PALF is cellulosic fiber, properties of PALF-thermoplastic composites could be enhanced by using established methods for other natural fibers. This includes the use of various types of commercial coupling agents such as silane coupling agent and maleic anhydride-grafted polymer. Addition of coupling agent in PALF thermoplastic systems provides the enhancement of mechanical properties of composites due to the improved interfacial interaction between fiber and matrix. Figure 16 shows the effect of silane coupling agent on properties of composites.

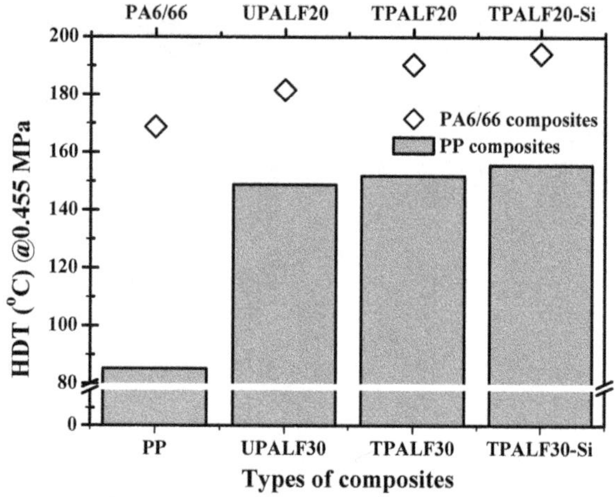

Fig. 16 Heat distortion temperature of PP, PP-PALF composites, PA6/66, and PA6/66-PALF composites (adapted from Kengkhetkit 2014; Nopparut and Amornsakchai 2016)

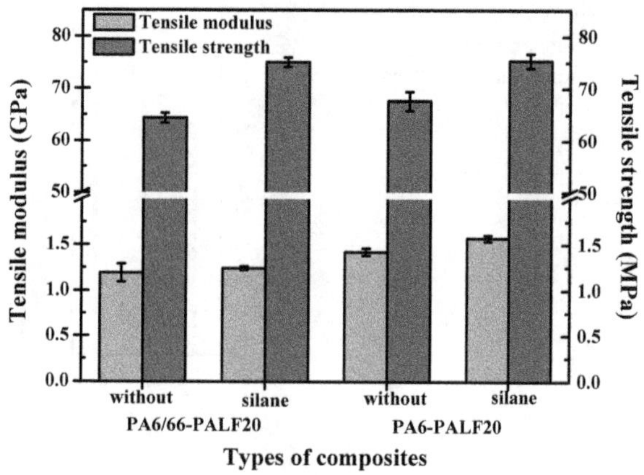

Fig. 17 Tensile modulus and tensile strength of PA6/66 and PA6 composites with and without silane coupling agent (adapted from Nopparut and Amornsakchai 2016; Panyasart et al. 2014)

Composites with silane coupling agent also have higher HDT than composites without silane coupling agent and neat polymers. In addition, the use of silane coupling agent in nylon composites provides the improvement in tensile strength in both PA6 and PA6/66 composites as shown in Fig. 17.

MAPP was used to improve the adhesion between filler and PP matrix in aligned PP-PALF composites. Enhancement of the performance of composite was observed (Kengkhetkit and Amornsakchai 2014). With the addition of 5%wt MAPP into PP-PALF composite, the flexural stress–stain curve of PALF composite changed significantly, and the composite exhibited greater toughness than that of composites without MAPP. This is shown in Fig. 18.

In plastic matrix composites, there can be an additional effect of chain orientation, in addition to PALF orientation, on mechanical properties of the composites. This has been demonstrated by Kalapakdee and Amornsakchai (2014). San, a PP-based thermoplastic elastomer, was used as the matrix. After aligning PALF,

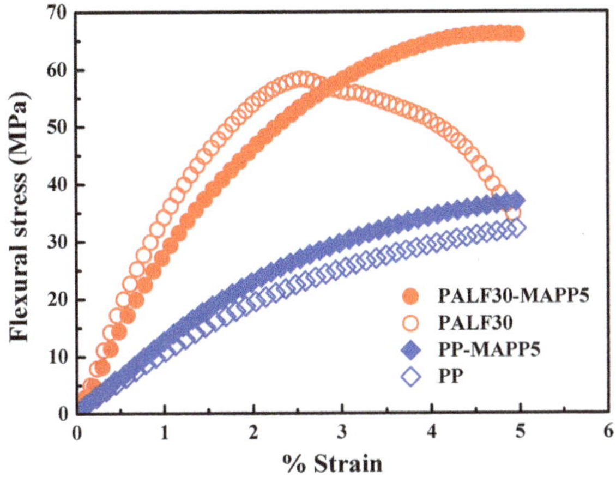

Fig. 18 Flexural stress–strain curve of PP and PP-PALF composites without and with 5%wt MAPP

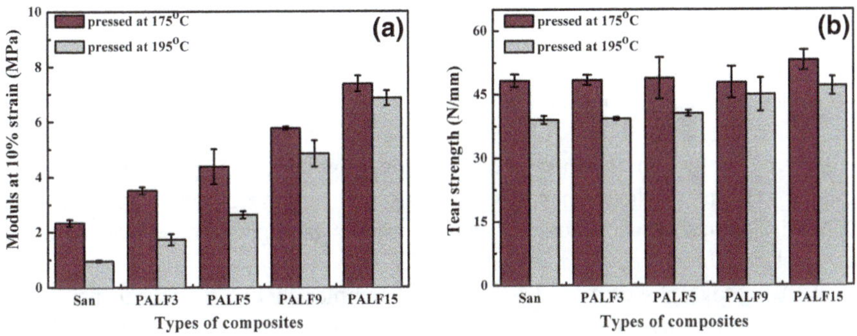

Fig. 19 Tensile modulus at 10% strain (**a**) and tear strength (**b**) of San-PALF composites compressed at two different temperatures (Adapted from Kalapakdee and Amornsakchai 2014)

crystalline orientation was also obtained. The orientation of PP is in the same direction as PALF. If subsequent compression molding is carried out at temperature low enough, such crystalline orientation can be retained, and this would provide additional reinforcement. Too high compression molding temperature would destroy the crystalline orientation, leaving only PALF orientation to reinforce the composite as shown in Fig. 19.

The combined effects of PALF alignment and matrix crystalline orientation on the mechanical properties of thermoplastic composite are also seen in PP with a greater degree of enhancement. Kengkhetkit showed that properties of PP-PALF prepreg that prepared according to the method described in Sect. 3.1 had relatively high mechanical properties (Kengkhetkit 2014). When the prepreg was compressed to form a thicker sheet of uniaxially aligned composites, the properties depended on compression molding temperature. This is illustrated in Fig. 20. Prepreg displays

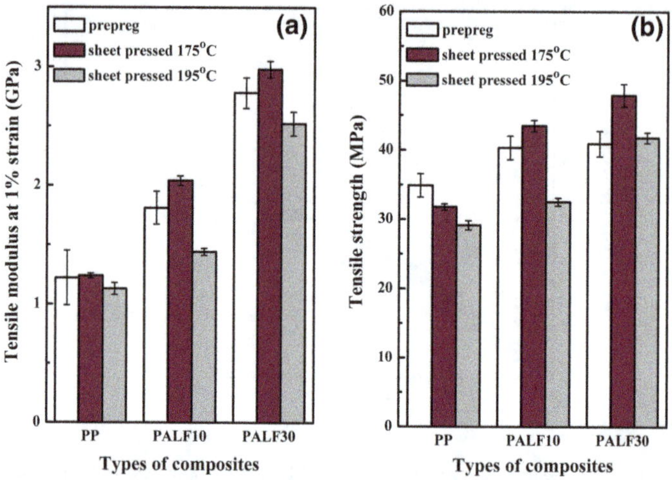

Fig. 20 Tensile moduli (**a**) and strengths (**b**) of PP-PALF prepregs and sheets containing different PALF content compressed at 175 °C and 195 °C

high mechanical properties due to the combined effect of PALF orientation and PP molecular orientation. During the subsequent compression molding, the prepreg was reheated again and molecular relaxation occurred. If the compression molding temperature is low enough just to melt the prepreg surface to give good bonding, limited molecular relaxation can occur and the composites only drop slightly. Otherwise, complete molecular relaxation occurs at high compression molding temperature. Under this situation, only PALF provides the reinforcement.

SEM photographs of etched PP-PALF30 samples in Fig. 21 reveal the strong difference of PP structure in prepreg and sheet pressed at 195 °C. Prepreg shows different morphology in parallel and perpendicular directions indication high molecular orientation while the sheet pressed 195 °C displays similar spherulitic structure in both directions.

In addition, the change in molecular orientation was confirmed with wide-angle X-ray scattering (WAXS). Prepreg and sheet compressed at 175 °C display crystalline and lamellar orientations. Azimuthal scans of the (110) reflections in Fig. 22 clearly indicate so.

It is clearly demonstrated, although from a limited set of examples, that PALF can be used to reinforce both commodity plastics like PP and high melting temperature engineering plastics like nylon. Mechanical properties and HDT of the plastic matrix can be significantly improved. It is important to note that improvement in HDT is not easily obtained from most mineral and cellulosic fillers and fibers. PALF reinforced plastics have several advantages over mineral filled plastics. These include lower density, lower carbon footprint, biodegradability, and incineratability (at the end of product's useful life). These composites can be used

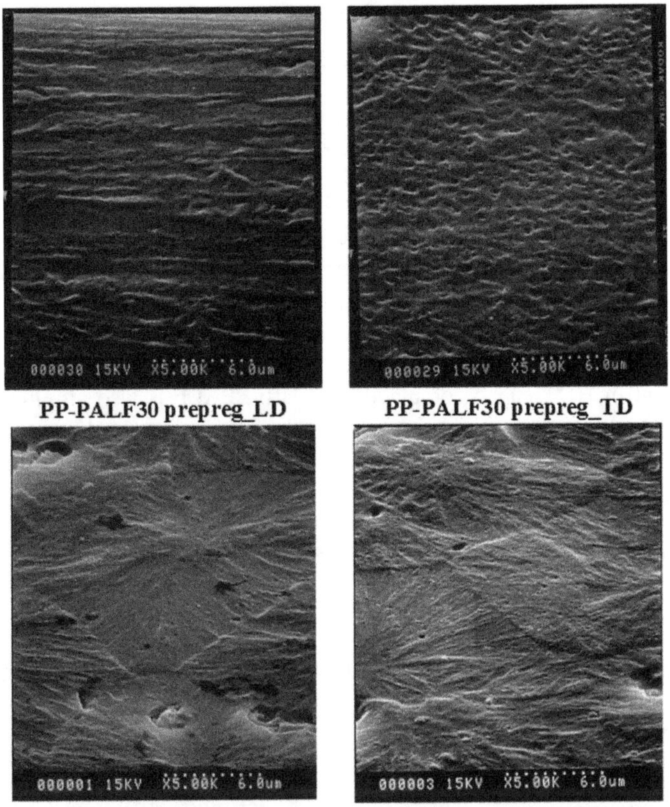

Fig. 21 SEM images of etched surface of PP-PALF30 prepreg (top row) and compressed sheet at 195 °C (bottom row) in the longitudinal (left column) and transverse (right column) directions

Fig. 22 Azimuthal scan of the (110) reflections for PP (o) and 30PALF-PP (x) prepregs (black) and sheets compressed at 175 °C (dark gray) and 195 °C (light gray)

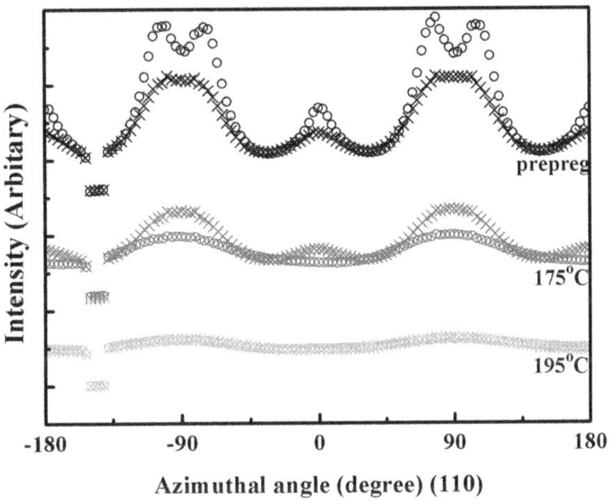

as automotive parts, engineering components, garden products, housewares, and containers.

4 Conclusions

The new concept of producing short and fine PALF as reinforcement for plastics and rubbers is very interesting and challenging. This will not only help farmers in dealing with their wastes but also make use of them as renewable materials to replace, in parts, petroleum-based materials. As-produced PALF provides very effective reinforcement for both commodity and engineering plastics. Significant improvement in HDT can be obtained. Similarly, PALF is also an effective reinforcement for rubbers. It provides rubbers with very distinct mechanical properties that cannot be obtained with common particulate fillers. PALF can be further treated to produce PALMF for even better reinforcement. The use of PALF as reinforcement for plastics and rubbers not only extends the applications of common plastics and rubbers to cover wider area but also makes them greener and more environmental friendly.

Acknowledgements We thank the following agencies for their kind financial support to the project: National Research Council of Thailand (NRCT), Thailand Textile Institute (THTI), and PERCH-CIC.

References

Chollakup R, Tantatherdtam R, Ujjin S, Sriroth K (2011) Pineapple leaf fiber reinforced thermoplastic composites: effect of fiber length and fiber content on their characteristics. J Appl Polym Sci 119:1952–1960

Devi LU, Bhagawan SS, Thomas S (1997) Mechanical properties of pineapple leaf fiber-reinforced polyester composites. J Appl Polym Sci 64:1739–1748

George J, Bhagawan SS, Prabhakaran N, Thomas S (1995) Short pineapple leaf fiber reinforced low-density polyethylene composites. J Appl Polym Sci 57:843–854

Kaewpirom S, Worrarat C (2014) Preparation and properties of pineapple leaf fiber reinforced poly (lactic acid) green composites. Fiber Polym 15:1469–1477

Kalapakdee A, Amornsakchai T (2014) Mechanical properties of preferentially aligned short pineapple leaf fiber reinforced thermoplastic elastomer: Effects of fiber content and matrix orientation. Polym Test 37:36–44

Kalia S, Kaith BS, Kaur I (2009) Pretreatments of natural fibers and their application as reinforcing material in polymer composites—a review. Polym Eng Sci 49:1253–1272

Karnani R, Krishnan M, Narayan R (1997) Biofiber-reinforced polypropylene composites. Polym Eng Sci 37:476–483

Kengkhetkit N (2014) Novel preparation and surface modifications of short pineapple leaf fiber for reinforcement of polypropylene. Ph. D. thesis, Faculty of Graduate Studies, Mahidol University, Thailand

Kengkhetkit N, Amornsakchai T (2012) Utilization of pineapple leaf waste for plastic reinforcement: 1. A new extraction method for short pineapple leaf fiber. Ind Crops Prod 40:55–61

Kengkhetkit N, Amornsakchai T (2014) A new approach to "Greening" plastic composites using pineapple leaf waste for performance and cost effectiveness. Mater Des 55:292–299

Leao AL, Cherian BM, Narine S, Souza SF, Sain M, Thomas S (2015) In: Faruk O, Sain M (eds) Biofiber reinforcements in composite materials. Elsevier, UK

Lopattananon N, Panawarangkul K, Sahakaro K, Ellis B (2006) Performance of pineapple leaf fiber–natural rubber composites: the effect of fiber surface treatments. J App Polym Sci 102:1974–1984

Lopattananon N, Jitkalong D, Seadan M (2011) Hybridized reinforcement of natural rubber with silane-modified short cellulose fibers and silica. J Appl Polym Sci 120:3242–3254

Lopez-Manchado MA, Arroyo M (2002) Short fibers as reinforcement of rubber compounds. Polym Compos 23:666–673

Luo S, Netravali AN (1999) Mechanical and thermal properties of environment-friendly "green" composites made from pineapple leaf fibers and poly(hydroxybutyrate-co-valerate) resin. Polym Compos 20:367–378

Mishra S, Mohanty AK, Drzal LT, Misra M, Hinrichsen G (2004) A review on pineapple leaf fibers, sisal fibers and their biocomposites. Macromol Mater Eng 289:955–974

Nopparut A, Amornsakchai T (2016) Influence of pineapple leaf fiber and it's surface treatment on molecular orientation in, and mechanical properties of, injection molded nylon composites. Polym Test 52:141–149

Panyasart K, Chaiyut N, Amornsakchai T, Santawitee O (2014) Effect of surface treatment on the properties of pineapple leaf fibers reinforced polyamide 6 composites. Energy Procedia 56:406–413

Payae Y, Lopattananon N (2008) Adhesion of pineapple leaf fiber to epoxy matrix: the role of surface treatments. Songklanakarin J Sci Technol 31:189–194

Pittayavinai P, Thanawan S, Amornsakchai T (2016) Manipulation of mechanical properties of short pineapple leaf fiber reinforced natural rubber composites through variations in cross-link density and carbon black loading. Polym Test 54:84–89

Prukkaewkanjana K, Thanawan S, Amornsakchai T (2015) High performance hybrid reinforcement of nitrile rubber using short pineapple leaf fiber and carbon black. Polym Test 45:76–82

Sapuan SM, Mohamed AR, Siregar JP, Ishak MR (2011) In: Kalia S et al (eds) Cellulose fibers: bio- and nano-polymer composites. Springer, New York

Satyanarayana KG, Pillai CKS, Pillai SGK, Sukumaran K (1982) Structure property studies of fibers from various parts of the coconut tree. J Mater Sci 17:2453–2462

Threepopnatkul P, Kaerkitcha N, Athipongarporn N (2009) Effect of surface treatment on performance of pineapple leaf fiber-polycarbonate composites. Compos Part B-Eng 40: 628–632

Wisittanawat U, Thanawan S, Amornsakchai T (2014a) Mechanical properties of highly aligned short pineapple leaf fiber reinforced - Nitrile rubber composite: Effect of fiber content and bonding agent. Polym Test 35:20–27

Wisittanawat U, Thanawan S, Amornsakchai T (2014b) Remarkable improvement of failure strain of preferentially aligned short pineapple leaf fiber reinforced nitrile rubber composites with silica hybridization. Polym Test 38:91–99

Wongpreedee T, Amornsakchai T (2015) Synchrotron X-ray diffraction study of pineapple leaf fiber reinforced natural rubber composites during stretching. Suranaree J Sci Technol 22: 253–263

Lightweight Wood Composites: Challenges, Production and Performance

Sandra Monteiro, Jorge Martins, Fernão D. Magalhães and Luísa Carvalho

Abstract Wood composites are materials made by bonding together wood and adhesives into a large material that can be used for different purposes. Nowadays, lightweight materials play an important role in several industries: aerospace, building and furniture. The reduction of weight is desirable for economic reasons (materials and transportation costs) and environmental reasons (resources, eco-efficiency). Once wood composites are employed in these industries, low density is a desired property. There are several options in the market to reduce the weight of composites, such as the use of low-density wood species, lower compaction of the wooden mat, incorporation of light fillers in the core layer of the panel, or use of sandwich panels with honeycomb core. All these strategies have challenges with respect to manufacturing, machinability (connections and lamination of the edges) and performance (physico-mechanical properties).

Keywords Lightweight · Wood-based panels · Sandwich panels

S. Monteiro · F. D. Magalhães · L. Carvalho (✉)
LEPABE—Laboratory for Process Engineering, Environment, Biotechnology and Energy, Faculdade de Engenharia da Universidade do Porto, Rua Dr. Roberto Frias, s/n 4200-465 Porto, Portugal
e-mail: lhcarvalho@estv.ipv.pt

S. Monteiro
e-mail: sandram@fe.up.pt

F. D. Magalhães
e-mail: fdmagalh@fe.up.pt

J. Martins · L. Carvalho
DEMAD—Departamento de Engenharia de Madeiras, Escola Superior de Tecnologia e Gestão de Viseu-IPViseu, Campus Politécnico de Repeses, 3504-510 Viseu, Portugal
e-mail: jmmartins@estgv.ipv.pt

S. Kalia (ed.), *Lignocellulosic Composite Materials*, Springer Series on Polymer and Composite Materials, https://doi.org/10.1007/978-3-319-68696-7_7

Contents

1 Introduction to Lightweight Wood-Based Composite

Wood composites are materials made by bonding together wood with different shapes and sizes and adhesives, forming a large material which can be used for a variety of products with the required physico-mechanical properties: furniture, flooring, roofing, palettes, decking, structural beams, marine applications, aircraft materials, etc.

The advantages of using wood-based composites, in relation to solid wood, are (1) use of small diameter trees, (2) use of waste wood from industrial processing and agriculture residues, (3) products with uniform and defect-free appearance, (4) better strength than solid wood, (5) products with unique shapes and dimensions and (6) products with special properties such as low thermal conductivity, fire resistance and better biological durability (Matuana and Heiden 2004).

Several factors have a huge impact on the mechanical properties of wood-based composites such as: wood species, forest management regimes (naturally regenerated, intensively managed), kind of adhesive used, shape (fibres, flakes, strands, particles, veneer, lumber) and the arrangement of wood elements in composite and the final composite density. The performance of wood composites can be characterized according to several engineering properties. Structural and non-structural applications are defined by the mechanical properties of the composite.

Static elasticity and strength are the key properties used to choose the materials or set up the design and the specifications (Cai 2012).

Nowadays, the use of lightweight materials becomes a topical issue in many developed industrial countries. The trend concerning lightweight materials is notorious in many industries and the furniture and wood-based panels industries do not miss out. The increase of the mobility of people demands that materials should be lightweight and should also have proper physico-mechanical properties. Generally, a lightweight material brings many advantages for producers, designers and consumers: (1) lower costs in consequence of mass reduction and transportation, (2) possibility of

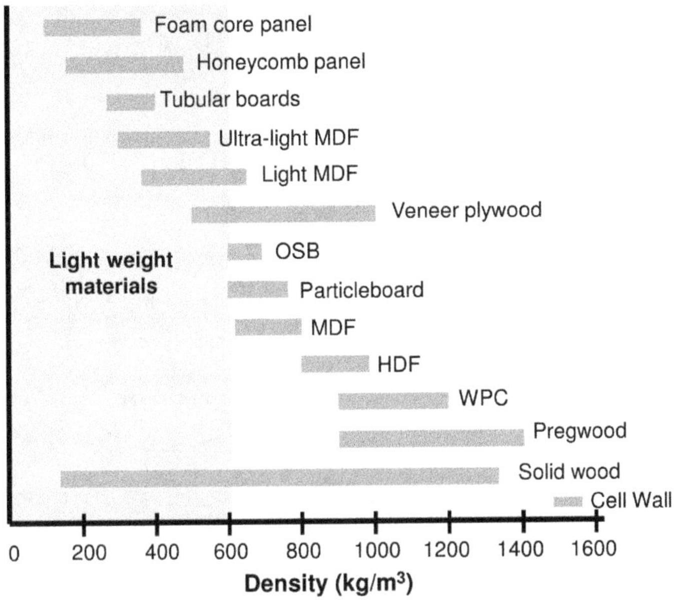

Fig. 1 Density of wood-based materials and for cell wall: foam core panel, honeycomb panel, tubular boards, ultra-light and light MDF (medium density fibreboard), veneer plywood, oriented strand board (OSB), particleboard, medium density fibreboard (MDF), high-density fibreboard (HDF), wood plastic composites (WPC), pregwood and solid wood adapted from Thoemen (2008)

obtaining thick elements, answering current design trends, (3) more rational use of wood, when its price is high or availability is scarce, (4) ease of transportation by producer and final costumer, with lower environmental impacts (Shalbafan 2013). According to CEN/TS 16368, panels with density below 600 kg/m^3 are classified as lightweight materials. Materials such as tubular board, honeycomb panel and panels with foam core are included in this group. Figure 1 shows the different ranges of density for wood-based materials, emphasizing those considered as lightweight.

The different strategies available for obtaining the lightweight wood composites already available in the market are summarized in Fig. 2 (Shalbafan et al. 2016b; Wang 2009; Dziurka and Mirski 2013).

Panels such as MDF and particleboards with low density are produced by using lower compaction of the wood mat. Another strategy is using very light wood species like *Ochroma* spp. (50–300 kg/m^3) or *Ceiba* spp. (300 kg/m^3) from North America and South America, respectively. *Populus* spp. (450 kg/m^3) (Trischler et al. 2014) from Europe is another type of lightwood also used in the production of lightweight particleboards. Extruded tubular boards are frequently used and the density achieved with this method may vary between 250 and 400 kg/m^3. Sandwich panels with foam core (made of polyurethane or polystyrene foam) or paper-based honeycomb structures are a major trend in the industry, achieving densities from 100 to 350 kg/m^3. However, lightweight panels have certain restrictions, which constitute challenges to overcome as follows:

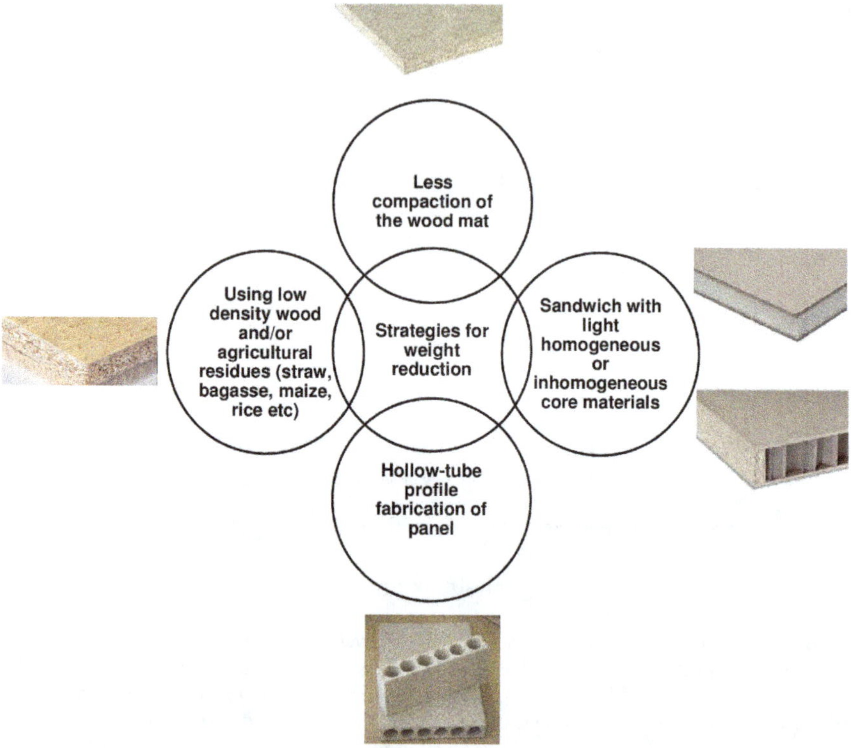

Fig. 2 Strategies for panel weight reduction

- Reduction of mechanical properties, which disallows the use of panels in applications requiring load-bearing capacity,
- The use of agricultural residues requires higher amount of resin,
- Difficult storage of non-wood bio-based materials such as agriculture crops,
- Limitations in terms of surface finishing and post-forming,
- Increase of production cost associated to the need of new additives due to lower bonding strength of the new raw materials (Balducci et al. 2008; Shalbafan 2013).

2 Fibreboard

Fibreboard is a wood-based panel with a nominal thickness of 1.5 mm or greater, manufactured from lignocellulosic fibres with the application of heat and/or pressure (EN 316). The bond is derived either from the felting of the fibres and their inherent adhesive properties or from a synthetic binder added to the fibres. Other additives can be included (EN 622-5).

The most commonly used lignocellulosic fibres are wood, but other plant fibres can be used such as bagasse and cereal straws. Thermo-mechanical pulping (TMP) is the process commonly used to make fibreboards. This process consists in breaking the bonds between the wood cells by combining the action of heat and mechanical energy. The lignin present in the wood has the ability to absorb small amounts of water; therefore, its softening temperature is dependent on moisture content. The TPM process uses high temperature (170–195 °C) and humidity (60–120%) that increase the probability of breaking the wood cells, when subjected to a mechanical energy during the refining process (Donaldson and Lomax 1989; Thoemen et al. 2010).

The methods of wood-based fibreboard manufacture are generally divided into wet and dry process. The definitions of fibreboards are formulated in the European standard (EN 622-5). Fibreboards are classified by their production process as follows:

- Wet process fibreboards (fibre distribution in water)
- Dry process fibreboards (fibre distribution in air).

Wet process boards are fibreboards having fibre moisture content of more than 20% at the stage of forming. Additionally, wet process boards are classified according to density, as follows:

–Hardboards (HB)	Boards with density ≥ 900 kg/m^3
–Medium boards (MB)	Boards with density ≥ 400 to <900 kg/m^3
–Softboards (SB)	Boards with density <400 kg/m^3

Dry process fibreboards (MDF) are fibreboards having a fibre moisture content of less than 20% at the stage of forming and having a density ≥ 450 kg/m^3. Medium density fibreboard (MDF) is one of the most rapidly growing composite panel products in forest products market.

Raw materials namely wood, binders and other additives and the production parameters are the factors that have a major effect on physico-mechanical properties of MDF (Akbulut and Ayrilmis 2004). When wood is considered as a raw material, wood fibre properties such as fibre structure and strength, anatomical and chemical properties of fibre, and fibre composition (percentages of whole and broken fibres and fines) are considered to be basic characteristics influencing fibreboard properties, since they occupy a large portion of the total panel volume. For marketing purposes, MDF of specific density range can be given different denominations. For example, the following density-related marketing terms for MDF have become established:

–HDF	MDF with density ≥ 800 kg/m^3
–Light MDF	MDF with density ≤ 650 kg/m^3
–Ultra-light MDF	MDF with density ≤ 550 kg/m^3

2.1 Wet Process Fibreboards: Softboard

Softboards are usually used in the interior of walls and floors, and as underlays in roofs, in order to provide heat insulation properties. The low density leads to a more energy efficiency caused by the many air voids in the panel (Rowell et al. 1995; Kollmann et al. 1975).

Softboards can be produced via a wet process (Fig. 3), which is really an extension of paper manufacturing technology.

Wood enters a chipper to be reduced in size (with sizes of approximately 25 and 5 mm thick). The chips are screened to remove undersized (<2 mm) and oversized (>50 mm) particles. The chips are washed to remove contaminants and to facilitate defibration, and then fed into the pre-heater or digester, where they are heated with overheated steam at 6–10 bar and at a temperature of 175–195 °C for 3–7 min. The steam treatment softens the lignin, which binds the fibres together allowing their separation in a pressurized discs' refiner under saturated steam at 8–10 bar. Here, the steamed chips are transformed into fibres. Some wet process boards are made without additional binders. If the lignocellulosic materials contain sufficient lignin, this compound can serve as a binder, as long as it is retained during the refining operation. Under heat and pressure, lignin will flow and act as a thermosetting adhesive, enhancing the naturally occurring hydrogen bonds. The fibres are discharged from the defibrator and then water is added to form a slurry (water is used as the dispersion medium for forming the fibres into a mat; this is the reason why the process is called wet process). This slurry is introduced into a wire mesh conveyer belt to form the board, losing its water prior to drying. The boards are lightly compressed and dried in an oven. The voids created after the evaporation provide the board its good thermal properties.

Contrary to other wood-based boards, fibre-based insulation materials are not required to display high strength. Surface appearance, internal bond and bending strength are also not of primary importance for softboards. The typical properties of these boards are presented in Table 1.

There are several insulation products available in the market. The board properties vary with production technologies, final application and specific requirements. Table 2 gathers examples of few commercial softboards and their properties.

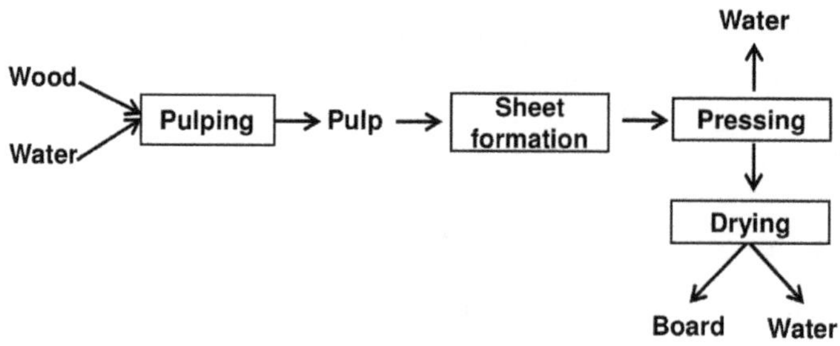

Fig. 3 Wet process manufacture adapted from Suchsland and Woodson (1986)

Table 1 Typical properties of softboards (Kollmann et al. 1975)

Property	Values
Density	170–280 kg/m^3
Flexural strength	1.0–2.7 MPa
Modulus of elasticity	80–400 MPa
Maximum thickness swelling	12–20%
Sound absorption of acoustical board (522 Hz)	50–85%

Table 2 Comparison of properties of different softboards available in market (Barbu and Pieper 2008)

Product	Application area	Thickness (mm)	Density (kg/m^3) EN 1602	Heat conductivity [W/(m K)]	Materials and additives
HolzFlex®	Intermediate rafter framework	40–200	40	0.038	–Wood fibres –Polyolefin fibres –Ammonium polyphosphate
Steico-flex	Intermediate rafter framework	40–200	50	0.038	–Wood fibres –Polyolefin fibres –Ammonium sulphate
Agepan THD N+F 230	Over rafter	40/60/80	230	0.047	–Wood fibres –Polyurethane resin
Thermosafe homogen	Universal insulating board	20–240	<110	0.037	–Wood fibres –diphenylmethane di-isocyanate (PMDI)

The major benefit of this production process is the low VOC emission level caused by the low amount of resin used and the drying time that is longer. The density (>150 kg/m^3) and the board thickness (<40 mm) are the two mainly restrictions in the wet process. Panels with higher thickness need a longer drying time which causes an increase in production costs (Barbu and Pieper 2008; Porter 2012).

2.2 *Dry Process Fibreboards: Light MDF and Ultra-Light MDF*

Light MDF and ultra-light MDF are widely used in the manufacture of furniture, doors and walls due to many excellent properties such as low density, low thermal conductivity and good sound absorption (Chen et al. 2016). These types of panels are manufactured by using what is called a dry process (Fig. 4).

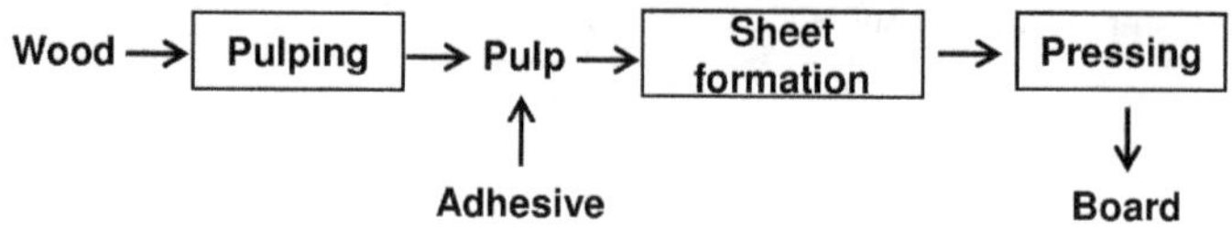

Fig. 4 Dry process manufacture adapted from Suchsland and Woodson (1986)

In this process, the fibres are produced in the same way as the wet process. After the defibration process, wood fibres are discharged into a pressurized blowline, a tube with 80–120 mm of diameter, which transports wet fibres to the dryer. In the blowline, when the fibre is blended under high pressure, the resin is more easily distributed due to the steam expansion that causes the separation of fibres (Thoemen et al. 2010).

Resins like polymeric diphenylmethane di-isocyanate (PMDI), urea-formaldehyde (UF) and phenol formaldehyde (PF) are used as adhesives (Suhaily et al. 2012). The adhesives, as well as other additives, are injected into the blowline through a 3–5 mm nozzle at high pressure (12–14 bars). The adhesive load is usually between 8 and 15% (resin solids/oven-dry wood). The wet resinated fibres are then blown through a flash tube dryer (with around 100 m long and 1–3 m in diameter) at around 30 m/s. At the end of the tube, cyclones separate the dried fibre from steam. After drying, fibres have approximately 8–12% moisture content. The dry fibre is conducted to mat formers. These are made with the purpose of distributing an even layer of fibres onto a continuously moving belt. Here, the panel thickness is controlled by monitoring the speed belt. The forming head spread fibre mattress homogenously. In the case of an 18 mm thick MDF board, the mat is around 680 mm high and has a bulk density of 23 kg/m^3. After passing a continuous prepress to reduce its height, the mat is then pressed in a continuous hot press with temperature in the range of 180–210 °C and pressure between 0.5 and 5.0 MPa (Thoemen et al. 2010). The pressing stage can be used to manipulate panel properties by changing the panel density profile. This operation has a major effect on the balance of properties of the resulting panel: a rigorous control of all processing variables is necessary to achieve appropriate product quality and to minimize pressing time (Carvalho and Costa 1998). Immediately after pressing the panels are fed into a star shape cooler where boards are stood on their edges to expose their faces to ambient air to facilitate rapid cooling. Finally, boards are subjected to finishing operations, which includes trimming, sanding and cut to size. Nowadays manufacturer also includes several surface treatments such as laminating, profiling and painting.

Light and ultra-light MDF are produced in the same way as MDF, but lower pressing pressure is used to attain lower density MDF. The main advantage of this process is the decrease of the density to 50 kg/m^3 and an extensive range of thicknesses (<200 mm). A disadvantage of the dry technology is the high-level amount of resins that is needed to strengthen the fibre bonds (Barbu and Pieper 2008).

Light MDF is used in situations where weight reduction is required in the range of 15–20% and where the machinability performance is not a limiting factor.

In the manufacturing of MDF, several important factors have direct effects on mechanical properties and stability of the panel, namely the resin system, moisture content, hot-pressing conditions and wood fibre characteristics. The structure and morphology of fibres have more influence than their mechanical characteristics on the board properties. Fibres with different morphologies have different specific surface areas (Xing et al. 2004). Short fibres need a higher amount of resin when compared with longer fibres, which generally results in better mechanical properties for the panels. When the resin content of the panels increases from 6 to 12%, an increase in mechanical properties (internal bonding, modulus of rupture and modulus of elasticity) is observed, while thickness swelling and water absorption decrease. In the thermo-mechanical pulping process, usually used to refine fibres, refining pressure and time are two important features that affect the final properties of the board. With the exception of water absorption, refining pressure has great influence in overall panel properties. On the other hand, refining time does not alter the properties of fibre and board (Nayeri et al. 2014). The effect of moisture content change on panel flatness is difficult to predict because of the characteristic variation of the density level throughout the thickness, or vertical density profile.

The panel properties depend on the compaction ratio of each layer that forms the panel. Generally, the difference between the layers is defined by the compaction ratio, since that no differences exist in geometry fibre and resin amount in the whole panel (Ganev et al. 2005).

The requirements for light MDF are different depending on its use in dry or humid conditions. Panels in humid conditions have to withstand the presence of water vapour in the surrounding air and casual contact with water on the surface. Table 3 specifies the requirements for light MDF boards for use in dry and humid conditions, particularly interior fitments including furniture.

Concerning ultra-light MDF, the requirements are different from light MDF. In this case, panels are divided into two groups:

- Ultra-light MDF Type 1 (UL1-MDF), these boards are typically used as insulation panels providing limited mechanical stiffness;
- Ultra-light MDF Type 2 (UL2-MDF), these boards are typically used as panels with stiffening function; they also can be used with fasteners. These panels have insulating properties.

The next table specifies the requirements for these two types of panels under dry conditions (Table 4).

The most important feature of low-density MDF is its homogeneous density profile along the thickness, which allows turning and edge profiling operations. The possibility of surface and edges finishing is one of the main advantages of this material over other wood composites.

There are several options of traditional light and ultra-light MDF in the market, in Table 5 are presented some examples.

Table 3 Requirements for light MDF boards for use in dry and humid conditions (EN 622-5)

Property	Conditions	Test method	Unit	Ranges of nominal thickness					
				>6–9	>9–12	>12–19	>19–30	>12–19	>45
Swelling in thickness 24 h	Dry	EN 317	%	20	16	14	12	11	11
	Humid			18	16	13	12	11	10
Internal bond	Dry	EN 319	N/mm^2	0.45	0.45	0.45	0.45	0.40	0.40
	Humid								
Bending strength	Dry	EN 310	N/mm^2	20	20	18	15	14	14
	Humid			20	20	18	16	16	14
Modulus of elasticity in bending	Dry	EN 310	N/mm^2	1700	1700	1600	1500	1400	1200
	Humid								

Table 4 Requirements for ultra-light MDF boards for use in dry conditions (type UL1-MDF and type UL2-MDF) (EN 622-5)

Property	Conditions	Test method	Unit	Ranges of nominal thickness				
				>9–12	>12–19	>19–30	>30–45	>45
Swelling in thickness 24 h	UL1-MDF	EN 317	%	18	14	13	12	12
	UL2-MDF							
Internal bond	UL1-MDF	EN 319	N/mm^2	0.15	0.15	0.15	0.13	0.13
	UL2-MDF			0.35	0.35	0.35	0.30	0.30
Bending strength	UL1-MDF	EN 310	N/mm^2	7.7	6.9	6	5.1	5.1
	UL2-MDF			18	16	14	12	12
Modulus of elasticity in bending	UL1-MDF	EN 310	N/mm^2	600	560	510	470	470
	UL2-MDF			1400	1300	1200	1100	1100

There is a new technology to produce low-density fibreboards different from traditional manufacturing. Dascanova® developed a unique mat modification technology that alters the compressibility of selected particles before pressing the final board. This allows redistribution of the density inside the board. Unlike conventional fibreboards that have homogeneous density in the core and higher density on the surfaces, Dascanova fibreboard has a vertical density profile that changes along the entire board (Fig. 5).

The ideal construction form is the triangular or wave structure. In comparison to standard technology, up to 30% less raw materials (incl. resin and other chemical components) are needed for producing Dascanova Fibreboard. The mechanical properties are similar to standard fibreboard, while internal bond strength (IB) is improved due to this special internal structure. The panels produced have an average density of 428 kg/m^3 with a bending strength and modulus of elasticity of 9.2 and 2.097 N/mm^2, respectively, (Denesi 2012).

Table 5 Mechanical properties of some light and ultra-light MDF available in the market for dry conditions. Information taken from the products' datasheets

Product	Application area	Thickness (mm)	Internal bond (N/ mm^2)	Swelling in thickness, 24 h (%)	Bending strength (N/mm^2)
Kronospan	Building material for a substitute for timber in non-load-bearing walls, ceilings, partitions	15–18	$\geq$0.80	12	45
Fibrapan	Loudspeaker boxes for sound equipments, doors, decorative mouldings	6–45	0.50–0.55	9–18	15–20
Caberwood	General-purpose joinery	12–25	0.45	12	20
Medite ultra-light	Furniture Ceilings Picture frames Mobile home components	12–18	0.35	10–15	15
Fibrapan ultra-light	Used in the construction industry and in the setting-up of exhibitions and fairs	9–28	0.50	13–17	14–18
Caberwood ultra-light	Theatre scenery Film sets and exhibitions	12–25	0.35	13–18	14–18

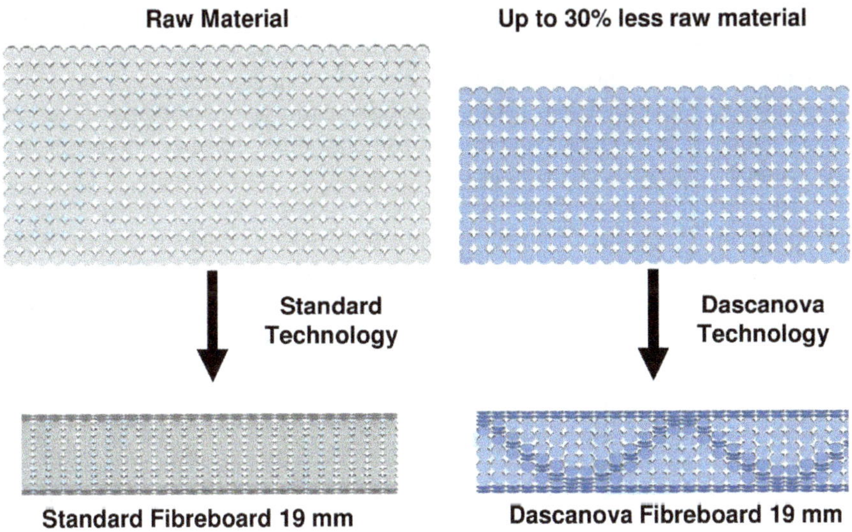

Fig. 5 Scheme of the comparison between the standard technology of fibreboard production and the developed Dascanova technology adapted from (www.wbpionline.com)

3 Particleboard

As defined in European Standard EN 309, particleboard is a 'wood-*based panel manufactured under pressure and heat from particles of wood (wood flakes, chips, shaving, sawdust and similar) and/or other lignocellulosic material in particle form (flax shives, hemp shives, bagasse fragments and similar) with the addition of an adhesive*'. As said before, particleboards with a density below 600 kg/m^3 are defined as lightweight particleboard (European technical specification CEN/TS 16368). These panels are commonly used in furniture, being overlaid with different materials for decorative purposes. Particleboard can also be employed in flooring systems, in house construction, for stair treads and as underlayment (Maloney 1989). These applications require high mechanical strength and good dimensional stability, once the panels are exposed to bending stress, static and impact forces. Besides that, particleboards must be resistant to water absorption (Nemli and Demirel 2007). Particleboard is the most popular wood-based panel in the market due to its lower cost, which results from the use of recycled wood, and to its intrinsic relatively low density compared with other boards.

3.1 Particleboard Manufacture

The raw materials used in particleboard manufacture consist of wood residues from sawmills, recycled cellulosic materials and plant residues such as wheat straw and bagasse. The initial step in particleboard production is the reduction of the wood into smaller sizes. For this, large particles are chipped in the Breakdown Machines, using hammer mills and flakers. The shape and integrity of the wood particles have a huge influence in the final properties of particleboard. The use of homogeneous raw material is very important to produce panels with good strength, smooth surfaces and similar swelling. After the particles are cut to the desired size, they are sent to the dryer in order to obtain the intended moisture content, usually in the range of 2–8%, depending on the resin chosen to produce the board. Higher values of moisture content can generate a large amount of water vapour during pressing, causing the lamination of the panel. The adhesive system must be prepared according to the formulation. Different amounts of adhesive are used in the surface and core layers. The adhesive content is calculated on a solid adhesive to oven-dry wood basis. Normally, particleboards have three layers, a core layer with coarser particles and a lower density and two surface layers with finer particles and higher densities (Maloney 1989). The smaller particles contribute to improve the surface quality and the larger one provides better mechanical properties like bending.

After being mixed with the resin, the particles are formed into a mat and then transferred to the hot press for the panel curing process. This step is critical and requires carefully controlled temperature, pressure and timing. The press time and temperature are very important since it is necessary to get a high enough temperature in the core to cure the adhesive, at the same time as the temperature at the surface should not be too high to avoid thermal degradation. Typical pressures are 2–4 MPa and temperatures 200–220 °C. After pressing, the boards are cut, sanded and finally, the edges are trimmed. Boards may also be subjected to different finishing. They can be veneered or overlaid with other materials, like impregnated paper laminates. Paint and lacquer can also be used in the board finishing (Nemli et al. 2005). A schematic example of particleboard production can be seen in Fig. 6.

The resins most commonly used in particleboard production are urea-formaldehyde (UF), melamine formaldehyde (MF) and phenol formaldehyde (PF), the last one being used to a much lesser extent. Other types can be used such as isocyanate resin (diphenylmethane di-isocyante—MDI and polymeric diphenylmethane di-isocyante—pMDI) (Dunky 1998; Stoeckel et al. 2013).

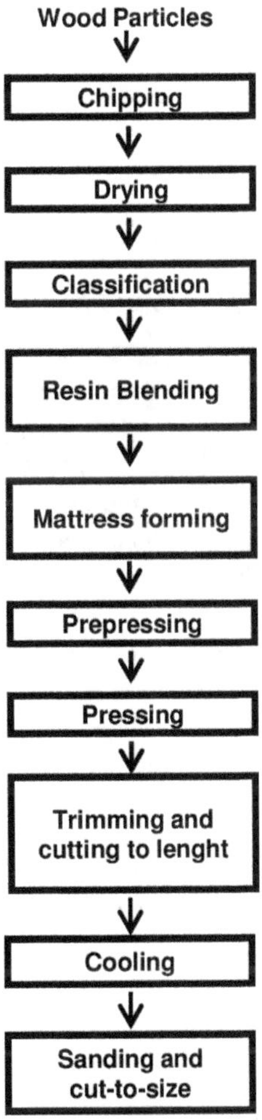

Fig. 6 Particleboard manufacture diagram

Table 6 General purpose lightweight boards LP1 and LP2 for use in dry conditions (CEN/TS 16368)

Type	Property	Test method	Unit	Requirement					
				Thickness range (mm, nominal)					
				>6–13	>13–20	>20–25	>25–32	>32–40	>40
LP1	Bending strenght	EN 310	N/mm^2	4.0	3.5	3.0	2.5	2.0	2.0
LP2				8.0	7.0	6.0	5.0	4.5	4.0
LP1	Modulus of elasticity in bending	EN 310	N/mm^2	550	500	475	450	400	375
LP2				1000	950	900	850	750	650
LP1	Internal bond	EN 319	N/mm^2	0.28	0.24	0.20	0.17	0.14	0.14
LP2				0.35	0.30	0.25	0.20	0.17	0.17

The advantage of UF resins when compared with the other ones is their high reactivity, fast curing, good adhesion to wood, and low cost (Costa et al. 2013).

Usually, physico-mechanical properties are evaluated according to European standards. The mechanical properties specifications for general-purpose lightweight boards (type one—LP1) and for use in dry conditions (type two—LP2) are presented in Table 6.

3.2 Particleboards Made from Light Wood Species

One of the interesting strategies used by the lightweight wood-based panel industry to achieve the density reduction of particleboard is the use of light available wood species. With this strategy, it is possible to achieve maximum reduction of one-third of the initial density. However, to maintain the mechanical properties such as internal bond and bending strength at acceptable values, producers had to increase the amount of used resin and optimize the board's density profile. On the other hand, the use of light wood species is strongly dependent on the availability and seasonality of wood, the global resources are limited and they are not sufficient to meet market needs. The production cost caused by the increasing of resin in the manufacture of boards also is a huge concern to the producers. Usually, the material costs related to this type of product are more than half of the production costs (Barbu et al. 2010).

In literature, there are a few examples of particleboards produced with wood light species such as poplar (*Populus spp.*) (Boruszewski et al. 2016) Seraya (*Shorea spp.*) (Kawai et al. 1986), paulownia (*Paulownia tomentosa*) (Kalaycioglu et al. 2005), kelempayan (*Neolamarckia cadamba*) (Lias et al. 2014). Figure 7 shows the internal section of particleboard made from *Populus* spp. wood.

Particleboards produced with these light wood species have a range of density between 300 and 600 kg/m^3. In these studies, an isocyanate adhesive is usually used as the binder, due to its ability to form covalent bonds with the hydroxyl

Fig. 7 Internal section of particleboard made from *Populus* spp. wood

groups on the wood surface. This high reactivity allows faster cure rates when compared with the amino and phenolic resins, and lower adhesive loads. In general terms, the properties of particleboards are significantly affected by the density of the raw materials; panels produced with light wood species may lead to panels with higher strength when compared with higher density species. This occurs because light species have higher compaction ratio, which allows a greater contact area between the wood particles during the pressing operation obtaining a good bonding. To obtain maximum bonding a certain amount of pressure is needed; the adequate pressure depends on the ratio between the board density and the wood species density. To produce particleboards with medium density, the required ratio is commonly higher than 1.3 (typically ratio in boards produced with amino resins). Boards produced with high-density species with admissible properties for use in most applications have high density. It is possible to obtain a reduction of density by mixing high and low-density species. On the other hand, the use of raw materials with different density can cause problems in some steps of the manufacturing process such as the following:

- milling operation—particles with different densities are milled differently;
- drying operation—loss of moisture content is affected by density;
- pressing operation—particle compressibility depends on density.

The amount of resin to be used can also be affected by the different characteristics of the wood material (Xu et al. 2004; Maloney 1989).

Boruszewski et al. studied the effect of technological parameters on mat compression during the producing of lightweight particleboards (500 kg/m^3) made of two low-density species (*Populus* spp. *and Pinus* spp.). Three series of panels were produced: one made with only pine, other made only with poplar, and the last one with a mixture of the two wood species. The results showed that the characteristics of the species used, such as heat transfer coefficients, and moisture content affect significantly the compression process. Lower density species require higher pressing times or higher heat transfer in the panel core layer. With regard to mechanical properties, the modulus of rupture (MOR) and the modulus of elasticity (MOE) had higher values for panels made with poplar wood. This is related to the low density of poplar wood, which leads to higher compaction. However, boards produced with mixed poplar-pine revealed the best performance in terms of the internal bond (IB). This result is explained by the density difference of the two wood species used, which requires the use of higher resin amount. Another reason is the higher extent of compaction of poplar face layers in the mixed boards, due to the lower density of this wood furnish (Boruszewski et al. 2016).

Studies carried out by Kalaycioglu et al. showed that particleboards produced with paulownia with a density of 550 kg/m^3 had MOE and MOR strength values that meet the requirements for general uses and interior fitments established in EN standards. As expected, the panels produced with a density of 650 kg/m^3 presented better mechanical properties (Kalaycioglu et al. 2005).

Another trend of lightweight wood-based panel industry is the use of alternative raw materials such as agricultural residues and annual/perennial plants to produce particleboards Dziurka et al. studied particleboards made from wood and rape straw particles to produce particleboards with density in the range of 350–550 kg/m^3. The results revealed that, as established, the reduction of density in both series of produced panels (wood and rape straw) lead to a reduction of bending strength and modulus of elasticity. Panels manufactured with rape straw showed a reduction of strength of 32% when the density of the boards is reduced from 550 to 350 kg/m^3. For the same density variation, strength reduction was 18% for particleboards produced from wood. The modulus of elasticity showed the same trend. The use of veneer on panel's surface improves significantly the bending strength. All produced boards meet the requirements of boards for interior decoration and furniture applications (lightweight boards LP2) (Dziurka and Mirski 2013).

3.3 Particleboards with Light fillers

In recent years, the competition for wood has been increasing, particularly due to the extensive use of this resource in the construction and furniture sectors. Therefore, the use of non-wood raw materials to reduce the weight of particleboards is an interesting approach. Materials like expanded polymeric particles, starch granulates or even popcorn (Kharazipour and Bohn 2012) have been mixed with wood in the core layer of panels. In low-density particleboards, the light fillers fill the voids between wood particles, allowing for improved machinability of panel edges (Shalbafan et al. 2016a).

BASF Company developed Kaurit Light, with a core made from a mix of expanded polystyrene foam (EPS) and wood particles (Fig. 8), with 30% lower weight than a conventional particleboard, while maintaining the same strength as standard (Peretolchin 2009). The size range of the EPS particles used is 0.3–0.8 mm.

To produce expanded polystyrene foam, deionized water, styrene and the mixture initiator are added to the reactor under agitation at ambient temperature. Then the stabilizer is added to the reactor. The stabilizer consists of insoluble inorganic salts such as magnesium carbonate, which prevent the monomer droplets from coalescing. It is during the heat cycle that occurs the development of droplet size. After polymerization, when the spherical particles are formed, the mixture is cooled and centrifuged to separate the water from the polymer beads and then the beads are dried. The beads and the water are added to the impregnation reactor. During heating, the reactor is pressurized with blowing agent, usually n-pentane, at 5–7% w/w with respect to the polymer. During this step, the blowing agent diffuses into the beads. At the end of this stage, the system is cooled down and the excess of the stabilizer is removed from the particle surface in a washer tank. The separation of polymer beads from water is made by centrifugation. Then the particles are dried.

Fig. 8 Internal section of particleboard produced with expanded polystyrene foam (EPS) mixed with wood particles in core layer

In the final processing, the EPS beads are heated, usually with steam that causes the beads to expand by foaming and their volume to increase (Kannan et al. 2007).

Another type of expanded material usually used in the core layer of particleboards is expandable microspheres (MS). This filler is commercially available under the name of Expancel and is produced by AkzoNobel company. The microspheres consist of a polymer shell (typically acrylonitrile or vinylidene chloride copolymerized with methyl methacrylate or methyl acrylate) encapsulating a gas (isobutene) (Magnus 2010). Increasing temperature above 85 °C causes the polymer shell to soften, which, in combination with the increased pressure of the encapsulated gas, results in sphere expansion. The particle size of MS is approximately 15 μm before expansion (Shalbafan et al. 2012), and 40 μm after expansion.

The main difference between MS and EPS is that the first is added to the core layer in the unexpanded state, having micrometric sizes. EPS particles are already expanded, having sizes of few millimetre. For this reason, MS can be more difficult to scatter in the core layer. Another important difference is the glass transition temperature (Tg): EPS particles have a higher Tg (103 °C) compared to MS (85 °C). This property affects the processing conditions of the panel. The inflammability of the blowing agents used in both fillers is the main limitation; this implies restrictions in some applications (Shalbafan 2013).

In the studies carried out by Shalbafan et al, the mechanical properties of multi-layered lightweight panels produced with MS core layer with different surface thickness (3, 4 and 5 mm) were evaluated. The target density of particleboards varied from 300 to 500 kg/m^3. The study revealed that the IB of the panels with MS in the core layer increased when the face layer thickness is raised. The same applies to bending strength. This is explained by the increase of density in the board core (120–180 kg/m^3), which promotes a better interaction between wood particles and MS. However, when compared with conventional panels, only particleboards with MS with higher density (500 kg/m^3) had IB higher than 0.24 N/mm^2 (IB requirement for lightweight particles boards' type 1) (Shalbafan et al. 2012).

In another study developed by Shalbafan et al., different amounts of expanded polystyrene particles (5, 10 and 15% based on the dry mass of wooden particles) were used to produce particleboards with two different densities (450 and 500 kg/m^3). The research showed that the use of these fillers affects the physical and mechanical properties of the particleboards. Fillers absorb lower amounts of resin when compared with wood particles, which improved the efficiency of the adhesive. As expected, panels with higher density had better flexural properties. Panels with a density of 500 kg/m^3 had higher IB when compared with conventional particleboards, due, according to the authors, to a bonding reinforcement effect caused by the fillers in panel core layer. Concerning physical properties, low-density particleboards had lower thickness swelling compared to conventional particleboards, mostly due to the presence of an hydrophobic filler. The same trend was observed for water absorption (Shalbafan et al. 2016a).

4 Extruded Particleboards

Extruded boards, also called tubular boards, are a type of particleboard produced by extrusion (Kreibaum process), while the conventional particleboards are manufactured by flat-pressing (Fahrni process) (Zhang and Lu 2015). In the extrusion process, wood residues and vegetable waste can be used if they are reduced to short particles. Glued wood particles are introduced in a hot press, compressed using a piston driven by an eccentric flywheel, and extruded through the hot section of the press, forming a board. Different types of hollow tubes are formed, depending on the diameters of the heated rods used (Fig. 9) (Wittmann and Wolf 2014).

Extruded boards are used in the production of furniture components such as low-density core for doors, tabletops and wall boards. They can also be employed in the production of prefabricated houses. Tubular boards have low strength in the longitudinal direction, higher linear swelling and lower perpendicular swelling when compared to the flat pressed boards. Usually, this type of boards is reinforced by using surface veneers (Kollmann et al. 1975; Borgin 1958).

5 Sandwich Panel

Sandwich panels are multilayer structures with two thin facings adhesively bonded to a thick core produced from lightweight material (Fig. 10). This type of arrangement leads to an increase of the strength and stiffness without much increase of the weight (Alessandro et al. 2013; Petras and Sutcliffe 1999).

This type of panel has several applications in fields where high performance (higher strength, damage tolerance and thermal resistance) and low weight are

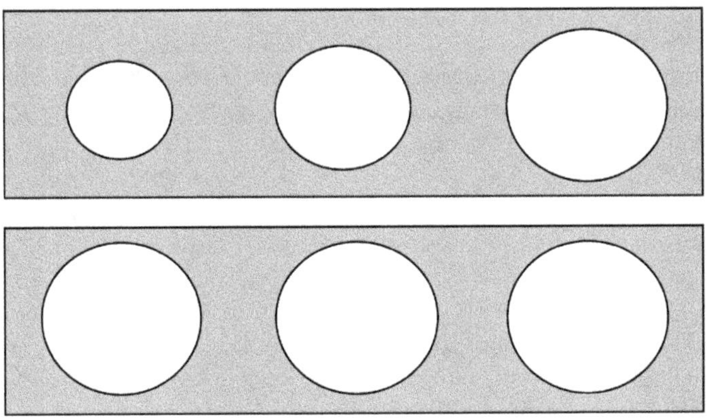

Fig. 9 Scheme of cross section of tubular boards with identical (top) and different (bottom) diameters of hollow tubes

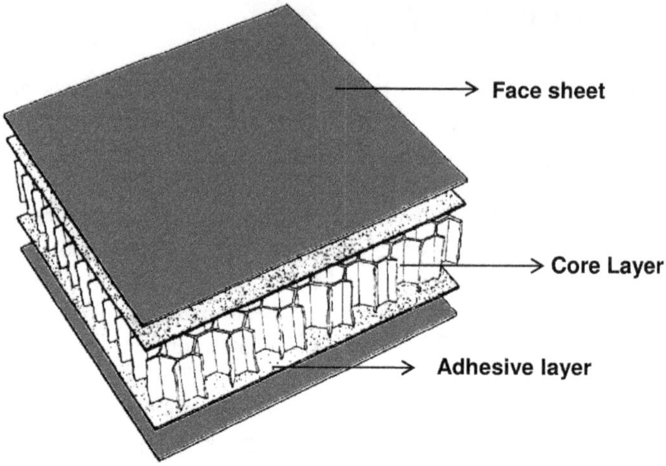

Fig. 10 Schema of sandwich structure adapted from Petras and Sutcliffe (1999)

required, such as aerospace and marine industries, furniture and automotive constructions (Schubel et al. 2005; Meraghni et al. 1999).

The appropriate selection of facing and core material enable obtaining a product with the desired characteristics. A waterproof facing can be used in several houses panels, a facing with abrasion resistance can be applied to floors. In addition, for decorative purposes, plastic facings could be applied to walls, doors and furnishings. The type of material used in the core can be selected to confer thermal insulation, fire and degradation resistance to the panel (Kollmann et al. 1975).

5.1 Sandwich Assembly

The crucial step in the assembly of the sandwich composite is gluing the faces to the core of the panel. This process could be carried out in a hot or cold press, in a continuous press or by moulding. Low pressures must be used to avoid crushing the core. In addition, the press plates must be precisely parallel; otherwise, the core can crush gradually.

Timber-based boards can be used as facing materials. However, this disallows use in continuous foaming lines and commonly the board's lengths are lower than 3.5 m. This type of facing is more suitable to the mould foaming and bonding methods. It is necessary to use appropriate joints to produce long panels. Wood-based chipboard or plywood can be applied to low humidity environments, however, if the panel is subject to long-term loading, core and face deformation may occur. Plywood is made from thin layers of wood veneers, designated plies that are glued and pressed to form the panel (Matuana and Heiden 2004). The usual thickness is in the range of 4–27 mm and the density varies between 550 and

700 kg/m^3. Chipboard is produced from wood chips with a thickness between 0.2 and 0.4 mm that are bonded together employing temperature and high pressure. The thickness is usually in the range of 2–25 mm with a density between 600 and 750 kg/m^3 (Davies 2001).

The preparation of cores in sandwich structures commonly involves cutting to adequate thickness and expanding honeycomb cores to obtain the intended cell size, if the core was acquired in the unexpanded form. Foams cores may be pre-formed into a block or foamed in situ, between the faces (Karlsson and TomasÅström 1997). An adhesive is used to bond the faces to the core. A low viscosity adhesive can be applied by spraying, in successive coats, while an adhesive with higher viscosity can be applied with a brush or hand roller (Kollmann et al. 1975).

5.2 *Types of Sandwich Core Materials*

Usually, the core of the panel is produced with low strength materials. However, the high thickness provides higher bending stiffness to the sandwich construction. Structurally, the core has two functions: (1) separate the faces and prevent deformation in the direction perpendicular to the face plane and (2) confer some degree of shear rigidity along the planes that are perpendicular to the face. Table 7 shows some properties of different materials that can be used as a core in a sandwich construction.

In the design of sandwich panels it is necessary to take into account the following factors: (1) the specific characteristics of the core, which can vary depending on the material used; (2) the effect of shear flexibility of the core; (3) the effect of the core on the buckling of the faces; (4) the effect of temperature in causing stresses and deflections and (5) the deformation behaviour (Davies 1993). Non-structural features like thermal conductivity, permeability, weight and moisture sorption should be considered in the design of the panel (Kollmann et al. 1975).

It is possible to distinguish two types of the core layer in sandwich panels as: homogeneous and non-homogeneous (Wang and Wang 2016). The first group includes cores made from hardwood strips (*Ochroma* spp., *Populus* spp., low-density *Pinus* spp.). Thermosetting or thermoplastic foams with open or closed cells are also used to produce a sandwich panel with the homogeneous core. The foams are produced from polymers like polyurethane, polystyrene and polyvinyl chloride (Pishan et al. 2014; Karagiozova et al. 2009; Bull and Edgren 2004). The group of non-homogenous cores includes textile, corrugated and honeycomb cores (Fig. 11). Honeycomb cores can be made of metal, plastic or paper.

Foamed core panels

Figure 12 shows a board with a foam core. There are several methods for manufacturing a sandwich panel with foam core.

In the batch process, the prefabricated layers are assembled together with the adhesive and the stack is pressed under temperature and pressure needed to cure the adhesive and form the final sandwich panel (Karlsson and TomasÅström 1997).

Table 7 Properties of different core materials (Kollmann et al. 1975)

Core material	Density (kg/m^3)	Elastic modulus (MPa)	Shear modulus (MPa)		Shear strength (MPa)	
			TL[a]	TW[b]	TL	TW
Balsa wood	80	1530	68	84	1.1	1.0
	112		103	145	1.6	1.4
	144		138	205	2.1	1.8
	176		174	264	2.6	2.2
Paper honeycomb	27	–	–	17	–	0.21
	32	–	–	–	–	–
	50	–	–	53	–	0.76
Polystyrene foam	28	8	8	–	0.22	–
	48	14	12	–	0.39	–
	72	21	18	–	0.39	–
Polyurethane foam	96	28	12	12	0.83	0.76
	144	83	25	28	0.65	0.69
	160	62	24	19	1.6	1.4

[a]TL-parallel to the core ribbon direction
[b]TW-perpendicular to the core ribbon direction

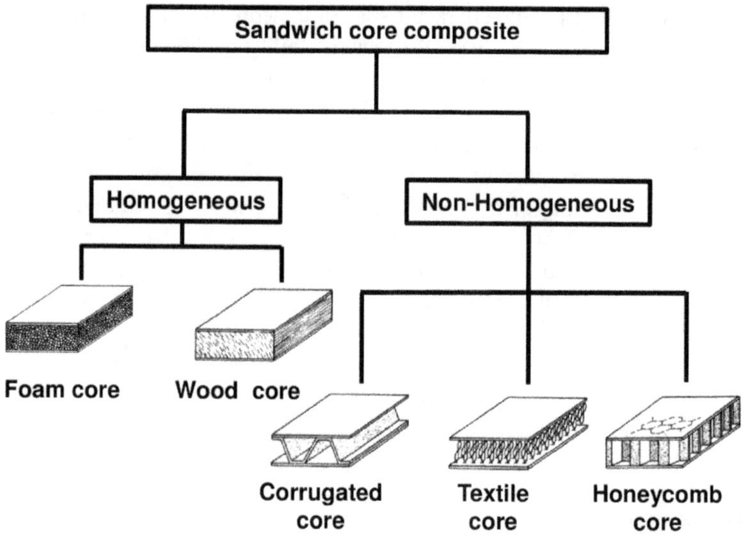

Fig. 11 Scheme of varies types of sandwich cores adapted from Pflug et al. (2002)

Fig. 12 Sandwich panel with a polystyrene foam core and wood-based particleboard in the facings

The high production costs necessary for this manufacturing process and the limited sizes countervail the weight reduction achieved with this method (Barbu et al. 2010). Another process is mould forming. In this case, an exact amount of foam is directly introduced into the cavity between two faces to fill the mould. It is a discontinuous process where the board is produced in a closed mould whose dimensions are the same as the final panel. The upper face is supported by spacers in order to maintain it in position. If edge profiles are required, these must be placed at this stage. The mould structure must be rigid enough to withstand the high pressures reached during foam formation. After the introduction of foam in the mould, the panel is left in the mould for a defined period of time and then is removed. The main advantage of this method is allowing the production of panels with complex shapes. However, this is a relatively slow process that can cause void spaces within the panel, due to the non-uniformity of foam flow (Davies 2001).

For the mass production of foam sandwich panel, a continuous process can be used. The face material run through the roll formers into coils, that creates the surface and edge profiles. In this process foils or impregnated paper are used for prefabricated facings and the core is usually foamed between the faces. Depending on the properties of the foam, an additional adhesive layer may be needed. For example, if the core layer is a polyurethane foam, the additional layer of adhesive is not used. The foam liquid is introduced by injection in the panel line. The foam expands and then the panel is cooled down and cut to the desired length (Fig. 13).

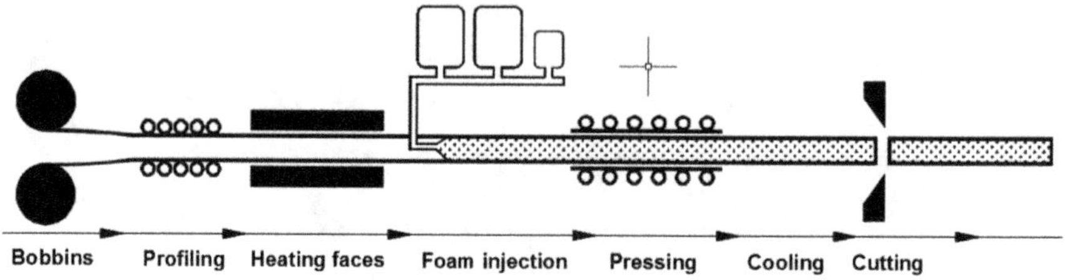

Fig. 13 Continuous line of foam production adapted from Davies (2001)

The main advantage of this process is to have less production steps, compared to the other process described above, which leads to a higher efficiency in time, material and human resources. This aspect had a huge effect on the final panel cost (Shalbafan 2013).

Honeycomb core panels

A honeycomb panel is constituted by a honeycomb-type core (typically with hexagonal cells) bonded to the surfaces (Nilsson and Nilsson 2002). Hexagonal cells are the most usual configuration; but there are others, such as the square, rectangular, corrugated, flex-core and reinforced cells shown in Fig. 14.

The type of surfaces chosen for a particular application is determined by functional requirements such as strength, stiffness, damage tolerance and appearance at minimum overall cost. In this kind of composite, the bending loads (compression and tension) are supported by the faces and the shear loads are supported by the core (Nilsson and Nilsson 2002). As said before, the core materials can be plastic, metallic or paper. Paper honeycomb panels (Fig. 15) have been widely used for the production of door leaves not only in the construction or furniture but also in aerospace and automobile sectors.

In general, any type of wood-based panel can be used for the faces, but the most common are particleboards and MDF, due to their low cost (Barboutis and Vassiliou 2005). Paper honeycomb can be produced by two different processes: expansion manufacturing and corrugation manufacturing (Fig. 16).

This process begins with cutting flat sheets of paper from a roll fixed on a rotating drum. As the sheets are stacked, alternating lines of glue are printed on each sheet. After the sheets are stacked the adhesive is cured. Slices of the unexpanded bock are cut with the required thickness (typically above 10 mm) and the sheets are pulled apart, causing the expansion of the hexagonal honeycomb core. After expansion, the residual stress is reduced by the controlled application of heat and moisture.

The corrugation process is another way to produce honeycombs (Fig. 17). Here the material is corrugated or folded into a certain shape and then the shaped layers are glued into a honeycomb block. The block can be then sliced into pieces of the required size (Pflug et al. 2002).

Honeycomb panels present good load-bearing behaviour, despite the weight reduction. However, they have some limitations as follows:

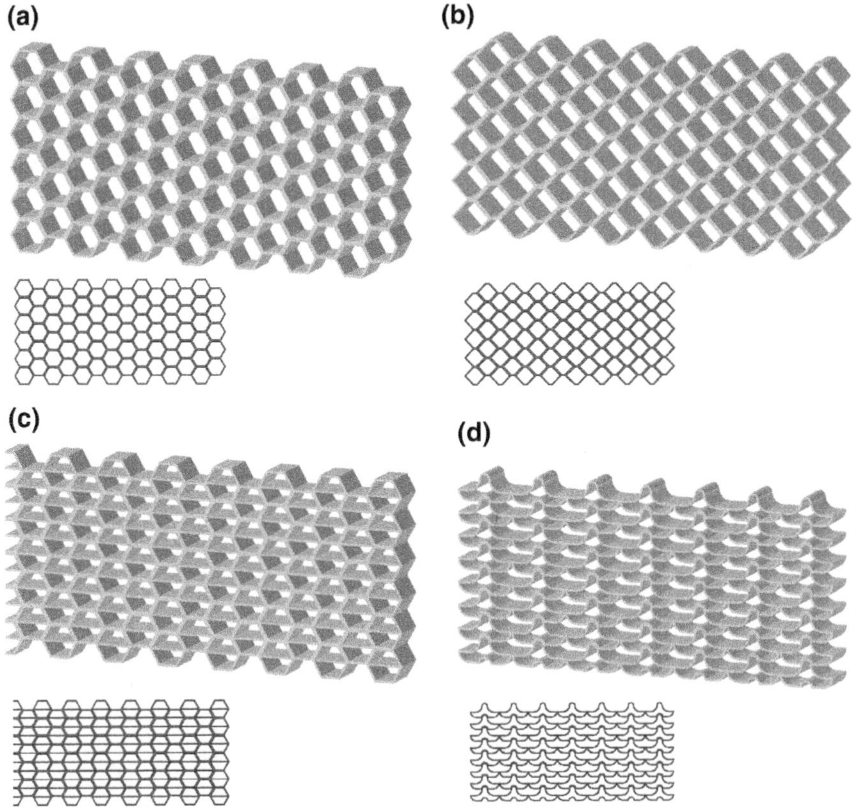

Fig. 14 Different honeycomb cells configurations: **a** hexagon, **b** square cell, **c** reinforced hexagon and **d** flex-core adapted from Bitzer (1997)

Fig. 15 Sandwich panel with paper honeycomb core

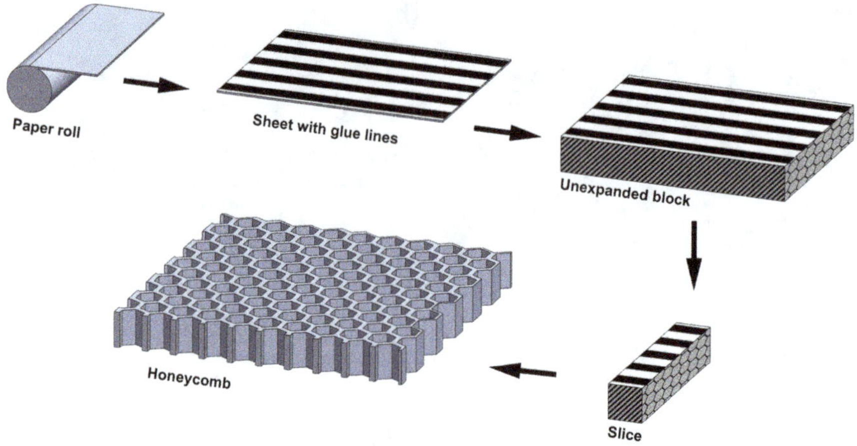

Fig. 16 Manufacture process by expansion of honeycomb adapted from Bitzer (1997)

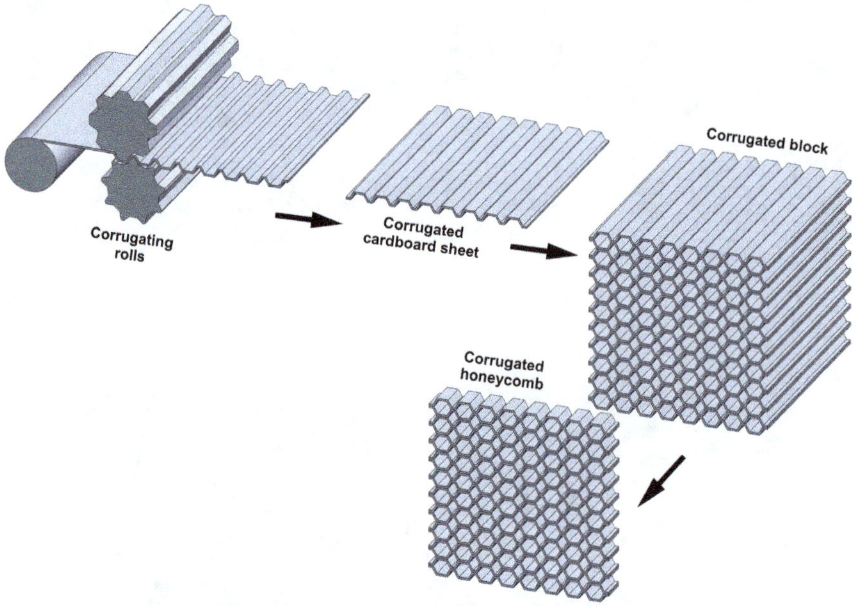

Fig. 17 Scheme of corrugated production process adapted from Bitzer (1997)

- mechanical properties of the structure without the frame are usually lower than for particleboards;
- it is necessary to develop specific fittings and edge connections to improve the mechanical integrity of the structure;
- edges lamination is very difficult;

- low resistance to parallel loads;
- low moisture resistance of the core;
- the board shows low resistance to highly localized external loads, as in case of impact;
- shear properties have different values if the tests are made along the length or across the face width (Aktay et al. 2008; Barboutis and Vassiliou 2005; Thoemen 2008).

6 Conclusion and Future Perspectives

Lightweight wood composites are in high demand for many different industries. A lightweight material can bring many advantages that appeal to manufacturers, designers and consumers. These include, lower raw material and transportation costs; new design possibilities, involving thicker elements; more rational use of wood, providing alternatives in case of wood shortages or high cost; easier transportation of the final product.

Different approaches have been adopted in the last decades to produce lightweight wood composites with varying density ranges and properties. The most commonly used products are lightweight fiberboards, particleboards and sandwich panels.

The biggest challenges faced by the manufacturers of these composites involve the reduction of mechanical performance that results from lowered density, the difficulties in edges connection/ lamination, and the low machinability. Further developments should provide strategies to overcome these limitations.

Acknowledgements This work was financially supported by: Project POCI-01-0145-FEDER-006939 (Laboratory for Process Engineering, Environment, Biotechnology and Energy—UID/EQU/00511/2013) funded by the European Regional Development Fund (ERDF), through COMPETE2020—Programa Operacional Competitividade e Internacionalização (POCI) and by national funds, through FCT—Fundação para a Ciência e a Tecnologia. Project NORTE-01-0145-FEDER-000005–LEPABE-2-ECO-INNOVATION, supported by North Portugal Regional Operational Programme (NORTE 2020), under the Portugal 2020 Partnership Agreement, through the European Regional Development Fund (ERDF).

Authors would like to thank Marcelo Oliveira, Department of Wood Engineering Polytechnic Institute of Viseu for helping with figures design.

References

Akbulut T, Ayrilmis N (2004) Influence of compression wood on physical and mechanical properties of medium density fiberboard. Wood Res 49:17–24

Aktay L, Johnson AF, Kröplin BH (2008) Numerical modelling of honeycomb core crush behaviour. Eng Fract Mech 75:2616–2630

Alessandro VD, Petrone G, Franco F, De Rosa S (2013) A review of the vibroacoustics of sandwich panels: models and experiments. J Sandw Struct Mater 15:541–582

Balducci F, Harper C, Meinlschmidt P, Dix B, Sanasi A (2008) Development of innovative particleboard panels. Drv Ind 59:131–136

Barboutis I, Vassiliou V (2005) Strength properties of lightweight paper honeycomb panels for the furniture. In: Proceedings of international scientific conference 10th anniversary of engineering design (Interior and furniture design). Sofia, pp 1–6

Barbu MC, Pieper O (2008) Light fibreboards for insulation purposes. In: Lightweight wood–based composites production, properties and usage. Proceedings of COST action E49 processes and performance of wood–based panels. Slovenia, pp 34–51

Barbu MC, Ludtke J, Thomen H, Welling J (2010) Innovative production of wood-based lightweight panels. Processing technologies for the forest and bio-based products industries. Kuchl, Austria, pp 115–122

Bitzer T (1997) Honeycomb technology. Springer, Netherlands

Borgin K (1958) Development of the particle board industry in europe. J South African For Assoc 32:56–71

Boruszewski P, Borysiuk P, Mamiński M, Czechowska J (2016) Mat Compression Measurements During Low-Density Particleboard Manufacturing. BioResources 11:6909–6919

Bull PH, Edgren F (2004) Compressive strength after impact of CFRP-foam core sandwich panels in marine applications. Compos Part B Eng 35:535–541

Cai Z (2012) Wood-based composite board. Wiley Encyclopedia of composites, pp 1–11

Carvalho LH, Costa CAV (1998) Modeling and simulation of the hot-pressing process in the production of medium density fiberboards (MDF). Chem Eng Commun 170:1–21

CEN/TS 16368, Lightweight Particleboards-Specifications

Chen T, Liu J, Wu Z et al (2016) Evaluating the effectiveness of complex fire-retardants on the fire properties of ultra-low density fiberboard. BioResour 11:1796–1807

Costa NA, Martins D, Pereira J et al (2013) C NMR Study of presence of uron structures in amino adhesives and relation with wood-based panels performance. J Appl Polym Sci 130:1–8

Davies JM (1993) Sandwich panels. Thin-Walled Struct 16:179–198

Davies JM (1rd) (2001) Lightweight sandwich construction. Blackwell Science Ldt, London

Denesi M (2012) Method for producing a particle-based element. US Patent 2012/032539 A1, 27 Dec 2012

Donaldson LA, Lomax TD (1989) Adhesive/fibre interaction in medium density fibreboard. Wood Sci Technol 23:371–380

Dunky M (1998) Urea-formaldehyde (UF) adhesive resins for wood. Int J Adhes Adhes 18:95–107

Dziurka D, Mirski R (2013) Lightweight boards from wood and rape straw particles. Drewno 56:19–31

EN 622-5, Fibreboards-Specifications-Part: 5 Requirements for dry process boards (MDF)

Ganev S, Gendron G, Cloutier A, Beauregard R (2005) Mechanical properties of MDF as a function of density and moiture content. Wood Fiber Sci 37:314–326

Kalaycioglu H, Deniz I, Hiziroglu S (2005) Some of the properties of particleboard made from paulownia. J Wood Sci 51:410–414

Kannan P, Biernacki JJ, Visco DP Jr (2007) A review of physical and kinetic models of thermal degradation of expanded polystyrene foam and their application to the lost foam casting process. J Anal Appl Pyrol 78:162–171

Karagiozova D, Nurick GN, Langdon GS, Chung Kim Yuen S, Chi Y, Bartle S (2009) Response of flexible sandwich-type panels to blast loading. Compos Sci Technol 69:754–763

Karlsson KF, TomasÅström B (1997) Manufacturing and applications of structural sandwich components. Compos Part A Appl Sci Manuf 28:97–111

Kawai S, Sasaki H, Nakaji M, Makiyama S, Morita S (1986) Physical properties of low-density. Wood Res 72:27–36

Kharazipour A, Bohn C (2012) Use of popcorn for timber and composite materials. US Patent 8,168,303 B2, 1 May 2012

Kollmann FFP, Kuenzi EW, Stamm AJ (1975) Principles of wood science and technology II wood based materials. Springer, Berlin

Lias H, Kasim J, Atiqah N, Johari N, Lyana I, Mokhtar M (2014) Influence of board density and particle sizes on the homogenous particleboard properties from kelempayan (Neolamarckia cadamba). Int J Latest Res Sci Technol 3:173–176

Magnus J (2010) Thermally expandable microspheres prepared via suspension polymerization-synthesis, characterization and application. Technical University of Denmark

Maloney TM (3 rd) (1989) Modern particle board & dry-process fiberboard manufacturing. Miller Freeman Publications, San Francisco

Matuana L, Heiden P (2004) Wood Composites. Encycl Polym Sci Technol 12:521–546

Meraghni F, Desrumaux F, Benzeggagh ML (1999) Mechanical behaviour of cellular core for structural sandwich panels. Compos—Part A Appl Sci Manuf 30:767–779

Nayeri MD, Tahir PM, Jawaid M, Ashaari Z, Abdullah LC, Bakar ES, Namvar F (2014) Medium density fibreboard made from Kenaf (Hibiscus cannabinus L.) stem: effect of thermo-mechanical refining and resin content. BioResour 9:2372–2381

Nemli G, Demirel S (2007) Relationship between the density profile and the technological properties of the particleboard composite. J Compos Mater 41:1793–1802

Nemli G, Ors Y, Kalaycioglu H (2005) The choosing of suitable decorative surface coating material types for interior end use applications of particleboard. Constr Build Mater 19:307–312

Nilsson E, Nilsson AC (2002) Prediction and measurement of some dynamic properties of sandwich structures with Honeycomb and foam cores. J Sound Vib 251:409–430

Peretolchin M (2009) Kaurit Light-the solution for lightweight furniture. In: International panel products symposium. Nantes, France, pp 139–143

Petras A, Sutcliffe MPF (1999) Failure mode maps for honeycomb sandwich panels. Compos Struct 44:237–252

Pflug J, Xinyu F, Vangrimde B, Verpoest I (2002) Development of a sandwich material with polypropylene/natural fibre skins and paper honeycomb core. In: Proceedings of the 10th European conference on composites materials. Belgium

Pishan S, Ghofrani M, Kermanian H (2014) Study on mechanical properties of lightweight panels made of honeycomb and polyurethane cores. Lignocellulose 3:59–68

Porter B (3rd) (2012) Carpentry and joinery. Routledge Taylor & Francis Group, New York

Rowell RM, Kawai S, Inoue M (1995) Dimensionally stabilized, very low density fiberboard. Wood Fiber Sci 27:428–436

Schubel PM, Luo J, Daniel IM (2005) Low velocity impact behavior of composite sandwich panels. Compos Part A Appl Sci Manuf 36:1389–1396

Shalbafan A (2013) Investigation of foam materials to be used in lightweight wood-based composites. University of Hamburg

Shalbafan A, Luedtke J, Welling J, Thoemen H (2012) Comparison of foam core materials in innovative lightweight wood-based panels. Eur J Wood Wood Prod 70:287–292

Shalbafan A, Tackmann O, Welling J (2016a) Using of expandable fillers to produce low density particleboard. Eur J Wood Wood Prod 74:15–22

Shalbafan A, Welling J, Luedtke J (2016b) Effect of processing parameters on mechanical properties of lightweight foam core sandwich panels. Wood Mater Sci Eng 7:69–75

Stoeckel F, Konnerth J, Gindl-Altmutter W (2013) Mechanical properties of adhesives for bonding wood—A review. Int J Adhes Adhes 45:32–41

Suchsland O, Woodson GE (1986) Fiberboard manufacturing Practices In the United States. USDA Forest Service, Virgina

Suhaily SS, Jawaid M, Abdul Khalil HPS, Mohamed AR, Ibrahim F (2012) A review of oil palm biocomposites for furniture design and applications: potential and challenges. BioResour 7:4400–4423

Thoemen H (2008) Lightweight panels for the furniture industry: some recent developments. In: Lightweight wood–based composites production, properties and usage, Proceedings of COST action E49 processes and performance of wood–based panels. Slovenia

Thoemen H, Irle M, Milan S (2010) Wood-based panels an introduction for specialists. Brunel University Press, London, England

Trischler J, Sandberg D, Thörnqvist T (2014) Evaluating the competition of lignocellulose raw materials for their use in particleboard production, thermal energy recovery, and pulp-and papermaking. BioResour 9:6591–6613

Wang D (2009) Impact behavior and energy absorption of paper honeycomb sandwich panels. Int J Impact Eng 36:110–114

Wang X, Wang Y (2016) Static analysis of sandwich panels with non-homogeneous soft-cores by novel weak form quadrature element method. Compos Struct 146:207–220

Wittmann O, Wolf F (2014) Wood-based materials. Ullmann´s Encyclopedia of Industrial Chemistry, Wiley, New York

Xing C, Riedl B, Cloutier A (2004) Measurement of urea-formaldehyde resin distribution as a function of MDF fiber size by laser scanning microscopy. Wood Sci Technol 37:495–507

Xu J, Sugawara R, Widyorini R (2004) Manufacture and properties of low-density binderless particleboard from kenaf core. J Wood Sci 50:62–67

Zhang H, Lu X (2015) Elastic properties of the extruded tubular particleboard. Wood Mater Sci Eng 10:185–188

www.wbpionline.com. Accessed 20 Sep 2016

Design and Fabrication of Kenaf Fibre Reinforced Polymer Composites for Portable Laptop Table

S. M. Sapuan, K. R. Purushothman, M. L. Sanyang and M. R. Mansor

Abstract Natural fibre is an emerging material that can help to reduce the dependency on non-renewable resources and sustains the ecological balance. This research focuses on the design and fabrication of a laptop table using kenaf fibre reinforced polymer composite. A composite portable laptop table was developed using standard design method and the manufacturing method used was hand lay-up. Market investigation, product design specification, conceptual design, detail design followed by fabrication were performed accordingly throughout this project. Stress distribution and deformation of the end product was tested using simulation software. This research has enabled to produce a strong, stable and aesthetic furniture which is made from composite material.

Keywords Kenaf · Biocomposites · Conceptual design · Hand lay-up Laptop table

Contents

S. M. Sapuan (✉) · K. R. Purushothman
Department of Mechanical and Manufacturing Engineering, Universiti Putra Malaysia, 43400 UPM Serdang, Selangor, Malaysia
e-mail: sapuan@upm.edu.my

S. M. Sapuan · M. L. Sanyang
Laboratory of Biocomposite Technology, Institute of Tropical Forestry and Forest Products, Universiti Putra Malaysia, 43400 Upm Serdang, Selangor, Malaysia

M. R.Mansor
Faculty of Mechanical Engineering, Universiti Teknikal Malaysia Melaka, 76100 Durian Tunggal, Melaka, Malaysia

S. Kalia (ed.), *Lignocellulosic Composite Materials*, Springer Series on Polymer and Composite Materials, https://doi.org/10.1007/978-3-319-68696-7_8

1 Background

The urge to maintain a sustainable and environmental-friendly raw resource came to light when scientists predicted that the fossil fuels primarily the petroleum can only last for another 20–30 years according to the current demand and growth of the world. Other natural resources such as iron, wood and coal also face the same problem which is depleting at an alarming rate. The need for a renewable resource alleviates following to global warming issues (results from carbon footprints) where the condition of the existing environment worsens. This occurrence awakens the conscience of many people that includes the government, non-government bodies and also the public regarding the dependency towards non-renewable resources and their effects towards the environment. The dilemma choosing the correct substituent to replace the petroleum-based thermoplastic materials which have excelled in

many factors have led both engineers and researchers to squeeze their brain in order to come up with a solution. It is then discovered the answer lays surrounding us; the natural fibre crops which are the products from any parts of the plant. These natural fibre crops actually have been cultivated ever since the prehistory era to make certain essential tools such as ropes and textiles. It can be classified into three groups namely plants, animals or minerals.

With the aid of engineering research and sustainable technology, natural fibres are able to produce lighter composites and lower cost compared to glass fibres and other synthetic polymers in various engineering needs. Natural fibres are biodegradable hence reverts the issue of limiting landfills besides does not release toxic fumes when burned. This biorenewable natural cellulosic polymer has remarkable mechanical characteristics such as relative high specific strength, less abrasive, favourable fibre aspect ratio, as well as good thermal, electric and acoustic insulating value (Asilah 2011).

However, there are certain disadvantages of natural fibres when it is used as reinforcement in polymer. The properties of resulting composites diminish when the dispersion of fibre is not uniform due to hydrophilic and intrinsic polar nature of the plant fibre combined with the nonpolar attributes of the most polymer matrix. On the other hand, another vital disadvantage of plant fibre is its hydrophilicity. It absorbs high moisture content and causes to swell and void at the interface. This impairs the dimensional stability of the composite. The properties of plant fibre vary according to cultivating environment whereby it may influence the shape, size and strength of its fibres. This directly alters the mechanical property of resulting natural fibre reinforced polymer matrix.

This overview tends to focus on kenaf fibre, one of the recognized constituents of natural fibre reinforced plastic composites throughout South East Asia region. It is a traditional crop came after wood and bamboo dominated the industrial purpose. It grows in warm and temperate region and related to hibiscus, cotton and okra. Kenaf is widely used in pulp and paper industries as a replacement for wood and currently serving as non-woven mats in the automotive industries. This natural fibre is discovered to have exceptional flexural strength combined with tensile strength that caught the attention for a vast range of moulded, extruded and non-woven products.

Manufacturing process selection is an important process of choosing the most suitable process to develop a product. The necessity of manufacturing process selection at an infant stage of producing a product has been addressed by many writers. In general, the selection process is carried out at the detail design stage. It leads to late spotting of errors at the final stage of the whole manufacturing process. At that point, the errors discovered in the manufacturing process will be lamented for redesigning and remanufacturing. It is also tough task to select a suitable fabrication process at early stage due to varying factors such as empirical knowledge and previous experience.

The primary aim of the current study is to carry out total design process as well as fabricate a laptop table from kenaf fibre reinforced unsaturated polyester composite.

2 Natural Fibre

The demand for natural fibre keeps growing as the environmental issues regarding synthetic polymer worsens. Natural fibre can be described as the product obtained from plants and animals that can be shaped into filaments, thread or even rope. They exist like hair-like materials which are uninterrupted filaments or in distinct elongated pieces. They can be used as one of the constituent materials in producing a component or a product. The main reason for natural fibre are much preferable compared to synthetic fibre such as glass is because they have wide ranges fibres and easily available besides being environmentally friendly. Natural fibres are also famous for their excellent properties such as low density, high specific strength and effectual costing (Asilah 2011; Faruk et al. 2012; Khalil et al. 2012; Sanyang et al. 2016).

Cultivated since prehistoric era, the natural fibre used then was flax and wool. Several vegetable fibres such as hemp were cultivated vastly and expanded all over Southeast Asia back at 4500 BC. In eighteenth and nineteenth century, Industrial Revolution forced the invention of machines processing natural fibre which produced an upsurge demand in natural fibre. In particular, natural fibre composites are much preferred in automotive industry for their minimal cost and low density. Glass fibre has a density of 2.6 g/cm^3 and comes in price between \$1.30 and \$2.00/kg. On the other hand, flax costs only \$0.22 to \$1.10/kg and has remarkable low density of 1.5 g/cm^3 (Jonn et al. 2000).

Unlike most synthetic fibres, all natural fibres are non-thermoplastic which means they do not soften when heated or flamed. At temperatures under the point they disintegrate, they show less response to dry heat, and there is no shrinkage or greater extensibility upon heating. Natural fibre does not become brittle if cooled to below freezing. Polypropylene has extensive melt flow index which is normally used to help in matrix adhesion and to obtain appropriate wetting of the fibres. In a study, samples with 40% fibre content of kenaf, coir, sisal, hemp and jute were manufactured. They were then piloted for tensile and impact test in order to differentiate their properties with glass fibre composites. The tensile strength matches the strength of glass fibre excluding coir. However, hemp obtained similar flexural strength as glass fibre. Furthermore, it was proven that kenaf fibre with greater fibre weight fraction improves ultimate strength, tensile modulus, and impact strength compared to glass mat composites. This study proves that natural fibre was able to outstand glass fibre in industries that do not involve heavy load (Wambua et al. 2003). A comparison between natural fibre and glass fibre (Table 1).

Natural fibre with these properties is very useful in reducing the greenhouse effect as it does not pollute the environment. Moreover, the processing of natural fibre is not as complex as processing synthetic fibre hence reducing the dependency on non-renewable sources such as petroleum. More extensive research will be needed on the relative environmental impacts of natural fibre composites and conventional composites in order to understand and extract further the better side of natural fibre (Joshi et al. 2003). Hence, the current work focuses on Kenaf fibre due to its rapid growth, high yield and excellent fibre quality.

Table 1 Comparison between natural fibre and glass fibre (Wambua et al. 2003)

	Natural fibres	Glass fibres
Density	Low	Double than that of natural fibre
Cost	Low	Comparatively higher than natural fibre
Renewability	Yes	No
Recyclability	Yes	No
Energy consumption	Low	High
Distribution	Wide	Wide
CO_2 neutral	Yes	No
Abrasion to machines	No	Yes
Health risk when inhaled	No	Yes
Disposal	Biodegradable	Non-biodegradable

2.1 *Kenaf Fibre*

Kenaf fibre with its scientific name *Hibiscus cannabinus* L. is a short day, an annual herbaceous plant that came from Africa (Akil et al. 2009). It has been then cultivated all around the world since it originates from Malvaceae, a species that is well known for economic and horticultural importance. Kenaf has a high growth rate, growing from 12–18 feet in about 4 months and yields 6–10 tonnes of dry weight per acre per year. Kenaf belongs to bast fibres category which consists of inner core fibre (60–75%) that are used for low-quality pulp. Meanwhile, the outer bast fibre (25–40%) produces better quality pulp (Akil et al. 2009).

Kenaf's main system can grow as deep as two metres in the soil. Similar to jute, kenaf exhibits adventitious roots on the base of the stems during water logging. That enables it to survive during harsh conditions. Apart from that, kenaf is primarily used for preparing ropes, canvas and sacking. Being a lignocellulosic fibre, kenaf is preferred for its economical and ecological benefits. Besides, it is well known for its non-abrasiveness during manufacturing, biodegradability, low density and greater specific mechanical properties (Bernard et al. 2011). In a study, kenaf was found to acquire carbon dioxide at an exorbitant rate as well as absorbing nitrogen and phosphorus from the soil (Rassmann et al. 2011).

2.2 *Characterists and Properties of Kenaf Fibre*

The composition of kenaf comprises of 75% cellulose and 15% lignin, indicates that it is able to degrade naturally (Karnani et al. 1997). It can be further classified as bast and core comprising 35 and 65%, respectively. Each fibre has its own distinctive usage, enabling zero wastage. Brief information on the chemical composition of different fractions of kenaf fibres is shown in Table 2.

Table 2 Chemical composition of different fractions of kenaf fibres (Nishino et al. 2003) [Waiting for permission]

	Kenaf whole (core + bast)	Kenaf core	Kenaf bast	Softwood	Hardwood
Extractive, %	6.4	4.7	5.5	0.2–8.5	0.1–7.7
Holocellulose, %	87.7	87.2	86.8	60–80	71.89
α-Cellulose, %	53.8	49.0	55.0	30–60	31–64
Lignin, %	21.2	19.2	14.7	21–37	14–34
Ash, %	4.0	1.9	5.4	<1	<1

3 Matrix

When fabricating a composite product, the adhesion of fibre-matrix is very crucial. In a particular product, matrix plays a big role in shifting the load to the stiff fibres through shear stress at the interface. To achieve this unique response, the bonding between polymeric matrix and fibre must be better. Otherwise, weak adhesion at the interface can defer the good qualities of the product, leaving it susceptible to environmental attacks as well as decreasing the life shell (Nishino et al. 2003).

Natural fibres are primarily made up of cellulose, where the base unit, anhydrous d-glucose, contains three hydroxyl (-OH) groups. As a result, these groups form intramolecular and intermolecular bonds, making the nature of plant fibres to be hydrophilic. Consequently, the adhesion ability of plant fibres generally reduced. To overcome this problem, many experiments have been conducted whereby alkaline solution is used to regenerate the damaged cellulose and to disintegrate the unused microscopic pits and cracks on the fibre, providing a better surface adhesion. On the other hand, the usage of coupling agents to repair the surface adhesion of fibre was discovered. It was designed to react with the hydroxyl group of the cellulose and to interact with the functional group of the matrix through distributing the stress transfer between the fibre and matrix (George et al. 2001).

3.1 *Biodegradable Polymer*

In common manufacturing process, the use of polymers such as polystyrene (PS) and polyvinyl chloride (PVC) has been used for ages. With the aid of research and technical development, the properties of regular polymer have been upgraded tremendously especially for short-term use such as party forks and spoon, plastic bags, plastic bottles and so on. Unfortunately, these polymers are too tough to be degraded in a matter of time. The products made from conventional polymers takes up majority of space in most landfills, provoking insufficiency in land space as well as health issues caused by the incineration of these products. Often, these products are discarded with food stains and residues hence making the degradation process becomes more complex (Vroman and Tighzert 2009).

Following to that concern, biodegradable polymer was brought to light. First introduced in 1980s, the main objective of this invention is to be degraded naturally by mean of microorganism. Normally, polymers are non-metallic compounds, solid and high in molecular weight. They are made up of reiterating macromolecules with changing characteristics where the purpose of the materials and mode of degradation depends highly on composition and processing method that is used (Kolybaba et al. 2003).

These polymers now can be divided into two groups namely natural and synthetic (Ikada and Tsuji 2000). The sources for natural polymers are abundant while the sources for synthetic polymers came from non-renewable petroleum. The usage of biodegradable polymers can be seen in the manufacturing process of products such as cling wrap, flushable sanitary products, sheet and non-woven packaging.

4 Factors Controlling Performance of Fibre Reinforced Composites

There are several factors controlling the performance of fibre-reinforced composites such as thermal stability and fibre hydrophilic nature.

4.1 Thermal Stability

To prevent the damage of surface properties, the manufacturing of natural fibre composite products must not be at high temperature (Keener and Stuart 2004). The main concern for the avoidance is because it can cause fibre degradation and volatile emissions that can affect the characteristics of the composite when exposed at high temperature. The ideal processing temperature is set to be around 200 °C even though the temperature can be increased but in short intervals. This causes a drawback that prohibits the processing of conventional thermoplastic such as polyethylene (PE), polyvinyl chloride (PVC), polypropylene (PP) and polystyrene (PS) (Rowell et al. 1999). However, Shibata discovered that the tensile strength of kenaf fibre decreased when exposed at 180 °C for 60 min. From their research, they concluded that fabrication of kenaf fibre composite must be below 200 °C (Shibata et al. 2006).

4.2 Fibre Hydrophilic Nature

The manufacturing process of kenaf fibre reinforced polymer composite became intricate because of the hydrophilic nature of both polymer matrix and natural fibre.

Even though both absorb moisture, natural fibre tends to absorb more moisture compared to the synthetic polymer. This causes instability in the structure of the composite product (Abdullah et al. 2007). The immanently polar and hydrophilic nature of lignocellulosic fibres as well as the nonpolar properties of most polymer results in non-uniform dispersion of fibres in a matrix, eventually causing distortion in properties of the resulting composite (Liu et al. 2007).

In order to overcome this problem, many research and studies have been carried out. To improve the surface properties and performance of composites, chemical preparations such as delignification, dewaxing, acetylation and chemical grating have been conducted (Sahed and Jog 1999). Alkaline treatment is used to improve the surface roughness of fibre to provide better mechanical interlocking and also to raise the amount of cellulose in order to facilitate more reaction sites. Alkaline treatment is able to upgrade fibre surface area in an experiment conducted to investigate the fibre-matrix interphase physicochemical interactions of a natural fibre reinforced plastic (Gonzaleza et al. 1999).

Another new eco-friendly technology has been developed named plasma treatment where it is able to change the surface properties of the natural fibre without affecting their bulk properties. In another separate study, the researcher was able to utilize SF6 plasma treatment to refine the hydrophobic property of silk fibre. The difference in hydrophobic property of the treated and untreated specimen was obvious (Chaivan et al. 2005).

5 Role of Natural Fibre in Global Industries

Malaysia is the ninth largest wood orientated commodities exporter in the world (www.miti.gov). Delivering up to 160 countries, the total value of the trade during the year 2011 alone exceeds RM 6.2 billion. Even though wood resources are still abundant in Malaysia, rampant illegal wood logging slowly reducing the advantage behold by National Timber Industry Policy (NATIP). At the current rate, excessive wood harvesting can cause a severe reduction in the area of forest up to 30% compared to the 19% value estimated by National Forestry Council (NFC). The consequence of this reduction does not only endanger the rich flora and fauna but also threatens the life of indigenous people besides risking pollution issues and global warming. Hence, in cooperating natural fibre in wood-based appliances have great potentials in dropping a number of trees logged. Natural fibres which can be commercially grown have faster growth and high yield. The main reason why natural fibre can be the suitable candidate to reduce wood consumption is because they are renewable, biodegradable and carbon neutral. The unique eco-balance of nature can be preserved and conserved in a much efficient way if natural fibre is used progressively (www.wildfibres.co.uk).

As for now, natural fibre has already begun contributing in automotive industry as a replacement for fibreglass in reinforcing components. Plant fibres are preferred as they are light and mechanically strong besides being easy for moulding process.

They cost less and reduce the car weight in a considerable amount. Car seats can now use coir as cushioning pads (Nor et al. 2010) and kenaf fibre composite polymer in parking brake lever (Mansor et al. 2014). On the other hand, aerospace industry took a leap in revolutionizing aircraft panels with natural fibre. Flax and ramie and hemp are being used to create internal aircraft parts especially cabin utilities (Nirmal et al. 2015). Weighing the developments in other major industries, natural fibre could play a larger role in creating sustainable furniture products that can perform as better as wood and aesthetically pleasing. In future run, more appliances in our daily life can be improvised with the use of natural fibre without harming the environmental balance.

6 Natural Fibre Composites Fabrication Techniques

There are various methods of fabrication of natural fibre composites namely hand lay-up, pultrusion, filament winding, compression moulding and resin transfer moulding.

6.1 Hand Lay-up

Hand lay-up or more known as wet lay-up can be regarded as the most established way to manufacture composite products. This method can produce products with amazing surface finishing with low tooling expense; however, it is only on the side where it touches the mould. It also needs exquisite labour skill as well as certain level of automated tool. This process requires an open mould where the inner surface will be leveled with the composite material. Some examples of parts produced using this method are boat, swimming pool and car bodies where the fibres will be applied on the inner surface of the mould. With the aid of releasing agent, the composite will not be stick on the mould, easing the process during removal. A gel coating is then applied for finishing touch and protection purpose (Mazumdar 2002). A study was conducted by Sapuan et al. (2006) to investigate mechanical properties of woven banana fibre reinforced epoxy composite. Hand lay-up technique was carried out on a rectangular plywood mould and pressed evenly for 24 h for curing before being tested.

6.2 Pultrusion

Pultrusion came from the combination of the word ‘pull’ and ‘extrusion’. It is used to fabricate natural fibre composite products by providing continuous fibrerovings on a creel and pulling it passing through a resin bath. This resin impregnation system normally contains epoxy or unsaturated polyester covers the fibre thoroughly and transfers the composite to a preform by the means of gripper. The preform determines the shape of the intended product before being heated at different temperatures for the curing process. Pultrusion can be used to produce solid or hollow objects of constant-profile sections (Mayer and Hancox 1993).

6.3 Filament Winding

Filament winding is similar to pultrusion in terms of providing continuous fibres (rovings) or tapes through a resin bath. This method differs from pultrusion as it is an automated process as it involves continuously rotating mandrel that wounds the fibre in helical, hoop and polar direction. Lilholt et al. used filament winding to produce a product from flax and hemp fibre reinforced polymer composite. The composite was wound circling the mandrel till reaching the intended length. The product can be removed upon completion. To ensure the placement of tow, layer thickness, and resin content, fibre tensioner comes in handy (Hyer 2009).

6.4 Compression Moulding

For mass production industries, compression moulding is preferred as it is fully automated and more efficient especially in automotive industries (Mazumdar 2002). It can be used to create flat or slightly curved parts using a heated die. Thermoplastics and thermosets undergo different steps in compression moulding. Thermoplastics require high heat and pressure because of its greater viscosity. The process begins when two open moulds containing the material is placed between heating plates. After being heated, required pressurized gas is supplied into the mould for it to take shape. The product will be removed after cooling (Biron 2012).

6.5 Resin Transfer Moulding

Also regarded as liquid transfer moulding, dry perform (composite) is placed in a closed mould that has the shape of the final product. Following it, the thermosetting resin will be injected into the closed mould and pre-impregnated onto the exterior to reform. Using a dispensing equipment, resin hardener, catalyst, colour and filler are injected prior to curing. Resin transfer moulding is preferred in automotive and aerospace industries because of its remarkable surface finish on both sides besides able to produce complex and large surface area objects. Mazumdar (Mazumdar 2002) found that this method was able to produce objects such as doors, hockey sticks, helmets and automotive panels.

7 Methodology

The possible sequence of procedures for overall project from the beginning is shown by a flowchart in Fig. 1. Details are further discussed in the following sections (Fig. 1).

7.1 Product Design Specification

Before starting the manufacturing process, a product design specification was needed to ensure the necessity of the product meets the project maturity. A basic design model chart is useful to guide the entire fabrication process as in Fig. 2.

The product design specification (PDS) was essential for the project progress as it has the necessary information regarding the product's result. The boundary conditions also must be realistic and prevent the forecast of the outcome. In addition, all areas related to the product foreword process should involve in its generation and only can be changed if entire related agrees. The PDS has been presented in written document as shown in Fig. 3.

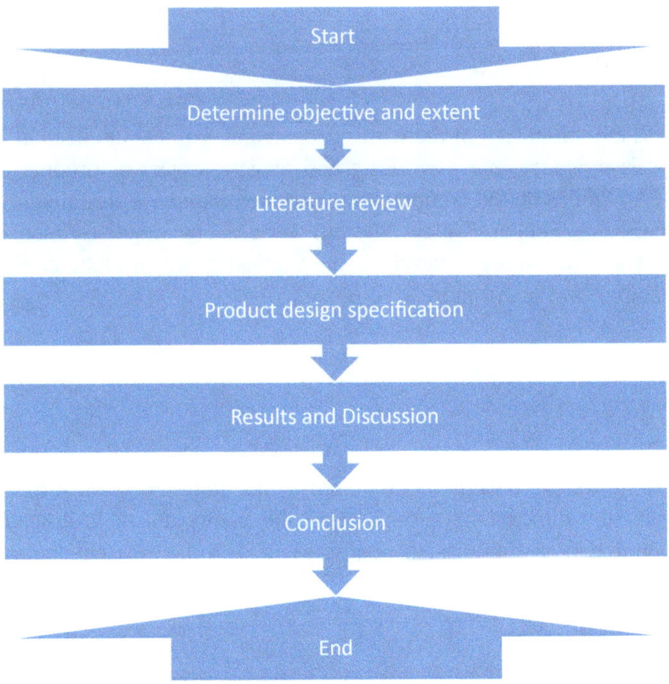

Fig. 1 Process flowchart for the entire project

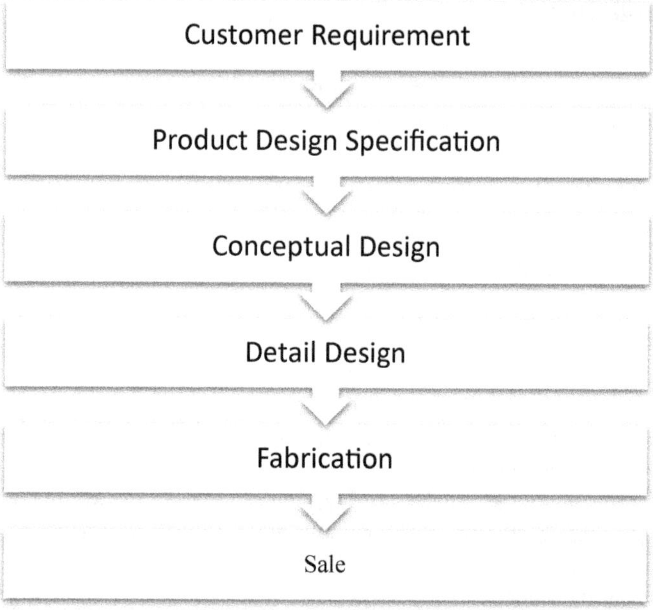

Fig. 2 Basic design model

7.2 Conceptual Design

This phase is the initial step of the whole project and requires intellectual process of shaping a research idea into a working and realistic object or product. This consumes time and relies on the complexity of product. Sapuan and Maleque (2005), in his research, have compared nine different designs of composite telephone stand and selected the best design based on the highest score in characteristics, ergonomics, stability and many more.

7.3 Detail Design

A detailed engineering design was required to facilitate the fabrication process as the details on the dimensions and shape of the composite table is available. An organized engineering drawing has been presented in the next chapter.

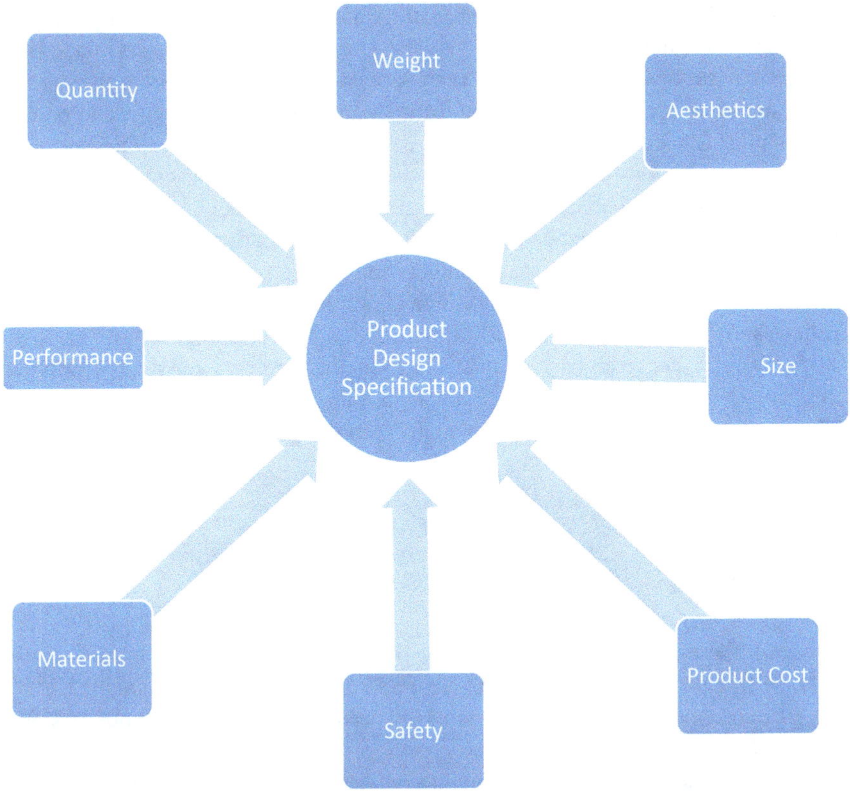

Fig. 3 Eight elements of Product design specification considered for composite laptop table

8 Fabrication of Composite Table

Fabrication of composite laptop table was carried out using hand lay-up technique. It is an open moulding form where it involves only one surface which follows the designated shape. A detailed drawing of the mould has been also prepared prior to the fabrication process. Mechanical fasteners are used to join the components of the table together and to ensure they hold together firmly to provide a steady and durable composite table.

8.1 Market Investigation

In the standard engineering design approach, all important information regarding the intended product design was gathered through market investigation. Through this step, various solutions and constraints can be identified that other designers may have included in their product (Salit 2014).

Working on a laptop is usually exhausting and painful. This is even prominent for anyone with health and mobility issues. Commonly, laptop users place their laptops on a solid and wide surface such as desk and tables. Prolong use on this surface cause neck and backbone pain. Moreover, laptops are being placed on laps as the name 'Laptop' derived; users find difficulty upon such use. The heat generated from the laptops brings discomfort and sores the thigh muscle. To overcome this, portable laptop table are widely used as they can be brought anywhere besides improving ergonomics instead of painfully sitting, overheating and cramping to use a laptop. It eliminates boredom and fatigue, improving productivity and comfort.

Based on market data analysis, portable laptop table are made from aluminium, plastics and wood. The average dimension of a portable laptop table is 500 × 300 mm and weighs less than 8 kg. In this project, kenaf fibre polymer composite is suitable to be used to fabricate a lightweight portable laptop table for its lightweight benefits and environmentally friendly properties such as recyclable and renewable.

8.2 Product Design Specifications (PDS)

In general, there are 32 elements of PDS specified in a design of a product (Pugh 1991). In this project, eight elements have been chosen to be implemented in the design of composite laptop table. Firstly, the performance of the composite laptop table must be comparable to with the conventional laptop tables which are made from wood or plastic. These tables are not meant to withstand extreme loads since the laptops are generally light. Performance can be seen when the table remains firm and stable after the laptop is placed on it. Secondly, the weight and size are two elements that must be included in this project. Laptop table does not exceed 500 mm in length and 800 mm in width. Following that, the weight is also not heavy measuring less than 8 kg. These elements are one of the key factors when users consider purchasing a laptop table.

On the other hand, material selection and product cost criteria are included in this project. Since this project encompasses on kenaf fibre reinforced polymer composite, the laptop table must be fabricated using this material. Combining parts made from natural fibre and wood creates a unique yet firm and strong product. The cost of the product must be considered since overpriced products are not marketable. Taking into consideration that natural fibre is slightly expensive than wood, the total fabrication cost for kenaf reinforced polymer composite laptop table must be regulated to excessive expenditure.

In addition, safety element was chosen because laptop table is used in close proximity. The distance body and table are close, so if the table topples, it should not harm or injure the user. An engineer or designer is responsible to design products that are safe to use, complying the code of ethics. To be marketable, the product must have attractable look. Aesthetic criteria behold the factor of a successful product. Besides being unique, the design and finishing of the product must be innovative and appealing. Lastly, to fabricate and assemble all the parts, the method of manufacturing must be considered. This element influences the outcome of the design and its manufacturability.

Detailed explanation of product design specification element is provided below;

(1) *Performance*

The laptop table should be able to withstand the weight of a conventional laptop as well as the force exerted by the hand on it. This combined force was a key factor in selecting the suitable design. To simulate this, the design of the portable laptop table was analyzed through Abaqus FEA (Finite Element Analysis)software. The stress and deformation of the design can be observed.

(2) *Weight*

The weight of the portable table should not exceed 8 kg and should be kept as light as possible without compromising the strength and quality of the product. Less weight means the user will be able to carry the product anywhere with less ease.

(3) *Product cost*

The total expenses of this product must be reasonable as too low expenditure constricts the use of quality components. Too high expenditure can also decrease the marketability of this product since average costs of a portable laptop table does not exceed RM 500.

(4) *Materials*

The raw material needed for this project is kenaf fibre and unsaturated polyester resin for the table platform. Meanwhile, plywood and silicon glue were needed to create the mould for the portable table components. The criteria of other components include easy fabrication, does not break or rot when exposed to the environment.

(5) *Safety*

The portable table does have smooth and regular surface. Any sharp edges trimmed to avoid unwanted injuries. The joints concealed carefully so that it does not protrude during usage.

(6) *Aesthetics*

The portable laptop table has decent finishing and looks presentable. Attractive presentation able to attract customers and increases its marketability.

(7) *Size*

The maximum length of the table is 500 mm and width did not exceed 800 mm.

(8) *Manufacturing Process*

The manufacturing of the composite portable laptop table was based on customer orders. The selected method of fabrication is using hand lay-up, the quantity produced be low because it depends heavily on labour skills and time.

8.3 Conceptual Design

In order to select a final design, design evaluation must be made based on several designs. By doing this, a final design can be obtained that satisfies all the design criteria. This enables the end product to have a quality and satisfying characteristics. There are many ways to do concept evaluation to select the best design.

Sapuan and Maleque (2005) and Sapuan et al. (2007) carried out their design concept evaluation by using the weighted objective method. In their project which was to fabricate telephone stand from banana pseudostem woven fibre reinforced epoxy composite and multipurpose table accordingly, they have successfully chosen the best design by comparing with several other design concepts. In this method, all vital elements of the design such as ergonomics, manufacturability and many more are evaluated and ranked appropriately. Each design will be given a weighted score and it was multiplied with the respective utility score which will be given by the designer. The design with highest total score will be selected for the final design.

Another method that was used to evaluate design concepts is by using Pugh evaluation method (Pugh 1991). In this method, several nominee designs were compared with a current design labeled as Datum. A positive score '+' was given if the concept performs better than Datum in the design criteria. Otherwise, negative score '−' was given if it yields lower than the Datum. If it yields equal performance as Datum, 'S' was given. Finally, the design with utmost difference between '+' and '−' was finalized.

In this project, Pugh evaluation method was used to choose the best design concept. Seven design ideas have been generated and by comparing to the base model, only one final design was selected through the evaluation method.

8.4 Concept Generation

The following design ideas are sketched and the following are its characteristics.

(a) *First Concept*

This laptop table was designed to have good stability and strength. Wider base area and short legs make this design to be resilient to various loads. Besides being easy to manufacture, this table has an added stand for mouse pad. But it was not portable since it has fixed legs and laptop pad. This design only offered a plane horizontal stand for laptop and cannot be adjusted to different viewing angle. The first concept is shown in Fig. 4.

(b) *Second Concept*

Second design concept shows a better performance in terms of gapped main-frame as it offered better ventilating advantage for the laptop. This design also possessed good strength that comes from its frame. However, this table was not portable and heavier. The second concept is shown in Fig. 5.

(c) *Third Concept*

Third concept was designed to have a fixed frame and a base. This was the simplest design of all and needs the minimum ease of manufacture. This design does not have additional space for mouse pad and not portable. The performance of this stand was the least of all. The third concept is depicted in Fig. 6.

(d) *Fourth Concept*

The fourth concept of the design process has yielded in a portable manner where the legs of the table can be folded. Besides, the performance of this design was improved by the adjustable laptop platform angle. The wide base area offers good

Fig. 4 First concept

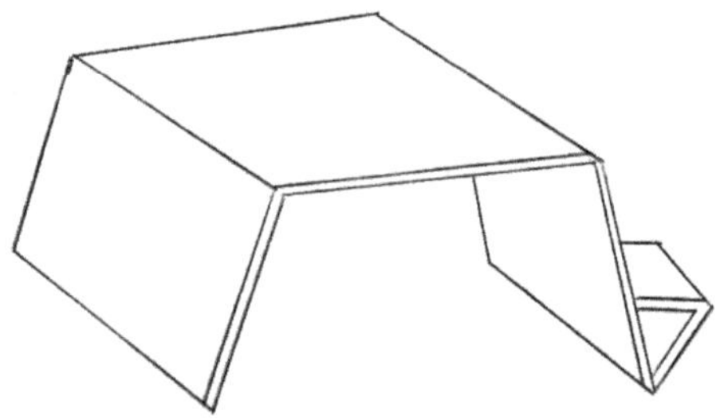

Fig. 5 Second concept

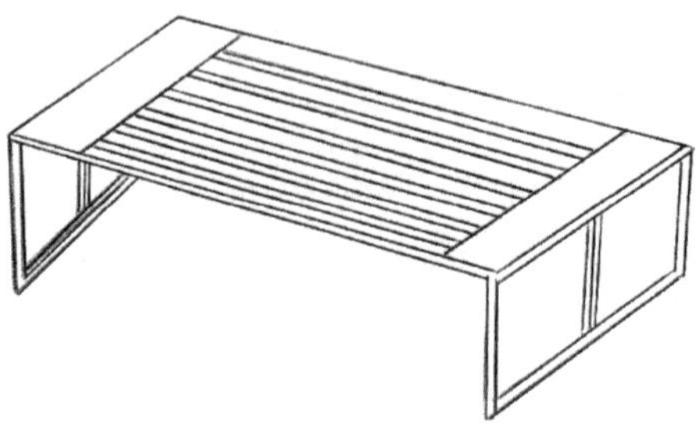

Fig. 6 Third concept

stability and the frame of the table provides good strength. It was easy to be fabricated within the reasonable weight specification. The fourth concept is shown in Fig. 7.

(e) *Fifth Concept*

The fifth design concept of the laptop table shows a portable product where it has holders on its sides so that the table can be lifted easily. The legs are wide and have a good stability which is foldable. The slanting leg was a big challenge for fabrication besides the cost of the manufacture was high since special hinge was required for the fitting process. The fifth concept is shown in Fig. 8.

(f) *Sixth Concept*

This design has good performance criteria since the legs and laptop platform was adjusted in various angle and height decision. It also weighs less as the legs most of the part consist of hinges. It was also portable and can be brought anywhere. The drawbacks are the compromised stability and high cost of manufacture. The sixth concept is shown in Fig. 9.

Fig. 7 Fourth concept

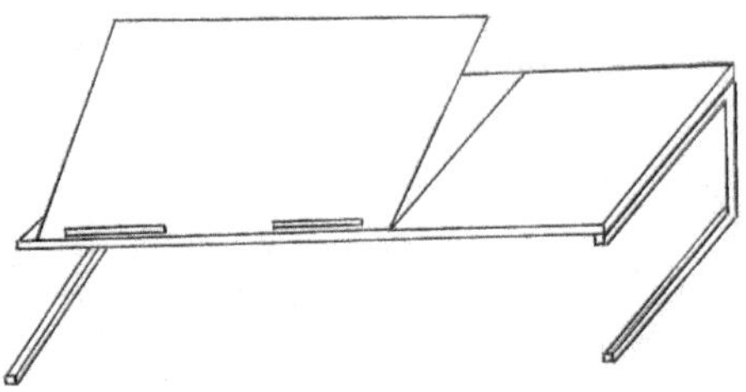

Fig. 8 Fifth concept

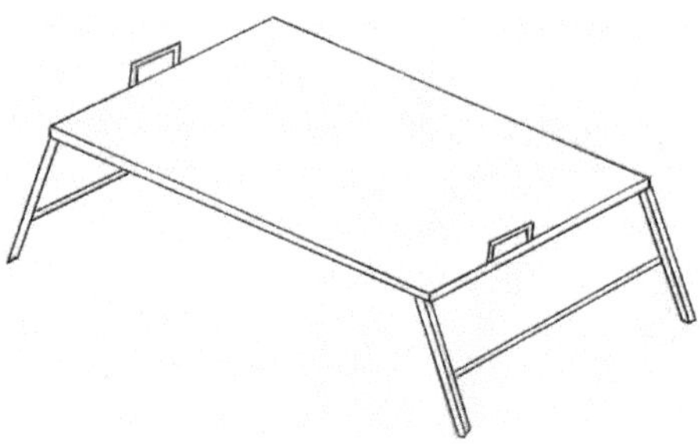

Fig. 9 Sixth concept

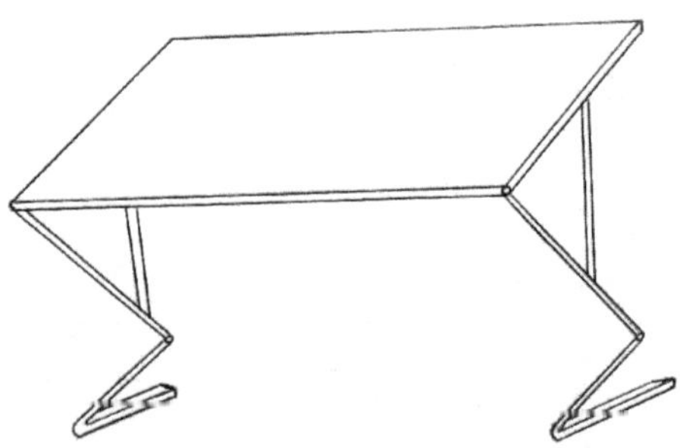

(g) *Seventh Concept*

The seventh concept shows a laptop stand with three legs and a single platform. This design offered good stability and has unique artistic value. It also has fairly good strength and performance. The manufacturability was difficult since it involves tight angle of operation. The cost to produce this table was accordingly high. The seventh concept is illustrated in Fig. 10.

8.5 Design Evaluation

By comparing the model design, seven design ideas were produced. Creative and innovative variations have been made in order to add value to the product besides fabricated using a natural fibre. In this step, one final design has to be selected for this project. Table 3 shows the selection of best design concept using Pugh evaluation method.

8.6 Detail Design

After selecting the final design concept, which was the fourth design concept, detail design was carried out to view the complete structure of the intended product, which are the dimensions, weight and structure of the laptop table. Computer-aided drawing software; CATIA (Computer Aided Three-dimensional Interactive Application) was used to design a three-dimensional view of the final product and its assembly (refer Appendix A). In order to identify the stress analysis when load is placed on the table, Abaqus FEA (Finite Element Analysis) simulation software was used after the completing the fabrication process.

8.7 Specification of the Table

To fabricate this design concept, a mould was needed to fabricate the composite material part, which was the laptop and mouse platform. Kenaf fibre was reinforced

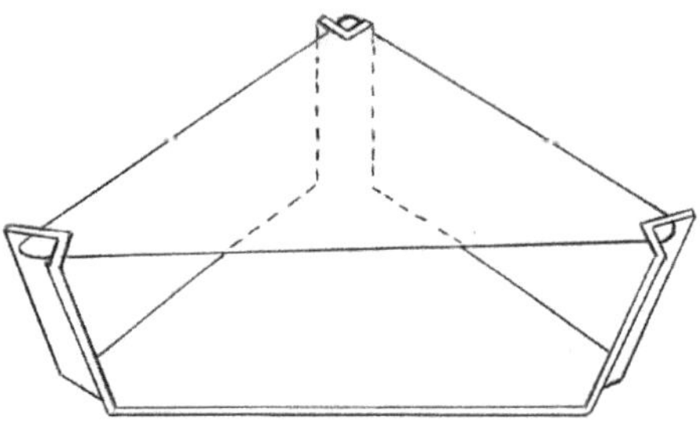

Fig. 10 Seventh concept

Table 3 Design concept selection using Pugh evaluation method for a portable laptop table

Concept	Baseline 	First	Second	Third	Fourth	Fifth	Sixth	Seventh
Performance	D	−	+	−	+	S	+	+
Stability	A	+	+	+	+	+	−	+
Strength	T U	−	+	+	S	+	−	+
Easy to manufacture	M	+	−	+	+	−	−	−
Portability		−	−	−	+	+	+	S
Weight		−	−	S	S	S	+	−
Cost of manufacture		S	−	S	+	−	−	−
Total score	$\sum +$ $\sum -$ $\sum S$	2 4 1	3 4 0	3 2 2	5 0 2	3 2 2	3 4 0	3 3 1
Positive difference	$\sum + (-) \sum -$	−2	−1	1	5	1	−1	0

Note Total score is obtained by obtaining the difference between '$\sum$+' and '$\sum$−'

with unsaturated polyester. The laptop platform measures length and width of 280 mm × 600 mm. Supporting both the platform is the main frame of the table. The length and width of the frame is 280 mm × 600 mm accordingly. Meanwhile, this table is supported by a pair of c-shaped legs. It comes at a height of 240 and 280 mm in length. Except for the laptop platform, rest of the table components are made from plywood which has uniform thickness of 11 mm. Overall, the table has a height of 296 mm, length and width measuring 280 and 600 mm, respectively. Figures 11, 12, 13 and 14 show the specifications of the table.

8.8 *Laptop Platform Design Consideration*

Before constructing the mould, there was a parameter to be resolved in forming the composite laptop platform. Firstly, the dimensions of the laptop platform are designed in such way so that it can accommodate laptops of all sizes. However, weight criteria remind that the entire table must not exceed 8 kg. By considering the weight of table frame, legs and hinges, the weight of the laptop platform must be between 2 and 5 kg. This consideration is because manipulating the platform weight was easier compared manipulating the wooden frame and leg which is by recalibrating the dimensions.

In general, natural fibre should not make up more than 40% of total volume of the composite. In this project, kenaf fibre occupies 28% of volume fraction, while unsaturated polyester set to be 70% in volume. Remaining 2% was the catalyst. The calculation of weight obtained is shown in Table 4.

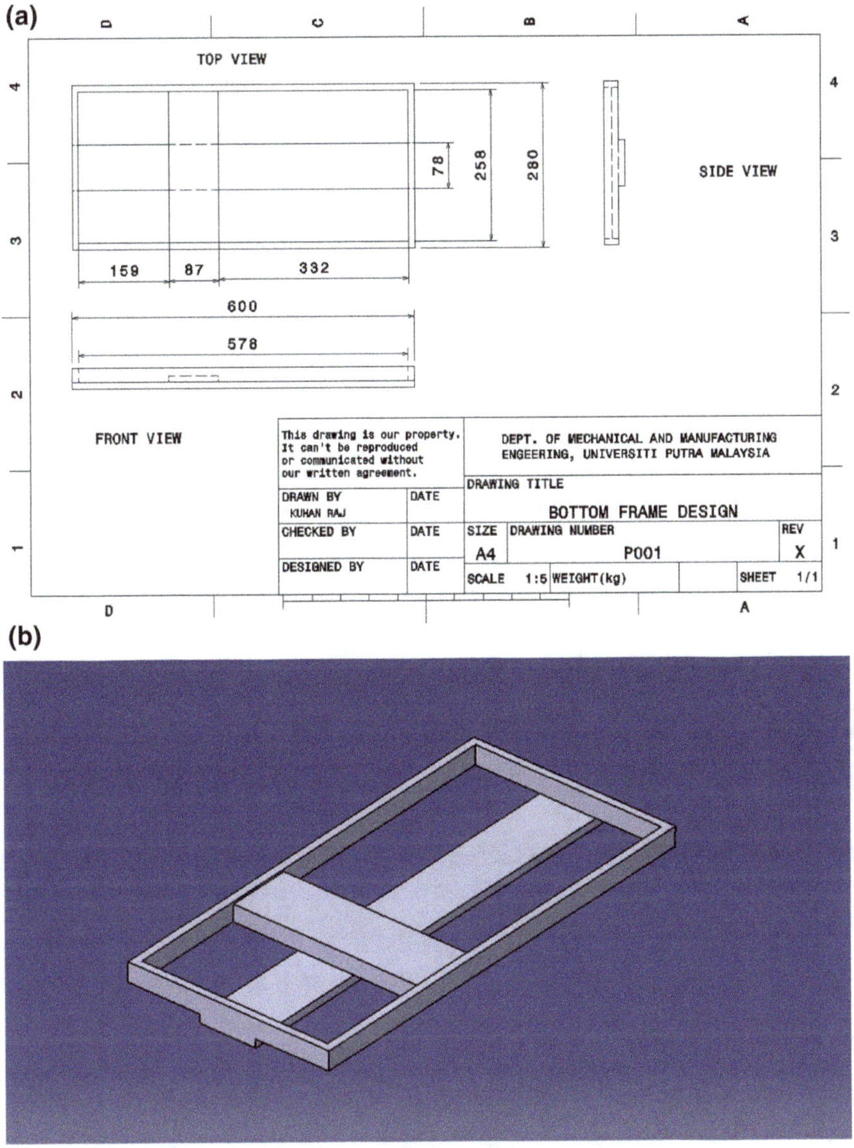

Fig. 11 **a** Drawing of table bottom frame (mm) **b** Design of table bottom frame

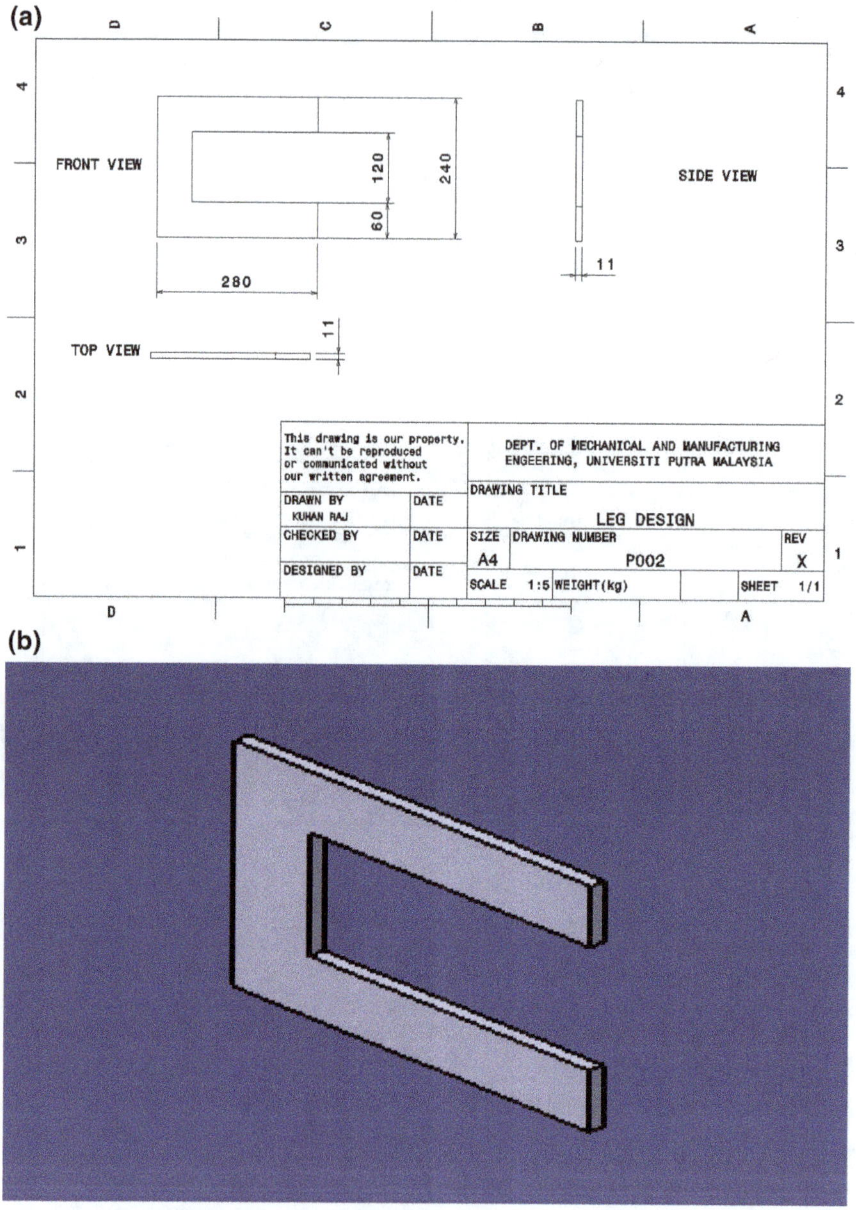

Fig. 12 **a** Drawing of table leg (mm) **b** Design of table leg

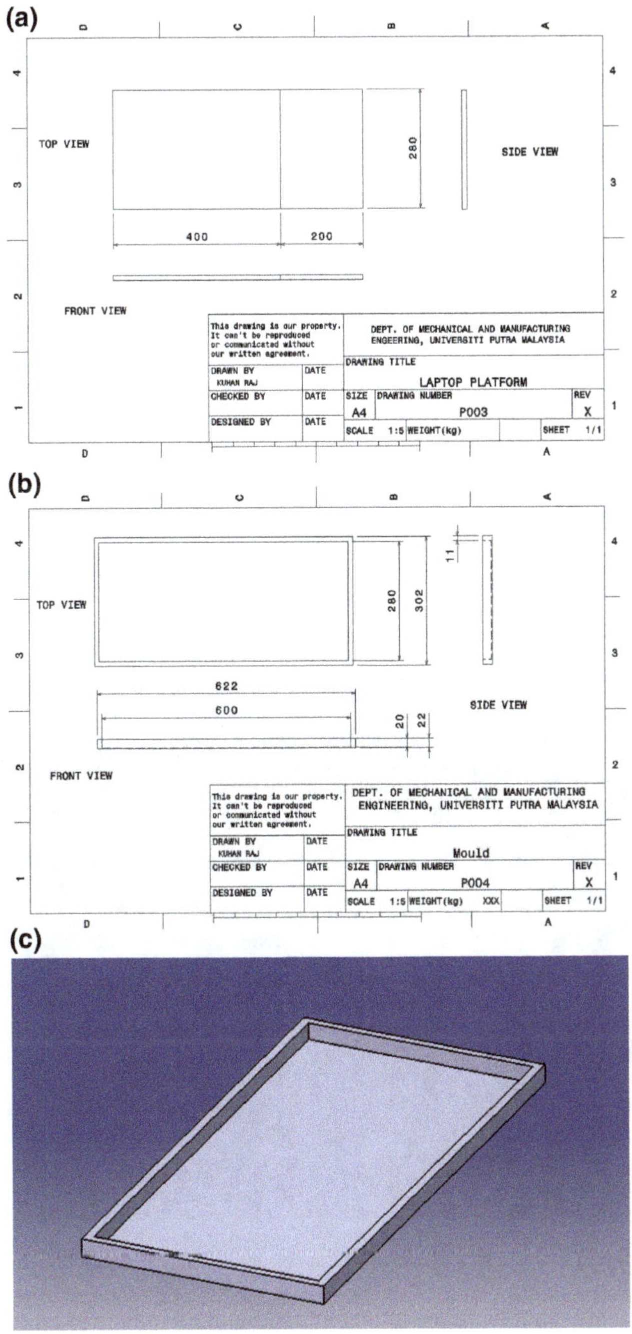

Fig. 13 **a** Laptop platform drawing (mm) **b** Drawing of mould (mm) **c** Mould design

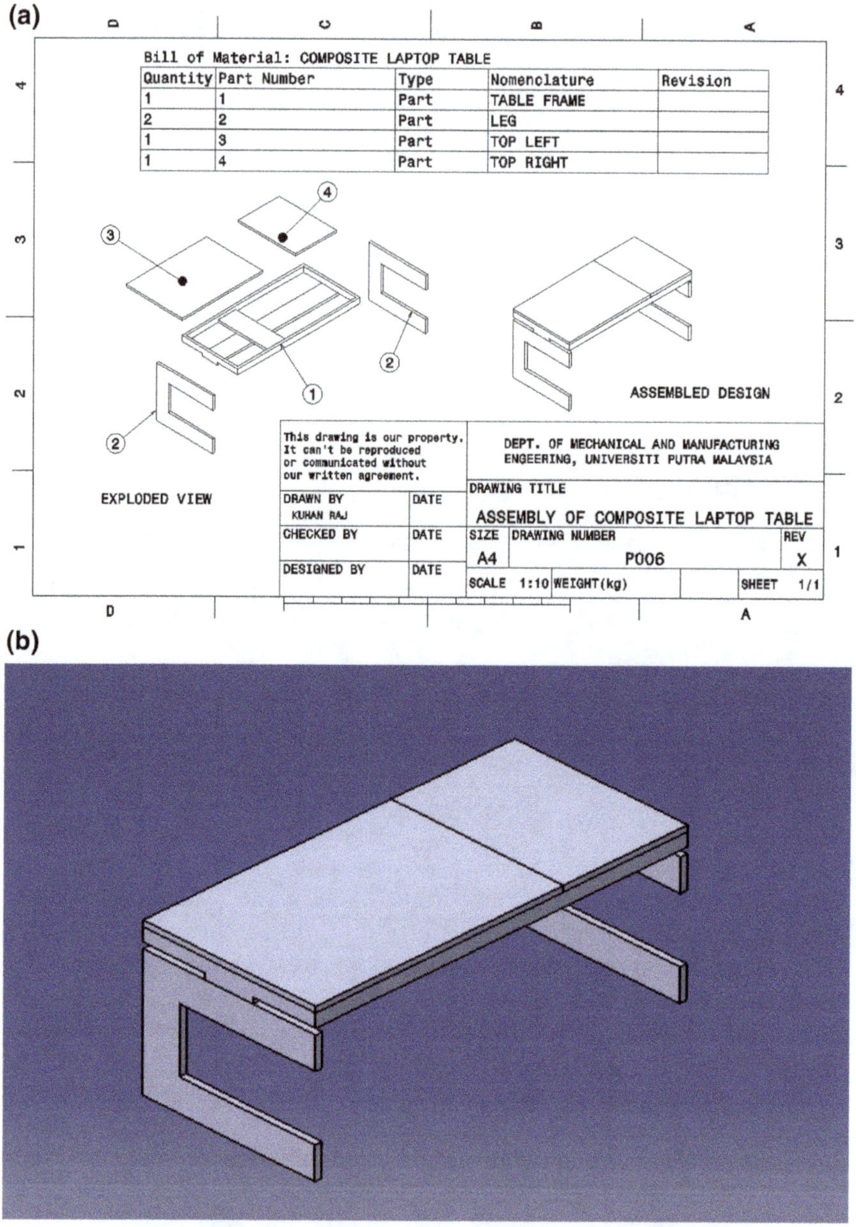

Fig. 14 **a** Drawing of assembly of composite laptop table (mm) **b** Assembly design of laptop table

Table 4 Calculation of required weight of matrix, catalyst and fibre

Volume of composite mould 70% of unsaturated polyester 2% of MEKP 28% of fibres	$=60\ \text{cm} \times 28\ \text{cm} \times 2\ \text{cm}$ $=3360\ \text{cm}^3$ $=0.7 \times 3360\ \text{cm}^3 \times 1.12\ (\text{g/cm}^3)$ $=2634.24\ \text{g}$ $=0.02 \times 3360\ \text{cm}^3 \times 1.55\ (\text{g/cm}^3)$ $=104.16\ \text{g}$ $=0.28 \times 3360\ \text{cm}^3 \times 1.4\ (\text{g/cm}^3)$ $=1317.12\ \text{g}$
Total weight	$=2634.24\ \text{g} + 104.16\ \text{g} + 1317.12$ $=4055.52\ \text{g}$

Fig. 15 Non-woven kenaf fibre mat

9 Fabrication of Composite Portable Table

The method of fabricating this composite portable table was hand lay-up as described earlier. This traditional method was used to produce products at minimum quantity and hence assuring good finishing and will be aesthetically attracting. There are several processes involved in fabricating the composite table. Firstly, it begun with fibre preparations, matrix preparation, followed by mould preparation, hand lay-up process and finally joining.

9.1 Fibre Preparation

The natural fibre chosen in this project was kenaf. It is one of the most common fibre used in trending green furniture industries as a replacement for wood and plastic. Non-woven kenaf fibre mat as shown in Fig. 15 was obtained from Innovative Pultrusion Sdn. Bhd, Malaysia.

9.2 Matrix Preparation

The preparation of matrix was carried out at Mechanical Lab in Faculty of Engineering. To bind the kenaf fibre, unsaturated polyester matrix was used together with methyl ethyl ketone peroxide (MEKP) as the catalyst. A total of 10 kg of resin were obtained from Berjaya Bintang Timur Sdn. Bhd., Cheras, Kuala Lumpur. The characteristics of resin are shown in Table 5. In order to obtain a good mixing property, both the solutions were measured prior to mixing. This is because incorrect mixing ratio will lead to poor bonding ability. Besides, it can cause the resin to harden too quickly even before the fabricating process finishes. The catalyst quantity was set to be 2% from the total weight of resin. The quantity of both chemicals is measured and poured into a metal container that was well cleaned. The solution was stirred gently so that no air bubbles form. This avoids porosity in the end product as shown in Fig. 16.

9.3 Mould Preparation

The laptop platform was fabricated using open mould method. Plywood was used as the mould since it has good even and clean surface. It was easy to handle and relatively cheap. Since the mould is in square shape, all the plywood pieces were

Table 5 Characteristics of unsaturated polyester resin

Characteristics	Value
Specific gravity	1.12
Viscosity-brookfield #3/60	450–600
Tensile strength, MPa	62
Elongation at break, %	2.2

Fig. 16 The mixing of unsaturated polyester and catalyst

cut in similar length of 900 mm. The thickness of the plywood is 20 mm and that set the thickness of the composite product. To join the plywood pieces, round head nails were used. The joining was done meticulously to avoid gaps that allow resin to seep out from the mould.

Once completed, moulding wax was applied on the base once an hour, continuously for 5 h. It was to make sure the composite can be removed easily from the mould surface after curing process. This wax also allowed the composite to have well even and shinny surface. The mould was prepared as shown in Fig. 17.

9.4 Hand Lay-Up Process

To begin with the fabrication process, the mould was cleaned and dried to remove unwanted dirt and maintain dry surface. The mould was then covered using a layer of aluminium foil. Next wax spray was applied for easy removal of the composite as well as to obtain excellent surface finishing. After allowing it to dry for several minutes, the resin and catalyst that were mixed earlier were poured in the mould. Then, the non-woven kenaf fibre mat was placed evenly on the mould. The fibre mat previously trimmed according to the mould shape so that it can fit evenly.

Following that, the resin mixture was applied again on the kenaf fibre mat using a brush and rolled firmly to disperse the resin mixture throughout the mould. The process was repeated with another two layers of kenaf fibre mat until it reached the desired thickness of 2 mm. Subsequently, the composite was allowed to cure under room temperature for duration of 24 h. Finally, the cured composite was detached from the mould as it is shown in Fig. 18. The excess and uneven edges were trimmed using bench saw and smoothened using grinder.

Fig. 17 Mould prepared for composite laptop table

Fig. 18 Cured composite laptop platform

Fig. 19 Constructing the table frame

9.5 Joining and Finishing Process

The final step in this project was the assembly or joining process as shown in Figs. 19 and 20. The trimmed composite was joined on the frame using L-bracket and slotted countersunk screw. Hinges were installed on the legs so that it can be foldable.

10 Results

To provide a better finishing look, the table was painted with shellac varnish. The wooden part of the table was painted thoroughly and gives a pleasingly attractive golden brown colour. To add better marketability characteristics, an adjustable height changing mechanism was added behind the laptop platform. It enables the user to adjust the height and angle of laptop platform to suit according to their comfort and use. Steel rod of 0.5 mm thickness was used for the locking device, while the locking rail was made up from wood. The illustrations of the finished product are shown in Figs. 21, 22, 23 and 24.

Fig. 20 Folding bracket for the legs

Fig. 21 Assembled composite laptop table

Fig. 22 Side view of the laptop table

Fig. 23 Front view when table is folded

Fig. 24 Functional laptop table

11 Discussion

In order to ensure the fabricated composite laptop table is stable and firm enough to withstand the load of the laptop combined with the force exerted by the user's hand, stress analysis was performed by using computer-aided simulation software, Abaqus. By using this software, the stress distribution and deformation of the table can be seen without performing any physical experiment. This kind of analysis saves material, time and costs besides pinpointing accurate data compared to conventional physical experiments (Table 4).

Table 6 Parameters used for simulation

Density	1.4 g/cm^3
Poisson's ratio	0.32
Young's modulus	42 MPa
Load applied	100 N

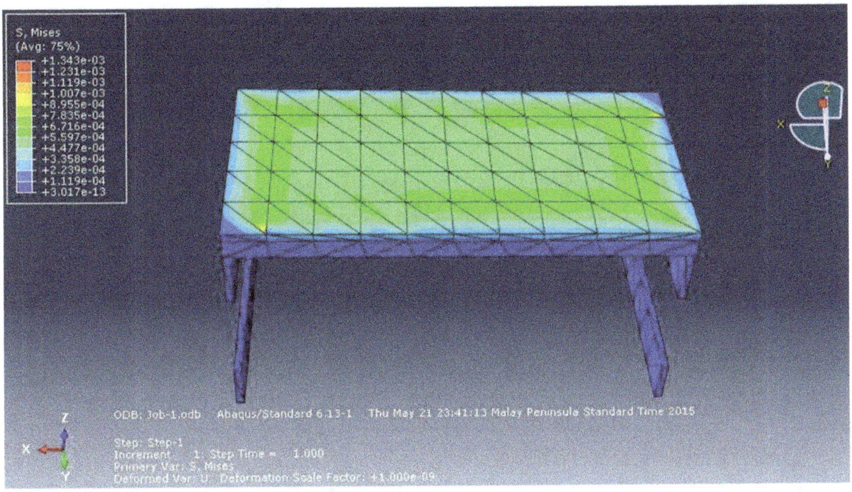

Fig. 25 Front view of laptop table

In this software, few important parameters were needed to simulate the stress analysis such as the density, Poisson's ratio and Young's modulus of kenaf fibre reinforced polyester composite. Since no physical experiments are conducted in this project, the Poisson's ratio and Young's modulus values were adapted from a similar project that involved the fabrication of kenaf fibre reinforced polyester composite (Kotresh et al. 2014). It was ensured that the density of kenaf fibre, the volume fraction of kenaf fibre, and type polymer used is similar to this project before adopting its parametric values. The values of parameters are shown in Table 6.

The results of the simulation are shown in Figs. 25 and 26.

The simulation reveals composite part of the table which is the laptop platform experience pressure of 559.7 Pa. Red colour indicates high-stress distribution while dark blue indicates low-stress distribution. The simulation proved that this table is safe to use since there is no red contour existing on the surface or legs of the table. The rate of deformation was minimum hence this table does not crack or break apart when the load was applied on it.

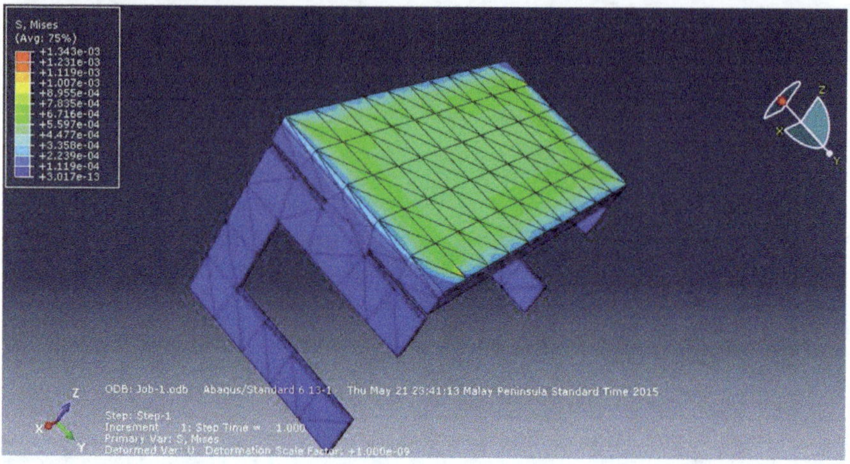

Fig. 26 Lower side view of laptop table

12 Conclusion

Overall, a composite portable laptop table was able to be produced in this project. The objectives were achieved successfully by manufacturing a composite laptop table using a systematic approach of the total design process. This composite laptop table can be brought anywhere by folding its leg inwards and can be carried easily using the holder. Besides serving as a laptop stand, the user can read books and newspaper on it.

In short, opting to natural fibre usage can reduce the dependency on non-renewable resources and help to conserve the green environment. Since composite furniture can be made from natural fibre, there are many chances they can be utilized for various purposes. Kenaf fibre reinforced polymer composite is one of the examples of the natural fibres diversity in becoming substituent for petroleum.

References

Abdullah A, Khalina A, Ali A (2007) Effects of fibre volume fraction on unidirectional kenaf/epoxy compositions: the transition region. Polym Plast Technol Eng 50:1362–1366

Akil HM, Cheng LW, Ishak ZM, Bakar AA, Rahman MA (2009) Water absorption study on pultruded jute fiber reinforced unsaturated polyester composites. Compo Sci Technol 69:1942–1948

Asilah CA (2011) Mechanical performance of kenaf fibre reinforced thermoplastic composite, B. Eng. Thesis, Universiti Malaysia Pahang, pp 24–38

Bernard M, Khalina A, Ali A, Janius R, Faizal M, Hasnah KS, Sanuddin AB (2011) The effect of processing parameters on the mechanical properties of kenaffibre plastic composite. Mater Des 32:1039–1043
Biron M (2012) Thermoplastics and thermoplastic composites: technical information for plastics users. Elsevier, Oxford
Chaivan P, Pasaja N, Boonyawan D, Suanpoot P, Vilaithong T (2005) Low-temperature plasma treatment for hydrophobicity improvement of silk. Surf Coat Technol 193:356–360
Faruk O, Bledzki AK, Fink HP, Sain M (2012) Biocomposites reinforced with natural fibers: 2000–2010. Prog Polym Sci 37:1552–1596
George J, Sreekala M, Thomas S (2001) A review on interface modification and characterization. Polym Eng Sci 41:1471–1485
Gonzaleza AV, Cervantes-Uca JM, Olayob R, Herrera F (1999) Effect of fiber surface treatment on the fiber–matrix bond strength of natural fiber reinforced composites. Compos B 30:309–320
http://www.miti.gov [Online] [Cited: May 8,2015.]
http://www.wildfibres.co.uk/why-natural-fibres.htm [Online] [Cited: May 8,2015.]
Hyer MW (2009) Stress analysis of fiber-reinforced composite materials. DEStech Publications, Inc
Ikada Y, Tsuji H (2000) Biodegradable polyesters for medical and ecological applications. Macromol Rapid Commun 21:117–132
Jonn A, Foulk Danny EA, Roy BD, David DM (2000) Flax fibre: potential for a new crop in the Southeast, 364
Joshi SV, Drzal LT, Mohanty AK, Arora S (2003) Are natural fiber composites environmentally superior to glass fiber reinforced composites? Resour Conservation Recycl 33:371–376
Karnani R, Krishnan M, Narayan R (1997) Biofiber reinforced polypropylene composites. Polym Eng Sci 37:476–483
Keener TJ, Stuart RK (2004) Maleated coupling agents for natural fibre composite. Compos A 35:357–362
Khalil HA, Bhat AH, Yusra AI (2012) Green composites from sustainable cellulose nanofibrils: a review. Carbohyd Polym 87:963–979
Kolybaba M, Tabil LG, Panigrahi S, Crerar WJ, Powell T, Wang B (2003) Biodegradable polymers: past, present, and future. In: ASAE meeting presentation, pp RRV03–0007
Kotresh, VK, Manjunatha (2014) Characterization and Investigation of tensile test on kenaf fiber reinforced polyester composite material. Int J Res Dev Eng Technol 2:132–136
Liu W, Drzal LT, Mohanty AK, Misra M (2007) Influence of processing methods and fiber length on physical properties of kenaf fiber reinforced soy based biocomposites. Compos B Eng 38:352–359
Mansor MR, Sapuan SM, Zainudin ES, Nuraini AA, Hambali A (2014) Conceptual design of kenaf fiber polymer composite automotive parking brake lever using integrated TRIZ–Morphological Chart-Analytic Hierarchy Process method. Mater Des 54:473–482
Mayer RM, Hancox N (1993) Design with reinforced plastics: a guide for engineers and designers. Springer Science & Business Media
Mazumdar SK (2002) Composites manufacturing: materials, product, and process engineering. CRC Press, Boca Raton
Nirmal U, Hashim J, Ahmad MM (2015) A review on tribological performance of natural fibre polymeric composites. Tribol Int 83:77–104
Nishino T, Hirao K, Kotera M, Nakamae K, Inagaki H (2003) Kenaf reinforced biodegradable composite. Compos Sci Technol 63:1281–1286
Nor MJ, Ayub M, Zulkifli R, Amin N, Fouladi MH (2010) Effect of compression on the acoustic absorption of coir fiber. Am J Appl Sci 7:1285
Pugh S (1991) Total Design, Integrated Methods for Successful Product Engineering. Addison-Wesley Publishers Ltd., Workingham
Rassmann S, Paskaramoorthy R, Reid RG (2011) Effect of resin system on the mechanical properties and water absorption of kenaf fibre reinforced laminates. Mater Des 32:1399–1406

Rowell RM, Sanadi A, Jacobson R, Caulfield D (1999) Properties of kenaf/polypropylene composites. Kenaf properties, processing and products. Mississippi: Agricultural & Biotechnology Engineering, pp 381–392

Saheb DN, Jog JP (1999) Natural fiber polymer composites: a review. Adv Polym Technol 18:351–363

Salit MS (2014) Tropical natural fibre composites: properties, manufacture and applications. Springer, Singapore

Sanyang ML, Sapuan SM, Jawaid M, Ishak MR, Sahari J (2016) Development and characterization of sugar palm starch and poly(lactic acid) bilayer films. Carbohydr Polym 146:36-45

Sapuan SM, Maleque MA (2005) Design and fabrication of natural woven fabric reinforced epoxy composite for household telephone stand. Mater Des 26:65–71

Sapuan SM, Leenie A, Harimi M, Beng YK (2006) Mechanical properties of woven banana fibre reinforced epoxy composites. Mater Des 27:689–693

Sapuan SM, Harun N, Abbas KA (2007) Design and fabrication of a multipurpose table using a composite of epoxy and banana pseudostemfibres. J Tropical Agric 45:66–68

Shibata S, Cao Y, Fukumoto I (2006) Lightweight laminate composites made from kenaf and polypropylene fibres. Polym Test 25:142–148

Vroman I, Tighzert L (2009) Biodegradable polymers. Materials 2:307–344

Wambua P, Ivens J, Verpoest I (2003) Natural Fibre: can they replace glass in fibre reinforced plastic? Compos Sci Technol 63:259–1264

Lignocellulosic Materials for Geotextile and Geocomposites for Engineering Applications

Jaideep Adhikari, Suvendu Manna, Sukanya Chowdhury Saha, Prosenjit Saha, Debasis Roy and Basudam Adhikary

Abstract Cost-effectiveness, environment-friendly nature of lignocellulosic (LC) materials have been utilized for geotechnical engineering for value-added end uses, however, so far the efficacy of such LC materials has not been widely accepted by the engineers due to lack of their long-term durability. Durability of LC materials can be enhanced by altering fibre chemistry or by applying a degradation-resistant surface coating. This chapter will present an extensive overview of such potential processes for enhancement of durability of LC materials followed by a comprehensive discussion on designing and testing parameters of geotextiles and geocomposites using LC materials. A special emphasis has been given to erosion control system and stabilization of slopes using jute fibres as a potential LC material.

Keywords Lignocellulose · Geotextile · Geomembrane · Geocomposites Erosion control · Jute

Abbreviation

ASTM	American Society for Testing Material
AOS	Apparent opening size
FoS	Factor of Safety
LC	Lignocellulosic
GT	Geotextiles
GSM	Gram per square metre

J. Adhikari · S. Chowdhury Saha · P. Saha
Dr. M.N. Dastur School of Materials Science and Engineering, Indian Institute of Engineering Science and Technology-Shibpur, 711103 Howrah, India

S. Manna · P. Saha (✉) · B. Adhikary
Materials Science Centre, Indian Institute of Technology, 721302 Kharagpur, India
e-mail: prosenjit@matsc.iiests.ac.in

D. Roy (✉)
Department of Civil Engineering, Indian Institute of Technology, 721302 Kharagpur, India
e-mail: debasis@civil.iitkgp.ernet.in

S. Kalia (ed.), *Lignocellulosic Composite Materials*, Springer Series on Polymer and Composite Materials, https://doi.org/10.1007/978-3-319-68696-7_9

JGT Jute geotextile
POA Percent open area
PWP Pore water pressure

Contents

1 Introduction

Geotextile and geocomposite are classified as important components of geosynthetics. American Society for Testing Material (ASTM) D 4439 defines geosynthetics as 'a planner product manufactured from polymeric material used with soil, rock, earth or other geotechnical engineering related material as an integral part of a human-made project, structure or system' (ASTM D4439—15a). Growing demand for geosynthetic is basically led by their five key functional attributes, such as, filtration, separation, drainage, reinforcement and protection. The class of geosynthetic comprises mainly of geotextiles, geogrid, geonets, geomembranes, clay liners, geopipe and geocomposite. Among them geotextile is widely used class of geosynthetic which is permeable in nature and serves typical applications like increment in the stability of soil; filtration, decrease the erosion caused by the wind, water. In general, woven, non-woven, knitted, knotted are the types of geotextiles that are used for geotechnical engineering. Lighter weight geotextiles are used in several applications viz. filtration, separation between soils, erosion control whereas heavier weight geotextiles are used in reinforcement purpose (Sen Gupta 1991). On the other hand, geocomposites are manufactured by combining geotextiles with either of geonet, geomembrane, geogrid, polymer core or combining geomembrane with geogrid, geosynthetics with soil, etc. Most of the geosynthetic market is dominated by four synthetic polymers such as polypropylene, polyethylene, polyester, polyamide. However, growing environmental concern leads us towards the development of natural materials. Lignocellulosic materials could be the ideal candidate to fulfil such environmental requirement. Natural fibres are suitable for their lower price and technical properties. The main drawback with lignocellulosic materials is their durability issue which can be modified tactfully by the experts according to the end applications.

1.1 Lignocellulosic Material

Lignocellulosic (LC) materials are any of the several natural, widely available, renewable resource material used in industrial purpose for the upliftment of sustainable global economy and constituting several polymers essentially found in the cell walls of a woody plant they are cellulose, hemicellulose and lignin. LC materials found their extensive uses in paper and pulp industries, in the production of biofuel mainly bio-ethanol, in natural fibre reinforced composite and in geotechnical applications. The mechanical properties of LC material are determined by its biomass (mainly governed by cellulose, hemicellulose and lignin). The chemical composition of these plant-based materials varies with geographical location, species, climate, processing condition, extraction methods, etc. The basic processing stage of LC materials deals with the formation of fibres and subsequently, they are processed in the form of non-woven mat, felt, yarn, woven or knitted fabrics for different applications. Cellulose molecules being linear in nature can easily arrange themselves in a periodic nature and forms the crystalline region of fibres. Overall cellulose molecules are semi-crystalline in nature. The crystalline part of cellulose provides strength to the fibre. Cellulose molecule contains a large number of hydroxyl groups which facilitate water absorption. Along with cellulose, hemicellulose also contributes to water absorption. Hemicellulose molecules are believed to be the main reason behind the hydrophilic nature of natural fibres as they are highly branched molecules and cannot crystalize hence forms an open structure. Amorphous nature of hemicellulose molecules contains plenty of hydroxyl groups which emanate from several sugar molecules. Lignin is also a highly branched molecule, amorphous in nature and exhibits the least influence in water absorption properties of a lignocellulosic material. Apart from cellulose, hemicellulose and lignin natural, fibres contain pectin (imparts flexibility), which is a wax-like substance. Pectins are basically hetero-polysaccharides comprises of poly galacturonic or pectic acid units. They are removed by treating the fibres in alkaline solution followed by washing with water. Waxes are compounds of alcohols that are insoluble in water and organic acids (Saha et al. 2016).

Natural fibres can be classified according to their extractive source.

Bast or phloem fibres are extracted from the stem of a dicotyledonous plant. By retting process, fibres are set to free from the cellular and other tissues. Ex.-Flax, Hemp, Jute.

Leaf fibres are extracted by scraping the pulp of leave from the fibres of a monocotyledonous plant. Ex. Abaca, Sisal.

Seed and fruit fibres are intended to protect the seed or fruit of a plant as these are the most vulnerable parts of a plant. Ex. Coir, cotton.

Some potential natural fibre used in different geotextile and geocomposite are tabulated in Table 1.

Table 1 Some potential natural fibres used in Geotextile and Geocomposites

Fibre	Cellulose content (wt%)	Hemicellulose content (wt%)	Lignin content (wt%)	Tensile strength (MPa)	Elongation (%)	References
Jute	61.0–71.5	13.6–20.4	12.0–13.0	393–773	1.16–1.5	Goda et al. (2008), Gupta et al. (2015)
Flax	71.0	18.6–20.6	2.2	345–1100	2.7–3.2	Goda et al. (2008), Chanda et al. (2008)
Sisal	67.0–78.0	10.0–14.2	8.0–11.0	511–635	3–7	Goda et al. (2008), Verma et al. (2013), Chanda et al. (2008)
Coir	36.0–43.0	0.2–0.3	41.0–45.0	131–175	4.7	Goda et al. (2008), Verma et al. (2013)
Ramie	68.6–76.2	13.1–16.7	0.6–0.7	400–938	1.2–3.8	Goda et al. (2008), Chanda et al. (2008)
Kenaf	72	20.3	9	743	–	Kumar et al. (2011), Komuraiah et al. (2014)
Bamboo	26–43	30	21–31	221–661	3.7	Kumar et al. (2011), Verma et al. (2013)
Abaca	56–63	20–25	7–9	430–760	–	Kumar et al. (2011), Chanda et al. (2008)
Palm	–	20.8	28.5	43–263	2.0–5.1	Chanda et al. (2008), Koba et al. (1990)

1.2 Approach Towards Durable Lignocellulosics

In a moist environment, the free hydroxyl groups of cellulose attract water molecules which cause the LC materials to swell. Cellulose chains of LC materials lose their integrity under exposure to several chemical environments such as acidity, alkalinity, salinity, UV irradiation, etc. by hydrolysis or oxidation. Biogenic enzymes also affect the cellulosic chains (Rowel 1996). Thus, the technologies related to the modification of cellulose chains for stability against degradation and mechanical strength indeed is a necessary factor to study. Hemicellulose and amorphous cellulose are more prone to any sort of enzymatic degradation caused by cellulolytic bacteria, fungi and insects. Oxidation, hydrolysis and dehydration reactions occur in the presence of acid and alkali environments, while mechanical damages of LC materials occurs in saline environments (Winandy and Rowel 2005). The main reason behind the mechanical damage is the surface swelling and shrinkage phenomena which are promoted by hygroscopic and salt-catalyzed hydrolysis action. UV irradiation causes oxidation of lignin and the degraded product leached out from the matrix. After delignification, the poorly bonded cellulose molecules also wear out from the surface of LC materials. Overall, the tensile strength of the materials falls significantly (Rowell and Stout 2007). Figure 1 describes the different type of degradation and the degrading agent which follows several mechanisms to attack the constituents of LCs. Susceptible components in order of their decreasing susceptibility can be identified from the figure.

The problem of degradation of LC materials can be overcome by three possible ways. The modifications are generally performed on the fibre whereas it can be done on the fabric and yarn also. The possible ways are

1. By formation of new covalent bond on the reactive sites of the LC fibres
2. Changing the fibre chemistry by radiation treatment
3. Coating of degradation-resistant chemicals on fibre surface.

Esterification, transesterification, etherification, silane treatment, grafting are the most common and popular chemical treatment method for making the fibres degradation resistant. Some biological agents such as laccase, protease enzymes are quite commonly used in chemical treatment methods. Cellulase, xylanase, pectinase natural enzyme, novozymes are used on LC fibres for enhancement in elongation at break, tensile strength and hydrophobicity of the materials.

The key mechanism is the removal of the amorphous materials from fibre surface and consequent fibrillation which results in enhanced mechanical strength and hydrophobicity.

Radiation method involves ion beam treatment for the alteration of surface properties of the LC materials. Gamma radiation and electron-beam methods are widely used modification methods. Gamma radiation is used where high penetrating efficiency and less treatment time is required whereas electron-beam treatment has been practiced for energy saving and controlled penetration-based process.

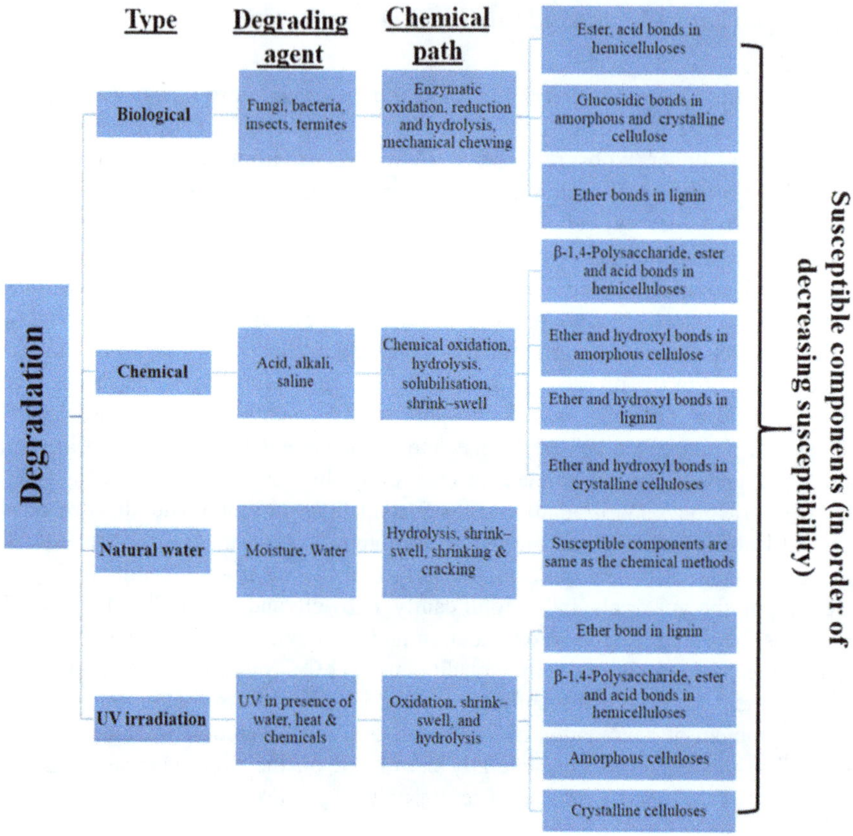

Fig. 1 Degradation reaction of LC materials (Saha et al. 2016)

LC materials either in the form of fibre, fabric or felt have been used to coat or block the sites with phenol–formaldehyde (PF) resin, boric acid–oil emulsion, neem seed extract, vegetable oils and other water repellent reagents to reduce their hydrophilicity. PF resins include melamine formaldehyde, cashew nut shell liquid–formaldehyde and polymerized cashew nut shell liquid, etc. Cross-linking occurs between LC materials and resin matrix through hydrogen bonds. Thus, the decrease in water uptake is attributed to the development of weakly bonded networks connected through secondary bonds (Saha et al. 2016).

2 Geotextiles and Geocomposites

Initially, Geotextiles were used as filtre fabrics and were envisioned to be a substitute of granular soil filtres. After that permeability, soil retention and other applications of GTs evolved and the textiles were tailored such a way that they can be used as suitably engineered materials. In the beginning, geotextiles were only intended to be the synthetic ones. Thus, for manufacturing a suitable geotextile type of polymer, fibre construction and weave design are the major concerned areas. For a synthetic textile material, the basic flow chart comprises

- Procurement of synthetic polymer
- Preparation of synthetic fibres by melt spinning/wet spinning/dry spinning technique. Most of the synthetic geotextiles are prepared by melt spinning process.
- Conversion of fibres into yarn (Monofilament, multifilament, staple fibre yarn, slit-film monofilament and slit-film multifilament)
- Preparation of the fabric by weaving the yarns

In the case of LC-based natural fibres, the fibres were extracted first from different sources and they were processed subsequently in the form of yarn and fabrics. The term yarn is used in textile where one or more fibres are drawn parallel and twisted into a single strand. Woven, Non-woven and knitted are the three choices for the formation of a fabric. For woven fabric weave design determines permeability, drapability and mechanical strength of the fabric and hence weave design is an important parameter for geotextiles and geocomposites. Weave design is determined by the sequence in which the longitudinal threads (known as Warp) are interlaced with the transverse pick (known as weft). Plain, basket and twill are the most common weave designs that have been already explored in the field of geotextiles. Each type of design also has several distinguishable separate patterns or formations, for instance, twill design has steep twill, broken twill, etc. patterns. Non-woven materials can be directly obtained after the fibre stage without converting them into yarn. Knitted fabrics can be stretched in any direction giving them much more elasticity than the woven fabrics. However, they are seldom used in geotextiles as they lack in mechanical strength. For the preparation of novel structures with knitted fabrics, yarns with high strength and low extension are inserted either in weft or warp size. High strength natural yarns are intended to be prepared with jute, flax, sisal, coir, abaca while soft yarns of viscose (regenerated cellulose), cotton, jute are designed for loop formation (Anand 2008).

Geocomposite material also conforms the basic philosophy of a composite material, i.e. combining different material in order to overcome the inadequacies of each material with respect to a specifically targeted application. The most conventional geocomposites are geotextile-geomembrane, geotextile-geonet, geotextile-geogrid and geomembrane- geogrid composites. Apart from the conventional geocomposites sometimes steel strands are also woven with geotextile materials to form a composite with superior mechanical properties. At times prefabricated holes are created with the aim of insertion of steel rods as an anchored material. Sometimes geotextiles are used in combination with the polymer core as an alternative to the traditional sand drains.

2.1 Lignocellulosic Materials for Geotextile and Geocomposites

Synthetic geotextile offers several advantages like durability, superior mechanical properties yet they lack in flexibility and drapability. Rigidness of synthetic materials prevents them from fitting with the uneven soil surfaces. So, when an inflexible GT is placed over this surfaces several wide openings are created between soil and the GT. Water molecules tend to flow between the openings and causes the GT to swell. Similarly, in the absence of water flow, the material tries to shrink. Thus, the reinforcement or stabilization of soils with GT gets hampered. Synthetic GT materials are also vulnerable to UV irradiation and overall stringent environmental policies drive the designers towards LC material for geotextile and geocomposite applications. As the LC materials are free from toxins and plasticizers, there is no possibility to add pollutants in the groundwater or to disturb the ecological system. LC materials are generally hydrophilic in nature and at the same time, their biodegradability rate is very much high. In some applications these properties are harmful nevertheless for application where the holding of moisture or water is necessary these properties are suitable. Moreover, the decomposed LC material adds organic nutrients to the soil. Due to the unique interlacement characteristics, woven geotextiles left with a satisfactory amount of open space for the plant to grow between the strands. The open mesh structure helps to pass sunlight, water-absorbing capacity helps to retain the moisture of the soil which in term helps in the consolidation of soil. Holding moisture and heat of the soil helps in the vegetation of plants. Thus, for this type of application open weave structure along with higher GSM fabric is preferred. Sometimes this environmental condition helps to grow several microorganisms which facilitate crop cultivation. Geotextiles are also used for weed controlling operation (mulching in agriculture/horticulture), preventing the plant from cold or wilting effect (loss of firmness of non-wooden part of the plant when the rate of loss of water is greater than absorption). Drapability can be defined as a material property having graceful folds under its own weight on retaining over a surface. Drapability of a GT, in the context of stabilization of slope, determines how a textile material will be able to adjust to the contour of slopes (Horrocks and Anand 2000). Degradation of lignocellulose-based GT is controlled by several parameters like type of soil (when exposed to the soil); type of asphalt, bitumen (during the road construction); depth of embedment (for reinforced material); climatic conditions (effect of moist, neutral, dry environment on agricultural applications), etc.

Among the natural LC materials, jute and coir geotextiles have been widely explored in India. Sisal, flax, ramie, palm and other LC materials are seldom used in this geographical location. However, in this part of the world, they are used in combination with Jute and coir. The application areas of these LC materials remain same as of jute and coir GT.

- Jute geotextiles found their extensive usage in road construction and pavement strengthening; under subgrade railway track; stabilization of embankment, slopes; consolidation of soils; prevention of erosion, etc. The use of heavy and coarse jute yarns makes them the ideal erosion control material for soil slopes under all climatic conditions. JGT is used in various structural civil engineering applications such as concrete slabs, cement mortar, etc. JGT coated with impervious synthetic non-woven are utilized for prevention of seepage from the structural civil engineered materials. JGT coated with bitumen enhances the lifetime of the material, nevertheless, bitumen coating makes JGT fabrics stiff and thus they are harsh to handle during installation. So chemically altered lignocellulosic for durability is preferred over the bitumen coating method. Some of the chemical treatments for enhancement of durability of Jute geotextiles are transesterification (Saha et al. 2012), coating with phenolic resin (Saha et al. 2012), etc. It is also used to protect or rather to cover dumps and heaps of solid wastes, granular materials (ex. fly ash in thermal power plants). JGT is extensively used in the construction of rural roads, pavements, railway tracks, concealed drains of hilly areas. Some common agricultural, horticulture and forestry sectors use of JGT are sunscreens, plant nets, harvesting nets, field-nets for protecting the crop from birds, windshield, weed management, mulching, etc. (Sen Gupta 1991, Ghosk 2014).
- Coir Geotextile-Due to their high lignin content, coir fibres are most durable among the natural fibres and extensively used in erosion control along the steep slopes. Despite the low cellulose content, coir has a very close fibre structure which helps them to possess suitable properties. It is primarily used for reinforcement of temporary clay structures, drainage applications along with prevention of soil erosion (Lekha 2004, 2006).

3 Designing of Geotextiles with Lignocellulosic Materials

According to the ASTM definition, GT is a permeable Geosynthetic exclusively made with textile material. For a particular application to design a GT several approaches are undertaken-

- Cost and availability
- Specifications
- Functions

Cost and availability method are very simple and basic approach yet it lacks with technicalities. Specification method deals with classification and utilization of the material according to their expected service life. In functional approach, the primary function of GT is evaluated and according to the purpose Factor of Safety (FoS) of that particular property of the material is calculated.

$$\text{FoS} = \frac{\text{Ultimate permisible property of the tested sample}}{\text{Required property(design) of the tested sample}} \tag{1}$$

If the factor of safety of a GT is sufficiently higher than the numerical value 1 then the GT material is adequate to be used for that particular application. In broader sense, ultimate allowable strength or property of the material must remain sufficiently higher than the required strength or property of the material. The designing of GT according to their function is discussed in the subsequent section.

3.1 Separation

A flexible porous geotextile (GT) segregates two dissimilar materials, preventing their intermixing and retaining their integrity and functionality. The purpose of separation is to maintain the unique performance of each individual layer and sometimes to improve the properties (Sarsby 2007). When coarse stone aggregates are placed over the fine soil particles two concurrent incident occurs over time.

The pressure exerts by the stone aggregates forced the soil particles to move into the voids of stone aggregates, thus the drainage capability is hampered.

While another process involves the reduction in strength of stone aggregate by trying to intrude into the soil particle.

A porous and flexible GT placed between subsoil and granular sub-base prevents the intermixing while permitting the free flow of water across the GT. In road construction, it restricts the aggregate to be punched down during initial compaction and during the dynamic loading of vehicles and, etc. Sometimes a temporary haul road is required during the construction of permanent works. This road is temporary and dug up after completion of work followed by disposal of it. LC materials without any surface modification are suitable for this kind of short-term application. Separation induces natural consolidation—formation of 'filtre-cake' in soil. Impact, puncture and tear propagation resistance are some of the mechanical properties required for separation function during installation, construction and for after use. Along with the mentioned properties Apparent Opening Size (A.O.S.), thickness, UV resistance, chemical stability, etc. of the LC materials are also required to have control over the designing of the GT. (http://www.engr.utk.edu/mse/Textiles/Geotextiles.htm).

GT used as a separator for unpaved, paved, railroads (between subgrade and stone base), for soils of roadways, earth and stone damps (between foundation and embankment), etc.

The basic construction method of a road is represented in Fig. 2. The granules placed over GT create many voids in between it. When a load is applied on top of the road, the stress is transferred from the surface to the aggregates and further through the GT it is conveyed to the underlying soil. The soil under stress pushes the GT upwards which finds it ways in the voids of the stones (Fig. 2). This field

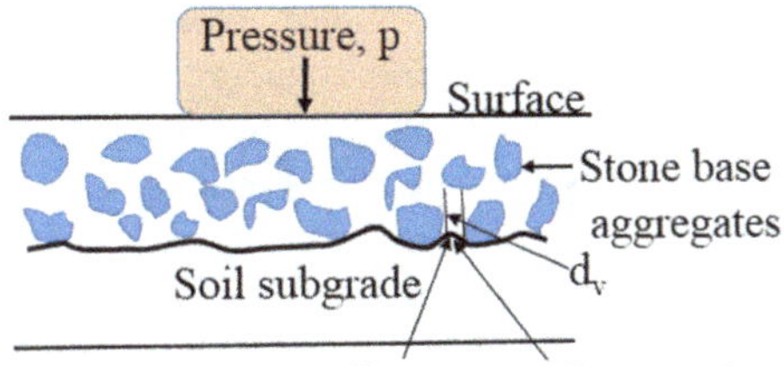

Fig. 2 Schematic representation of GT used for separation purpose

situation can be imitated according to ASTM D 3786 burst test where a GT is stressed over a gradually increasing hemispherical shape until the GT fails. Thus, for designing a GT to be used as a separator evaluation of burst resistance of GT material is important.

From the field situation Girould suggested a mathematical formula-

$$T_{\text{reqd.}} = \frac{1}{2} p' d_v [f(€)] \tag{2}$$

where $T_{\text{reqd.}}$ is the strength required for GT, p' is the stress on GT which is less than the stress (p) implied by the tire at the ground surface, d_v is the maximum void diameter of the stone which can be approximated by 0.33 $d_a.d_a$ is the average stone diameter and $[f(€)]$ is the strain function of the deformed GT.

For estimating the burst strength of GT the modified form of the above formula

$$T_{\text{ult}} = \frac{1}{2} p_{\text{test}} d_{\text{test}} [f(€)] \tag{3}$$

where T_{ult} is the ultimate strength GT, p_{test} is the pressure at the time of failure of GT and d_{test} is the diameter of the burst test device.

Factor of safety can be calculated as

$$\text{FoS} = \frac{T_{\text{allowable}}}{T_{\text{reqd.}}}, \; T_{\text{allowable}} = \frac{T_{\text{ult.}}}{\text{cumulative reduction factor}} \tag{4}$$

Another infield analogy was taken into account while designing geotextile, i.e. assessment of grab strength. A GT is sandwiched between two layers of stone aggregates. So, when the stone aggregate above the GT is forced to the stone aggregate lying below the GT, a lateral stress is generated within the GT. The lower aggregates must remain in direct contact with the geotextile. Due to the applied stress, the stone base tries to spread latterly and thus the lignocellulose-based GT also tries to expand accordingly. Therefore, the analogy to estimate the grab strength of the material can be visualized. The maximum strain is measured that a GT is able to withstand during the setting down process of the upper granules to the level of GT. Mathematical model is used to evaluate this property (Korner 1993).

The other two common parameters while designing GT for separation are puncture resistance and impact resistance. During the installation period and later at service period a GT is exposed to the pressure of several sharp stones, miscellaneous debris, roots, etc. These items can remain either in the beneath or top of the material. These materials are able to puncture the GT during the backfilling or at the time of traffic load imposed on the road. Impact or tear resistance is evaluated to have the assessment on the extent of material strength subjected to withstand the sudden free fall of any construction equipment or stones.

3.2 Reinforcement

Reinforcement is basically a type of strengthening mechanism. GT as a reinforcement to soil makes them superior with modified engineered properties. The soil has much higher compressive strength compared to its tensile strength. Under load, soil expands and in order to have sufficient shear strength, a strong soil/fabric interface is required. To have a strong interface and to avoid slippage higher frictional coefficient is required between soil and the fabric. Soil and reinforcement must experience same extension such that the implied stress is shared by each of the reinforcing element and then redistributed in the soil. In this way, stress is confined internally. Due to the requirement of the high frictional coefficient and high tensile strength, heavy grades of geotextiles are used to reinforce earth structures. In normal condition, if a soil is stressed vertically it tries to expand laterally. In the case of reinforcement with GT, internal confinement of stress prevents the expansion of soil laterally. GT exerts confining action on soil and absorbs a part of the stress causing shear failure. The method of reinforcing soil can also be utilized for stabilizing slopes and embankment. GT acts against rotational slides, lateral dispersion and vertical subsidence when inserted within a soil body at appropriate levels at initial stages. In the case of road reinforcement, GT helps to reduce deflections, rutting and fatigue cracking and overall help to strengthen the pavements. Sisal, Coir (Mwasha 2009), Jute GT are used for soil reinforcement purpose.

In developing countries, for construction of inexpensive short-term projects, utilizing manpower and the abundance of renewable material make senses of utilizing the LC materials. The stability of such material is monitored periodically and the material is replaced when it is unable serve as reinforcement. Furthermore, from the environmental concern point of view they enrich the soil without introducing harmful residues.

- Groundwater fills the pores in the soil particles. While constructing the embankments of soft compressible soil base a GT is used. The soft soil does not have sufficient strength to endure the shear stresses and generally fails under three modes—splitting, circular and basal. GTs are placed in the vertical increment in an embankment between the embankment fill and the underlying soft soil. A lateral stress is produced which resists the embankment from

splitting. A uniform compressive forces act on the underlying soil surfaces. Water pressure of pores in the soil is maximum at centre of the embankment while compared to the extremities. The embankment loading increases this water pressure in the pores. Pore water pressure (PWP) dissipates with time in the underlying soft soil. A pressure gradient is set up which enables the water to migrate sideways beneath the embankment and thus the PWP reduces with time. Hence, the stability of the embankment is improved.
As the PWP decreases with time, the force required for stabilization of the embankment also diminishes with time. The stabilizing force to the embankment is provided by the GT. This decreasing rate of the required stabilizing force can be designed with the deterioration rate of the LCs. As LC material loses its strength due to degradation over a period of time, this feature can be utilized to design LC for the above-mentioned purpose. Sometimes, consolidation drains are used to enhance the dissipation of excess PWP.

- Another short-term use of the LC materials is found in the establishment of temporary road/ structures by the armed forces. The armed forces have to construct temporary roads very quickly when dealing with the confrontations. However, these structures are designed to be demolished by machinery or explosive when the soldiers evacuate the place. The idea is to avoid leaving any resources which might get exploited by the enemies. By employing indigenous lignocellulosic fibre materials this situation is tackled as the natural geotextiles left to rot in the soil or set on fire.

Use of GT in road construction (Functions in asphalt overlays)

Stress relief: Restraining of soils subjected to hydrodynamic forces while allowing passage of water.

Barrier: Paving fabric, properly installed in between an existing road surface and a new asphalt overlay. Hence provides a stress relieving function, as the bitumen-saturated fabric is allowed for slight differential movements between the layers.

Reinforcement: The use of stress–strain behaviour of a geotextile to improve the mechanical properties of soil.

The main reason behind the generation of relative stress cracking in the pavement is due to repeated thermal expansion and contraction and also due to the lack of bearing capacity under load. Figure 3 represents the two mechanisms involved behind such kind of stress cracking.

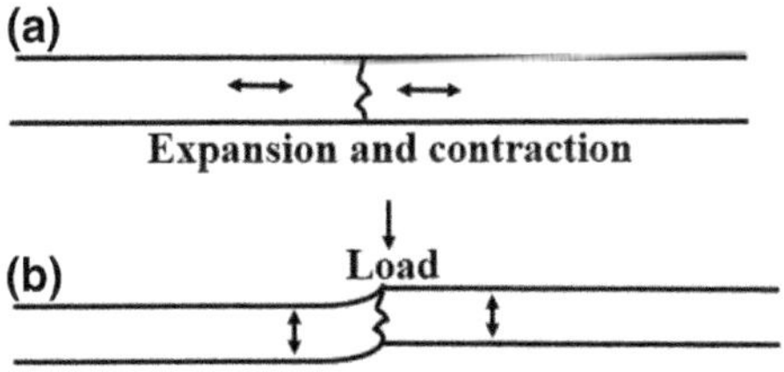

Fig. 3 Schematic representation of **a** thermal induced fatigue, **b** Traffic induced fatigue

The full-width GT used for this applications are coated with asphalt cement or asphalt emulsion and then overlaid with final bituminous. GT should be waterproofed and durable for this kind of applications. Hence, use of LCs for this kind of applications is very infrequent in the literature.

Assessment of base failure of GT as reinforcement is evaluated by the shear strength of the bonding system while top failure and slope failure are assessed by determining the tensile strength of GT. Permeability, creep, chemical and degradation resistant, etc. are some of the other important properties required for the reinforcement function with GT (http://www.engr.utk.edu/mse/Textiles/Geotextiles.htm).

3.3 Drainage (Fluid Transmission)

A GT can transport fluids or gasses within its plane. Normally the strength of a soil is increased by the decrease in its water content and vice versa. Excess PWP can be dissipated by the drainage system and thus consolidation of soil takes place. To reduce the drainage path temporary drains with GT is used for limited life spans. These drains are required until the soil consolidation has taken place. Thus, in this application naturally degraded GT is used.

In highways, airfields and railroads GT is used around the perforated pipe underdrains. Figure 4a shows protection of stones by GT around the perforated pipe, Fig. 4b shows GT is used directly to protect the perforated pipes with sand in the backfill and Fig. 4c represents GT wrapped stone with no perforated pipe. Proper designing of geotextile makes it free from soil contamination. Adequate flow capacity and adequate soil retention capacity are the two main parameters while designing such kind of trench drains.

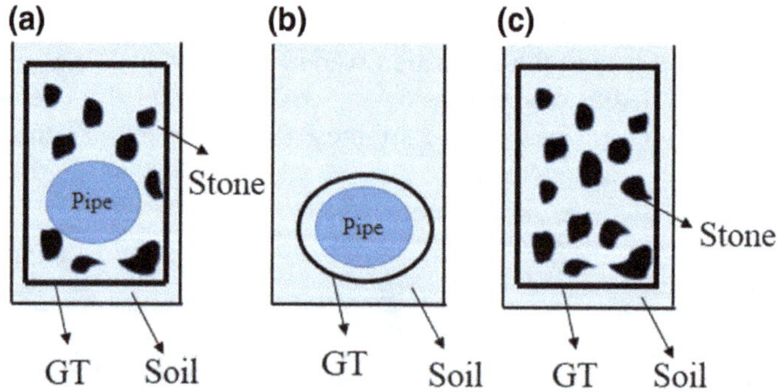

Fig. 4 Construction of underdrains pipe **a** GT around the stone, **b** GT wrapped pipe, **c** GT wrapped stone

Drainage behind the reinforced concrete wall:

Behind the conventional reinforced concrete wall, a vertical drainage system is needed. The water from backfill soil must drain out to reduce the generation of hydrostatic pressure. This hydrostatic pressure along with the horizontal soil pressure is able to cause failure of the walls. The sand used in this drainage system must stay free from clogging for the lifetime. Thus, the sand must be protected with GT filtre. GT can be used in several types of arrangements such as a rigid wall with underdrain facility, a rigid wall with weep holes, a temporary wall with open sheet, etc. In all type of the retaining walls, GT is used where it serves drainage as a primary function and protecting the wall as a secondary function.

3.4 Filtration

GT enables to perform two contrasting functions simultaneously such as soil retention and permeability of water through and along the GT. Open pore size of geotextile is selected to avoid several interrelated phenomena such as blocking, blinding and clogging. An open geotextile structure is needed to pass the liquid through the filtre (GT) in order to avoid excess hydrostatic pore pressure and at the same time, a close structure is needed in the upstream for soil retention. As the amount of liquid flow through soil is related to the particle size, this kind of contradictory design is possible. A common empirical relationship between permeability coefficient and soil particle size is used in this regard.

$$k = \mathrm{C}\mathrm{d}_{10}^{2} \tag{5}$$

where k is the permeability coefficient, C is the site-specific constant and d_{10} is the effective soil particle size at which 10% of the soil is fine.

Large particle size relates to the high flow conditions requiring a GT with sufficiently larger pore size and smaller particle size relates to low flow conditions requiring the GT to be designed with smaller pore size.

Clogging of the geotextile is a process in which fine silt particles settle on or in the geotextile or at the interface between the geotextile and the soil. Clogging reduces the flow of water.

When the major fraction of the actual soil close to the geotextile is fat clay or fine silt particles, incorporation of a thin layer of filtre sand between the soil and the geotextile will be advantageous. It is accepted, that a certain part of the soil is smaller than the pore openings of the geotextile, and there will be a loss of the smallest soil particles. However, this will stop after a period of time due to the formation of a natural filtre.

3.5 Erosion Control System and Stabilization of Slopes

- **Stabilization of steep slope**

This application differs from the others in the sense that GTs are directly laid over the top of the surface without submerging it in the soil. Slopes erode due to the migration of surficial soil particles. The surface run-off is caused by the flow of water and severe wind. Slopes can be managed by restricting migration of surficial soil particles. Different soil erosion zones are described in Fig. 5. Geotextile material controls erosion of soil by reducing direct impact of rain droplets, creating mini check dams and arrest migration of surface soil particles. The mesh-like structure acts as mini check dams and traps seeds and soil particles during free water flow. Subsequently, by holding moisture, it assists in creating a microclimate encouraging the germination of seeds and growth of vegetation for the conservation of soil.

Mechanism of surface erosion in slopes

Precipitation, surface flow, subsurface flow and storage can be correlated as-

$$\text{precipitation} = \text{surface flow} + \text{through flow (sub} - \text{surface flow)} + \text{storage}$$

- if throughflow and storage can be increased then surfaceflow gets reduced
- through flow depends on soil permeability
- if precipitation is high, surfaceflow will dominate
- if soil permeability is high, through flow will be dominant
- the problem gets complicated as the infiltration rate varies with time
- the aspect of storage can be facilitated by creating micro barriers on the slope. Such micro barriers also curb the intensity of surface flow.

GT, when inserted within an embankment in appropriate layers, can prevent rotational slides. Erosion of top soil either on a flat ground or on a slope can be

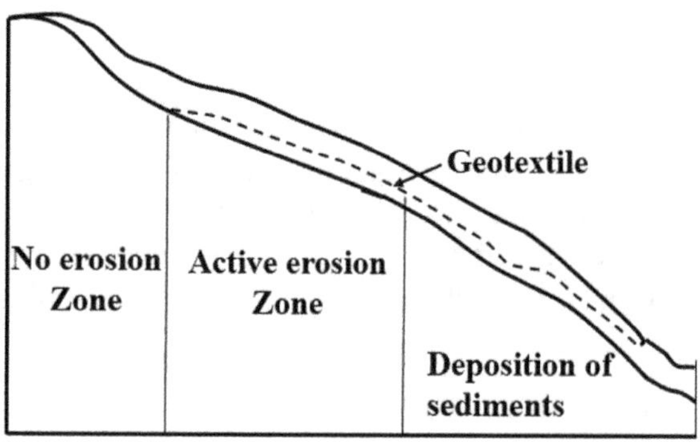

Fig. 5 Schematic representaion of different zones of erosion

effectively prevented by open weave GT. Three-dimensional construction of open weave GT helps to reduce the velocity of surface run-off by interposing successive micro barriers to the direction of flow and entrap the soil particles; dissociated by the kinetic energy of raindrops. It has also been reported that LC-based GT add micronutrients to the soil and do not add nitrogenous product on biodegradation. Moreover, these GT residue is beneficial as it enhances the hydraulic conductivity of the soil. LC material-based GT helps to retain the moisture and soil fertility, therefore helps in crop cultivation process of steep slopes (Vishnudas et al. 2012).

- **Stabilization of soil utilizing biofuel co-product:**

Lignin or lignin derivative products can reinforce soil directly or by forming humic acids. Biofuel industry utilizes several lignocellulosic-based biomass. The coproducts from biorefinery are used to reinforce sandy soil (Ceylan et al. 2015).

- **Slit fencing:**

Slit fences are used to prevent the run-off of sediment into the downstream of rivers. Sometimes a geogrid is also associated with GT and used for this purpose. Slit fences are anchored in a trench. As the material is exposed to sunlight it must be UV stabilized along with other GT properties. In this application GT is first used as a filtre then it clogs with soil particles in the lower part.

Protection of river and canal banks:

Bitumen treated woven JGT has performed satisfactorily in controlling erosion of the river and canal banks in different trials. Woven JGT can serve as a better and cost-effective substitute of the conventional granular filtre. Availability of granular aggregates often poses difficulty, apart from the difficulties encountered in exercising quality control. A layer of woven JGT treated with a suitable water repellent additive may replace the layers of granular aggregates. An armour layer over the fabric is, however, necessary to prevent the fabric displacement and its exposure to weather (Ghosh et al. 2009). The schematic diagram of the mechanism is represented in Fig. 6 and the different installation sequences are described in Fig. 7.

Control of surface soil detachment by GT:

- can prevent rain splash detachment of surficial soil particles and arrest their migration. Act as mini check dams.
- reduces surface run-off
- reduces erodability coefficient of soil
- protects soil and seeds from direct exposure
- controls dehydration of soil
- allows air and light
- adds nutrients to the soil and acts as mulch on biodegradation (natural GT).

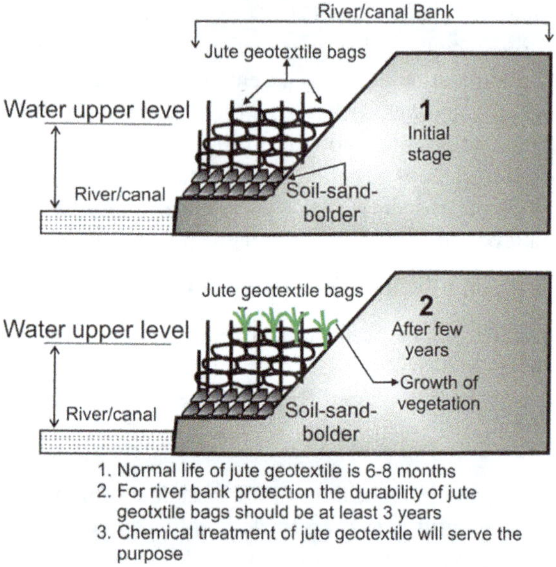

Fig. 6 JGT for rever/canal bank-mechanism

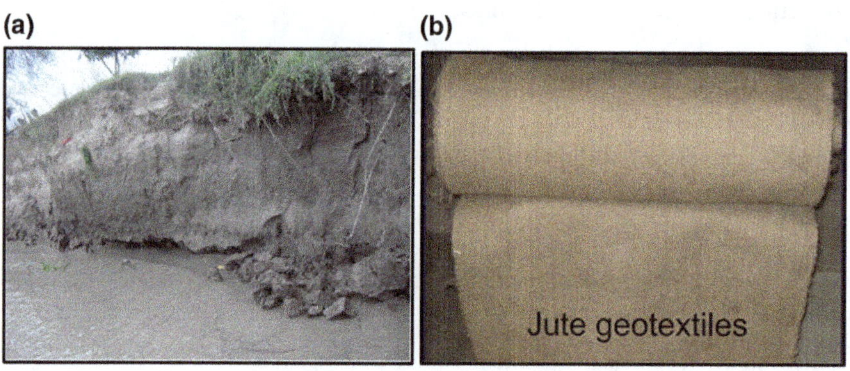

Fig. 7 Installation of Jute geotextile (JGT) for prevention of soil erosion **a** river embankment **b** JGT woven fabric

Market estimation of Jute Geotextile (JGT) for erosion control application

In Indian scenario JGT has been widely explored especially in the eastern region and coir-based GT is explored in the southern region. Again needless to say that selection of material depends on the availability of the material in that particular geographical location to make the product commercially viable. Acceptable specification that generally used for erosion control applications are presented in Table 2.

Table 2 General working specifications of jute geotextiles (JGT)

Type of fabric	Weight (gsm)	Width (cm)	Threads		Thickness (mm)	Quality ratio = Tensile strength of yarn/count of the yarn x 100	Tensile strength (KN/m)		Water absorption capacity (%)	Indicative price (Rs/m^2)	Application areas
			MD	CD			MD	CD			
B-Twill woven type 1	650	40	10	10	0.5	0.55 (wd), 0.50 (wed)	20	18	80	18–20	River embankment
B-Twill woven type 2	680	42	11	12	0.45	0.65 (wd), 0.55 (wed)	21	20	70	18–20	Do

MD machine direction, *CD* cross machine direction, *wd* warp direction and *wed* weft direction

The estimated market demand of geotextiles only for erosion control application for low/medium income countries has been calculated by considering the market share of the countries such as India and Bangladesh and the estimated JGT market for high-income countries has been calculated with respect to USA and Canada. Initially, the areas were selected around the world where the annual rainfall is more than 600 ml, after that the areas where the probability of flood is in higher side (more than 15 flood in last 30 years) were identified. For each case the total length of flood-prone river dikes in both coastal and non-coastal areas was calculated as per the information obtained from (http://www.banglapedia.org/docs/HT/R_0207.HTM), Central Water Commission, India, Draft recommendation for a National Levee Safety Program, 15 January, 2009. Then the total length was estimated considering the length of right bank and left bank of river side as shown in column 3 of Table 3. The percentage of potential embankment length out of total length was estimated as per the report. (http://www.banglapedia.org/docs/E_0049.HTME), Draft recommendation for a National Levee Safety Program, Jan 15, 2009, (www.textileworld.com/Articles/2011/January/Jan_Feb_issue/Geotextiles.html). Percentage of length out of total embankment length has been estimated as 25% (Ghoshal and Som 1993) and the potentially JGT requirement out of the length for geotextiles was considered as 10% (Chand 2008).

The most potential areas considered as having emerging markets for JGT all over the World are Pacific Asia, USA, Japan, Europe, Canada, Australia, South Africa and South America (http://www.the-infoshop.com/report/fd90373-eosynthetics.html). Region wise global market share of JGT is presented in column 2 of Table 4. Overall world geotextiles market potential was estimated from the annual reports on the construction industry prepared by Business Monitor International (2008), The US market for technical textiles, 2002, report from David Rigby Associates (2008) (www.textileworld.com/Articles/2011/January/Jan_Feb_issue/Geotextiles.html) and Crosth-waite and Connaughton (2008) considering the US demand for geotextiles in column 4 of Table 4 Following this the market for rest of the countries and rest of the world was calculated and presented in Table 4. The same ratio as India and Bangladesh was applied for low-middle income countries and the ratio of USA-Canada was applied for other higher income countries. During the selection of the rest of the world, the regions with the rainfall less than 500 mm were not considered for the market estimation of the JGT.

Table 3 Estimated market for Jute Geotextile for erosion control (2012–2013)

Country	Flood-prone river length (km)	Estimated repairable length of river dikes (km)	Geotextiles needed (km)[a]	JGT market (Sqm)[b]	Market (crore INR)
India	24,122	7960[c]	1990	3,184,104	3.18
Bangladesh	11,307	7462[c]	1865	1,492,524	1.50
USA	92,339	16,621[d]	4155	6,648,400	6.64
Canada	9233	1662[d]	415	664,840	0.66

[a]Considering that 25% of the repairable length can be captured by geotextiles [b]Length of geotextiles needed × 0.1 × 1000 × 16 [c]estimates 33% of the length of flood-prone river [d]estimates 9% of the length of flood-prone river

Table 4 Global market estimation of Jute Geotextile for erosion control (2012–2013)

Area	%Market share	Geotex market (INR Crore)	JGT market (INR Crore) @10% in 2012–2013
Asia Pacific	41	155.8	15.58
USA	24	91.2	9.12
Canada	2.4	9.12	0.912
S America	6	22.8	2.28
Japan	10	38	3.38
Europe	11.6	44.08	4.408
Australia	2	7.6	0.76
Africa	1	3.8	0.38
Rest of world	2	7.6	0.76
Total INR (2012–2013)	100	380	38.00

Stability of slope by wave flume test:

The stability of the slope is predicted by an actual field analogy-based wave flume test. Wave flume test consists of two apparatus they are wave generator and deformation recording system. A metal panel driven by electric motor through connecting rod and flywheel is used to generate waves of different energies and frequencies. Thus the actual condition in the site is simulated. The measurement system actually measures the deformation of revetment surface by longitudinally flexible measurement bar and wire rollers. At the upper part of the revetment, wave exerts excess pore water pressure and results in soil collapse. At middle part soil eroded by up and down drag force of the flow and erodes along the inclined bank. These two mechanisms can be explained by seepage along the parallel and perpendicular direction. Downward seepage parallel to the slope occurs when the water level drops in the front of the slope. Similarly, upward flow normal to the slope also causes erosion. In the up rush flow, pore water pressure increases and carries soil upwards. Eroded soil can pass through the GT, leading to an increase in pore water pressure or it can migrate along the gaps by flowing downwards and finally deposited on the surface. A thick GT with a low number of constriction is appropriate for protection from the cyclic flow (Faure et al. 2010).

4 Designing of Geocomposites with Lignocellulosic Materials

Erosion control and revegetation:

To offer higher performance and diverse application range with a single product, attributes of different materials are combined. GT alone can be used for erosion control and revegetation purpose. However, to withstand sunlight, rain, wind and other environmental constraints, specialty geocomposites are beneficial instead of the lower GSM fabric-based GT. Erosion process caused by water or wind medium can be described by three sequence-detachment, transportation and deposition. The flexible composites must be anchored with the soil system either wise poor functioning will cause the system to fail. Materials can be of two type's temporary erosion controllable and permanent erosion controllable. Permanent erosion controllable geocomposite materials facilitate revegetation which eventually becomes entangled with the root system of the vegetation and acts as a reinforcement to the soil. Natural biodegradable materials are very frequently used for the temporary erosion controllable product. Jute geonet in combination with non-woven geotextile mat is used for mulching application (Ahn 2002).

Reinforcement:

Natural LC materials used as GT have several advantages, however, they lack in strength compared to the conventional GT materials. Thus, to overcome the inadequacies of a GT, geocomposites are tried as reinforcement. Sometimes lignocellulosic GT materials are combined with the conventional geosynthetic materials. A modified GT can be prepared by combining different polymers, or a strengthened GT can be prepared by reinforcing it with fibreglass or steel. The basic purpose of geomembrane is the transportation of fluid or vapour, however, single geomembrane lacks in tensile strength. Thus, geonet and geogrids are utilized to strengthen the geomembranes and then these material acts as geocomposites. Fibre, meshes, mattresses are used in conjunction with several geosynthetics as reinforcement to the soil. To overcome the brittle nature of different structural materials, fibres are used as reinforcement. Fibre reinforced building material such as concrete offers greater resistance against cracking, shrinkage. LC fibre reinforced concrete allows designing thinner section retaining sufficient strength (Koerner 1993).

Filtration:

In railroad construction, GTs are used and they experience higher abrasion at the upper surfaces. At the same time, filtration is also required for these materials at the lower side. Coarsening of GT or increasing the open pore size of GT does not provide any solution as they introduce several complexities. To overcome this situation a common method is to attach different types of geosynthetic family members together. The lower part is made of non-woven material and the upper part is prepared by giving an abrasive resistant coating on the GT material. GTs are often combined with geonets. A GT is placed over the geonets to avoid the extrusion of clay through the geonet. Non-woven GTs are used in such kind of applications.

Drainage:

The principle of drainage remains same a discussed in the earlier section. However, for wide and larger planer areas geonets/geotextile composite is used.

The in plain flow capacity of a geonet/geotextile composite is much higher than the flow capacity of a very thick GT.

Containment or liquid vapour barrier:

This basic mechanism of this type of containment by geocomposite is utilizing a GT which provides strength and a bitumen which provides the moisture barrier. The majority of this type of product deal with various synthetic materials like fibre glass, high-density polyethylene, polyester film, etc.

5 Tests and Parameters for Evaluating the Performance of Geotextiles and Geocomposites

To evaluate the performance of Geotextiles most of the test methods are adapted from the textile industry. This includes testing of physical and mechanical properties of the GT nonetheless the modification came in the assessment of properties like porosity, environmental factors and degradation rates (Koerner 1993).

5.1 Physical Properties Testing—The Index Tests for GT

Specific gravity:

While designing GT for a specific purpose establishing a relation between volume and density is needed, thus specific gravity of the material needs to be evaluated. According to the ASTM D 792 or ASTM D1505 specific gravity is defined as the ratio of material's unit volume to the volume of distilled, de-aired water at 4 °C.

Mass per unit area:

It gives the proper idea about the weight of a textile material. Gram per square metre (GSM) is the most widely used unit in this regard. The test is performed according to ASTM D 5261 and ISO 9864.

Thickness:

For filtration, separation type of application measurement of thickness in unavoidable. The thickness is measured at a specific pressure and the tolerance level of accuracy should remain within 0.02 mm. ASTM D 5199 and ISO 9863 stands deal with measuring the thickness of a geotextile material. Thickness range between 0.25 and 3 mm is used for most of the common GT materials.

Stiffness:

The term stiffness relates to the bending stiffness of the material. Free fall of material under its own weight is used to perform this kind of test. A strip of 25 mm fabric is slide horizontally over a surface lengthwise. The material bends gravitationally under its own weight and touches a plane inclined at an angle of 41.5°. This bending length is used for measuring bending modulus and flexural rigidity. The common Shirley apparatus is widely used for calculating the bending length. ASTM D1388 is related to the stiffness (Savile 1999).

$$C = L\left(\frac{\cos\frac{1}{2}\theta}{8\tan\theta}\right)^{\frac{1}{3}} \tag{6}$$

where C is the bending length and L is the length of the fabric projecting and θ is the angle with which the fabric bends.

Flexural rigidity

$$G = M * C^3 * 9.807 * 10^{-6}\,\mu\text{NM} \tag{7}$$

M is the fabric mass per unit area GSM.

Bending modulus

$$\text{Bending modulus} = \frac{12G10^3}{T^3}\text{N/m}^2 \tag{8}$$

T is the fabric thickness (mm).

Stiffness provides a suitable working surface for the installation of GT. High stiffness is required for soft soils.

Drape:

Knitted fabrics are floppy and easily follow the typical curvatures of the base material. However, woven textiles are preferred due to their superior mechanical properties over the knitted ones. Woven fabrics are relatively stiff over the knitted fabrics. Hence, sometimes drape measurement is also carried out after stiffness measurement. How the material follows the contour of soil can be evaluated by this test. In this test, a circular fabric is placed over a disk and hangs under its own weight. The diameter of the fabric should be greater than that of the disk. The fabric folds around the disk under its own weight and the shadow is cast on the paper. Then it is cut along the shaded path. From this draped trace paper the area of the supporting disk is also left out. Finally, the mass of the draped part is measured. In another tracing paper the mass of the strip, between fabric diameter and the disk diameter is measured. The drape coefficient is calculated from the ratio of two masses.

5.2 Mechanical Testing

Tensile testing:

Tensile strength is the most important parameter while designing as it directly relates to the endurance capacity of a material. For reinforcement purpose, it is a primary function while for separation, drainage, filtration it acts as a secondary function. Tensile strength of single fibre, fibre bundle, yarn and fabric all can be evaluated by different tensile testing instruments. The need of testing either fibre, yarn or fabric depends on the type of application. Testing (generally for yarn) instruments generally worked under three mechanisms

CRE Constant rate of extension
CRL Constant rate of loading
CRT Constant rate of traverse

CRE and CRL type of instruments are preferred to compute this kind of scientific work as precise control over testing speed, gauge length, etc. are possible. In CRE principle rate of increase of specimen length is uniform with time and in the case of CRL principle increase of the load is uniform with time and rate of extension is dependent on the load-elongation characteristics of the specimen. In CRE principle specimen experiences a high rate of loading while in CRL principle specimen experiences a high rate of extension.

To perform grab tensile test larger width of fabric than the jaw width is used while in strip test fabric width exactly similar to the jaw width is used. Grab tests are performed to measure the fabric assistance of lateral threads, sometimes the effect of selvage's are also approximately estimated with this kind of test. When the wider width fabric is stretched under tension the fabric experiences severe poison's ratio effect and it tries to rope up giving higher value of strength (Booth 1996).

Seam strength:

Often in practical purposes, the end of a GT is needed to be stitched together with another GT for making it suitable for application purposes. However, stress transfers from one GT to another through the seam. Several types of stitching are utilized by the textile industry to improve the efficiency of the stitched part. Needless to say that the seam part of the fabric remains much weaker than the overall GT and hence it is needed to be evaluated before using it. Laboratory grade seam strength measurements for GT follows ASTM D4884 and ISO 13426 standards. Seam efficiency can be calculated by

$$\text{Seam efficiency}(\%) = \frac{T_{\text{seam}}}{T_{\text{GT}}} * 100 \tag{9}$$

where T_{seam} and T_{GT} are the wide width tensile strength of seam and overall geotextile, respectively.

Burst strength:

The need of burst strength is mentioned above in the separation function of GT. According to ASTM D 3786, a 30 mm diameter-based hemisphere made of rubber is used to deform GT until the failure occurs.

Tear test:

The commonly used tear tests for GT are tongue tear test, trapezoidal tear test, Elmendorf tear test. During installation of GT, a material undergoes a different type of tear stresses. The chance of a fabric failure occurs from a small tear which acts as an initiator. To mimic this situation, a pre-cut is made in the sample according to the test requirements. Sample with initiation cut is loaded on the instrument and the strength or energy required to tear the sample along the path of the initiation cut is measured. For trapezoidal tear test, the samples are loaded on a tensile testing instrument itself. The widely used testing method is Elmendorf type, where a pendulum lever system is used tear apart the fabric along the line of initiation cut. Tear tests are performed generally on the woven fabrics only (Booth 1996).

Compressibility:

The compressibility of a geotextile is relatively low and for heavier woven GT it does not have any direct consequences on designing. However, for needle-punched non-woven fabric, used for filtration and transportation of liquid, compressibility is an important parameter for designing. The transfer capacity of the non-woven GT decreases under compression.

Fatigue testing:

Fatigue resistance is the ability of a material to sustain the repeated cyclic loading until the failure of a material. A typical fatigue test consists of several types of stress cycles where a GT of wide width variety is stressed (generally in CRE principle) to a predetermined load below its ultimate strength and then returned back to a lower or zero load. This stress cycle is repeated until the failure occurs and the number of cycles required to fail the material is calculated. Several types of stress cycles are reversible stress cycle, repeated stress cycle, random stress cycle. It can be referred easily that the lower value of predetermined load gives a higher number of loading cycles. Thus, when a GT is exposed to a certain load the resulting lifetime of that material can be correlated with this kind of test. So, increasing the maximum allowable exposure strength of a GT will reduce the lifespan of the material and vice versa.

Impact test:

Resistance offered by a material under sudden load is estimated in this type of test. Several construction materials during installation, rocks, stones, etc. can readily cause puncture and tear to the material. Thus fabric with a sufficiently high impact resistance is required for field applications. An impacting cone attachment with Elmendorf-type tear test can directly give the energy required for fabric breakage. However, this type of attachment is performed to test fabrics with low impact strength. Impact pendulum type devices are used for higher strength fabrics. Drop-cone type of test is another modified impact strength apparatus where a fabric is punctured first followed by tearing of the material.

Puncture test:

Impact resistance measures the performance of a material under dynamic conditions. However, the behaviour of a textile material against the impression of stones, rocks are predicted by the puncture test. A simple method is used where a fabric is clamped against a cylinder and a steel rod is forced at a prescribed rate to puncture the material. The resistance to puncture is measured in force units. ASTM D 4833 standard is associated with this test. The small size of the above-mentioned test apparatus causes large deviation in the test result of the same material between its low density and high-density region. So, to avoid the discrepancy of the test, larger size puncture test apparatus has been developed. This type of test incorporates the conventional soil testing apparatus—CBR (California bearing ratio) plunger and mould, used in civil engineering.

Frictional behaviour

A sufficiently high frictional coefficient between GT and soil is required to prevent the slippage or sliding away of material under normal atmospheric condition. A common shear test apparatus in which GT is firmly attached to the one-half of apparatus and soil remains on the other half. The upper half attached with GT is placed over the soil part. The required amount of shear force needed for sliding is recorded. The response between GT to soil and soil to soil is compared, and according to that shear strength efficiencies of GT on the cohesion behaviour of soil are measured.

Pull-out test:

Pull-out test estimation is required for GT used in anchored or reinforcement applications. In these case, GT is covered with soil particles on either side. This strength depends on the normal force applied to the soil. This test greatly resembles the direct shear test method used for frictional behaviour. However, they are different as in pull-out test they are deformed heavily. Thus the resistance offered by a material in this type of test is lower in value than that of the shear test resistance values.

5.3 Hydraulic Properties Testing

Porosity:

Porosity is the ability of liquid flow through or within the GT. It is dependent upon the pore size. Porosity of geotextile is defined as the ratio of void volume to the total volume. Yet, in practical situation, it is indirectly measured by the following formula

$$n = 1 - \frac{m}{\rho t} \tag{10}$$

where '*n*' is the porosity, '*m*' is the mass per unit are (GSM), 'ρ' is the density (g/m^3) and '*t*' is the thickness of fabric (m).

Apparent opening size (AOS) is estimated according to ASTM D 4751 standard in which glass beads of different diameter is sieved through the material. The known diameter glass beads are passed through until the weight of the bead passed through the specimen remains at 5% by weight. This defines the O_{95} opening size standard. Another test method calculating percent open area (POA) is utilized for monofilament woven geotextile. A light source is passed through the fabric to measure POA. The overlapping fibre in non-woven restricts the light source to pass hence this method is not applicable for the non-woven.

Permeability:

Filtration through geotextile involves transportation of fluids perpendicularly. To avoid any hindrance in the path of flow of water, cross-plane permeability under zero stress must be enumerated. The cross-plane permeability values of a fabric change under applied stresses. So, to make the testing, performance oriented 2–5 layers of GT is placed in an open mesh-like grid made of stainless steel. Normal pressure on top of the specimen is applied by implying ceramic balls. So, variation in fabric thickness under normal condition (for LCs) or under compression (for all type of material) is an issue of major concern for the estimation of in-plane permeability of a GT. All of the above permeability calculation involves Darcy's formula

$$q = k_p i A \tag{11}$$

$$q = k_p i (W * t) \tag{12}$$

$$k_p t = \theta = \frac{q}{iw} \tag{13}$$

where '*q*' is the flow rate (m^3/sec), 'k_p' is the permeability (m/sec), '*i*' is the hydraulic gradient, '*W*' is the width of fabric (*m*) and '*t*' is the thickness of fabric (m).

Soil retention:

Soil retention test is obligatory for a GT used in underwater slit curtains, slit fencing purposes. A known amount of slit uniformly mixed in water is allowed to pass through the GT and flow rate, the velocity of slit passing through the GT is calculated. It indicates the void spaces and clogging behaviour of a GT. Fresh water is constantly added in the mixture of slit to obtain a uniform head during testing.

5.4 *Serviceability Tests*

Creep tests:

Creep is the elongation of a material under application of prolonged constant stresses. Synthetic polymers are very much sensitive to creep hence this test is unavoidable for synthetic GT. However, the performance of natural lignocellulosics under the prolonged application of load needs to be estimated also.

Abrasion tests:

Abrasion tests are basically wearing behaviour of a material where a GT is abraded against another surface. Several ASTM methods are associated with the abrasion test. Comparative tensile properties are evaluated between the abraded and non-abraded test specimen. After abrasion of a specific number of cycles, the percentage retention of tensile strength is reported from this kind of tests. Disk kind of test apparatus is very simple (use of abrasion wheels), however, to make the test, field-performance oriented tumbler type of test is preferred where a basalt stone grade is used for abrasion.

Clogging test

Clogging test is important for prediction of retaining long-term flow capability of a GT. It is a process in which fine silt particles settle on or in the GT. The soil particles from the site are procured and placed directly over the GT material which is subjected to the clogging test. A constant flow head of slit is maintained throughout the process of investigation. For natural LC materials, biological clogging under leachate (a typical heterogeneous liquid material takes of several constituents while passing through the solid material) flow is an area of concern. Biological clogging measurements are very much necessary for drainage and filtration purpose. The permeability of a GT decreases with time as several acidogenic conditions promote bacterial growth. Biological clogging is a complex phenomenon and several types of research are going for the estimation of clogging under constant leachate flow (Palmeira et al. 2008).

Degradation behaviour:

Degradation or loss of material occurs due to the molecular chain scission, bond breaking and extraction of components. Degradation can be caused by several factors such as sunlight (UV degradation), temperature, hydrolytic action, oxidation (polyolefin are mostly susceptible to this kind of degradation), chemical reaction, biological attack (through microorganisms), ageing of material and from several other mechanisms. Intensities and spectrum of sunlight vary with time of the year. Geographical location, atmospheric pollution, cloud cover, humidity, wind, etc. also affects the UV degradation of LC materials over the time. Laboratory tests should simulate this factors while considering sunlight degradation. Under high temperature, the temperature dependent degradation rate of a polymeric material accelerates.

GT is used as a base material for most of the geocomposites. Hence the tests for evaluating the properties of geocomposites can be estimated with the discussed test methods commonly applicable for geotextile.

6 Applications

Application of geotextile and geocomposites has been discussed already according to their function (Ghosh et al. 2009). The basic five type of application are as follows:

i. Filtration—restraining of soils subjected to hydrodynamic forces while allowing passage of water.
ii. Separation—prevention from intermixing of dissimilar soils.
iii. Drainage—collecting and transporting water in the plane of a geotextile.
iv. Reinforcement—use of stress–strain behaviour of a geotextile to improve the mechanical properties of soil.
v. Protection—preventing or limiting local damage by the use of a geotextile.

7 Conclusions

Geotextiles and geocomposites are subjected to various kind of exposures (chemical, mechanical and physical) and the required functionalities primarily lead the designing of geotextiles. Designing is the process where an ideal material is selected by following an optimized parameter to meet the demands of specific end uses. This selection of candidate depends on several properties which are required to be estimated depending upon material's durability under potential exposure conditions to the field. However, the property estimation of a material is based on typical experimental results.

Prime importance is given to cost-effectiveness and availability of the material while designing geotextiles and geocomposites. For lignocellulosic (LC) material, the property of the same fibre varies with geographical location. So, selection of an LC material for designing geotextile and geocomposite should consider exposure conditions for end-use application. LC-based geotextiles are suitable in soil erosion prevention, revegetation, restoration and reinforcement applications. Natural fibres have superiority in many aspects over the synthetic ones such as cost, availability, drapability, etc. Drapability of natural fibre-based mat enables the material to fit in the contour of the slopes perfectly and thus, it helps in strengthening and stabilizing stream banks, slopes, wetlands, hillside soils, etc. The degraded product helps in adding nutrients to the soil. However, durability of these LC materials are the most concerned area of research. Chemical alteration in fibre chemistry or degradation-resistant surface coating of fibres are the two most commonly used method for durability enhancement of natural fibres.

References

Ahn TB, Cho SD, Yang SC (2002) Stabilization of soil slope using Geosynthetic mulching mat. Geotext Geomembr 20:135–146

Anand S (2008) Designer natural fibre geotextiles—a new concept. Indian J Fibre Text Res 33:339–344

ASTM D4439—15a Standard terminology for geosynthetics ASTM international Sept 2015

Booth JE (1996) Principles of textile testing: an introduction to physical methods of testing textile fibres. In: Yarns, Fabrics 3rd edn. CBS Publishers & Distributors Pvt. Ltd

Ceylan H, Kim S, Uzer AU, Yang B (2015) Strength performance of Iowa soils stabilized with biofuel industry co-product. Procedia Eng 125:317–323

Chand V (2008) Market and marketability of jute geotextile. Proceedings international workshop on jute geotextiles—technical potential and commercial prospects

Chand N, Fahim M (eds) (2008) Tribology of natural fiber polymer composites–chapter 1 natural fibers and their composites

Crosthwaite D, Connaughton J (2008) World construction 2007-2008. Davis Langdon Management Consulting, London, UK

Faure Y-H, Ho CC, Chen R-H, Le Lay M, Blaza J (2010) A wave flume experiment for studying erosion mechanism of revetments using geotextiles. Geotextiles Geomembranes 28:360–373

Ghosh M, Choudhury PK, Sanyal T (2009) Suitability of natural fibres in geotextile applications. IGC Geotide, pp 497–501

Ghosh SK, Bhattacharyya R, Mondal MM (2014) A review on jute geotextile—part 1 International Journal of Research. Eng Technol 3(2):378–386

Ghoshal A, Som N (1993) Geotextiles and geomembranes in India-state of usage and economic evaluation. Geotext Geomembr 12:193–213

Goda K, Takagi H, Netravali AN (2008) Chapter 13-Fully biodegradable green composites reinforced with natural fibers- Natural fibre reinforced polymer composites: from macro to nanoscale Sabu Thomas. Laly A, Potha

Gupta MK, Srivastava RK, Bisari H (2015) Potential of jute fibre reinforced polymer composites: a review international journal of fiber and textile research 5(3):30–38

Horrocks AR, Anand SC (2000) Handbook of technical textile. Wood head Publishing Limited

Koba Y, Ishizaki A (1990) Chemical composition of palm fiber and its feasibility as cellulosic raw material for sugar production agric. Biol Chem 54(5):1183–1187

Koerner RM (1993) Designing with Geosynthetics 5th edn

Komuraiah A, Kumar NS, Prasad BD (2014) Chemical composition of natural fibers and its influence on their mechanical properties. Mech Compos Mater 50(3):359–376

Kumar R, Obrai S, Sharma A (2011) Chemical modifications of natural fiber for composite material pelagia research library. Der Chemica Sinica 2(4):219–228

Lekha KR (2004) Field instrumentation and monitoring of soil erosion in coir geotextile stabilized slopes—a case study. Geotext Geomembr 22(5):399–413

Lekha KR, Kavitha V (2006) Coir geotextile reinforced clay dykes for drainage of low-lying areas. Geotext Geomembr 24(1):38–51

Mwasha A (2009) Using environmentally friendly geotextiles for soil reinforcement: a parametric study. Mater Des 30:1798–1803

Palmeira EM, Remigio AFN, Ramos MLG, Bernardes RS (2008) A study on biological clogging of nonwoven geotextiles under leachate flow. Geotext Geomembr 26:205–219

Rowell RM (1996) Chemical modification of non wood lignocellulosics, chapter 9. In: Hon DNS (ed) Chemical modification of lignocellulosic materials. Marcel Dekker, New York

Rowell RM, Stout HP (2007) Jute and Kenaf fibers Chapter 7. In: Lewin M (ed) Handbook of fiber chemistry, 3rd edn. CRC Press, Boca Raton

Saha P, Roy D, Manna S, Ba Adhikari, Sen R, Roy S (2012a) Durability of lignocellulosic fibers treated with vegetable oil–phenolic resin. Carbohyd Polym 87:1628–1636

Saha P, Roy D, Manna S, Adhikari BA, Sen R, Roy S (2012b) Durability of transesterified jute geotextiles. Geotextiles Geomembranes 35:69–75

Saha P, Chowdhury Sukanya, Roy D, Adhikari B, Kim JK, Thomas S (2016) A brief review on the chemical modifications of lignocellulosic fibers for durable engineering composites Polym. Bull. 73(2):587–620

Sarsby RW (2007) Use of 'Limited Life Geotextiles' (LLGs) for basal reinforcement of embankments built on soft clay. Geotext Geomembr 25:302–310

Savile BP (1999) Physical testing of textiles, 1st edn. CRC Press, Boca Raton

Sen Gupta AK (1991) Geotextiles: opportunities for natural-fibre products. International Trade Forumv 27 10(6)

Verma D, Gope PC, Shandilya A, Gupta A, Maheshwari MK (2013) Coir fibre reinforcement and application in polymer composites: a review. J Mater Environ Sci 4(2):263–276

Vishnudas S, Savenije HHG, Zaag PV, Anil KR (2012) Coir geotextile for slope stabilization and cultivation—a case study in a highland region of Kerala. South India Phy Chem Earth 47–48:135–138

Winandy JE, Rowell RM (2005) Chemistry of wood strength. In: Handbook of wood chemistry and wood composites. CRC Press, Boca Raton

www.textileworld.com/Articles/2011/January/Jan_Feb_issue/Geotextiles.html

www.engr.utk.edu/mse/Textiles/Geotextiles.htm

www.the-infoshop.com/report/fd90373-geosynthetics.html

www.banglapedia.org/httpdocs/HT/R_0207.HTM

www.banglapedia.org/httpdocs/E_0049.HTME

Lignocellulosic Fibres-Based Biocomposites Materials for Food Packaging

H. Angellier-Coussy, V. Guillard, E. Gastaldi, S. Peyron and N. Gontard

Abstract Current requests in the field of food packaging lead to the reasoned design of materials able to improve the global environmental balance of the food/packaging system by minimizing the negative environmental impact of the packaging material while improving its positive role in the food wastes and losses reduction that strongly impact our environment. This means to simultaneously control food degradation reactions while limiting undesirable migrations of additives from packaging towards in respect of our health and remaining economically competitive. The substitution of oil-based materials by ones issued from renewable and non-food resources (e.g. issued from bioconversion of agro-food wastes, for example) and furthermore, fully biodegradable in natural conditions is also a necessity and represents a significant breakthrough from the research in the field of food packaging. In this context, increasing attention is given to full-biocomposites, i.e. composite materials based on constituents all biosourced and biodegradable. Developing full-biocomposites for food packaging requires taken into account numerous factors, and this is even more important for complex biodegradable materials due to the gap in knowledge on their behaviour and potentialities in usage conditions. The objective of this chapter is to decipher the state of the art on full-biocomposites by considering the specific stakes relative to the food packaging application. After the first part of introduction, the second part will present the role

H. Angellier-Coussy (✉) · V. Guillard · E. Gastaldi · S. Peyron · N. Gontard
Unité Mixte de Recherche Ingénierie Des Agropolymères et Technologies Emergentes, INRA/Université de Montpellier/Montpellier SupAgro/CIRAD, 2 Place Viala, Cedex 1, 34060 Montpellier, France
e-mail: helene.coussy@umontpellier.fr

V. Guillard
e-mail: guillard@univ-montp2.fr

E. Gastaldi
e-mail: emmanuelle.gastaldi@umontpellier.fr

S. Peyron
e-mail: peyron@univ-montp2.fr

N. Gontard
e-mail: gontard@univ-montp2.fr

S. Kalia (ed.), *Lignocellulosic Composite Materials*, Springer Series on Polymer and Composite Materials, https://doi.org/10.1007/978-3-319-68696-7_10

of packaging to ensure food quality and safety and how it should be designed in such a way to reduce food waste and losses. The third part will present the window of mass transfer properties of full-biocomposites, which is the main functional property when considering the food packaging application. The fourth part will consider the economical competitiveness of full-biocomposites, the fifth part will treat the safety issues and the sixth of the different options of end of life and waste management.

Keywords Biocomposites · Food packaging · Mass transfer properties
Food safety

Contents

1 Introduction

All long their life cycle, packaging materials consume resources (most of them being non-renewable): materials, water and energy, leading to waste and pollutants accumulation in air, soils and ocean (e.g. the Great Pacific Garbage Patch, collection of plastic debris in the North Pacific Ocean). Eight million tons of plastics, principally packaging, leak into the ocean each year, equivalent to dumping the contents of one garbage truck into the ocean every minute. If no action is taken,[1] in 2050, this is estimated to increase to four garbage trucks per minute that will lead to find 1 kg of plastic per 1 kg of fish in the ocean. Significant efforts have been developed since the beginning of 2000s to encourage the minimization of resources uses (light-weighted 'just-necessary' packaging materials) the maximization of material recycling through decontamination or composting (rise of biodegradable packaging materials). The substitution of oil-based materials by ones issued from renewable and non-food resources (e.g. issued from bioconversion of agro-food wastes, for example) and furthermore fully biodegradable in natural conditions[2] is a necessity and represents a significant breakthrough from the research in the field of food packaging. The challenge relies nowadays on the improvement of the global environmental balance of the food/packaging system which is expected to become much more positive than it is now by minimizing the negative environmental impact of the packaging material while improving its positive role in the food wastes and losses reduction that strongly impact our environment. One should keep in mind that when a food product is lost or wasted, a packaging product is also discarded at the same time creating some additional waste management issues.

Among innovative technologies throughout the food supply chain, food packaging is a particular key player to improve food preservation, quality, safety and retailing/consuming conditions, and thus to reduce food losses in all countries (FAO 2011; Wikström and Williams 2010; Angellier-coussy et al. 2013a). Packaging is usually wrongly considered as an additional economic and environmental cost rather than an added-value for food losses and wastes' reduction. However, efficient packaging solutions could be a key factor for sustainable food consumption by improving food preservation and thus reducing food losses, by balancing the issue of energy consumption and/or food quality degradation related to conventional food processing (for instance, balancing cold chain issues by the use of modified atmosphere packaging technologies).

Reducing the overall environmental impact of the food/packaging system means to simultaneously consider the environmental impact of the packaging material itself, including waste management issues, in addition to its contribution to the reduction of environmental impact of food waste and losses. That means the development of eco-friendly packaging by reaching a compromise to answer the sometimes contradictory concerns of environmental, economic and social sustainability of food. This goal could not be achieved without the use of knowledge engineering and artificial intelligence methodologies to answer these multi-criteria and contradictory concerns of all stakeholders of the food chain.

---

[1]World Economic Forum, Ellen MacArthur Foundation and McKinsey & Company, The New Plastics Economy—Rethinking the future of plastics (2016) http://www.ellenmacarthurfoundation.org/publications.

[2]Biodegradable in natural land conditions or home composting conditions in opposition to industrially compostable materials such as PLA.

Ensuring the economic sustainability of food also means ensuring the economical competitiveness of the packaging. Thus, packaging materials should be manufactured using conventional techniques such as compounding, extrusion, injection or compression moulding by applying the industrial production rates. The performance/cost ratio should also always be kept in mind. Finally but not least, the health of the consumer has to be ensured by controlling the mass transfer of components from the packaging towards the food and vice versa.

In this context, lignocellulosic-based biocomposites appear as innovative and promising materials of future for the food packaging sector. In the objective of reducing the overall environmental impact of the food/packaging system, increasing attention is given to full-biocomposites, i.e. composite materials based on constituents all biosourced and biodegradable (Berthet et al. 2015a). Full-biocomposites are based on a biosourced and biodegradable polymer matrix, i.e. bio-polyesters (e.g. polyhydroxyalcanoates) and agropolymers (e.g. proteins, polysaccharides, etc.), and lignocellulosic particles or fibres as fillers. The cheapest and most environmentally virtuous lignocellulosic fillers would be those derived from agricultural wastes, forestry by-products or food industry solid by-products. The drawbacks hampering currently implementation of full-biocomposites at industrial scale are mainly their limited processability, unstability in usage conditions, high production cost (e.g. bio-polyesters) or controversial environmental claims. Developing full-biocomposites for food packaging requires taking into account all the previously presented stakes: reducing the overall environmental impact of the food/packaging system by (i) designing materials with functional properties that allow ensuring food quality and reducing food waste and losses, and (ii) reducing the environmental impact of the packaging material itself by an appropriate choice of resources and material end of life, while maintaining an economical competitiveness using conventional shaping processes at industrial scale, and ensuring food safety and health of the consumer.

The objective of this chapter is to decipher the state of the art on full-biocomposites by considering the specific stakes relative to the food packaging application. The first part will present the role of packaging to ensure food quality and safety and how it should be designed in such a way to reduce food waste and losses. The second part will present the window of mass transfer properties of full-biocomposites, which is the main functional property when considering the food packaging application. The third part will consider the economical competitiveness of full-biocomposites, the fourth part will treat of the safety issues and fifth of the different options of end of life and waste management.

2 The Role of Packaging to Reduce Food Waste and Losses

2.1 Importance of Mass Transfer in the Food/Packaging System

The main functions of packaging are to protect and stabilize food until its consumption. Food preservation could be defined in terms of reduction of degradation reactions, namely physical–chemical and microbial reactions for non-living products and also physiological reactions for living products. Food quality and safety are dependent on the initial quality of the raw materials and on the rate of degradation reactions, which occur during the processing of the food, storage, distribution and in-home preparations steps (Steele 2004). The rates of degradation reactions of food are function of temperature, which is controlled by the cold chain, light transmission and atmosphere composition around the food, which is controlled by packaging.

Indeed, packaging can play a major role by defining around the food, a headspace atmosphere whose composition is controlled via the mass transfer through the packaging materials from the environment towards the headspace and the food. It concerns the transfer of gases and vapours such as O_2, CO_2 or water vapour, which are key factors of food quality preservation by controlling food degradation reactions (such as oxygen influencing oxidation of vitamins, essential fatty acids, etc.). This technology is called Modified Atmosphere Packaging (MAP) and could be passive or active (Fig. 1). Passive MAP relies on the modification of internal packaging atmosphere due to an equilibrium between production/consumption of gases by respiring products such as fresh fruits and vegetables and gas permeation through the packaging material while active MAP relies on the gas depletion or gain by using gas flush associated with barrier packaging film or by using gas or vapours absorbing or emitting packaging materials (Floros and Matsos 2005; Ozdemir and Floros 2004; Guillaume et al. 2008; Raouche et al. 2011).

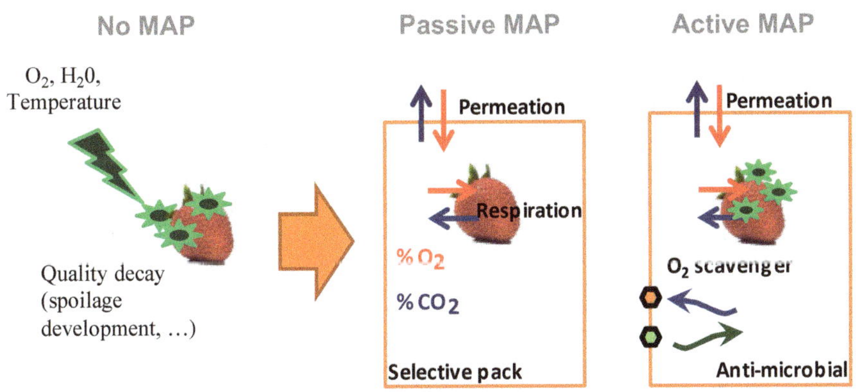

Fig. 1 Illustration of the MAP technology

In addition to the transfer of gases and vapours from the environment towards the headspace and the food, mass transfer can also occur from the foodstuff towards the packaging, which can be detrimental to the food sensory properties (e.g. scalping of aroma compounds), and also from the packaging towards the food (e.g. undesirable migration of potentially toxic molecules). All these transfers have an impact on food quality and safety. The mathematical modelling of these mass transfers is the basis for a better design and dimensioning of the food packaging system. Thus, tailoring of an adequate packaging material means designing it to control all these transfers precisely.

2.2 Modelling Tools and Decision Support Systems

Nowadays, choice of a packaging material is based on a lot of empiricism without any (or rarely) use of rational, requirement-driven approaches. Requirement-driven approaches consist in characterizing the needs and requirements of the food and then translating them into mass transfer properties. Once they are known, they are considered as a constraint in the choice or conception of the packaging. This optimized approach is based on the use of mathematical modelling tools and avoids over-testing and over-sizing of the packaging material and overall a better food quality and safety by adequately targeting the appropriate material and packaging dimensioning to the food requirements (Cagnon et al. 2013; Chaix et al. 2015). Beyond material permeability, the design or selection of packaging includes numerous other factors such as cost, industrial feasibility, environmental impact, safety and stability, waste management of the packaging material throughout all over the food/packaging life cycle. As these factors cannot be all simultaneously optimal, stakeholders are faced with the difficulty of finding a compromise. For instance, the food manufacturer may have the following query: '*I want optimal gas permeability in order to guarantee product quality throughout its shelf life and I would like a transparent packaging material made from renewable resources with a cost for raw material less than 3 €/kg …*'. This query could be completely different from the point of view of another stakeholder, for instance, the consumer who will favour food quality and safety.

Recent development in the field of knowledge engineering and artificial intelligence has permit to carry out a multi-criteria analyses of argued constraints and preferences of the different stakeholders of the food chain and to provide an answer to multi-criteria queries in packaging decision (Guillard et al. 2015; Tamani et al. 2015).

For example, in the case of modified atmosphere packaging (MAP) of fresh fruits and vegetable, MAP modelling tools have been integrated into a complete Decision Support System (DSS) that takes into account the preferences, acceptances and needs of the different stakeholders of the food packaging chain. This DSS proposes some argued and ranked packaging solutions for a given product (Guillard et al. 2015). Modelling tools are based on the fact that in MAP, gas concentration in the headspace changes according to an interplay between gas permeation through the packaging material, O_2 consumption and CO_2 production by the fresh food that respires, and until an equilibrium is reached. In order to extend the storage of the produce, this atmosphere must be as closed as possible to the optimal and recommended atmosphere. Such gas transfers in food/packaging system can be modelled by combining mathematical equations representing gas permeation through the material (such as Fick's law) and gas consumption and production during respiration (such as Michaelis–Menten-based equations, with possible carbon dioxide inhibition). Such models can be used either to predict gas evolution, i.e. partial pressures of O_2 and CO_2, within the food packaging knowing all the other parameters and/or to determine optimal O_2 and CO_2 permeability of the material, knowing all the other parameters and considering optimal recommended O_2 and CO_2 partial pressures as a target.

3 Mass Transfer Properties of Biocomposites for Food Packaging

Biosourced and/or biodegradable materials could display some mass transfer properties more suitable for fresh produces than classically used synthetic ones (Guillaume et al. 2010; Cagnon et al. 2012; Guillaume et al. 2013). In addition, recent outcomes have demonstrated that incorporation of nano- or microparticles such as lignocellulosic fibres into biopolymers permits to modulate their gas and water vapour transfer properties (Berthet et al. 2015b; Berthet et al. 2015c; Angellier-Coussy et al. 2013b) in order to match a specific application. Combining biopolymers with relevant active features based on biomolecules is also a solution to overcome their limitations for developing new application in the food sector (antimicrobial packaging for instance), without limiting sustainability. To achieve a tailored design of these eco-efficient food packaging solutions, it is not only necessary to understand the functional properties of the material as a whole but also those of each raw constituent, in particular, their mass transfer properties.

3.1 Basic Knowledge on Mass Transfers

Considering a composite material as a homogeneous material, mass transfer is supposed to obey to Fick's law which related the flux (J) to the gradient of concentration $\left(\frac{\partial c}{\partial x}\right)$ through a proportionality coefficient, (D):

$$J = -D\left(\frac{\partial c}{\partial x}\right) \tag{1}$$

where D is the coefficient of diffusion or diffusivity ($m^2\ s^{-1}$) which characterizes the mobility of a molecular species in the material.

Equation (1) could be easily simplified in case of monodirectional transfer by:

$$J = -D\left(\frac{\Delta C}{\Delta x}\right) \tag{2}$$

where ΔC is the difference of concentration on both sides of the plane composite material and Δx the thickness of the material.

Henry's law gives the correspondence between the concentration (C) and the partial pressure (p) for the dilute system:

$$C = S \times p \tag{3}$$

where S is the solubility coefficient (mol $m^{-3}\ Pa^{-1}$), a thermodynamic parameter which characterizes the number of moles of a molecular species sorbed into the material.

By combining Eqs. (2) and (3), the first Fick's law could be expressed for a monodirectional flux through a plane sheet of material as follows:

$$J = \frac{DS(p_1 - p_2)}{\Delta x} \tag{4}$$

where p_1 and p_2 are the pressure on the two sides of the film.

The mass transfer phenomenon in the biocomposite could be therefore represented by a sorption–diffusion mechanism (PlasticsEurope 2014) described by the three coefficients, i.e. solubility (S), diffusivity (D) and permeability (P). The permeability coefficient combines the effects of diffusion and solubility as follows:

$$P = D \times S \tag{5}$$

The permeability coefficient is conventionally used to characterize the barrier properties of plane sheet materials in several fields of science such as membrane science, packaging science, etc., and by extension those of composites considered as homogeneous materials.

By the addition of permeable particles in a polymer matrix, the P of the composite material was a result of the contribution of both permeability of the neat matrix and that of the particles themselves. The composite material is then considered as a bi-phasic material. A huge part of the work done on mass transfer properties of composite material containing permeable particles have focused on the predictive modelling of the composite permeability from the permeability in the neat matrix and that in the particle itself.

3.2 Gas and Water Vapour Transfer Properties in Raw Constituents

Lignocellulosic fillers Due to their chemical composition, lignocellulosic fibres naturally display a hydrophilic behaviour, allowing the sorption and the diffusion of water vapour. Main studies are devoted to explore mechanisms of moisture sorption, which are quite complex because it often involves a continuous change of the structure of the particles in the case of swelling (Okubayashi et al. 2004). With increasing water uptake, the lignocellulosic structure gets looser, favouring the invasion by water molecules at higher water vapour pressures. Generally, moisture sorption isotherms of cellulose-rich fibres can be fitted using a parallel exponential kinetics (PEK) model suggesting that moisture exchange relies on two different mechanisms occurring simultaneously in slow and fast processes, as observed on textile fibres such as lyocell (Okubayashi et al. 2004), cotton (Okubayashi et al. 2004) or viscose (Okubayashi et al. 2005), and cellulosic nanoparticles prepared from sisal fibres (Belbekhouche et al. 2011). The fast and slow sorption sites can be related to different types of amorphous regions, external/internal fibre surfaces and direct/indirect sorption. Water molecules are known to absorb directly and fast on hydroxyl groups of external surfaces and amorphous regions, while it is more slowly on inner surface of micro-cavities and crystallites (even if crystalline regions are often considered to be impermeable to vapour molecules) (Okubayashi et al. 2004; Luduena et al. 2012). Additional water molecules can absorb on the water molecules binding the fibre directly. In the case of fibre treatment, removing amorphous components (e.g. cellulose crystals isolation) and hydrophilic groups (e.g. surface treatment) may result in a lower water sorption.

Belbekhouche et al. (2011) observed on cellulosic nanoparticles prepared from sisal fibres that the first-half sorption–diffusion coefficient, corresponding to the diffusion of water through the surface of particles, was higher than the second-half sorption–diffusion coefficient, more representative of the diffusion at the core. It is an indication that the diffusion of water is rather controlled by the surface than by the core, probably because of a barrier effect related to the presence of water at the surface during sorption kinetics. Indeed, the Park model generally well fitting the moisture sorption isotherms of cellulose-rich (Okubayashi et al. 2004) and lignocellulosic fibres (Bessadok et al. 2009a), it can be considered that water molecules condense layer by layer on the adsorption sites.

All published moisture sorption kinetics are performed on either cellulose-rich fibres (Okubayashi et al. 2004; Belbekhouche et al. 2011; Lavoine et al. 2012; Mannan 1997) or lignocellulosic long fibres (length of the order of several millimetres) (Bessadok et al. 2009a; Célino et al. 2013; Gouanvé et al. 2007). Furthermore, very few studies are dealing with the identification of moisture diffusivity in lignocellulosic fibres, with diffusion coefficients ranging from around 2.2×10^{-8} m^2 s^{-1} (Bessadok et al. 2009b) to 3.4×10^{-13} m^2 s^{-1} (Mannan 1997), depending on the substrate and the used methodology. Currently, no reliable methodology has been published to evaluate moisture diffusion in short lignocellulosic fibres (sizes ranging from 15 to 300 μm) that are, for example, obtained by successive dry grinding of raw lignocellulosic biomass such as wheat straw (Berthet et al. 2015b; Ghizzi et al. 2011). Indeed, Wolf et al. (Wolf et al. 2016). showed through modelling tools that it was not possible to extrapolate the moisture diffusivity obtained for a piece of native wheat straw internode to one of ground particles, due to huge differences of structure. In this perspective, future research should focus on a methodology to identify the moisture diffusivity in short lignocellulosic fibres. Finally, to the best of our knowledge, no data are available concerning the diffusivity and sorption of gas such as O_2 and CO_2 in lignocellulosic fibres.

Biosourced and biodegradable matrices. On the first hand, agro-polymers (e.g. starch, proteins, cellulose, lignin, chitosan), which are directly extracted from the plant and are available at a reasonable cost (around 1€/kg), generally display a high intrinsic reactivity and relatively poor moisture and gas barrier properties (Berthet et al. 2015a). As an example, the water vapour permeability (WVP) values of protein-based films (values ranging from 6×10^{-11} to 5×10^{-12} mol m^{-1} s^{-1} Pa^{-1} for wheat gluten-based films depending on the film preparation process (Angellier-coussy et al. 2011)) are normally 2–4 orders of magnitude higher than that of low-density polyethylene (LDPE) (WVP = 0.05×10^{-12} mol m^{-1} s^{-1} Pa^{-1}) (Series 1995). This is due to the presence of free hydroxyl groups in the matrix, which interact strongly with migrating water molecules. Relative humidity considerably affects the WVP and gas permeability of protein films, due to the plasticization of the protein network. This effect is even more pronounced for CO_2 than for O_2, which is particularly interesting for the packaging of respiring products such as fruits and vegetable (Mujica-Paz and Gontard 1997).

On the opposite, polyesters such as microbial polyesters (polyhydroxyalcanoates) and polylactic acid, which are currently available at a cost that remains high (between 2 and 5 €/kg), display excellent barrier properties and water resistance, close to classical petro-sourced polymers.

3.3 *Gas and Water Vapour Transfer Properties in Biocomposites*

Studies dealing with gas transfer properties in biocomposites are very rare (Valdés García et al. 2014; Sonia and Dasan 2013), but much more numerous concerning water vapour permeability (Berthet et al. 2015b; Berthet et al. 2015c; Luduena et al. 2012; Valdés García et al. 2014; Sonia and Dasan 2013; Berthet et al. 2016; Sirviö et al. 2014; Sanchez-garcia et al. 2008). Available literature highlights that mass transfer properties can be significantly modulated by the incorporation of lignocellulosic fillers, resulting in either an increase or a decrease of permeability values, depending on the respective intrinsic properties of the raw constituents, the filler content and overall on the structural changes of the material induced by the

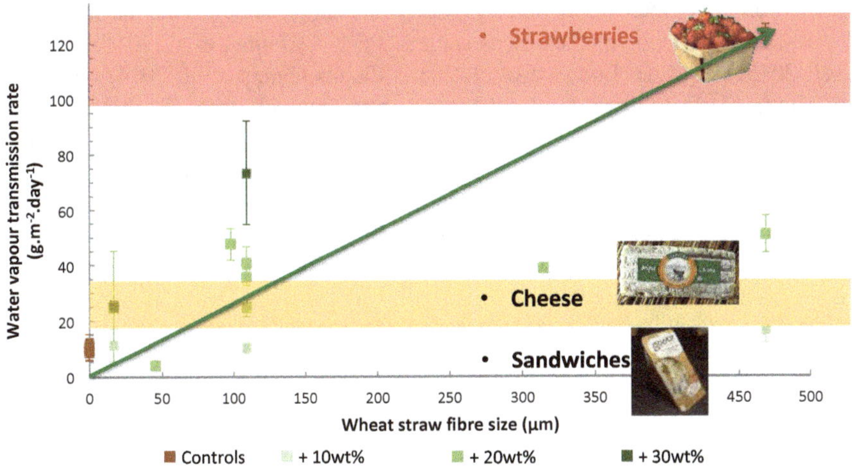

Fig. 2 Correlation of water vapour transmission rate of PHBV/wheat straw fibres-based biocomposites with requirements of some food products

presence of fillers, including changes in crystallinity, the occurrence of a plasticizing effect, the creation of a tortuous pathway preventing the diffusion of molecules or the creation of defects and microvoids at the filler/matrix interface favouring the diffusion of molecules. As demonstrated by Berthet et al. for PHBV/wheat straw fibre biocomposites (Berthet et al. 2015b; Berthet et al. 2015c), such a modulation in mass transfer properties could very interesting for food packaging purposes. For example, they showed that modulating filler size (from a d_{50} of 17 µm up to 450 µm) and content (up to 30 wt%) allowed to fulfil the requirements of respiring fresh produces such as strawberries, thus enabling to preserve them in a better way than currently used polyolefins (Fig. 2) (Berthet et al. 2015c).

It is clear that lignocellulosic fibres cannot be considered as impermeable to water vapour, which makes more complex the interpretation of data (Berthet et al. 2015a). Competitive phenomena are often described to explain the evolution of water vapour permeability in full-biocomposites. A decrease in water vapour permeability could be observed in the case of highly hydrophilic matrices displaying very low water vapour barrier properties such as polysaccharides (Sirviö et al. 2014) and/or highly crystalline cellulose, at least for low filler contents (Sanchez-garcia et al. 2008). This was mainly ascribed to a tortuosity effect, i.e. the creation of a more tortuous pathway for diffusing molecules. In the case of alginate-based biocomposites, the increase of the matrix cross-linking induced by the presence of cellulose was also in favour of a reduction of water vapour permeability (Sirviö et al. 2014). Increasing fibre content was seen to result in fibre agglomeration and it was thought to cause a reduction in matrix homogeneity and cohesion and lead to preferential penetrant paths and to detrimental effects in barrier properties (Sanchez-garcia et al. 2008).

Most often, an increase in water vapour permeability is observed, particularly in the case of apolar polymer matrices such as PHBV (Berthet et al. 2015b; Berthet et al. 2015c) or PCL (Luduena et al. 2012; Valdés García et al. 2014) combined with lignocellulosic fillers (and not highly crystalline cellulose). Such an effect is generally exacerbated by increasing filler contents and large size fibres (Berthet et al. 2015c). An increase in water vapour permeability is first ascribed to a thermodynamic effect, i.e. an increase of the solubility coefficient, due to the hydrophilic character of lignocellulosic fillers as compared to the polymer matrix. This is also related to structural changes of the materials induced by the presence of fillers that result in the creation of a preferential pathway for the diffusion of water vapour, especially at the filler/matrix interface.

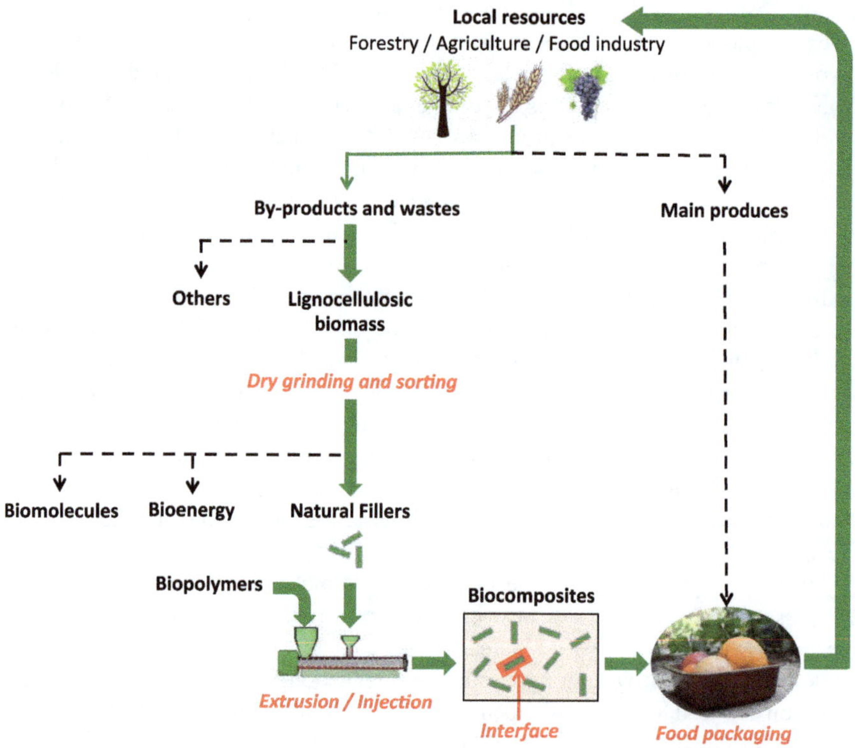

Fig. 3 Eco-conversion of lignocellulosic biomass into fillers for biocomposites

4 Economical Competitiveness

4.1 *Raw Materials: Towards the Valorization of Agro-Residues*

A key feature for enabling biocomposites to be applicable as food packaging relies on developing tailor-made materials from both non-food and low-cost resources. The cheapest and most environmentally virtuous lignocellulosic biomass that could be used as raw materials for the production of fillers is the one obtained from either agricultural, urban, forestry or food industry solid by-products and wastes (Fig. 3). The conversion of lignocellulosic by-products or wastes as fillers in biocomposites would be in favour of waste reduction while adding value to the agro-food chain.

Recently, efforts have been carried out to produce fillers with controlled properties (size, morphology, surface reactivity) using solvent-free processes. In the framework of the European project EcobioCAP, biodegradable and functional fibres with sizes from the order of several micrometres up to the order of the millimetre were simply achieved by successive dry grinding (cut, impact and ball milling) of wheat straw without any pretreatment (Berthet et al. 2015b; Berthet et al. 2015c) (Fig. 3). The size reduction of the lignocellulosic biomass upon successive grinding also results in changes of morphology and surface reactivity. As an example, successive grinding of wheat straw leads to a reduction of fibre aspect ratio and an increase of the dispersive component of surface energy (Berthet et al. 2015c). This is explained by a demasking of hydrophobic compounds such as lignin at the surface of particles. It was thus deduced that the compatibility between fine particles and apolar polymer matrices would be favoured. However, natural fibres prepared solely with these kinds of processes are generally characterized by broad size and composition distributions (Berthet et al. 2015c; Averous and Le Digabel 2006). This polydispersity is induced by the complex and heterogeneous structure of the available lignocellulosic biomass, especially due to differences in histological tissues and anatomic parts throughout the plant. Therefore, a batch of lignocellulosic fibres prepared only by dry grinding might contain particles with sizes ranging from the order of the millimetre (coarse particles) down to few microns (fines or dusts). Different processes are industrially available to sort fibres by separating coarse fibres from dusts according to size criterion. The most interesting of them, when considering composite applications, are dry sorting processes, which allow to prevent alteration of the bulk characteristics of particles. They mainly include sieving and air classification.

4.2 *Shaping Processes*

In order to be widely used for food packaging purposes, materials should be manufactured using techniques conventionally used for plastics from fossil resources, i.e. mainly extrusion, blowing, injection or compression moulding, by applying the industrial production rates.

Among the disadvantages associated with classical thermal treatments, the main risk is the thermal degradation of either the bioplastic used as matrix or the lignocellulosic fibres. Indeed, most biopolymers are especially sensitive to temperature, their melting temperature being close to their degradation temperature (e.g. PHBV (Leroy et al. 2012)). In the case of proteins such as wheat gluten, the temperature of degradation is even below their glass transition temperature, imposing the use of external plasticizers to avoid thermal degradation reactions such as Maillard reactions (Montaño-Leyva et al. 2013). Lignocellulosic fibres are also sensitive to thermal treatments as degradation can occur that weakens the fibres. The inherent moisture that is naturally bound to the hemicellulose component can change depending on temperature and exposure time. Furthermore, heat treated biomass is known to globally become darker and brittle, due to the loss of its tenacious and fibrous nature resulting from the decomposition of hemicelluloses and partial depolymerization of lignin and cellulose. Several publications have investigated these issues and have generally found that exposure to temperatures at or below 170 °C did not have a significant effect on the strength of the fibres (Gassan and Bledzki 2001). Thermal treatment of lignocellulosic biomass, i.e. mild torrefaction, can be voluntary and positively used for reinforcing applications. Indeed, it enables to hydrophobize lignocellulosic fillers, as shown for wheat straw, resulting in a better fibre/matrix adhesion (Berthet et al. 2016).

Regarding natural fibres, it is also well known that processes involving shearing such as extrusion result in a degradation of fibre length and morphology (Zini et al. 2007; Beaugrand and Berzin 2013). The effect is particularly pronounced for long fibres (initial length of several millimetres) (Grande and Torres 2005) and almost negligible for short fibres with sizes lower than 300 μm in the case of soft extrusion conditions (Bourmaud et al. 2016) and 80 μm for severe processing conditions (Teuber et al. 2016). Depending on the processing conditions, the dispersion level of the fibres within the composite material might also be affected, thus impacting its resulting mechanical and barrier properties. Finally, the homogeneous feeding of melt extrusion devices could be difficult with some fibres, as reported for brewing spent grains, flax or hemp fibres, thus imposing the use of a force feeder equipment. Attention must also be paid to the moisture content of lignocellulosic fibres at the moment of the manufacturing process. It is generally advised to carefully dry natural fibres before compounding to avoid any alteration of the intrinsic mechanical properties of fibres, a potential hydrolytic degradation of biopolymer matrices or a bad fibre/matrix adhesion. However, it was recently shown that in spite of structural differences induced by the introduction of moist wheat straw fibres in PHBV, tensile properties of resulting biocomposites were not significantly affected, leading to the conclusion that the initial fibre moisture content was not a predominant factor for controlling the mechanical properties of PHBV/wheat straw fibres composites (Berthet et al. 2015d). It should be thus studied case by case.

5 Safety Issues of Biocomposites

5.1 Regulation and Food Contact Ability

To ensure food safety, the chemical migration of components from the packaging towards the food should be controlled. In the European Union, an important effort of harmonization on Food Contact Material (FCM) regulations has been undertaken since 1976, when the Framework Directive was adopted laying down the general principles of FCM safety. European regulation 1935/2004 (European Commission 2004) is the basic community legislation that sets the general guidelines for the compliance of all FCMs. The regulation 10/2011/EC translates the requirements of framework regulation 1935/2004 to plastics laying down the limits and conditions that must be respected for migration tests. The requirement of inertia for every FCM is assessed by (i) an overall migration limit (OML), i.e. the total mass released by the packaging material during a migration test, which is set to 60 $mg.kg^{-1}$ of food and (ii) a specific migration limit (SML) defined for each substance listed in a positive list on the basis of toxicological data. Interactions between the material packaging and the food depend on the characteristics of the food, i.e. its water activity, acidity and fat content. So far, for evaluating the chemical safety of packaging materials in contact with food, materials shall be tested with food simulants. The recommended conditions migration tests must be applied according to worst case scenario and are in practice applicable only to water-resistant materials even though the use of water-sensitive materials, such as paper, board or bioplastics, is particularly widespread.

The introduction of new materials, such as bioplastics and lignocellulosic fibres, has deeply changed the panorama of food packaging. In absence of specific regulation, bioplastics should lie within the scope of European regulation (EC) 1935/2004. So far, very few studies are referenced about the safety assessment of biodegradable biopolymer-based packaging. Among the biopolymers issued from bioconversion, the poly(3-hydroxybutyrate-co-3-hydroxyvalerate) (PHBV) is the unique biopolyester obtained by means of bacterial fermentation that may be used as FCM with a SML of 0.05 mg/dm^2 for the crotonic acid (depolymerization product) according to the specification of Annex I (10/2011/EC). In the case of water-sensitive materials intended only to contain dry foodstuffs, migration values tend to be particularly high in common migration tests designed for water-resistant plastic materials and using food-simulating liquids (Gällstedt and Hedenqvist 2004; Mauricio et al. 2010). These results highlighted the difficulty to transfer the control provision of synthetic polymers to biodegradable materials and then the inaccuracy of recommended migration tests for such materials. A possible alternative test that could be recommended for such materials should be derived from either paper testing methods or active packaging. Paper and board materials are among the oldest biodegradable packaging materials but are not yet regulated by a specific Directive at the European level. The test condition to evaluate migration from paper containing potential organic migrants and dedicated to be employed in contact with dry food is based on the use of Tenax ™ [modified poly (phenylene oxide); Varian, Houten, the Netherlands] according to the recommendation set by the European Committee for Standardization norm CEN/TC 172 (2014), as successfully demonstrated by Nerin and Asensio (2007). However, such a limited approach fails to reproduce the variety of foodstuff packed in these materials and only imitates dry fatty and solid products. Furthermore, the use of such an approach rises up the problem of contact at the interface with the materials since Tenax™ is in the form of powder. In the case of active packaging such as oxygen scavenger, a dedicated test was developed to perform a more realistic overall migration experiment with filter paper soaked with simulant (Dainelli et al. 2008). This method showed much lower realistic overall migration than with the conventional test with liquid simulants.

One critical key point of biodegradable packaging is the sensitivity balance in the respect that the material should be durable enough for protection but still biodegradable. The structural and physical–chemical stability of packaging materials under the whole life cycle of packaging materials are the first keys to ensure their ability to preserve food safety within defined limits and recommended usage conditions. If moisture, temperature and mechanical stress are well-known parameters, other ones such as food acidity, water activity or fat content are also able to have an influence on biodegradable materials' properties and stability. Substances arising from physical–chemical material degradation as well as breakdown products and/or intentionally added additives could migrate during the contact of the materials with the food. A recent study demonstrated that, as evidenced for other polyesters such as PET, ethanol 95% (v/v) was the most severe food simulant for PHBV films, with a strong impact on their physical–chemical and chemical stability (Chea et al. 2015). In the case of ethanol, a high sorption value was noticed, together with a significant plasticizing effect and an increase in the water vapour permeability, which was mainly explained by a decrease in both the molecular weight and the crystallinity degree of PHBV films. However, it was shown that the occurrence of structural changes, intimately related to the affinity between the polymer and the FSL, was not the unique parameter governing the physical–chemical stability and the inertness of PHBV films. Indeed, in the case of water, a decrease in the polymer molecular weight and crystallinity were observed with no impact on functional properties and overall migration, meaning that such structural changes do not fully predict the physical–chemical stability and chemical safety of PHBV films. Thus, an in-depth interest should be also given to the affinity between potential migrants and FSLs and their migration behaviour.

To conclude, knowledge should still be increased on the dependency between chemical safety and the complex physical–chemical and biological instability of the material, i.e. potential degradation of polymeric network, breakdown products formation (non-intentionally added substances), enhanced migration.

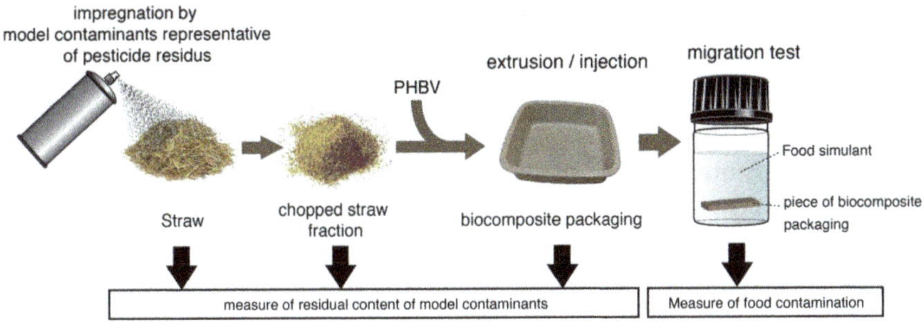

Fig. 4 Approach proposed for the safety assessment of food contact biocomposite materials

5.2 *Undesirable Migrations from Biocomposites and Challenge Tests*

The few studies on the ability of biocomposite materials to be used as FCM have been conducted on PLA (Yang et al. 2016; Fortunati et al. 2012) and PHA (Yu et al. 2014; Scarfato et al. 2015; Dhar et al. 2015) including cellulose nanocrystals with investigations exclusively focused on the impact of the filler inclusion on overall migration level. On this issue, no drastic detrimental effect of cellulose nanocrystals was evidenced with migration level within standard limits set by the legislation. However, the production pathway of biocomposites comes along with new concerns related to the presence of potential toxicological substances able to migrate towards food, which could be natural contaminants (mycotoxines, phycotoxines), pesticide residues (including herbicides, insecticides and fungicides) and ubiquitous environmental pollutants. Environmental pollutants include polycyclic aromatic hydrocarbons, but also additives used in all life products materials, and especially plasticizers like phthalates that are commonly used in the formulation of packaging materials. The chemical nature of these contaminants is variable and the question of their persistence after the processing steps of biocomposites remains a critical point.

A special attention should be thus paid to evaluate the decontamination efficiency of the respective production and extraction processes used to prepare lignocellulosic fibres. For that purpose, "challenge tests" could be used, which consists of voluntary enriching materials with surrogates (molecules representative of the contaminants) and quantify their residual amount at each step of the preparation process, as performed in the frame of the EcoBioCAP project (Fig. 4).

The objective of such an approach is to place the decontamination efficiency in the centre of the human health risk assessment. As conducted in the case of the evaluation process of recycled materials proposed by the European Food Safety Agency, the strategy is based on (i) the selection of molecules, which are representative of the potential contaminants and for which toxicological data are available including Maximum Residue Limit (MRL), Acute Reference Doses (ARfD) and Acceptable Daily Intake (ADI), and (ii) a realistic impregnation of raw materials.

6 Environmental Impact of Biocomposites: End of Life

Food packagings are often accused of representing significant amounts of wastes because of their very short life duration and single-use and difficulty. In spite of the improvement of the techniques of recycling, the environmental impact of plastics still raises problems due to their fossil origin and by consequence, the fossil carbon which enters their composition may be released in the atmosphere in the form of CO_2 at their end of life. To put in perspective, we shall call back, however, that plastics use for their synthesis only 4% of the world consumption of crude oil (Duval 2014).

To reduce the carbon footprint of packaging materials, the most often followed strategy is the reduction of the amount of raw materials used, which is generally translated by a lightening and/or a decrease of the quantity of used packagings. This strategy may achieve some limits, especially in terms of processability, imposing to consider all the life cycle of packaging materials. Solutions can be mainly brought at two steps, i.e. the production of raw materials and the end of life process, and should be set up in relation to the targeted application and the infrastructure of waste management.

6.1 Options of Waste Management for Biocomposites

The possible options for waste management, i.e. landfill, energy recovery, recycling, reuse, composting, anaerobic digestion, are strongly dependent on the implemented policy on a local scale. To make a choice, it is necessary to take various criteria into account, including the energy consumption, the potential risks to the health, the transport of the materials, the necessity of a constant supply in quantity and in quality, etc.

Landfill Landfill remains the most used treatment for plastics wastes in the majority of countries, reaching an average value of 38% in Europe (PlasticsEurope 2014). In the conditions of burying (darkness and anaerobic conditions), non-biodegradable materials remain relatively stable. Nevertheless, the European directive encourages to limit the use of landfill for biodegradable materials, including biocomposites, because of the release of methane into the atmosphere in absence of oxygen (European Commission 1999). Methane is characterized by a potential of global reheating approximately 25 times higher than CO_2 (on a time-scale of 100 years).

Table 1 Characteristics of materials for food packaging

Material	Carbon content (%)	Renewable carbon content (%)	Calorific capacity (MJ/kg)	Biodegradability	Ref.
Paper	44.4	100	19.7	+	(ECN 2009)
Starch	44.4	100	17.4	+	(ECN 2009)
PLA	50.0	100	18.7	±	(Vink et al. 2007)
PHA	70.4	100	36.6	+	(Patel et al. 2006)
PHB	55.6	100	23.8	+	(Patel et al. 2006)
PHBV	60.0	100	27.4	+	(Patel et al. 2006)
PCL	64.0	0	28.2	+	(Hermann et al. 2011)
PBAT	29.2	0	29.2	+	(Hermann et al. 2011)
PE	81.9	0	46.2	–	(Babrauskas 1992)
PP	68.9	0	46.4	–	(Babrauskas 1992)
PS	88.5	0	42.5	–	(Babrauskas 1992)
PET	56.4	0	22.2	–	(Babrauskas 1992)

Energy recovery The vapour produced during incineration is transformed into electricity thanks to an alternator with an efficiency of about 25% or directly used to warm houses with an efficiency of 60%. The interest of this option is that the produced electric and thermal energies can be used to substitute energies stemming from electric and natural gas productions. Energy recovery is connected to the carbon content of the material and to its calorific capacity (Table 1). In the EU, only the processes answering criteria of energy efficiency (efficiency > 60 or 65%) can be considered as energy recovery.

Recycling The requirement for recycling is the existence of a functional network integrating the various steps as the collection, the sorting and the transformation in a recycled matter. Among the recycling technologies, the mechanical recycling is the only one applicable to the industrial scale. In reality, plastics are still little recycled with regard to the other raw materials with a 23% recycling rate against 71% for the glass and 88% for paper and cardboards. The difficulties related to the mechanical recycling of plastics are mainly due to the fact that plastic materials considerably differ by their nature and structure, what makes them incompatible and leads to mechanical properties often very lower than those of virgin materials. This requires laborious sorting and cleaning processes. Other reasons limiting their recycling are in connection with degradation reactions (thermal, UV and mechanical) that can occur upon processing and usage. As regards biocomposites, one key point in favour of their recyclability is the development of materials from a thermoplastic matrix rather than a thermoset matrix (Faruk et al. 2012). Lignocellulosic filler-based biocomposites can be reprocessed by means of mechanical processing, such as successive injection moulding cycles, as shown for example for PLA/flax fibres biocomposites (Le Duigou et al. 2008). The recyclability is usually guaranteed up to several reprocessing cycles when considering the preservation of mechanical properties. However, different degradative effects can occur over reprocessing, such as fibre length reduction and separation of fibre bundles (Bourmaud et al. 2016) and chain scission in the synthetic thermoplastic matrix, which leads to a decrease in the molecular weight and an increase in crystallinity (Vilaplana et al. 2010). Grinding a biocomposite and using resulting milled composite particles as reinforcements is another method of recycling, as shown by Grozdanov et al. for PLA/rice husks and kenaf fibres biocomposites that were reused into polymer mortars and concrete structures for the low-cost building industry (Grozdanov et al. 2010). Generally, full-biocomposites have less recyclability than their neat thermoplastic matrix, because composites are more sensitive to thermomechanical degradation. That is why composting is supposed to be the most appropriate waste management route for such materials.

Reuse The reuse of the material stemming from packagings is also a way of interesting valuation, which could be perfectly integrated into a concept of the circular economy by allowing the worn product to become a new resource. However, it involves upstream steps, including selective sorting and collection, which are not adapted to all the materials. As an example, the paper-cardboard sector is particularly concerned by this way of valuation with a recycling rate close to 80%.

Composting Composting is an aerobic biological process involving bacteria and moulds, which allows converting the organic matter in biofertilizers. Various standards can be used to guarantee that a material is compostable or not. In comparison with home composting, industrial composting imposes higher temperatures (50–60 °C) and a frequent mixture of the biomass, what results in a higher rate of degradation. The gas released is mainly CO_2, methane being exceptional and that of protoxide of nitrogen is very low compared with home composting. Composting may be the most appropriate waste management route when a biodegradable matrix is used.

Anaerobic digestion (AD) It is a process of biological degradation of the organic matter in the presence of a consortium of bacteria and in absence of oxygen, leading to the production of biogas. Biogas has low economic value due to its high CO_2 content that decreases energy content and limits direct uses to heat production or co-generation. AD digestates, resulting from the incomplete degradation of materials into bioenergy, can be used as biofertilizers. But this still raises concerns about possible hygienic (pathogens), environmental (antibiotics, metals) hazards and storage (degradation and GHG emission) issues, while storage is necessary because of the limitation of land disposal imposed by EU nitrogen regulation (Scaglia et al. 2015). The use of two-phase AD processes allows increasing conversion yields and allow the concurrent production of bio-hythane, a biogas enriched in H_2 (up to 10–20%) with lower impact on GHG emission than conventional catalytic methods

(Micolucci et al. 2014). The VFAs rich residual effluents generated by the 2 stages AD are promising precursor for further conversion to PHAs and derivatives (Reis et al. 2011), which have numerous potential uses as specialty biopolymers for medical, packaging applications (Chen 2009). AD applicability is still poor for lignocellulosic-rich waste streams whose conversion yields are not high enough (Yuan et al. 2014; Sawatdeenarunata et al. 2014).

In conclusion, the environmental footprint of a packaging is closely correlated at his end of life. Landfill and energy recovery are compatible with all the materials while composting and anaerobic digestion require that the material is biodegradable in certain conditions. In every case, the end of life of biocomposites will result in a release of carbon into the atmosphere but according to the origin of this carbon, either fossil or biogene, the environmental impact of these emissions will be more or less significant. The production of added-value products such as bioenergy, biomolecules or biofertilizers, should be taken into account in the assessment of the overall environmental impact of a packaging material.

6.2 *Nature of the Polymer Matrix: BioSourced and/or Biodegradable*

Among biosourced or partially biosourced plastics, some are realized from polymers possessing a structure identical to their counterparts of origin fossil as the polyethylene stemming from some sugar cane for example. While others as PLA or PHA are characterized by a new and original structure which distinguishes them conventional petrochemical polymers and present besides the advantage to be biodegradable (in certain conditions). As their conventional counterparts, biosourced polyolefins do not have the capacity to biodegrade. The resulting wastes are thus preferentially incinerated or recycled. During their incineration, biosourced polymers release CO_2 in the atmosphere but contrary to fossil polymers, their impact on the global warming is considered as unimportant because it is a carbon of short cycle stemming from renewable resources.

It should be specified nevertheless that the production of biosourced polymers is far from being neutral from an environmental point of view since it is dependent on fossil resources for the production of agricultural resources as well as for the transformation of raw resources and the shaping of materials. Additionally, each agricultural production needs water, which is negatively considered in life cycle assessment. Concerns are also related to the availability of the agricultural resources, the competition between food and non-food usages, the risks of deforestation as well as to the potential depletion of water. The current researches are focusing on the conversion of wastes (either liquid or solid) and algaes.

The development of biodegradable packaging materials, either biosourced such as PHBV or from fossil origin such as PBAT or PBS allows envisaging new ways of waste management. Packaging materials, which are often soiled by food, could be treated as biological wastes by composting or anaerobic digestion without needing to be separated from the wet organic waste, which they may contain.

Thus, the production of biosourced plastics is not neutral from an environmental point of view. It is thus advisable to limit them to certain specific uses.

7 Conclusion and Perspectives

Full-biocomposites composed of bioplastics combined to lignocellulosic fibres have been widely studied over the last decade. Main applications areas of biocomposites are still focused on automotive, building and sport. Consequently, current available scientific literature mainly deals with the related improvement of mechanical properties, for example by attempting to favour percolation phenomena and a strong fibre/matrix interfacial adhesion. Since food packagings are now identified as a future application area of biocomposites, for example, for the production of rigid food trays, it becomes necessary to investigate mass transfer properties as key properties of food preservation and food safety.

It was evidenced that lignocellulosic fibres often act as defects, which impact both mechanical and permeability properties, degrading the firsts and increasing the seconds. Increasing material permeability by choosing the appropriate formulation seems to be a valuable approach, especially in the packaging of respiring products such as fruits and vegetables. But mechanical properties should remain sufficient to preserve the integrity of packaging materials, thus ensuring their fundamental role of container and protection towards physical damage. Furthermore, owing to the large diversity and complexity of composition and structure of the lignocellulosic biomass, conclusions are often system dependent. Thus, in order be able to dissociate phenomena in the balance, lignocellulosic particles with controlled characteristics should be produced.

Finally, whatever the targeted sector of application, the stability of biocomposites all over their life cycle is a *sine qua non* condition for sustainability. One other bottleneck to the use of full-biocomposites for packaging is their intrinsic reactivity towards thermomechanical treatments and environmental conditions. We are faced with unstable systems characterized by a dynamical evolution of structure and properties. A better knowledge and control of their stability upon usage conditions is still necessary, especially in the case of food contact.

References

Angellier-coussy H, Guillard V, Guillaume C, Gontard N (2013a) Agro. Food Ind 23(4):15–19

Angellier-Coussy H, Gastaldi E, Correa F, Gontard N, Guillard V (2013b) Eur Polym J 49:299–306

Angellier-coussy H, Guillard V, Guillaume C, Gontard N (2011) In Biopolymers—new materials for sustainable films and coatings pp 213–232

Averous L, Le Digabel F (2006) Carbohydr Polym 66:480–493

Babrauskas V (1992) In Babrauskas V, Grayson S (eds) Heat release in fires London, pp 207–223

Beaugrand J, Berzin F (2013) J Appl Polym Sci 128(2):1227–1238

Belbekhouche S, Bras J, Siqueira G, Chappey C, Lebrun L, Khelifi B, Marais S, Dufresne A (2011) Carbohydr Polym 83:1740–1748

Berthet M-A, Angellier-Coussy H, Guillard V, Gontard N (2015a) J Appl Polym Sci 42528:1–18

Berthet M, Angellier-coussy H, Machado D, Hilliou L, Staebler A, Vicente A (2015b) Gontard N 69:110–122

Berthet M-A, Angellier-Coussy H, Chea V, Guillard V, Gastaldi E, Gontard N (2015c) Compos. Part A Appl Sci. Manuf

Berthet M, Gontard N, Angellier-coussy H (2015d) Compos Sci Technol 117:386–391

Berthet M, Commandré J, Rouau X, Gontard N, Angellier-coussy H (2016) Mater Des 92:223–232

Bessadok A, Langevin D, Gouanvé F, Chappey C, Roudesli S, Marais S (2009a) Carbohydr Polym 76:74–85

Bessadok A, Roudesli S, Marais S, Follain N, Lebrun L (2009b) Compos Part A Appl Sci Manuf 40:184–195

Bourmaud A, Akesson D, Beaugrand J, Le Duigou A, Skrifvars M, Baley C (2016) Polym Degrad Stab 128:77–88

CEN/TC 172 (2014) Practical Guide for Users of European Directives, Unit "Chemical and physical risks; surveillance" of the Health & Consumer Protection Directorate-General of the European Commission

Cagnon T, Guillaume C, Guillard V, Gontard N. (2012) Packag Technol Sci (Paper Pres)

Cagnon T, Méry A, Chalier P, Guillaume C, Gontard N (2013) Innov Food Sci Emerg Technol 20:288–298

Chaix E, Couvert O, Guillaume C, Gontard N, Guillard V (2015) Compr Rev Food Sci Food Saf 14:1–21

Chea V, Angellier-Coussy H, Peyron S, Kemmer D, Gontard N (2015) J Appl Polym Sci 41850:1–8

Chen Guo-Qiang (2009) Chem Soc Rev 38:2434–2446

Célino A, Fréour S, Jacquemin F, Casari P (2013) J Appl Polym Sci 130(1):297–306

Dainelli D, Gontard N, Spyropoulosc D, Zondervan-van den Beukend E, Tobbacke P (2008) Trends Food Sci Technol 19:99–108

Dhar P, Bhardwaj U, Kumar A, Katiyar V (2015) Polym Eng Sci 55(10):2388–2395

Le Duigou A, Pillin I, Bourmaud A, Davies P, Baley C (2008) Compos. Part A Appl Sci Manuf 39 (9):1471–1478

Duval C (2014) In: Hamaide T, Deterre R, Feller JF (eds) Impact environnemental des matières plastiques pp 57–69

EC Commission Regulation (EC) No 10/2011 of 14 January 2011 on plastic materials and articles intended to come into contact with food

ECN (2009) The composition of biomass and waste. Phyllis database, version 4.13. Energy research Centre of the Netherlands (ECN), http://www.ecn.nl/phyllis/

European Commission (1999) EU Landfill Directive. Council Directive 1999/31/EC, European Commission, http://eur-lex.europa.eu/LexUriServ/LexUriServ.do?uri.CELEX:31999L0031:EN:HTML

European Commission (2004) Off J Eur Union 47 (L338):4–17

FAO (2011) Appropriate food packaging solutions for developing countries

Faruk O, Bledzki A, Fink H-P, Sain M (2012) Prog Polym Sci 37(11):1552–1596

Floros JD, Matsos KI (2005) Introduction to modified atmosphere packaging. In: Jug H. Han (ed) Innovations in food packaging. Elsevier Academic Press New York, USA, pp 159–172

Fortunati E, Peltzer M, Armentano I, Torre L, Jimenez A, Kenny JM (2012) Carbohydr Polym 90(2):948–956

Gassan J, Bledzki AK (2001) J Appl Polym Sci 82:1417–1422

Ghizzi D, Silva G, Guilbert S, Rouau X (2011) Powder Technol 208:266–270

Gouanvé F, Marais S, Bessadok A, Langevin D, Métayer M (2007) Eur Polym J 43:586–598

Grande C, Torres FG (2005) Adv Polym Technol 24(2):145–156

Grozdanov A, Avella M, Buzarovska A, Gentile G, Errico M (2010) Polym Eng Sci 50(4):762–766

Guillard V, Buche P, Destercke S, Tamani N, Croitoru M, Menut L, Guillaume C, Gontard N (2015) Comput Electron Agric 111:131–139

Guillaume C, Chalier P, Gontard N (2008) In: Chiellin E (ed) Environmentally compatible food packaging Boca Raton. CRC Press, pp 396–418

Guillaume C, Schwab I, Gastaldi E, Gontard N (2010) Innov. Food Sci Emerg Technol 11(4):690–696

Guillaume C, Guehi D, Gontard N, Gastaldi E (2013) J Food Eng 119(2):362–369

Gällstedt M, Hedenqvist MS (2004) J Appl Polym Sci 91:60–67

Hermann B, Debeer L, De Wilde B, Blok K, Patel M (2011) Polym Degrad Stab 96:1159–1171

Lavoine N, Desloges I, Dufresne A, Bras J (2012) Carbohydr Polym 90:735–764

Leroy E, Petit I, Luc J, Colomines G, Deterre R (2012) Polym Degrad Stab 97:1915–1921

Luduena L, Vazquez A, Alvarez V (2012) Carbohydr Polym 87(1):411–421

Mannan KM (1997) Talukder M. A I Polymer (Guildf) 38(10):2493–2500

Mauricio M, Guillard V, Peyron S, Gontard N (2010) J Appl Polym Sci 166(5):2526–2535

Micolucci F, Gottardo M, Bolzonella D, Pavan P (2014) Int J Hydrogen Energy 39:17563–17572

Montaño-Leyva B, Ghizzi Da Silva G, Gastaldi E, Torres-chávez P, Gontard N, Angellier-coussy H (2013) Ind Crops Prod 43:545–555

Mujica-Paz H, Gontard N (1997) J Agric Food Chem 45:4101–4105

Nerin C, Asensio E (2007) Anal Bioanalitical Chem 389:589–596

Okubayashi S, Griesser UJ, Bechtold T (2004) Carbohydr Polym 58:293–299

Okubayashi S, Griesser UJ, Bechtold T (2005) Cellulose 12:403–410

Ozdemir M, Floros JD (2004) Crit Rev Food Sci Nutr 44(3):185–193

Patel M, Crank M, Dornburg V, Hermann B, Roes L, Hüsing B et al (2006) Medium and long-term opportunities and risks of the biotechnological production of bulk chemicals from renewable resources—the potential of white biotechnology. The BREW project. Final project report. Utrecht: Utrecht University, http://www.chem.uu.nl/b

PlasticsEurope (2014) Plastics–the facts 2013. An analysis of European latest plastics production, demand and waste data

Raouche S, Mauricio-Iglesias M, Peyron S, Guillard V, Gontard N (2011) Innov. Food Sci Emerg Technol 12(4):426–434

Reis M, Albuquerque M, Villano M, Majone M (2011) Mixed culture processes for polyhydroxyalkanoate production from agro-industrial surplus/wastes as feedstocks. Comprehensive biotechnology, Environmental biotechnology and safety, vol 6, pp 669–683

Sanchez-garcia MD, Gimenez E, Lagaron JM (2008) Carbohydr Polym 71:235–244

Sawatdeenarunata C, Surendraa KC, Takaraa D, Oechsnerb H, Khanala SK (2014) Bioresour Technol 178:178–186

Scaglia B, Pognani M, Adani F (2015) Sci Total Environ 514:314–321

Scarfato P, Di Maio L, Incarnato L (2015) J Appl Polym Sci 132(48)

Series PH (ed) (1995) Permeability and other film properties of plastics and elastomers; Plastics Design Library (PLD). Norwich, NY, USA

Sirviö JA, Kolehmainen A, Liimatainen H, Niinimäki J, Hormi OEO (2014) Food Chem 151:343–351

Sonia A, Dasan KP (2013) J Food Eng 118:78–89

Steele R (2004) Understanding and measuring the shelf-life of food, 1st ed Steele R (ed) Woodhead Publishing Limited: Cambridge, UK

Tamani N, Mosse P, Croitoru M, Buche P, Guillard V, Guillaume C, Gontard N (2015) Comput Electron Agric 113:174–192

Teuber L, Militz H, Krause A (2016) Compos Part A Appl Sci Manuf 84:464–471

Valdés García A, Ramos Santonja M, Beltrán Sanahuja A, Garrigós Selva M (2014) del C. Polym Degrad Stab 108:269–279

Vilaplana F, Strömberg E, Karlsson S (2010) Polym Degrad Stab 95(11):2147–2161

Vink E, Glassner D, Kolstad J, Wooley R, O'Connor R (2007) Ind Biotechnol 3(1):58–81

Wikström F, Williams H (2010) Packag Technol Sci 23(7):403–411

Wolf C, Guillard V, Angellier-Coussy H, Ghizzi D, Silva G, Gontard N (2016) J Appl Polym Sci 43329:1–10
Yang W, Fortunati E, Dominici F, Giovanale G, Mazzaglia A, Balestra GM, Kenny JM, Puglia D (2016) Int J Biol Macromol 89:360–368
Yu H, Yan C, Yao J (2014) RSC Adv 4(104):59792–59802
Yuan X, Wen B, Ma X, Zhu W, Wang X, Chen S, Cui Z (2014) Bioresour Technol 154:1–9
Zini E, Focarete ML, Noda I, Scandola M (2007) Compos Sci Technol 67(10):2085–2094

Lignocellulosic Fibres Reinforced Polymer Composites for Acoustical Applications

Elammaran Jayamani and Muhammad Khusairy Bin Bakri

Abstract In this chapter, an overview of various composites for sound absorption applications were reported. It includes composites made of a polymer matrix reinforced with synthetic fibres and natural fibres. New developments dealing with composites made of lignocellulosic fibre were discussed in detail and the merits and demerits of composite made of synthetic fibres and natural fibres have been discussed. In this chapter, procedures for estimating the sound absorption coefficients of various sound absorbing natural fibres were discussed with a mathematical model. Factors that may influence the sound absorption coefficients of porous materials, such as fibre size, porosity, flow resistivity, thickness, tortuosity and density were described. Empirical models to predict the flow resistivity and sound absorption coefficient were also discussed. This chapter also examines the critical issues and scientific challenges that require further research and development of polymer composite materials for their increased acceptance in the modern world for sound absorption purpose.

Keywords Lignocellulosic · Fibres · Reinforced · Polymer · Composites
Acoustical · Applications

Contents

E. Jayamani (✉) · M. K. B. Bakri
Faculty of Engineering, Computing, and Science, Swinburne University of Technology
Sarawak Campus, Kuching, Sarawak, Malaysia
e-mail: elammaranj@gmail.com

S. Kalia (ed.), *Lignocellulosic Composite Materials*, Springer Series on Polymer and Composite Materials, https://doi.org/10.1007/978-3-319-68696-7_11

1 Introduction

The sound is mechanical waves that resulted from the vibration in the molecules. The sound of the frequency audible to human ears is approximately in the range of 15–16 kHz, which is known as the normal audible frequency range. Furthermore, the human ear is sensitive mostly at frequencies between 3 and 6 kHz (Ver and Beranek 2006). Noise and sound are interrelated, where both are actually audible pressure fluctuations in the air. However, noise is usually considered to be an 'unwanted sound' due to its loudness and randomness in vibration. The presence of noise in our daily life will definitely have an impact on our health. For example, noise can cause deafness, sleep interference, slack in performance and behaviour, memory acquisition and mental health as well annoyance (Mechel 2008). Due to the health risks posed by noise, it is important for us to investigate and study how noise can be controlled or absorbed using various sound absorbers. Something that can be heard by the auditory system or diagnosed by the receivers is commonly defined as sound (Talty 1988). The auditory parts covered the sensory system and organ (i.e. the ears, eardrumetc.), while receivers covered the electrical input (i.e. microphone, diaphragm sensor etc.). Scientifically, the changes in pressure in a medium that composed of different sets of frequency causes the variation of oscillation wave that created a sound (Cory 2005; Crocker and Arenas 2007; Talty 1988). Sound can be transmitted, transferred and travelled in a different medium, such as air, water and solid and from one another medium except in a vacuum. In a solid medium, sound move in a transverse wave, while in another medium, such as air and water, sound move in longitudinal waves and sometimes with a mixture of a transverse wave. The speaker is one of the most common sources of sound. It creates vibration motion and shock wave on the surface area of speaker solid diaphragm that forces the air molecules to move and vibrate. According to Talty (1988), the up and down (or back and forth) fluctuation pressure in the surrounding air was created by the speaker diaphragm. Pulsating gases in-car exhaust system and flowing water or moving fluids in a river also created a similar fluctuating wave (Wakefield Acoustic Ltd 2004).

The inner and middle parts of the structural mechanisms of the ears are known to be very sensitive to the change of fluctuating air pressure. It caused the vibration in the inner and middle parts of the ears that is called as hearing sensation. It produced signals that being transcoded and transmitted to the brain to be evaluated and categorized as sound. Sound can travel in any directions radiated from its sources. For example, through the motions, the stronger the sound created from source, the bigger the pressure waves, hence, the louder the sound, the faster the vibration at and from source, and the higher the sound frequency. Figure 1 shows the typical structure of the human ear.

Approximately, a normal young person can receive 15–16 kHz range of audible frequency. Unfortunately, the majority of normal hearing sensitivity can detect audible frequency up to approximately 3–6 kHz at a pressure of 20 μPa. While comparing the audible frequency at atmospheric pressure (101.3 kPa), the variation fractional may vary approximately up to 2×10^{-10} (Vér and Beranek 2006). Sound frequency is measured in terms of cycles per second or Hertz (Hz), while the sound level is expressed in term of Decibel (dB) (Crocker and Arenas 2007). It is known that the speeds of the sound wave that enter mediums varied accordingly. As the sound wave enters different mediums, the speed will vary accordingly. Depending on the medium density that the sound enters, the sound that move, transmitted or

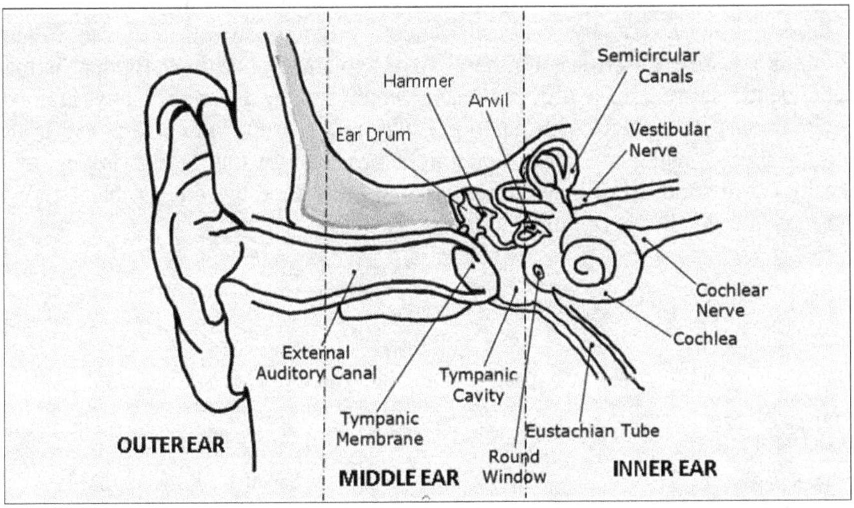

Fig. 1 Typical structure of human ear

transfer either become very slow or fast. Table 1 shows the speed of sound in different mediums.

Through simple understanding, noise is an undesirable sound that can be heard by human or animal. The definition is completely dependent on the hearer level of understanding of noise. Scientifically, in term of mechanical and physical understanding, discordant sound resulted from noise is due to the non-periodic and change of vibration in a medium (Cheremisinoff 1996). The psychological and emotional feeling is created as a response to noise, which is also depending on the consciousness and sub-consciousness to the hearer feeling. Changes in the certain level and variation of sound may psychologically and physiologically causes the hearer to feel anger, frustrated, stressed and distracted. Thus, the noise effect can even affect our health, behaviour, communication, environment, etc.

The long exposure of noise for a certain intensity and duration normally may lead to hearing damage. Depending on the sensitivity and the situation of the hearer, hearing loss may either be permanent or temporary. It is believed that the significant temporary hearing loss or threshold shift due to exposure to loud noise may lead to a permanent hearing loss for recurring or prolonged exposure. However, it could not be clarified the exact relationship between temporary and permanent hearing loss. At high-frequency ranges, hearing loss might seriously affect the understanding and way of speech. Thus, at higher frequencies, it is very important to distinguish the speech consonants that carried a lot of information.

There should be more concerns regarding on this matter, even though the evidence to support the non-auditory health effects cases may not be as ample as the hearing loss cases. The reason behind this may be due to noise able to alter the normality function of the neurological, endocrine and cardiovascular systems. The noise alternations affect the in-equilibrium disorder in the body system and may cause a rise in blood pressure, constriction of blood vessels and change in heart rhythm. As studies performed on the animal to demonstrate such phenomena, it is noticed that noise may produce similar effects, or compound effect of, other more common stresses. Unfortunately, there is not enough support to show that permanent due to recurring noise is due to change in physiology and psychology. In noisy environments, the effects of behavioural may range from dulling response up to certain specify auditory signals that might produce frustration and falling to a sensitizing of annoyances that normally would be ignored by some. Table 1 shows the typical noise levels in different environments.

Table 1 Typical noise levels in different environments

Source of noise	Sound level (dB)	References
Iron factory	100.4	Ali (2011)
Power plant	91.5	Ali (2011)
Residential areas (in city)	55	Ali (2011)
School yard	55	Sarantopoulos et al. (2014)
Roadside	70	Sarantopoulos et al. (2014)

In work environment, noise can magnify the minor exasperation. It is also known that human performance would also affected by the presence of noise, especially those who involved in gathering, processing and analyzing information or data. Other than that, prolonged exposure to noise would also affect the health. The issue of prolonged exposure to noise and its undesirable effect on health is not only limited to the industrial and office workplace, but also at home and outdoor environment. Technological advances have provided a lot of new design and engineering system and indirectly becoming one of the major contributions that produced a lot of noise to the environment, such as engines, tyres and subways. Sometimes, noise sources from household too can be as severe as industrial sources. Example of household noise sources most probably produced by small machinery, such as chainsaw, lawn mowers, televisions, hand tools, fans, air conditioners and stereos. Sometimes, if these noise levels are left uncontrolled and unmonitored, it may subsequently affect the unprotected and unregulated occupants. Thus, significantly, these noise levels would indirectly cause hearing impairment.

2 Sound Absorbers

There are three types of the sound absorber, namely porous absorber, Helmholtz resonator and membrane absorber. In the application of noise control or sound absorption, the porous absorber is the most widely being used as the sound absorbing material. Porous absorber is a solid material, where its inner structure consists of numerous cavities, channel or interstices, which enable sound waves to enter through them. Porous absorber can be classified into cellular, fibrous and granular (Rey et al. 2012). Thus, this encourages the focuses on more research on the fibrous porous material. While, Helmholtz resonators are commonly used in controlling narrow band low-frequency noise. The reason behind this was due to the Helmholtz resonator having a cavity connected to the main duct through a neck in which the cavity and neck are connected in series that leads to two resonances at low frequency. Unfortunately, due to the narrow band performance of the resonator, the Helmholtz resonator is not very effective in controlling and absorbing noises which do not have substantial tonal characteristics (Tang et al. 2012).

A membrane absorber has only the vibrating metal membranes on a lightweight metal honeycomb structure and does not contain porous materials or fibre free. While,membrane absorber has a smooth cover membrane that helps protect against humidity and solid particles. This characteristic promotes the use of membrane absorber in silencers or as surface treatments in architectural applications (Kang and Fuchs 1999). When sound is absorbed by the porous material, heat is generated as a result of the dissipation of acoustical energy. Sound, as a pressure wave, causes air molecules to vibrate with the frequency of the sound wave when sound approaches a porous material. When a sound wave enters through the interface of a rigid or flexible wall or the skeleton of the porous or fibrous sound absorbing material, the drag force caused by the interface and fluid (air molecules) will produce friction.

The flow of sound waves that expand and contract when flowing through irregular pores as well as a change in the direction of flow direction causes the sound wave to lose momentum (Cox and D`Antonio 2009).

A good sound absorber will keep the reflection of sound at the surface as low as possible. The acoustical energy should be dissipated after entering the absorber before it reaches the surface of the absorber again due to being reflected from the rigid backing. This implies a need of sufficient thickness of porous material (Ver and Beranek 2006). Resonant absorbers involve a mass vibrating against a spring. Losses occurred by damping to remove the sound energy. Damping is at its highest efficiency by placing porous absorbent in the neck of the Helmholtz resonator. Alternatively, viscous losses within the neck can be used to gain absorption for Helmholtz resonators with small openings. This is a technique that allowed resonators without porous absorbent to be produced (Cox and D`Antono 2009).

3 Sound Absorbing Materials

The sound absorbing materials with different structure have their unique sound absorbing characteristics upon interacting with a variety of sound frequency and intensity. It includes hollow structure, porous structure, solid composite materials multi-dimensional size or length structure. Therefore, a brief review based on the sound absorbing materials and the reason it was replaced with more advanced alternatives were needed to be discussed in order to understand the development of sound absorbing materials. The first materials that being used as sound insulating or soundproofing materials were asbestos. Considered as fibrous mineral materials, asbestos is not easily burned, corrode and generally a good electrical insulator. However, it is easily deformed with other materials.

Asbestos is the term name for a specific mineral that takes place in a certain natural occurring mineral fibre (National Toxicology Program 2011). The natural occurring mineral fibre contains in asbestos are *actinolite, amosite, anthophyllite, chrysotile, crocidolite* and *tremolite* (National Toxicology Program 2011). The asbestos first application is reported in the United States in the early 1800s along the lines of the industrial revolution in the country (National Toxicology Program 2011; National Park Service 1999). In the 1940–1975 eras, asbestos managed to reach its highest peak demand, although at the first stage introduction of it was very rare and challenging. Asbestos was widely used in building such as school, office, houses and plant. It is usually designed for soundproof product and fire insulator for ceiling and roof.

Unfortunately, throughout the introduction of the asbestos period to industry, there is a lack of understanding on the harmfulness of asbestos towards environmental, animal and human health. The risk of asbestos started to reveal after a decade because of the advancement of the technologies and research. Reacted with water or air in an extensive period of time, the asbestos may release the insignificant amount of poisonous fibrous particle. It is usually from the fibre parts of the asbestos which is highly volatile and yet insoluble. In other words, the released of the poisonous fibrous particle are unreactive and degraded slowly in the environment (Clement International Corporation, United States Public Health Service, and United States Agency for Toxic Substance and Disease Registry 2002).

The released of small asbestos particle and when it is inhaled for a long period of time may cause a health problem. This will cause the small growth deformation called cancer. Asbestos (especially *tremolite*) is inflammatory, extremely strong scouring action when inhaled and bio-persistence. It is considered to be the related main contribution to mesothelioma and lung cancer. It can also exist as an impurity embedded phase within non-asbestos materials and other asbestos fibres such as *chrysotile*, for instance, vermiculite and talc (Churg et al. 1984; McDonald and McDonald 1997; Bernstein et al. 2005).

Most of asbestos applications and usage have been reduced in all over the countries due to the on-going awareness on the bad effect posed by it to the health. While other countries within the European Union (EU) have taking an initiative by completely banned the use of asbestos (Kogel et al. 2006). Eventually, factories have to halt asbestos production and terminate their operation due to the low demand on asbestos due to the tight importing and exporting restriction. However, some company in the industry succeeds to pursue new alternative material made from artificial materials such as synthetic fibre materials. As a result, the artificial fibres or synthetic fibres related industry has expanded as the main substitute for asbestos fibres during their introduction. There are more than hundred types of synthetic fibre that it is known in the world today.

Often manufactured with cellulose, synthetic fibres are used as a starting material modified from natural polymers (Rouette 2001). Due to urgent need and large demand to replace asbestos material, lack of research on this area lead to the unknown worst effect possessed by synthetic fibre. In the lab, as where the research conducted, it was found that the synthetic fibre was having characteristic, behaviour, hazardous and other properties almost similar to those asbestos material. Specifically, it may cause the formation of growth called cancer and injury in the lung, (Su and Cheng 2009). Because of it slow degradation and deterioration rate, the usage of synthetic fibre is bad for the environment. Because of many cases regarding on asbestos and synthetic fibre reported, especially concerning on the health and environmental, this had causes emerging of alternative sound absorbing based on the natural fibre materials research and development. The replacement with natural fibre materials can help to reduce this problem. It is known that natural fibre is affordable, biodegradable, processable, recyclable, renewable and sustainable (Fragomeni and Venkatesan 2010).

Up till now, most of the reported acoustic behaviour obtained did not consider the result of other properties (i.e. morphological, chemical, mechanical etc.) and what affected it in details, although a lot of researches were done on it. Further explanation and interpretation can be done especially on the structure-properties relationship either by mechanical means, chemical functionality group, and fibre polymer interfaces morphologically. Therefore, a study on using natural fibre composites as replacements in certain applications, such as soundproof insulation with constructive structure is becoming more important.

4 Mechanism of Sound Energy Absorption in Lignocellulosic Fibre Composites

It is known that the air molecules vigorously vibrated when sound is produced. The vibrations of the air molecules were due to the transferable energy properties between itself, two or more and different sizes of molecules in the air. The sound energy will be transferred to absorber materials such as lignocellulose fibre composites from air medium to solid medium. As the molecular particle of air and absorbing materials collided, the sound energy was transferred from the air to the absorbing materials. The vibration and collision between two different medium and different molecular particles cause friction and vibration in both air and solid medium. This friction and vibration in both air and solid medium also generate heat energy. Unfortunately, not all the heat energy and sound energy generated is absorbed by the lignocellulose fibre composites materials because usually some of them were reflected back in the air. Figure 2 shows the reflection and absorption of sound absorbing materials.

Internally, the sound energy and heat energy were widely spread inside the absorbed materials. Cox and D'Antonio (2009) stated that when flowing through irregular pores, the flow of sound wave expands and contracts. It will cause the sound wave to lose it momentum as there is a change in the flow direction. Almost all the noise created will absorb by the perfect sound absorber and very minimal quantity of sound would be reflected. After being absorbed by the sound absorbing

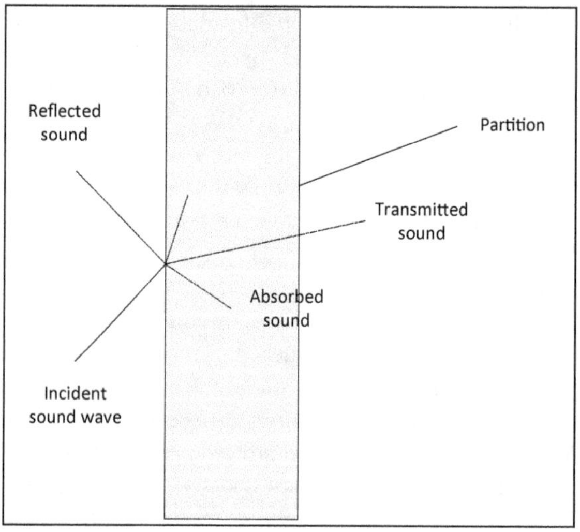

Fig. 2 Interaction of sound wave on material

material, the sound energy degenerated before it reaches a surface due to reflection from the rigid backed of the material (Vér and Beranek 2006). This also indicates that porous materials used should be necessarily thick so that the reflected noise can be reduced.

5 Factors Affecting Sound Absorption Coefficients of Lignocellulosic Fibre Reinforced Composites

Sound absorption takes place when a wave of sound passes through a material and results in a reduction in the acoustic energy. The materials capable of reducing acoustic energy significantly are also called as sound absorptive materials. The changes in the parameters (i.e. porosity, density etc.) are important. This needs to be understood deeply, because it influences the properties of the materials, which dictates the outcome of sound absorption properties of a material, such as the absorption coefficient. Sound absorbing materials were typically used to control an acoustic environment into a certain level of an enclosed volume by reducing the amplitude of the reflected waves. Common factors such as the fibre size, porosity, flow resistivity, density, thickness, etc. affect sound absorption of the materials that will be investigated and discussed.

5.1 Fibre Size

Fibre sizes vary depending on different materials being used. The changes in the fibre size imply the change in the fibre diameter or the fibre length itself or both. It is also one of the main factors that influence the sound absorption properties of porous materials. The effect due to changes in fibre size can be observed through related studies reported in the literature. Lee and Joo (2003) managed to conduct an experiment based on the usage of three different types of non-woven fabric materials that were made of recycled polyester fibres which had the same fibre length but different denier. The fibre deniers used were 1.25, 2 and 7 den in diameter with a 38 mm in length. For the binding purpose, low-melting point polyester staple fibre with a 6 den in diameter with 42 mm in length was used. Denier (den) is the unit measurement of the fibre linear mass density of fibre size. Lee and Joo (2003) concluded that the fine fibre content increases the sound absorption coefficient of the materials due to superior airflow resistance by mean of friction of viscosity through the vibration of the air.

Koizumi et al. (2002) also reported that a decrease in fibre diameter size leads to an increase in the sound absorption coefficient. The reason behind this was that sound waves move within the thin fibre more easily when compared to their motion through thick fibres. Whereas, according to Sun et al. (1993), by using fine fibre denier, more fibres are needed to fulfil or reach the equivalent of its original fibre content or same volume density. This results in higher airflow resistance and more tortuous path. A study by Koizumi et al. (2002) also showed that the fine fibre denier that range from 1.5 to 6 dpf (denier per filament) performed acoustically superior compared to the coarse denier fibres. Furthermore, it is also reported that the micro-denier fibres (less than 1 dpf) can provide a dramatic increase in the acoustical performance of sound absorption in the materials (Koizumi et al. 2002).

5.2 *Porosity*

The simplest definition of porosity is an empty void or space in materials. According to Allard et al. (1989), the porosity of materials is defined as the ratio of the volume of the voids in the porous materials to its total volume. In terms of mechanical and mathematical perspective, porosity is the fraction of the volume of medium voids (i.e. air). Porosity usually traps gas (i.e. air), liquid (i.e. water) or solid (i.e. starch) form of materials. The porosity in materials is higher, especially for materials with foam type structure or open pores, such as porous silicon and porous rubber. Furthermore, according to Shoshani and Yakubov (2001), an absorber of non-woven webs has high-sound absorption coefficient. The design of such a structure, the absorber should have increasing porosity along the direction of propagation of the sound wave. Based on Canham (1997), porosity can be identified in different ways and it depends on the shape and size packed inside a material. The formula for porosity can be identified as below:

$$\text{Porosity, } \phi = \frac{V_{\mathrm{v}}}{V_{\mathrm{t}}} \tag{1}$$

Whereas, V_{v} is the volume of medium voids (i.e. air) in the acoustical materials (m^3), and V_{t} is the total volume of the acoustic materials (m^3).

According to Sagartzazu et al. (2008), the porosity values are defined within the range of 0–1. In other words, it is measured from 0 to 100%. It is known that the porosity of some porous absorbers, such as polymer foams and fibrous polymer, lie between the ranges of 0.95–0.99. The influence of porosity has been well presented through the research conducted by Knapen et al. (2003). In their research, the porosity was manipulated and the effect on the sound absorption coefficient against frequency was investigated. The high porosity in a material causes high contact between the material surface and sound waves. This leads to more acoustical energy being dissipated. In addition to that, the number, size and types of pores are important factors that should be considered while studying sound absorption mechanism in porous materials. Thus, to allow sound dissipation by friction, the sound waves have to enter the porous materials. In other words, there should be enough pores within the material and on the material surface for the sound to pass through and get dampened.

5.3 Flow Resistivity

The determination of how much the air enters a porous absorber and the resistance to the air flow from the entering the structure is called flow resistivity. Thus, flow resistivity is one of the most important aspects that influence the characteristics of sound absorption in materials. It is also known that the specific flow resistance per unit thickness of the materials influences the sound absorbing properties of fibrous materials. Interlocking within the fibres in non-woven materials provides the friction that resists acoustic wave motion. In general, when a sound wave enters these materials, the amplitude will decrease due to friction when the waves travel through the tortuous and rough passages. The acoustical or sound wave energy was converted into heat energy as sound waves energy pass through the frictional rough and tortuous passage of the materials (Conrad 1983). Friction quantity is expressed by the resistance of the materials to airflow that is called airflow resistance and it is defined as the following:

$$\text{Flow Resistivity, } R_{\mathrm{f}} = \frac{\Delta P}{\Delta T u} \tag{2}$$

Where, ΔP is the differential sound pressure across the thickness of the sample (Pa), ΔT is the incremental thickness (m), and u is the particle velocity through the sample (ms^{-1}).

Based on Crocker and Arenas (2007), the high flow resistivity materials had a high resistance of air flow through the materials. This showed that materials with high flow resistivity sometimes tend to reflect more sound it absorbs. Furthermore, according to Ren and Jacobsen (1993), impedance and propagation constant characteristics describe the acoustical properties of porous materials. Hence, the flow resistance of the materials are controlled to a great extent. For a porous material, the airflow resistivity per unit thickness is proportional to the shear viscosity coefficient of the fluid (i.e. air) involved and inversely proportional to the square of pore size characteristics of materials. This implies, with given porosity of a fibrous material, the flow resistivity per unit thickness is inversely proportional to the square of fibre diameters (Uno Ingard 1994). According to Cox and D'Antonio (2009), flow resistivity may vary between common porous absorber, thus, this can imply that it is one of the most important factors in the research.

5.4 Density

Density is one of the important parameters that influence the sound absorption coefficient behaviour in materials. According to the studies done by Koizumi et al. (2002), for middle and high range frequencies, sound absorption increases with increase in density. Similar results were found by Ballagh (1996) research while working with wool materials with varying bulk density. Based on the both results, it showed that the low-density type of material absorbs more sound at the low frequency. In contrast, higher density materials normally absorb more in the middle and high frequencies. Furthermore, the material starts to perform well with a big margin or gap range in higher density at 500 Hz and above. This is due to the increase in the energy loss as the surface friction increases in a denser material, which leads to an increase in the sound absorption coefficient. Moreover, when the apparent density is high, the quantity of fibres per unit area increases. The sound absorption coefficient increases as the energy loss increase when the surface friction increases. The density of a material composite can be determined according to the formula:

$$\text{Density, } \rho = \frac{W_a}{W_a + W_w - W_b}(\rho_{water}) \tag{3}$$

Where, W_a is the specimen weight in air (kg m^{-3}), W_w is the partly immersed wire specimen holder weight (kg m^{-3}), W_b is the specimen weight fully immersed in distilled water along with the partly immersed wire specimen holder (kg m^{-3}) and ρ_{water} is the density of distilled water at testing temperature (kg m^{-3}).

5.5 Thickness

There is an enormous amount of study done for porous materials, especially on the sound absorption behaviour and the effect of its thickness. According to Coates and Kierzkowsld (2002), an effective sound absorption of a porous absorber was achieved when the thickness of the materials is around one-tenth of its incident sound wavelength by following the rule of thumb. Whereas, according to Hirabayashi et al. (1995), peak absorption may occur at one-quarter incident sound wavelength at the resonance frequency. Thus, these showed that there is a relationship between thickness either at low, medium or high-frequency sound absorption. According to Ibrahim and Melik (1978), as the material increases in the thickness, at the low frequency, the sound absorption will increase as well. However, at a certain high frequency, the sound absorption had an inconsistent effect.

The effects of thickness against sound absorption were also included in Hakamada et al. (2006) report. According to Hakamada et al. (2006), based on their experiment on the three samples of material with different thickness at 1, 2 and 20 mm, the result clearly showed the effect of thickness difference on the sound absorption coefficient. It showed that the thicker the material, the more sound it absorbs. Material thickness can enhance the acoustic impedance for porous materials (Huang et al. 2008). It can be seen clearly in Fatima and Mohanty (2011) results, where the natural rubber latex jute composites showed higher sound absorption coefficients in thick composites when compared with thin composites.

Meanwhile, Jiang et al. (2012) also reported that the chlorinated polyethylene (CPE) and seven-hole hollow polyester fibres (SHPF) composites show improved sound absorption when the thickness increased. Ersoy and Küçük (2009) also reported that a linear increase in sound absorption coefficient can be observed in the sample results, as the thickness increased.

5.6 *Tortuosity*

Tortuosity is the measurement of the elongated passage way through the pores in certain thickness of the sample. However, there are many other definitions describing tortuosity. According to Yamashita et al. (2009), tortuosity is defined as a parameter in the internal structure of porous materials that represent the quantity of sound waves that enter the materials air space. However, according to Knapen et al. (2003), tortuosity is the internal structures that influence the acoustical properties of the material. According to Wassilieff (1996), tortuosity is the measurement of how far the pores deviate from the normal, or twist bout the material axis. Tortuosity equation can be defined as below:

$$\text{Tortuosity, } \tau = \phi \frac{R_s}{R_f} \tag{4}$$

Where, ϕ is the porosity, R_s electrical resistivity of the saturated sample (N.s. m^{-4}) and R_f is the resistivity of the fluid used (N.s.m^{-4}).

Based on Sakagami et al. (1996) report, it appears that the optimum sound absorption is achieved by the less tortuous materials rather than by the materials that are more tortuous. Knapen et al. (2003) stated that the more tortuous the materials, the lower absorption coefficient it has. Higher tortuosity causes larger fluctuations in the sound absorption coefficient. The fluctuations will start manifesting from the lower frequencies as the tortuosity increased. However, Sakagami et al. (1996) stated that tortuosity only had a small effect on the noise transmission coefficient. Horoshenkov and Swift (2001) reported that the tortuosity only affects the location about quarter wavelengths of its peaks, whereas the flow resistivity and porosity may involve in affecting the height and width of the peaks. Furthermore, the value of tortuosity might determine the high-frequency response of sound absorbing porous materials.

5.7 *Compression*

The effect of compression plays the key role in the field of sound absorption or acoustical absorption. The results of manipulating porous materials such as fibrous materials or cellular materials create diverse variants of physical parameters that influence sound absorption. The compression of materials causes the sound absorption to differ in distinct porous materials. According to Castagnede et al. (2000) and Wang et al. (2008), the sound absorption properties decreased as the fibrous mats or porous materials being compressed. On compressing, the various fibres in the mat were brought nearer to each other without deformation (without any change in fibre size) (Castagnede et al. 2000). The compression also resulted in a decrease in thickness and indirectly cooperates with other physical variation (Castagnede et al. 2000).

Based on the study by Castagnede et al. (2000), compression led to an increase in flow resistivity and tortuosity, and indirectly decreased the shape factor (include characteristic length) and porosity. Hence, this occasionally caused inconsistencies in sound absorption behaviour. According to Wang et al. (2008), the large compression rate caused the effect on the sound absorption at the lower frequency, as the final thickness of porous materials is invariable. Unfortunately, these physical parameter variations in compressed materials are not reason for the drop in sound absorption directly because it is typically related to the decrease in sample thickness based on the experiment Castagnede et al. (2000) conducted. According to Keshavarz and Ohadi (2013), depending on the compression method, it may either give negative or positive effects on sound absorption coefficient.

5.8 Surface Impedance

Jayaraman (2005) has investigated the influence of surface impedance on sound absorption by using air laid samples with different configurations. It can be seen that the sound absorption coefficients increased when the surface impedance of the materials were altered. The surface impedance influences the sound absorption significantly by altering the tortuous path and porosity of the material. Higher sound resistivity was estimated when the material with a given layer of thickness produced higher dissipation. The reason behind this was due to the surface impedance that increased at the same time with a resistivity that caused a greater amount of reflection, especially on the surface layer that indirectly produces a lower absorptive capability. Moreover, according to Simón and Pfretzschner (2004), the whole process must be frequency dependent, as the lower frequency bands on the necessary layer thickness increases, the resistivity will decrease.

5.9 Placement/Design

According to Everest (2001), the sound absorption of materials may be influenced by the different types of sound absorbers used in different places such as sides, ends and ceilings at different axial modes (can either in transverse, vertical or longitudinal axis). Sound absorbing materials placed near corners or along edges of the room surfaces typically demonstrated effective sound absorption, especially in a rectangular room. For example, in the case of speech studios, the absorbing materials must be placed at head height on the walls, so that it will be more effective at higher audio frequencies. At a certain point, the sound absorption can be twice more effective as the same material placed elsewhere if it were applied at lower portions of high walls (Everest 2001). Moreover, the material surfaces should never face each other.

If there is an air cavity behind the materials, the materials will act as a membrane of a certain mass. This design is frequency dependent. The air inside the materials cavity is analogous to mechanical spring. Significantly, the sound absorption of materials is enhanced by embedding an air gap between the rigid back surface and the sample (Hakamada et al. 2006). In the research work by Hakamada et al. (2006), it was observed that the air gap increase when the thickness decrease. Aso and Kinoshita (1965) concluded that the greatest absorption coefficient in a thin sample of large porosity slightly decreased as in the depth of the air space increased. Besides, their study revealed that the absorption coefficient in a thick sample is almost constant, independent with the profundity of its air space.

5.10 Temperature

Srivastava et al. (2006) and Harris (1966) examined the effect of temperature on the mineral sound absorption and the outcomes showed that sound absorption was affected when there was a change in temperature. Furthermore, Srivastava et al. (2006) had established a theoretical connection between the thermal conductivity using a least-square regression method and the noise reduction coefficient. These relations were done under the distinctive temperature condition.

6 Methods to Measure Sound Absorption Coefficient of Composites

There are two methods or techniques to obtained sound absorption coefficient. They are reverberation and impedance tube. In extend, impedance tube was divided into two sub-methods or sub-techniques. They are standing wave and two-microphone transfer function impedance tube. All of these methods or technique usually use concentration ratio of absorbed sound energy which is the reflected energy to incident energy.

6.1 Reverberation Method

A method that uses the outcome of reflections of sound in receiving and hearing echoes at certain source locations in a compartment or room is called reverberation. Reverberation in a room is usually calculated in terms of reverberation time. Factor that affected reverberation time is usually room size and room interior surface sound absorption. Chen (2011) stated that the reverberation time was determined at the time required for reflections of the direct sound to attenuate by 60 dB below the level of the direct sound. According to Jambrosic et al. (2008), the testing procedure of reverberation is to calculate the empty room reverberation time. The reverberation time was repeatedly taken in an empty room where the specimen was placed. Average reverberation time was used to calculate sound absorption performance. Figure 3 shows the reverberation room method.

6.2 Standing Wave Method

The first method to measured sound absorption by using sound speed in a tube was the standing wave method (Oldfield 2006). One of the most popular methods for standing wave is using standing wave ratio (SWR). The magnitude and the sequential minimum and maximum location of the standing wave pattern in the tube was determined using traversing microphone. Thus, these will produce the acoustic reflection coefficient and impedance. Fundamentally, standing wave method creates a standing wave from the transmitting sound sources in the tube before it is reflected at the end side of the mounted sample. To minimum and maximum of the standing wave were obtained by utilized the measurement of the amplitude by moving the microphone location. The calculation of sound absorption coefficient was in relation

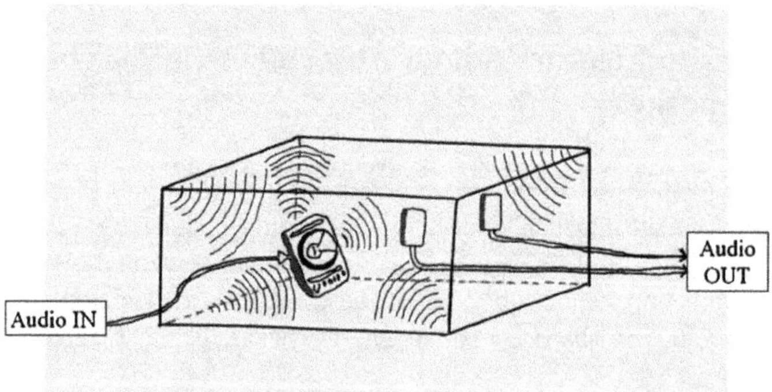

Fig. 3 Reverberation room method

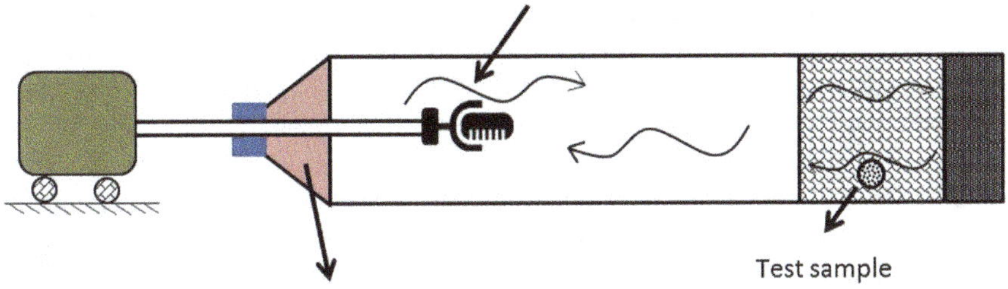

Fig. 4 Standing wave method

to the change in frequency. Standing wave method normally standardized in accordance with ASTM C384-04 (2011) and ISO 10534-1 (1996). Figure 4 shows the standing wave method.

Based on standing wave method concept, a standing wave is created when the source of sound being switched on. Along the tube, moving back and forth (up and down) by using a microphone, the pressure point for the minimum and maximum were obtained. To get the reading, the microphone is connected to a voltmeter. The phase and the magnitude of sample location based on the standing wave data were computed to obtained sound absorption coefficient. The standing wave method will consume a lot of time due to the varieties of value and position needed to be distinguished independently for each related frequency. However, the results are easy to be interpreted and still reliable.

6.3 *Two-Microphone Transfer Functions Method*

Alternatively, the sound absorption can also be calculated experimentally by using the two-microphone transfer function impedance tube method (Huiping et al. 2006). Since the introduction of the two-microphone transfer function impedance tube method, it has replaced the popularity of standing wave method due to it several advantages such as accuracyetc. Act as a waveguide cylinder, impedance tube was used to generate plane wave from the sound source. The two microphones were located at two locations near the sample to measure the sound pressure as shown. This method frequency is depending on the distance between the positions of the microphone and the tube diameter size (Suhanek et al. 2008). Along a waveguide (Lefebvre et al. 2007), the measurement was taken at two different locations connected to the object and the impedance sound was taken and evaluated.

In their reported work, Putra et al. (2013) test the coefficient of sound absorption for biodegradable sample based on sugar cane fibres by using the two-microphone method. Before being dried heated at 80 °C for five minutes to remove the remaining water, the sugar cane was and cut off. During fabrication of test specimens, different compositions of sugar cane fibre were bind with polyester and polyurethane. Then, the different pressure of hot-pressed is used on the mould to create different porosity levels. Regardless of different type of the binder in the sugar cane, the result showed that no effect on 40% binder of the total composition on the performance of the absorption. The fibres started to show a good result of the sound absorption coefficient of 0.5 after the frequency of 3500 Hz and keep increasing to 0.78 at 4500 Hz. Research made by Putra et al. (2013) provided a lot of information of bio-composites material sound absorption properties and reliability of two-microphone method to characterize composite material sound absorption properties. Figure 5 shows the two-microphone transfer function impedance tube method.

The calculation for sound absorption for the two-microphone method is based on the one-dimensional planar and pressure wave propagation mathematical theory. While, Seybert (1988) explained that in the tube, this method is driven by random sound. By advantage of using small samples, as a change in parameters, this method creates fast measurement of normal incidence wave and also it is easy to be disassembled and assembled. Additionally, the results demonstrated by Horoshenkov et al. (2007) shows that the result attained varied at different laboratory conditions. The benefit of the transfer function impedance tube is that the apparatus is smaller and therefore it is more useful for small test specimens. Conversely, it allows the surface impedance to be measured in addition to the sound absorption (Oldham et al. 2011). However, the shortcomings are that the sound absorption can only be measured at normal incidence wave and uncertainties exist while measuring the sound absorption of materials having a porous structure. Thus, other related characterisation needed to be done to reduce the uncertainties.

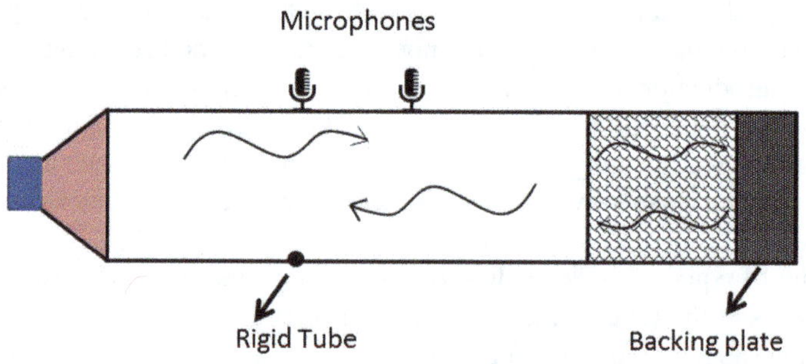

Fig. 5 The two-microphone transfer function impedance tube method

7 Fabrication of Lignocellulosic Fibre Reinforced Polymer Matrix Composites

Several researchers (Khedari et al. 2003; Khedari et al. 2004; Zulkifli et al. 2008) successfully reported composite made from particle board using agricultural wastes. One of the researchers found out that the sound absorption properties of glass wool fibres were similar to those bamboo fibre (Koizumi et al. 2002). Even though it had almost similar density, the enclosed surface fibreboard bamboo materials were found to have higher sound absorption properties than those plywood materials. Yang et al. (2003) found out that in the range of 500–8000 Hz, randomly cut rice straws and wood particle composite board exhibits higher sound absorption compared with other particleboard, plywood and fibreboard. With further research, it shows that natural fibre composite has becoming more popular due to it superiority such as combustible, biodegradable, weighable and recyclable, when compared with synthetic fibre composites. It is ranked among high-performance composites due to the good physical properties and advantages in the environment and economy (Avella et al. 2000).

According to Simón and Pfretzschner (2004), materials for sound absorbing are chosen based on it materials, types, dimensions and the frequency range it can control. Taking the example of ramie fibre poly (l-lactic) acid composites, it sounds absorption coefficient is from 0.089 to 0.353 at the frequency of 250–1600 Hz (Chen et al. 2010). Wheat straw polypropylene composites also showed high sound absorption coefficient from 0.03 to 0.2 at the range of 300–1800 Hz when compared with jute fibre polypropylene composites (Zou et al. 2010). Furthermore, Reddy and Yang (2011) showed high sound absorption coefficient form 0.06 to 0.8 at the range of 1000–5000 Hz for jute zein composites as compared with jute polypropylene composites. Rice straw and wood particle commercial urea formaldehyde composite boards also showed better sound absorption coefficient in the frequency range of 500–8000 Hz as compared to particleboard, plywood, and fibreboard (Yang et al. 2003). Moreover, Yang et al. (2004) also showed high sound absorption coefficient for commercial rice straw and waste tyre particle polyurethane composites at the frequency range of 2000–8000 Hz, whereas its performance is superior than fibreboard, particleboard, and rice straw-wood particle composite boards. Table 2 shows the summary of acoustical properties of natural fibre.

Today, agricultural raw waste materials are spread widely and being used as replacement for synthetic fibres and wood-based acoustic absorption materials purposes. Plentiful of agricultural waste in Malaysia can be used in various materials applications even in the near future, for example, rise husk, coconut, and oil palm fibre (Mohd Nor et al. 2004). Furthermore, it is easy to crush agricultural lignocellulose fibres such as rice straw, wheat straw, and oil palm frond into particles and chips compared with wood particles or synthetic fibres. These fibres are cheaper, nonabrasive, abundant, renewable and less toxic. Thus, through research, current and future applications based on the renewable resources for acoustical

Table 2 Summary of acoustical properties for natural fibre

Materials	Fabrication methods	Sound absorption coefficient	Specimen parameters	References
Bamboo	Compression	0.02–0.4 (at 3000 Hz)	Thickness of 50 mm	Koizumi et al. (2002)
Wood	Compression	0.1 (at 4000 Hz)	Thickness of 25 mm	Wassilieff (1996)
Cellulose (isofloc)	–	1.10 (at 3000 Hz)	Thickness of 60 mm	Asdrubali (2006)
Sheep wool		0.80 (at 3000 Hz)		
Mineral wool		0.90 (at 3000 Hz)	Thickness of 40 mm	
Arenga pinnata	–	0.40–0.80 (at 5000 Hz)	Thickness of 10–40 mm and diameter of 28 mm	Ismail et al. (2010)
Cotton	Compression	0.50 (at 500 Hz)	Thickness of 50 mm	Oldham et al. (2011)
Kapok	–	0.67–0.99 (at 4000 Hz)	Thickness of 20–60 mm, density of 8.3–41.7 kg/m^3, and combined diameter of 30 and 60 mm	Xiang et al. (2013)
Coir	Compression	0.78 (at 4000 Hz)	Thickness of 20 mm and combined diameter of 100 and 28 mm	Fouladi et al. (2013)
Sugarcane				
Grass		0.70 (at 4000 Hz)		
Corn		0.90 (at 4000 Hz)		

materials can lead to viable alternatives (Nick et al. 2002; Asdrubali 2006). Table 3 shows the summary of acoustical properties of natural fibre composites.

8 Sound Absorption Simulations

8.1 Empirical Model for Sound Absorbing Materials

The properties of sound absorbing material can be determined by phenomenological and microstructural models. Unfortunately, at certain condition the validity is very limited. As in formulation of empirical models, generally, the theoretical point of view were not being considered, neither the structure inside the material taken were also into account (Ballagh 1996).

In the past decades, several models were developed by researchers (Qunli 1988; Dunn and Davern 1986) built on the principles and model made by Delany and Bazley (1970). Subsequently, these models were extendedly modified by Mechel

Table 3 Summary of acoustical properties of natural fibre composites

Materials	Fabrication methods	Sound absorption coefficient	Specimen parameters	References
Rice straw/ wood urea formaldehyde composites	Hot compression (500 psi 140 °C)	0.05–0.5 (at 8000 Hz)	Thickness of 10 mm, diameter of 25 mm, and fibre loading of 0–30 wt%,	Yang et al. (2003)
Luffa epoxy composites	Compression (1000 psi)	0.1 at 6000 Hz	Thickness of 5 mm, diameter of 25 mm	Jayamani et al. (2014)
Hemp brick (66%)	–	0.50 (at 2000 Hz)	Thickness of 300 mm	Asdrubali (2006)
Polystyrene (PST)		0.20 (at 3000 Hz)	Thickness of 40 mm	
Recycled latex + coco		0.50 (at 3000 Hz)		
Rami poly (L-lactic acid) composites	Hot compression	0.01–0.02 (at 1800 Hz)	Thickness of 3 mm, diameter of 110 mm and various grade of PLLA.	Chen et al. (2010)
Wheat straw/ polypropylene composites	Hot compression	0.03–0.23 (at 3000 Hz)	Thickness of 3.2 mm, fibre loading of 40–80 wt%, and diameter of 63 mm	Zou et al. (2010)
Polyester fibres polyethylene composites	Hot compression	0.03–0.35 (at 10,000 Hz)	1st condition: Thickness of 1–3 mm and fibre loading of 0–20 wt%, 2nd condition: Thickness of 20 mm and diameter of 60 mm	Jiang et al. (2012)
Flax/epoxy	Compression (laminated)	0.11 (at 2000 Hz)	Thickness of 3 mm and diameter of 100 mm	Prabhakaran et al. (2014)
Glass/flax/epoxy	Compression (laminated)	0.15 (at 2000 Hz)		
Oil palm/zein composites	Hot compression	0.11 (at 6000 Hz)	Thickness of 5 mm and diameter of 25 mm	Jayamani et al. (2016a)
Banana/epoxy composites	Compression	0.13 (at 6000 Hz)	Thickness of 5 mm and diameter of 25 mm	Jayamani et al. (2016b)

and Ver (1992). By normalized it with a desired frequency and referred it to the flow resistivity of the materials, the impedance and propagation coefficient characteristic of the material can be predicted. Another prediction can be made from the complex quantities for acoustic properties. Natural wool materials are sometimes outside of the limits available empirical models, although this method was used extensively. According to Ballagh (1996), in some of the cases, microstructure models are best to used. It is because of the consideration of the internal structure and physical properties of the material were taken to calculate and predict the acoustic performance. Normally, the models used were correlated with parameters like porosity, tortuosity, flow resistivity, and shape factors. Over the past decades, different models were development by Zwikker and Kosten (1949), Rayleigh (1896), Attenborough (1982), and Biot (1956). Other researchers (Allard and Champoux 1992; Johnson et al. 1987), used shape factors modelled complex model structure.

8.2 *Empirical Model for the Flow Resistivity*

In a material sound absorption, the influenced of flow resistivity is significant. The fluctuation of sound absorption coefficient decreased and increased is profoundly related with the flow resistivity that changes in the materials. Thus, in modelling of sound absorption, flow resistivity plays a big role. In modelling of sound absorption coefficient, either as constant or variable, the flow resistivity can be manipulated in the formula. To calculate the flow resistivity, Bies and Hansen (1980), Garai and Pompoli (2005) and Mechel (2008) model can be used. Normally, the flow resistivity is derived based on the fibre radius, porosity and the bulk density from related calculation and measurement of the materials.

8.2.1 Mechel Model

Based on Mechel (2008) models, by having the same radii for flow resistivity, the relationship between the fibre and the sound incidence derivation were made. Normally, based upon the fibre radius and bulk density or material porosity of the materials, the equations calculation was made. The equations based on a model created by Mechel (2008) are as below:

$$\text{Flow Resistivity, } \sigma = 10.56\left(\frac{\eta}{r^2}\right)\left(\frac{\mu^{1.531}}{r^2(1-\mu)^3}\right) \tag{5}$$

and

$$\text{Flow Resistivity, } \sigma = 6.8\left(\frac{\eta}{r^2}\right)\left(\frac{\mu^{1.296}}{r^2(1-\mu)^3}\right) \tag{6}$$

Where, η is the viscosity of the medium (i.e. air viscosity for $1.84 \times 10^{-5} Pa.s$), r is the radius of the fibre (m) and μ is the massitivity ($\mu = 1 - \varepsilon$, where ε is the porosity of the material). Equations 5 and 6 can be simplified as below:

$$\text{Flow Resistivity, } \sigma = \frac{10.56\eta(1-\varepsilon)^{1.531}}{r^2\varepsilon^3} \tag{7}$$

and

$$\text{Flow Resistivity, } \sigma = \frac{6.8\eta(1-\varepsilon)^{1.296}}{r^2\varepsilon^3} \tag{8}$$

Where, ε is the porosity of the material and r is the radius of fibre (m).

From 6 to 10 μm fibre diameter range, Eq. 5 was used and from 20 to 30 μm fibre diameter range, Eq. 6 was used. Oldham et al. (2011) stated that no closed cell or a small amount of binder material, the porosity, ε is given by the equation below:

$$\text{Porosity, } \varepsilon \approx 1 - \frac{\rho_b}{\rho_m} \tag{9}$$

Where, ρ_b is material bulk density (kg m^{-3}) and ρ_m, is the material matrix density (kg m^{-3}).

8.2.2 Bies and Hansen Model

The model expressed has the same matrix materials with the bulk density as the main functions of porosity (Bies and Hansen 1980). The flow resistivity was expressed as below (Bies and Hansen 1980):

$$\text{Flow Resitivity, } \sigma = \frac{K}{d^2\rho_m^{-1.53}} \tag{10}$$

Where, K is the constant value (3.18×10^{-9}), d is the fibre diameter (m), and ρ_m is the bulk density of the material (kg m^{-3}). Oldham et al. (2011) revised the model and is given as below:

$$\text{Flow Resistivity, } \sigma = \frac{3.18 \times 10^{-9}}{4r^2\rho^{-1.53}} \tag{11}$$

Where, r is the radius of fibre (m). The above expression was obtained and measured by using fibre mineral and fibreglass with approximately 2600 kg m^{-3} material matrix density (Oldham et al. 2011).

8.2.3 Garai and Pompoli Model

Oldham et al. (2011) stated that Garai and Pompoli model is improved based on the Bies and Hansen model that influence with a variety of the fibre diameter (i.e. between 18 and 48 μm), a mean value (i.e. around 33 μm) with a matrix density material (approximated 1350 kg m^{-3}). The equation on Garai and Pompoli (2005) is as below:

$$\text{Flow Resistivity}, \sigma = \frac{28.3 \times 10^{-9}}{4r^2\rho^{-1.404}} \quad (12)$$

Where, r is the radius of fibre (m). The equation revised by Garai and Pompoli (2005) is based on the Bies and Hansen model below:

$$\text{Flow Resistivity}, \ \sigma = A\rho_{\mathrm{m}}^{B} \quad (13)$$

Where, ρ_{m} is the bulk density of the material (kg m^{-3}), and both A and B are the free parameters of the materials. Thus, an ideal value for A and B needed to decide. To obtain A and B values, Garai and Pompoli (2005) used the least-square best-fitting method and managed to get the values of 25.989 and 1.404.

8.3 *Empirical Model for the Sound Absorption Coefficient*

Garai and Pompoli (2005) and Delany and Bazley (1970) were two models used to predict sound absorption coefficients. Garai and Pompoli (2005) model is improved version based on Delany and Bazley (1970) model.

8.3.1 Delany and Bazley Model

By a simple power-law relations, Delany and Bazley model were obtained by best-fitting a range of fibrous porous absorbers with a big amount of experimental data (Oldham et al. 2011). Based on the dimensional and represented in power-law relations variable, the data obtained were normalized (usually the flow resistance and the frequency). The empirical formula below:

$$\text{Characteristic Impedance}, \ Z_{\mathrm{c}} = \rho_0 c\left(1 + 0.0571\left(\frac{\rho_0 f}{\sigma}\right)^{-0.754} - j0.087\left(\frac{\rho_0 f}{\sigma}\right)^{-0.732}\right) \quad (14)$$

$$\text{Complex Wave Number}, \ k_{\mathrm{c}} = \frac{\omega}{C}\left(1 + 0.0978\left(\frac{\rho_0 f}{\sigma}\right)^{-0.754} - j0.187\left(\frac{\rho_0 f}{\sigma}\right)^{-0.732}\right) \quad (15)$$

Where, ρ_0 is the air density (kg m^{-3}), σ is the flow resistivity (Ns m^{-4}), f is the frequency (Hz), and $\omega = 2\pi f$ is the angular frequency (rads^{-1}). Based on the equation, the range is limited in $0.01 < \frac{\rho_0 f}{\sigma} < 1$. The sound propagation was influenced by two complex quantities, coefficient propagation $\gamma = \alpha + j\beta$ and characteristic impedance, $Z = R + jX$ (Delany and Bazley 1970).

8.3.2 Garai and Pompoli

Delany and Bazley improved formula based on Garai and Pompoli model. It is known that the empirical formula made by Delany and Bazley (1970) unlikely suitable for other material such as polyester fibre material (Garai and Pompoli 2005). For polyester fibre materials, with new impedance model (NMI), equations are derived from the best-fit model's value. As shown in Table 4, results of polyester material are given and tabulated with a comparison between Delany and Bazley the eight coefficient value.

Based on the power-law relationship, new equation was rewritten from data obtained by Garai and Pompoli (2005), where it can be fit in the range of $0.05 < \frac{\rho_0 f}{\sigma} < 8.4$ (Oldham et al. 2011). Below is the simplified equation:

$$\text{Real Part of Characteristic Impedance, } Z_R = \rho_0 C_0 \left[1 + C_1 \left(\frac{\rho_0 f}{r} \right)^{-C_2} \right] \tag{16}$$

$$\text{Imaginary Part of Characteristic Impedance, } Z_I = -\rho_0 C_0 \left[C_3 \left(\frac{\rho_0 f}{r} \right)^{-C_4} \right] \tag{17}$$

$$\text{Real Part of Propagation Constant, } \gamma,\ \alpha = \frac{2\pi f}{C_0} \left[C_5 \left(\frac{\rho_0 f}{r} \right)^{-C_6} \right] \tag{18}$$

$$\text{Imaginary Part of Propagation Constant, } \gamma,\ \beta = \frac{2\pi f}{C_0} \left[1 + C_7 \left(\frac{\rho_0 f}{r} \right)^{-C_8} \right] \tag{19}$$

Where, ρ_0 is the air density (kg m^{-3}) and f is the frequency (Hz).

Table 4 Value of the eight coefficient equation from new impendence model (NMI) compare with Delany and Bazley model value (Garai and Pompoli 2005)

Model	C_1	C_2	C_3	C_4	C_5	C_6	C_7	C_8
Delany and Bazley	0.057	0.754	0.087	0.732	0.189	0.595	0.098	0.700
NMI	0.078	0.623	0.074	0.660	0.159	0.571	0.121	0.530

Based on the equation, the sound absorption coefficient for a rigid backed fibrous layer at normal incidence can be obtained by using the equation below:

$$\text{Surface Impedance}, Z_l = (Z_R + iZ_I)[\coth(\alpha + i\beta)l] = Z_{lR} + iZ_{lI} \tag{20}$$

$$\text{Sound Absorption at Normal Incidence}, \ \alpha_n = \frac{4Z_{lR}\rho_0 C_0}{|Z_l|^2 + 2\rho_0 C_0 Z_{lR} + (\rho_0 C_0)^2} \tag{21}$$

Where, l is the layer thickness (m) and ρ_0 is the density (kg m^{-3}). In most applications that usefully for noise control purposes, the simplified procedure can be used where it is best fit based on the values of both characteristic impedance and propagation constant (Delany and Bazley 1970; Dunn and Davern 1986). In analogous ways, according to Gardner et al. (2003), the measured values of the absorption coefficient and surface impedance neural network was obtained and optimized on the sound absorption coefficient.

9 Conclusions

Over the last decade, the uses of composites have been rapidly evolving as due to the shortage of mineral materials, in which commonly used as a sound absorbing material. Thus, a significant research and development were done in order to improve the sound absorption, especially in the field of polymer matrix composites. Furthermore, the natural fibre reinforced with polymer matrix composites had gained enormous attention toward engineering, industrial, technologies, and manufacture due to its potentials and advantages for sound absorption applications and also due to its various environmental concerns. Specifically, with polymer matrix materials, natural fibres have the ability to function in many ways, lightweight, renewable, low density, produce less CO_2, and environmentally friendly. Furthermore, it is easy to fabricate, manufacture and have properties comparable to those other materials used for sound absorption. Further research and innovation, especially in material design and fabrication are increasing due to it revolutionize and realization for sound absorption applications especially in industry and automotive. Thus, this showed that the future sustainability and growth of composites and polymer from renewable sources are reliant on continued research.

References

Ali SA (2011) Industrial noise levels and annoyance in Egypt. Appl Acoust 72(1):221–225
Allard JF, Champoux Y (1992) new empirical equations for sound propagation in rigid framed porous materials. J Acoust Soc America 91(6):3346–3353
Allard JF, Depollier C, Guignouard P (1989) Free field surface impedance measurements of sound-absorbing materials with surface coatings. Appl Acoust 26(3):199–207
Asdrubali F (2006) Survey on the acoustical properties of new sustainable materials for noise control. EuroNoise 30(1):1–10
Aso S, Kinoshita R (1965) Maximum sound absorption coefficient of a fiber assembly. J Text Mach Soc Japan 11(3):81–87
ASTM C384-04 (2011) Standard test method for impedance and absorption of acoustical materials by impedance tube method. In: ASTM International. West Conshohocken, PA
Attenborough K (1982) Acoustical characteristics of porous materials. Phys Rep 82(3):179–227
Avella M, Rota GL, Martuscelli E, Raimo M, Sadocco P, Elegir G, Riva R (2000) Poly (3-Hydroxybutyrate-Co-3-Hydroxyvalerate) and wheat straw fibre composites: thermal, mechanical properties and biodegradation behaviour. J Mater Sci 35(4):829–836
Ballagh KO (1996) Acoustical properties of wool. Appl Acoust 48(2):101–120
Bernstein DM, Chevalier J, Smith P (2005) Comparison of Calidria chrysotile asbestos to pure tremolite: inhalation biopersistence and histopathology following short-term exposure. Inhalation, Toxicol: Int Forum Respir Res 17(9):427–449
Bies DA, Hansen CH (1980) Flow resistance information for acoustical design. Appl Acoust 13 (5):357–391
Biot MA (1956) Theory of propagation of elastic waves in a fluid-saturated porous solid. I: low-frequency range. J Acoust Soc America 28(2):168–178
Canham LT (1997) Properties of porous silicon, 1st edn. Institution of Engineering and Technology, London
Castagnede B, Aknine A, Brouard B, Tarnow V (2000) Effects of compression on the sound absorption of fibrous materials. Appl Acoust 61(1):173–182
Chen Z (2011) Investigation on simulation and measurement of reverberation for small rooms. Int J Adv Comput Sci Appl 2(7):69–73
Chen D, Li J, Ren J (2010) Study on sound absorption property of ramie fiber reinforced poly (L-Lactic acid) composites: morphology and properties. Compos A Appl Sci Manuf 41 (8):1012–1018
Cheremisinoff NP (1996) Noise control in industry: a practical guide, 1st edn. William Andrew Inc./Noyes Publications, New Jersey
Churg A, Wiggs B, De Paoli L, Kempe B, Stevens B (1984) Lung asbestos content in chrysotile workers with mesothelioma. Am Rev Respir Dis 130(6):1042–1045
Clement International Corporation, United States Public Health Service and United States Agency for Toxic Substance and Disease Registry (2002) Toxicological profile for asbestos, 1st edn. United States Department of Health and Human Services, Washington, DC
Coates M, Kierzkowsld M (2002) Acoustic textile-lighter, thinner and more sound absorbent. Tech Text Int. International Newsletters
Conrad JH (1983) Engineering acoustics and noise control, 1st edn. Prentice Hall, New Jersey
Cory W (2005) Fans and ventilation: a practical guide, 1st edn. Elsevier, Amsterdam
Cox TJ, D'Antonio P (2009) Acoustic absorbers and diffusers: theory, design and application, 2nd edn. Taylor & Francis, London
Crocker MJ, Arenas JP (2007) Handbook of noise and vibration control, 2nd edn. Wiley, New Jersey
Delany ME, Bazley EN (1970) Acoustical properties of fibrous absorbent materials. Appl Acoust 3 (2):105–116
Dunn IP, Davern WA (1986) Calculation of acoustic impedance of multi-layer absorbers. Appl Acoust 19(5):321–334

Ersoy S, Küçük H (2009) Investigation of industrial tea-leaf-fibre waste material for its sound absorption properties. Appl Acoust 70(1):215–220

Everest FA (2001) The master handbook of acoustics, 4th edn. McGraw-Hill, New York

Fatima S, Mohanty AR (2011) Acoustical and fire-retardant properties of jute composite materials. Appl Acoust 72(2–3):108–114

Fouladi MH, Nassir MH, Ghassem M, Shamel M, Peng SY, Wen SY, Xin PZ, Mohd Nor MJ (2013) Utilizing Malaysian natural fibers as sound absorber. Modelling and measurement methods for acoustic waves and for acoustic microdevices, InTech, pp 161–170

Fragomeni S, Venkatesan S (2010) Incorporating sustainable practice in mechanics and structures of materials, 1st edn. CRC Press, Boca Raton, Florida

Garai M, Pompoli F (2005) A simple empirical model of polyester fibre materials for acoustical applications. Appl Acoust 66(12):1383–1398

Gardner GC, O'Leary ME, Hansen S, Sun JQ (2003) Neural network for prediction of acoustical properties of polyurethane foams. Appl Acoust 64(2):229–242

Hakamada M, Kuromura T, Chen Y, Kusuda H, Mabuchi M (2006) High sound absorption of porous aluminum fabricated by spacer method. Appl Phys Lett 88(25):106–254

Harris CM (1966) Absorption of sound in air versus humidity and temperature. J Acoust Soc America 40(1):148–159

Hirabayashi T, McCaa DJ, Rebant RG, Rusch P, Saha P (1995) Application of noise control and heat insulation materials and devices in the automotive industry. SAE Technical Paper

Horoshenkov KV, Khan A, Bécot F-X, Jaouen L, Sgard F, Renault A, Amirouche N, Pompoli F, Prodi N, Bonfiglio P (2007) Reproducibility experiments on measuring acoustical properties of rigid-frame porous media (round-robin tests). J Acoust Soc America 122(1):345–353

Horoshenkov KV, Swift MJ (2001) The effect of consolidation on the acoustic properties of loose rubber granulates. Appl Acoust 62(6):665–690

Huang J, Sun B, Dai G-c (2008) Sound absorption properties of lightweight reinforced thermoplastic laminates. J Funct Polym 21(1):177

Huiping T, Ping T, Zhengping X, Qiangbing W (2006) Research progress of sintered porous metal. Rare Met Mater Eng 35(2):428–432

Ibrahim MA, Melik RW (1978) Physical parameters affecting acoustic absorption characteristics of fibrous materials. In: Proceedings of the mathematical and physical society of Egypt, pp. 125–130

Ismail L, Ghazali MI, Mahzan S, Ahmad Zaidi AM (2010) Sound absorption of arenga pinnata natural fiber. World Acad Sci Technol 43(1):804–806

ISO 10534-1 (1996) Acoustics—determination of sound absorption coefficient and impedance in impedance tubes—part 1: method using standing wave ratio. International Organisation for Standardization, Geneva

Jambrosic K, Horvat M, Domitrovic H (2008) Reverberation time measuring methods. J Acoust Soc America 123(5):3617

Jayaraman KA (2005) Acoustical absorptive properties of nonwoven. North Carolina State University

Jayamani E, Hamdan S, Rahman MR, Soon KH, Bakri MKB (2014) Processing and characterization of epoxy/luffa composites: investigation on chemical treatment of fibers on mechanical and acoustical properties. BioResources 9(3):5542–5556

Jayamani E, Hamdan S, Ezhumalai P, Bakri MKB (2016a) Investigation on dielectric and sound absorption properties of banana fibers reinforced epoxy composites. Jurnal Teknologi 78(6–10):97–103

Jayamani E, Hamdan S, Soon KH, Rahman MR, Bakri MKB (2016b) Acoustical, thermal, and morphological properties of zein reinforced oil palm empty fruit bunch fiber bio-composites. J Appl Polymer Sci:1–8

Jiang S, Xu Y, Zhang H, White CB, Yan X (2012) Seven-hole hollow polyester fibers as reinforcement in sound absorption chlorinated polyethylene composites. Appl Acoust 73 (3):243–247

Johnson DL, Koplik J, Dashen R (1987) Theory of dynamic permeability and tortuosity in fluid-saturated porous media. J Fluid Mech 176(1):379–402

Kang J, Fuchs HV (1999) Effect of water-films on the absorption of membrane absorbers. Appl Acoust 56(2):127–135

Keshavarz R, Ohadi A (2013) Effects of compression on the sound absorption of transversely isotropic fibrous materials at obligue incidence. Appl Acoust 74(1):383–395

Khedari J, Charoenvai S, Hirunlabh J (2003) New insulating particleboards from durian peel and coconut coir. Build Environ 38(3):435–441

Khedari J, Nankongnab N, Hirunlabh J, Teekasap S (2004) New lowcost insulation particleboards from mixture of durian peel and coconut. Build Environ 39(1):59–65

Knapen E, Lanoye R, Vermeir G, Lauriks W, Van Gemert D (2003) Sound absorption by polymer-modified porous cement mortars. In: 6th International conference on materials science and restoration, vol. 1, no. 1, pp. 347–358

Kogel JE, Trivedi NC, Barker JM, Krukowski ST (2006) Industrial minerals and rocks. In: Society for mining, metallurgy, and exploration, 7th edn. Littleton, Colorado

Koizumi T, Tsujiuchi N, Adachi A (2002) The development of sound absorbing materials using natural bamboo fibers. High Perform Struct Compos 4(1):157–166

Lee Y, Joo C (2003) Sound absorption properties of recycled polyester fibrous assembly absorbers. Autex Res J 3(2):78–84

Lefebvre A, Scavone GP, Abel J, Buckiewicz-Smith A (2007) A comparison of impedance measurements using one and two microphones. Proceedings of ISMA 1(1):1–8

McDonald JC, McDonald AD (1997) Chrysotile, tremolite and carcinogenicity. Ann Occup Hyg 41(6):699–705

Mechel FP (2008) Formulas of acoustics, 2nd edn. Springer-Verlag, Berlin Heidelberg, New York

Mechel FP, Ver IL (1992) Sound absorbing materials and sound absorbers. In: Beranek L, Ver IL (eds) Noise and vibration control engineering: principles and applications. Wiley, pp. 203–243

Mohd Nor MJ, Jamaludin N, Tamiri FM (2004) A preliminary study of sound absorption using multi-layer coconut coir fibers. Electron J Tech Acoust 3(1):1–8

National Park Service (1999) Health and safety risk of asbestos: conserve O gram, 1st edn. Department of the Interior, Washington, DC

National Toxicology Program (2011) Report on Carcinogens, 12th edn. United States Department of Health and Human Services, North Carolina

Nick A, Becker U, Thomas W (2002) Improved acoustic behavior of interior parts of renewable resources in the automotive industry. J Polym Environ 10(3):115–118

Oldfield R (2006) Improved membrane absorbers. The University of Salford

Oldham DJ, Egan CA, Cookson RD (2011) Sustainable acoustic absorbers from the biomass. Appl Acoust 72(6):350–363

Prabhakaran S, Krishnaraj V, Kumar MS, Zitoune R (2014) Sound and vibration damping properties of flax fiber reinforced composites. Procedia Engineering 97(1):573–581

Putra A, Abdullah Y, Efendy H, Farid WM, Ayob MR, Py MS (2013) Utilizing sugarcane wasted fibers as a sustainable acoustic absorber. Procedia Eng 53(1):632–638

Qunli W (1988) Empirical relations between acoustical properties and flow resistivity of porous plastic open-cell foam. Appl Acoust 25(3):141–148

Rayleigh JWS (1896) Theory of sound II, 1st edn. Macmillan and Co., London

Reddy N, Yang Y (2011) Novel green composites using zein as matrix and jute fibers as reinforcement. Biomass Bioenerg 35(8):3496–3503

Ren M, Jacobson F (1993) A method of measuring the dynamic flow resistance and reactance of porous materials. Appl Acoust 29(4):265–276

Rey Rd, Alba J, Arenas JP, Sanchis VJ (2012) An empirical modeling of porous sound absorbing materials made of recycled foam. Appl Acoust 73(6–7):604–609

Rouette H-K (2001) 'Jute', encyclopedia of textile finishing, 1st edn. Springer, Berlin Heidelberg, Cambridge, UK

Sagartzazu X, Hervella L, Pagalday JM (2008) Review in sound absorbing materials. Arch Comput Methods Eng 15(3):311–342

Sakagami K, Kiyama M, Marimoto M, Takahashi D (1996) Sound absorption of a cavity-backed membrane: a step towards design method for membrane-type absorbers. Appl Acoust 49 (3):231–241

Sarantopoulos G, Lykoudis S, Kassomenos P (2014) Noise levels in primary school of medium sized city in Greece. Sci Total Environ 482–483(1):493–500

Seybert AF (1988) Two sensor methods for the measurement of sound intensity and acoustic properties in ducts. J Acoust Soc America 83(6):2233–2239

Shoshani Y, Yakubov Y (2001) Use of nonwovens of variable porosity as noise control elements. Int Nonwovens J 10(4):23–28

Simón F, Pfretzschner J (2004) Guidelines for the acoustic design of absorptive devices. Noise Vib Worldwide 35(1):12–21

Srivastava RL, Dhabal RL, Suman BM, Saini A, Panchal P (2006) An estimation of correlation on thermos-acoustic properties of mineral wool. J Sci Ind Res 65(3):232–236

Su W-C, Cheng YS (2009) Deposition of man-made fibres in human respiratory airway casts. J Aerosol Sci 40(3):270–284

Suhanek M, Jambrosic K, Horvat M (2008) A comparison of two methods for measuring the sound absorption coefficient using impedance tubes. In: ELMAR 2008, 50th International symposium, IEEE, vol.1, no.1, pp. 321–324

Sun F, Banks-Lee P, Peng H (1993) Sound absorption in an anisotropic periodically layered fluid-saturated porous medium. Appl Acoust 39(1–2):65–76

Talty JT (1988) Industrial hygiene engineering—recognition, measurement, evaluation and control, 2nd edn. William Andrew Inc./Noyes Publications, New Jersey

Tang SK, Ng CH, Lam EYL (2012) Experimental investigation of the sound absorption performance of compartmented Helmholtz resonators. Appl Acoust 73(9):969–976

Uno Ingard K (1994) Notes on sound asborption techology, 1st edn. Noise Control Foundation, New York

Vér IL, Beranek LL (2006) Noise and vibration control engineering—principles and application, 2nd edn. Wiley, New Jersey

Wakefield Acoustic Ltd (2004) City of vancouver noise control manual, 1st edn. Sound Smart, Victoria, B.C

Wang C-N, Kuo Y-M, Chen S-K (2008) Effects of compression on the sound absorption of porous materials with an elastic frame. Appl Acoust 69(1):31–39

Wassilieff C (1996) Sound absorption of wood-based materials. Appl Acoust 48(4):339–356

Xiang H-F, Wang D, Liua H-C, Zhao N, Xu J (2013) Investigation on sound absorption properties of kapok fibers. Chin J Polym Sci 31(3):521–529

Yamashita T, Suzuki K, Adachi H, Nishino S, Tomota Y (2009) Effect of microscopic internal structure on sound absorption properties of polyurethane foam by x-ray computed tomography observation. Mater Trans 50(2):373–380

Yang H-S, Kim D-J, Kim H-J (2003) Rice straw-wood particle composite for sound absorbing wooden construction materials. Biores Technol 86(2):117–121

Yang H-S, Kim H-J, Son J, Park H-J, Lee B-J, Wang T-S (2004) Rice-husk flour filled polypropylene composites; mechanical and morphological study. Compos Struct 63(3–4):305–312

Zou Y, Huda S, Yang Y (2010) Lightweight composites from long wheat straw and polypropylene web. Biores Technol 101(6):2026–2033

Zulkifli R, Mohd Nor JM, Mat Tahir MF, Ismail AR, Nuawi MZ (2008) Acoustic properties of multi-layer coir fibres sound absorption panel. J Appl Sci 8(20):3709–3714

Zwikker C, Kosten CW (1949) Sound absorbing materials, 1st edn. Elsevier, New York

GPSR Compliance
The European Union's (EU) General Product Safety Regulation (GPSR) is a set of rules that requires consumer products to be safe and our obligations to ensure this.

If you have any concerns about our products, you can contact us on

ProductSafety@springernature.com

In case Publisher is established outside the EU, the EU authorized representative is:

Springer Nature Customer Service Center GmbH
Europaplatz 3
69115 Heidelberg, Germany

www.ingramcontent.com/pod-product-compliance
Lightning Source LLC
Chambersburg PA
CBHW071722010526
44275CB00003B/116
9783319686950